The Active Female

THE ACTIVE FEMALE

HEALTH ISSUES THROUGHOUT THE LIFESPAN

Edited by

JACALYN J. ROBERT-MCCOMB

Texas Tech University
College of Arts and Sciences
Health, Exercise, and Sport Sciences
Lubbock, TX

REID NORMAN

Texas Tech University Health Sciences Center
School of Medicine
Pharmacology and Neuroscience
Lubbock, TX

MIMI ZUMWALT

Texas Tech University Health Sciences Center
School of Medicine
Orthopaedic Surgery and Rehabilitation
Lubbock, TX

HUMANA PRESS ✸ TOTOWA, NEW JERSEY

Editors

Jacalyn J. Robert-McComb
Texas Tech University
College of Arts and Sciences
Health, Exercise, and Sport Sciences
Lubbock, TX

Reid Norman
Texas Tech University Health Sciences Center
School of Medicine
Pharmacology and Neuroscience
Lubbock, TX

Mimi Zumwalt
Texas Tech University Health Sciences Center
School of Medicine
Orthopaedic Surgery and Rehabilitation
Lubbock, TX

ISBN: 978-1-58829-730-3   e-ISBN: 978-1-59745-534-3

Library of Congress Control Number: 2007928340

©2008 Humana Press, a part of Springer Science+Business Media, LLC
All rights reserved. This work may not be translated or copied in whole or in part without the written permission of the publisher (Humana Press, 999 Riverview Drive, Suite 208, Totowa, NJ 07512 USA), except for brief excerpts in connection with reviews or scholarly analysis. Use in connection with any form of information storage and retrieval, electronic adaptation, computer software, or by similar or dissimilar methodology now known or hereafter developed is forbidden.

The use in this publication of trade names, trademarks, service marks, and similar terms, even if they are not identified as such, is not to be taken as an expression of opinion as to whether or not they are subject to proprietary rights.

While the advice and information in this book are believed to be true and accurate at the date of going to press, neither the authors nor the editors nor the publisher can accept any legal responsibility for any errors or omissions that may be made. The publisher make no warranty, express or implied, with respect to the material contained herein.

Printed on acid-free paper

9 8 7 6 5 4 3 2 1

springer.com

Preface

Medical practitioners and health care educators must be continually vigilant of the growing and ever-changing health issues related to girls and women who lead an active lifestyle and participate in sports and exercise. There have been landmark legislations that have changed the social perception that girls and women not only can, but should be physically active. With any changing social milieu, there are evolving health issues associated with the journey. Continuing medical education for physicians, nurses, allied health professionals, health educators, and certified professionals in sports medicine is vital to the economic and public health care system. Education has been recognized as the most important tool that we can use to prevent disease and illness.

In 1972, Congress passed Title IX of the Educational Amendments Act, assuring that girls and women would have equal opportunity to participate in interscholastic and intercollegiate sports. The effect has been an increase in the participation of women in interscholastic sports from approximately 300,000 to greater than 2.2 million in 1998 *(1)*.

Participation in recreational exercise for fitness and health, from young girls to elderly women, has substantially increased in the last four decades and has become a more prominent part of public life than ever before *(2)*. Physical activity has been recognized as a therapeutic means to decrease illness and increase health and well-being for girls and women of all ages and racial groups. In the US Public Health Services release, "Healthy People 2000," one of the recommendations was to increase the physical fitness of all women in an effort to reduce the health disparities between men and women and among different ethnic and racial groups *(3)*.

What makes women's health issues unique? Girls and women are different from boys and men, not only physiologically but also psychologically. Body image issues are more prominent in young girls than young boys and body dissatisfaction seems to start very early in life. Collins et al. *(4)* reported that 42% of a sample of 6-year-old to 7-year-old girls indicated a preference for body figures different and thinner than theirs. Thompson et al. *(5)* found that 49% of 4th-grade females indicated that their ideal figure would be thinner than their current figure. Young girls' bodies begin changing at puberty. This may be a hindrance to sport performance. Internal and external pressures placed on girls and women to achieve or maintain unrealistically low body weight may affect the normal female life cycle. Menstrual cycling, childbearing and menopause are experiences that are unique to the female life cycle. Lack of menstrual cycling caused by energy deficiency may even seem desirable to young females, yet there are long-term health consequence that are not so obvious to the ill-informed.

In 1992, The Female Athlete Triad was the focus of a consensus conference called by the Task Force on Women's Issues of the American College of Sports Medicine *(6)*. The three components of the Triad are disordered eating, amenorrhea, and osteoporosis. However, these are not elite disorders, these disorders are not limited to athletes, and these disorders are seen in young girls and elderly women who have never participated in collegiate or intercollegiate sports. These disorders represent a growing health concern for girls and women of all ages and physical skill levels.

Recognizing the lack of inclusion of women in health research and realizing that many health issues are unique to women, the US National Institutes of Health (NIH), established the Office for Research in Women's Health in September of 1990. The charge of this office was to improve women's status across the lifespan through health biomedical and behavioral research *(7)*. More recently, The Female Athlete Triad Coalition was formed in 2002 as a group of national and international organizations dedicated to addressing unhealthy eating behaviors, hormonal irregularities, and bone health among female athletes and active women. The Female Athlete Triad Coalition represents key medical, nursing, athletic, health educators, and sports medicine groups, as well as concerned individuals who come together to promote optimal health and well-being for female athletes and active women (http://www.femaleathletetriad.org).

INSTRUCTIONAL MATERIALS

The instructional materials that accompany *The Active Female: Health Issues Throughout the Lifespan* include a companion CD with PowerPoint lecture notes for each chapter. These slides are intended to be a resource for lectures, seminars, and other presentations. Also included on the CD are figures from each chapter. Other instructional materials are provided at the end of the book and include the appendices and multiple choice review questions. These enhancements are designed to reinforce and enliven the richness of the text for the student, the professor, and the professional user of the book.

We believe the instructional materials and the content in this book are ideal for one- or two-day workshops, focused conferences on women's health issues, or college and university classes. Since PowerPoint lecture notes and multiple choice review questions are provided for each chapter, this textbook is ideal for the development of traditional and on-line courses in women's health issues that meet the qualifications for CEC and CME credits set by licensing and certifying organizations.

Jacalyn J. Robert-McComb
Reid Norman
Mimi Zumwalt

REFERENCES

1. Bunker LK: Psycho-physiological contributions of physical activity and sports for girls. *President Council on Fitness and Sports Res Digest* 1998;3:1–8.
2. Garrett W, Lester G, McGowan J, Kirkendall D. *Women's Health in Sports and Exercise*. Rosemont, IL, American Academy of Orthopaedic Surgeons, 2001.
3. Public Health Service: Healthy People 2000: National Health Promotion and Disease Prevention Objectives. Full Report, With Commentary. Washington, DC, US Department of Health and Human Services, 1990, DHHS publication (PHS) 91-50212.
4. Collins LR, Lapp W, Helder L, Saltzberg J. Cognitive restraint and impulsive eating: insights from the Three-Factor Eating Questionnaire. *Psychol Addict Behav* 1992;6:47–53.
5. Thompson SH, Corwin S, Sargent RG. Ideal body size beliefs and weight concerns of fourth-grade children. *Int J Eat Disord* 1992;21:279–284.
6. Yeager K, Agostini K, Nattiv A, Drinkwater B. The female athlete triad. *Med Sci Sports Exerc* 1993;25:775–777.
7. Klimis-Zacas D, Wolinsky, I. *Nutritional Concerns of Women*. Boca Raton, FL, CRC Press, 2004.

Acknowledgments

I would like to acknowledge the following people for their technical contribution in helping me toward completion of my book chapters. Without their dedication and hard work, the task would have been much more difficult for me: Dr. Herb Janssen, Ms. Meadow Green, Ms. Sabrina Eckles, Ms. Barbara Ballew, Ms. Tara Vega, Ms. Jennifer Askew, and Ms. Alicia Niemeyer. I also want to thank Gail Branum, my nurse; Al Rosen, my friend; Mich Zumwalt, my brother; Francoise Sullivan, my mother; and most of all Demi and Miko, my children for all their love and undying moral support for me always!

Mimi Zumwalt

Contents

Preface	v
Acknowledgments	vii
List of Contributors	xiii
List of Appendices	xv

PART I: FOCUSING ON ACTIVE FEMALE'S HEALTH ISSUES: UNIQUE GENDER-RELATED PSYCHOLOGICAL AND PHYSIOLOGICAL CHARACTERISTICS OF FEMALES

1. Body Image Concerns Throughout the Lifespan
   Jacalyn J. Robert-McComb 3

2. Reproductive Changes in the Female Lifespan
   Reid Norman 17

3. Considerations of Sex Differences in Musculoskeletal Anatomy
   Phillip S. Sizer and C. Roger James 25

PART II: PREOCCUPATION WITH BODY IMAGE ISSUES AND DISORDERED EATING ISSUES IN THE ACTIVE FEMALE

4. Body Image and Eating Disturbances in Children and Adolescents
   Marilyn Massey-Stokes 57

5. The Female Athlete Triad: *Disordered Eating, Amenorrhea, and Osteoporosis*
   Jacalyn J. Robert-McComb 81

6. Disordered Eating in Active Middle-Aged Women
   Jacalyn J. Robert-McComb 93

7. Eating Disorder and Menstrual Dysfunction Screening Tools for the Allied Health Professional
   Jacalyn J. Robert-McComb 99

8. Education and Intervention Programs for Disordered Eating in the Active Female
   Jacalyn J. Robert-McComb 109

PART III: REPRODUCTIVE HEALTH

9 The Human Menstrual Cycle
   Reid Norman 123

10 Abnormal Menstrual Cycles
   Reid Norman 131

11 Psychological Stress and Functional Amenorrhea
   Reid Norman 137

12 Effects of the Menstrual Cycle on the Acquisition of Peak Bone Mass
   Mimi Zumwalt 141

PART IV: PREVENTION AND MANAGEMENT OF COMMON MUSCULOSKELETAL INJURIES IN ACTIVE FEMALES

13 Prevention and Management of Common Musculoskeletal Injuries in Preadolescent and Adolescent Female Athletes
   Mimi Zumwalt 155

14 Prevention and Management of Common Musculoskeletal Injuries in the Adult Female Athlete
   Mimi Zumwalt 169

15 Prevention and Management of Common Musculoskeletal Injuries Incurred Through Exercise During Pregnancy
   Mimi Zumwalt 183

16 Prevention and Management of Common Musculoskeletal Injuries in the Aging Female Athlete
   Mimi Zumwalt 199

17 Osteoporosis and Current Therapeutic Management
   Kellie F. Flood-Shaffer 213

PART V: SAFE EXERCISE GUIDELINES THROUGHOUT THE LIFESPAN

18 Physical Activity Recommendations and Exercise Guidelines Established by Leading Health Organizations
   Jacalyn J. Robert-McComb 227

19 Exercise Guidelines for Children and Adolescence
   Jacalyn J. Robert-McComb and Chelsea Barker 241

20 Exercise Precautions for the Female Athlete: *Signs of Overtraining*
   Jacalyn J. Robert-McComb and Abigail Schubert 247

21 Exercise Guidelines and Recommendations During Pregnancy
   Jacalyn J. Robert-McComb and Jessica Stovall 253

Contents

22 Mindful Exercise, Quality of Life, and Cancer:
*A Mindfulness-Based Exercise Rehabilitation Program
for Women with Breast Cancer*
Anna M. Tacón *261*

23 Exercise Guidelines for the Postmenopausal Woman
Shawn Anger and Chelsea Barker *271*

PART VI: NUTRITION, ENERGY BALANCE, AND WEIGHT CONTROL

24 Estimating Energy Requirements
Jacalyn J. Robert-McComb *279*

25 Nutritional Guidelines and Energy Needs for Active Children
Karen S. Meaney, Kelcie Kopf, and Megan Simons *287*

26 Nutritional Guidelines and Energy Needs for the Female Athlete:
*Determining Energy and Nutritional Needs to Alleviate the
Consequences of Functional Amenorrhea Caused by Energy
Imbalance*
Jacalyn J. Robert-McComb *299*

27 Ergogenic Aids and the Female Athlete
Jacalyn J. Robert-McComb and Shannon L. Jordan *311*

28 Nutritional Guidelines and Energy Needs During Pregnancy
and Lactation
Jacalyn J. Robert-McComb *323*

29 Nutritional Guidelines, Energy Balance, and Weight Control:
Issues for the Mature Physically Active Woman
Jacalyn J. Robert-McComb *335*

Appendices *345*
Index *439*

LIST OF CONTRIBUTORS

PRIMARY AUTHORS

Jacalyn J. Robert-McComb, PhD, FACSM
Professor at Texas Tech University, Department of Health, Exercise, and Sport Sciences, Texas Tech University, Lubbock, TX, Adjunct professor in the Department of physiology, Texas Tech University Health Science Center, Certified by the American College of Sports Medicine as an Exercise Test Technologist, Exercise Specialist and Clinical Program Director

Reid Norman, PhD
Professor and Chairman, Pharmacology and Neuroscience, Texas Tech University Health Science Center School of Medicine Lubbock, TX

Mimi Zumwalt, MD
Attending Orthopaedic Surgeon, Associate Professor of Orthopaedic Surgery, Director of Sports Medicine, Team Physician, Texas Tech University Health Sciences Center School of Medicine, Clinical Associate Professor of Rehabilitation Sciences, School of Allied Health, Lubbock, TX, Certified by the American College of Sports Medicine as an Exercise Leader

INVITED GUEST AUTHORS

Kellie F. Flood-Shaffer, MD, Fellow of the American College of Obstetricians and Gynecologists, Department of Obstetrics and Gynecology, Texas Tech University Health Science Center School of Medicine, Lubbock, TX

C. Roger James, PhD, FACSM, Center for Rehabilitation Research, School of Allied Health Sciences, Texas Tech University Health Science Center, Lubbock, TX.

Phillip S. Sizer Jr, PT, PhD, OCS, FAAOMPT, Professor & Program Director, ScD Program in Physical Therapy; Director, Clinical Musculoskeletal Research Laboratory; Department of Rehabilitation Sciences, School of Allied Health Sciences, Texas Tech University Health Science Center, Lubbock, TX

Marilyn Massey-Stokes, EdD, CHES, Associate Professor in Health, Exercise, & Sport Sciences, Texas Tech University, Lubbock, TX

Anna M. Tacón, PhD, Associate Professor in the Department of Health, Exercise, & Sport Sciences, Texas Tech University, Lubbock, TX

Karen S. Meaney, EdD, Associate Professor in the Department of Health, Exercise, & Sport Sciences, Texas Tech University, Lubbock, TX

Shawn Anger, MS, NSCA-CPT, Physical Therapy Today, Lubbock, TX

Chelsea Barker, MS, NASM-CPT, Physical Therapy Today, Lubbock, TX

Shannon L. Jordan, MS, Department of Health, Exercise, & Sport Sciences, Texas Tech University, Lubbock, TX

Kelcie Kopf, MS, Department of Health, Exercise, & Sport Sciences, Texas Tech University, Lubbock, TX

Jessica Stovall, BS, Department of Health, Exercise, & Sport Sciences, Texas Tech University, Lubbock, TX, MS Graduate Student

Abigail Schubert, BS, Department of Health, Exercise, & Sport Sciences, Texas Tech University, Lubbock, TX

Megan Simons, BS, Department of Health, Exercise, & Sport Sciences, Texas Tech University, Lubbock, TX

List of Appendices

Appendix 1:	Body Image Quality of Life Inventory	347
Appendix 2:	Body Image Concern Inventory	349
Appendix 3:	Physical Appearance State and Trait Anxiety Scale: Trait	351
Appendix 4:	The SCOFF Questionnaire	352
Appendix 5:	Eating Attitudes Test (EAT-26)	353
Appendix 6:	Bulimia Test—Revised (BULIT-R)	358
Appendix 7:	Student-Athlete Nutritional Health Questionnaire	364
Appendix 8:	Female Athlete Screening Tool	366
Appendix 9:	Eating Disorder Organizations and Resources	369
Appendix 10:	Determining Moderate and Vigorous Exercise Intensity Using the Heart Rate Reserve (HRR) Method	371
Appendix 11:	Determining Moderate and Vigorous Exercise Intensity Using the Borg Rating of Perceived Exertion (RPE) Scale	372
Appendix 12:	The Physical Activity Readiness Questionnaire (PAR-Q) C. 2002	373
Appendix 13:	General Organizational Guidelines for Exercise in Children and Adolescents	374
Appendix 14:	American College of Sports Medicine Guidelines for Resistance Training with Children	375
Appendix 15:	Kraemer's Age-Specific Exercise Guidelines for Resistance Training	376
Appendix 16:	Sample Exercise Resistance Program for Postmenopausal Women: 4-week, 6-week, 8-week, and 12-Week Programs	377
Appendix 17:	Illustrations of Exercises from Sample Resistance Program for Postmenopausal Women	379
Appendix 18:	Physical Activity Level Categories and Walking Equivalence	385
Appendix 19:	Estimated Energy Expenditure Prediction Equations at Four Physical Activity Levels	386
Appendix 20:	Estimated Calorie Requirements (in Kilocalories) for Specific Age Groups at Three Levels of Physical Activity Using the Institute of Medicine (IOM) Equations	389
Appendix 21:	Nutrition Questionnaire with 3-Day Recall	390
Appendix 22:	Food Frequency Questionnaire	393
Appendix 23:	US Department of Health and Human Services and the US Department of Agriculture, 2005 Dietary Guidelines for Americans	395

Appendix 24:	MyPyramid Food Intake Patterns at Varying Calorie Levels with Discretionary Calories	396
Appendix 25:	Dietary Reference Intakes (DRIs): Recommended Intakes for Individuals, Macronutrients	398
Appendix 26:	Dietary Reference Intakes (DRIs): Acceptable Macronutrient Distribution Ranges	399
Appendix 27:	Dietary Reference Intakes (DRIs): Estimated Average Requirements for Groups	400
Appendix 28:	Dietary Reference Intakes (DRIs): Recommended Intakes for Individuals, Elements	402
Appendix 29:	Dietary Reference Intakes (DRIs): Recommended Intakes for Individuals, Vitamins	403

I Focusing on Active Female's Health Issues: Unique Gender-Related Psychological and Physiological Characteristics of Females

1 Body Image Concerns Throughout the Lifespan

Jacalyn J. Robert-McComb

CONTENTS

    1.1   LEARNING OBJECTIVES
    1.2   INTRODUCTION
    1.3   RESEARCH FINDINGS
    1.4   CONCLUSIONS
    1.5   SCENARIO WITH QUESTIONS AND ANSWERS

1.1. LEARNING OBJECTIVES

After completing this chapter, you should have an understanding of the following:

- The difference between normal body image concerns, body dissatisfaction, and the preoccupation with body image concerns, or a pathological concern for thinness.
- Mediating factors that contribute to body image dissatisfaction in females.
- Prepubertal, adolescent, young adult, midlife, and older adult body image concerns.
- Clinical assessment tools for the evaluation of body image.
- Effective body image education and management programs referenced in the scientific literature.

1.2. INTRODUCTION

Although there is little agreement as to the exact definition of body image, there is little disagreement that body image is a multidimensional construct [1]. Thompson et al. [2] suggested that "body image" has come to be accepted as the internal representation of your own outer appearance. However, this may be an oversimplistic notion, given the complexity of the body image construct. Concerns about body image range from a normal desire to look attractive, body dissatisfaction, to a pathological concern with thinness or perfection [3].

There are medical issues that may arise from body dissatisfaction at both ends of the weight continuum ranging from anorexia nervosa to obesity [4,5]. In fact, the absence of refined measures developed for the use in the assessment, prevention, and

From: *The Active Female*
Edited by: J. J. Robert-McComb, R. Norman, and M. Zumwalt © Humana Press, Totowa, NJ

treatment of body image concerns associated with medical disease has been termed the "single-most neglected area in the study of body image"*(6)*. It is well known that body dissatisfaction plays a role in the development and maintenance of eating pathology *(7)*; however, body image concerns are pertinent to other psychiatric disorders, and these disorders are seen in all ages of patients. Negative body image and disordered eating behaviors in children and youth are common *(8)*; however, these attitudes and behaviors do not simply stop at adolescence. These unhealthy attitudes and behaviors many times carry on into adulthood *(9,10)* and are seen even in the older adult *(11)*.

Awareness of the etiology and the development of body image disturbances, knowledge of body image assessment techniques *(5,12,13)*, and effective prevention and management programs *(14)* are important for clinicians and health care educators to understand so that they may be able to educate and guide those they have in their care *(15–17)*. It is also important that those in the caring industry become aware of their own perceptions of body image and how these perceptions may influence patient care. Even physicians and mental health professionals are influenced by their patient's appearance and may treat unattractive individuals differently *(1)*.

1.3. RESEARCH FINDINGS

1.3.1. The Difference Between Normal Body Image Concerns, Body Dissatisfaction, and the Preoccupation with Body Image Concerns or Body Image Disturbances

As the concept of body image disturbances and the related pathology transcends fields in medical and allied health fields, there is increasing research devoted to the study of body image—*Body Image: An International Journal of Research* has been devoted to this topic. Body image is highly individualized and clinicians must recognize the subjectivity inherent in the development of body image. A host of factors, both developmental and proximal, combine to shape an individual's body image experience *(18)*. These factors have been grouped into current/proximal and historical/developmental categories. Developmental influences include sociodemographic factors, peer and familial influences, internalization of cultural ideal, and personality attributes. Proximal factors refer to everyday experiences, how they are interpreted, and their effects on mood and behavior.

Perhaps, the most perplexing issue related to body image is its definition *(14)*. Commonly used terms include body dissatisfaction, negative body image, body dysphoria, body image distortion, body esteem, body image disturbance, and body image concerns. Body image concerns are best conceptualized as occurring along a continuum. At one end of the continuum is body dissatisfaction and at the other end is body image distortions/disturbances. Reports of body dissatisfaction alone do not constitute body image disturbances. Body image dissatisfaction is a common psychological problem affecting many Westernized women *(4)*. Body dissatisfaction refers to the negative subjective evaluation of one's physical body, such as figure, weight, stomach, and hips. Body dissatisfaction should also be differentiated from the overemphasis placed on weight and shape in determining self-worth, which is a symptom of both anorexia and bulimia nervosa *(19–20)*. "Disturbance" typically denotes a clinical problem, characterized by persistent and chronic distress that may also interfere with interpersonal, psychosocial, or occupational functioning, and consequently may warrant consideration for treatment *(14)*. The most recognized and codable diagnoses with body image disturbances have been in eating disorders and body dysmorphic disorders *(6,20)*; however, these disorders are not limited to these pathologies alone. Body dissatisfaction must also be distinguished from body image distortion *(7)* wherein the individual perceives their body to be significantly larger than it really is, which is a symptom of anorexia nervosa *(19,20)*.

1.3.2. Mediating Factors that Contribute to Body Image Concerns in Females

As one reads the research literature, the most prominent mediating factor precipitating body dissatisfaction seems to be the media *(8)*. The media influences young women about what their bodies should look like, suggesting that the ideal body is extremely thin *(21)*. Field et al. *(22)* found that negative attitudes about weight and shape were strongly related to the frequency of reading fashion magazines. Baker et al. *(23)* found that visually impaired women had a less-negative body image than sighted women suggesting that the media contribute to these images. There is also an increase in the Internet websites promoting anorexia (pro-anorexia) and bulimia (pro-imia); these disorders include body distortion or disturbance as a diagnostic criterion *(8,24)*.

However, the notion that the media somehow "cause" weight and shape concerns seems oversimplistic given that the media primarily reflect beliefs and attitudes in the minds of the consumer *(15)*. The images that are portrayed by the media must be internalized as the images that are relevant to the culture that you identify with, the culture that you consider yourself belonging to, or even desire to belong to.

African-American females have reported that the thin ideal portrayed in the media relates more to Caucasian females *(25)*. There is no evidence to suggest that African-American and Caucasian females internalize media representation of the female body image in the same way *(4)*. To the contrary, an African American's susceptibility to advertisements depicting Caucasians has been associated with the strength of the African American's own ethnic identity *(26)*.

Therefore, it is vital for clinicians to understand the concept of culture when assessing body image, because of the subjectivity inherent in the internalization of what is acceptable in that specific culture. Yet, culture is not easily defined.

There is no single definition of culture, nor is there a consensus among scholars as to what the concept should include *(27)*. The definition of culture most relevant to traditional health implies that culture is a "metacommunion system," wherein not only spoken words have meaning, but every object of perception has meaning as well *(28)*. Spector *(29)* has proposed that the following are characteristics of culture: (1) the medium of personhood and social relationships, (2) consciousness, (3) an extension of biological capabilities, (4) an interlinked web of symbols, (5) the potential to create and limit human choices, and (6) a duality of existence-culture can simultaneously exist both in a person's mind and in the environment.

The body image construct also has considerable lability within "culture." There is a dynamic and fluid relationship between situational factors, goals, and body image experiences *(14,30)*. For athletes, body dissatisfaction and negative effect most emerged when considering their bodies within the social context, where femininity is defined consistent with the Victorian ideal *(31)*. Social messages purport that the acceptable female body is small and toned; yet, the athletic body is large and muscular. There seems to be a conflict between a female body for sport and a socially acceptable female body. This is particularly true in sports where a muscular body is beneficial (e.g., softball, basketball, and body building).

Other mediating factors that contribute to body image concerns are family attitudes and beliefs. Hill and Franklin *(32)* concluded that mothers have an important role in the transmission of cultural values regarding weight, shape, and appearance. Shoebridge and Gowers *(33)* found that an overprotective or "high concern" of parenting is common in children who subsequently develop anorexia nervosa that has body image disturbances as a specific criteria for this disorder.

Social class may also be a mediating factor in the development of body image concerns. Body dissatisfaction has also been shown to be common in middle-aged women *(34,35)*. In general, research has demonstrated that for a given body size, socioeconomically advantaged women are more dissatisfied with or concerned about their bodies than socioeconomically disadvantaged women *(36)*. However, not all researchers have found this to be true *(37)*.

1.3.3. Prepubertal, Adolescent, Young Adult, Midlife, and Older Adult Body Image Concerns

It seems that concerns about weight and dieting are appearing in younger children. Shapiro et al. *(38)* showed that dieting and exercise were used to control weight in as many as 41% of girls aged 8–10. Even children as young as 5 years of age are expressing fears of becoming fat and having body image concerns *(39)*. Davison and colleagues *(40)* found that at ages 5–7, girls who participated in aesthetic sports (e.g., dance, gymnastics, and cheerleading) reported higher weight concerns than girls who participated in nonaesthetic sports (e.g., soccer, volleyball, and tennis).

Nonetheless, adolescence is viewed as the stage of greatest risk in the development of body image and weight concerns (generally thought to be from age 11 to 19) *(15)*. Peer pressure, bullying, and teasing about weight has been identified as a precipitating factor in body dissatisfaction in adolescent females. Cooper and Goodyer *(41)* showed that from age 11 to 16, there was a steady increase in weight and shape concerns in females in a community sample: 11–12 years (15.5%), 13–14 years (14.9%), and 15–16 (18.9%) experienced body image concerns. Packard and Krogstrand *(42)* found that more than one-half (52%) of rural white women between age 8 and 17 reported weight concerns and that this pattern increased with age. Other researchers have found that children between the age of 9 and 14 became constant dieters when thinness was important to their fathers *(43)*.

Surprisingly, pregnancy seems to help women's viewpoint on body image concerns. Robb-Todter *(44)* employed a qualitative research paradigm to investigate women's experience of weight and shape changes during pregnancy and in the early postpartum period. She found that women did not loose interest in their weight, shape, or appearance as a result of pregnancy but that it seemed less important than in the past. Concern about the baby's well-being superseded women's concerns about weight and shape. Based on the experiences of these women, she concluded that pregnancy and motherhood may have the potential to help women put weight issues and eating habits in perspective, if only temporarily.

For the most part, research has also shown that adult women are dissatisfied with their body *(4)*. Potts *(10)* utilized the Body Shape Questionnaire *(45)* to assess body dissatisfaction in women ($n = 171$) aged 35–50 ($M = 41$). They found that 87% wanted to be thinner, yet only 35% were actually overweight. They also found that the women in their study (age 35–50) were more dissatisfied with their body than women 20 years younger. McLaren and Kuh *(35)* used self-report data from 912 54-year-old women to analyze body dissatisfaction adjusting for body mass index. They found that women from the nonmanual working classes as adults were more dissatisfied with their body than those from the manual class as adults, and they also found that higher educational qualifications were associated with more dissatisfaction with weight and appearance, and education appears to be more important than occupationally defined social class in explaining body dissatisfaction. In another research publication *(34)*, they stated that weight dissatisfaction was reported by nearly 80% of a sample of 1026 54-year-old women even though 50% were of normal weight (BMI < 25). Additionally, women were more dissatisfied with their bodies in the fifties than they had been in the forties. This same trend has been found in other studies, suggesting that women in midlife have incorporated society's image of the ideal female, and not measuring up to that ideal, they are dissatisfied with their bodies *(10)*.

1.3.4. Clinical Assessment Tools for Body Image

As the instruments to assess body image concerns are discussed, it is important to emphasize the complexity of the body image construct *(46,47)*. Of the various theories and approaches to assessment are considered for body image, a cognitive behavioral approach has received the most empirical attention *(14)*.

1.3.4.1. Cognitive Behavioral Approach

A cognitive behavioral approach to assessment entails identifying factors that precipitate and maintain body image concerns. The primary goals of a cognitive behavioral assessment are to (1) contextualize body image concerns in a way that will increase patient awareness of precipitating and maintaining factors and (2) provide a guide for treatment goals and planning based on this assessment. Current/proximal factors which should be considered during body image assessment are (1) impact of body image concerns; (2) patient's investment in appearance (meaning of attractiveness to one's sense of self, perceived discrepancy between self and ideal, and internalization of appearance ideals); (3) activating events/triggers (external cues); (4) cognitive and emotional processing (internal dialogue, cognitive distortions, core beliefs, and stress reactivity); (5) behavioral strategies/self-regulatory behaviors (coping style, reassurance seeking, avoidance behaviors, social comparison, and repetitive checking/grooming); and (6) goals and obstacles to treatment (expectations, motivation, social support, and medical or psychiatric comorbidity) *(14)*. Historical/developmental factors to consider during assessment are (1) sociodemographic factors (family origin/ethnicity, gender, and age); (2) cultural/socialization factors (interpersonal experiences and familial, authoritarian and peer influences); (3) physical characteristics of attribute (age of onset, body mass index, acquired versus congenital, and ability to control attribute); (4) personality attributes (strictness/perfectionism and self-worth); (5) history of treatment attempts (successful, unsuccessful, surgical history, and weight loss attempts); and (6) comorbidity (medical illnesses and Axis I disorders) *(14)*.

Self-monitoring is an integral part in assessment in the cognitive behavioral model. The patient should be instructed to record any situation that triggers experiences related to body image, appearance-related beliefs and thoughts, and their effects on mood and behavior. In addition to providing a foundation to guide treatment planning, monitoring allows assessment of treatment progress and outcomes. For a more thorough understanding of the cognitive behavioral approach in the assessment of body image disturbances, we refer the readers to *A Handbook of Theory, Research, and Clinical Practice* by Cash and Pruzinsky *(18)* and to the *Handbook of Eating Disorders and Obesity* by Thompson *(48)*.

1.3.4.2. COMMONLY USED BODY IMAGE ASSESSMENT SCALES AND QUESTIONNAIRES

Table 1.1 lists some common scales that have been reported in the research literature that have an internal consistency rating and test-retest reliability rating of at least 0.70 *(49–56)*.

Caution is recommended in adopting scales without psychometric properties. Even when psychometric properties are listed, judgment must be used in generalizing the standardized sample to your target sample. Banasiak et al. *(57)* cautions that although extensive research has been conducted on body image concerns in adolescence, many of the instruments used to assess these concerns in adolescence have been validated using adult samples. However, they did find that many of the measures developed on adults can be applied to middle adolescent girls when care is taken to ensure that the girls understand the terms used in the assessment instrument.

Examples of body image questionnaires that have been validated for college-age women and that have internal consistency and test-retest reliability scores above 0.70 can be found in Appendices 1–3: Body Image Quality of Life Inventory by Cash and Fleming *(49)*, Body Image Concern Inventory by Littleton et al. *(16)*, and Physical Appearance State and Trait Anxiety Scale: Trait by Thompson *(58)*.

1.3.5. Effective Body Image Education and Management Programs Referenced in the Scientific Literature

1.3.5.1. BODY IMAGE EDUCATION

Results of a body image program for adult women developed at the University of Alberta *(59)* suggested that participation in the program had a significant positive impact on women's body image. Based on common themes derived from a needs assessment, a 12-session program was developed for noneating disordered women aged 20–60 years. As the program's goal was to promote women's acceptance of their bodies regardless of their weight, weight loss strategies were excluded from the program. The program was based on Social Cognitive Theory. Sessions ranged from 90 to 120 min in length and were structured as follows: (1) review of previous session's homework and feedback, (2) presentation and discussion of selected topics, (3) individual and/or group exercises to develop skills or concepts, and (4) assignment of journal and homework

Table 1.1
Instruments for assessing body image with high internal consistency and test-retest co-efficients (> 0.70)

Author	Test Name	Description of Test	Reliability	Standardization Sample
Cash and Fleming (49)	Body Image Quality of Life Inventory	A 19-item instrument designed to quantify the impact of body image on aspects of one's life. Participants rate the impact of their own body image on each of the 19 areas using a 7-point bipolar scale from −3 to +3.	IC: 0.95 college-age women TR: 0.79 college-age women	116 college-age women ($M = 21.3 \pm 5.1$)
Littleton et al. (16)	Body Image Concern Inventory	A *brief* instrument for assessing dysmorphic concern only takes a few minutes to answer. Despite its brevity, the BICI provides an assessment of body dissatisfaction, checking and camouflaging behavior, and interference due to symptoms such as discomfort with and avoidance of social activities (*see* Appendix 2).	IC: 0.93 (Cronbach's alpha) 184 undergraduates TR: none given	184 undergraduates at a medium-sized southeastern university, approximately 89% were women
Reed et al. (50)	Physical Appearance State and Trait Anxiety Scale	Participants rate the anxiety associated with 16 body sites (8 weight relevant and 8 non-relevant); trait and state versions available.	IC: Trait: .88–.82, State: .82–.92 TR: 2 weeks, .87	Female undergraduate students

(Continued)

Table 1.1
(Continued)

Author	Test Name	Description of Test	Reliability	Standardization Sample
Garner and Olmstead (51); Garner (52)	Eating Disorder Inventory (EDI and EDI-2). Body Dissatisfaction Scale	9-item subscale assesses feelings about satisfaction with body size; items are 6-point, forced choice; reading level is 5th grade	IC: Adolescents (11–18) Females = 0.91 Males = 0.86 Children (8–10) Females = 0.84 TR: None given	610 males and females aged 11–18 (55) 109 males and females age 8=9 8–10 (56)
Shisslak et al. (53)	McKnight Risk Factor Survey III (MFRS-III)	Participants use 5-item subscale assesses concern with body weight and shape	IC: Elementary = .82 Middle school = .86 High school = .87 TR: Elementary = .79 Middle school = .84 High school = .90	103 females, 4th–5th grade; 420 females, 6th–8th grade; 66 females, 9th–12th grade
Wooley and Roll (51)	Color-A-Person Body Dissatisfaction Test	Participants use five colors to indicate level of satisfaction with body sites by masking on a schematic figure	IC: .74–.85 TR: 2 weeks (.72–.84), 4 weeks (.75–.89)	102 male and female college students, 103 bulimic individuals

IC, Internal consistency; TR, test-reset.

exercise. Topic themes were as follows: (1) introducing body image, (2) influences on body image, (3) relaxation and desensitization, (4) discovering our body image distortions, (5) changing self-defeating body image behaviors, (6) doing what is best for you, (7) the truth about fat and dieting, (8) listening to your body, (9) learning to love body movement, (10) reviving your friendship with your body, (11) the natural process of aging, and (12) evaluation and wrap-up.*

1.3.5.2. MANAGEMENT OF BODY DISTURBANCES

There are two primary approaches for treating body image disturbances that have been referenced in the research literature and are supported by clinicians: the cognitive behavioral and the feminist approach. These approaches can be considered as a treatment option for a wide variety of clinical populations *(2)*.

The cognitive behavioral strategy as developed by Cash *(60)* has eight components. The first component has been discussed in section 1.3.4.2 and is a comprehensive body image assessment. The second component involves body image education based on the findings from the initial assessment. The third component is body image exposure and desensitization. During this component, clients develop relaxation skills and use them to manage their body image distress. The fourth step is identifying and challenging appearance assumptions and the problems produced by these maladaptive core beliefs about appearance. In the fifth step, clients dispute negative appearance assumptions through audio taping corrective thinking dialogues and keeping a diary. The sixth component targets both avoidant behaviors and compulsive patterns by modifying self-defeating body image behaviors. The seventh component involves the development of body image enhancement activities, such as dancing, and the client is instructed to expand the number of positive body-related experiences. In the eighth component, clients evaluate their progress, set future goals, and develop strategies for coping with setback.

The feminist approach differs from the cognitive behavioral approach in three primary ways *(2)*. First, the feminist approach criticizes approaches that focus on treating body image problems by changing a woman's appearance (diet and exercise). They are of the view that woman should not be defined by their appearance. Second, feminist therapy relies on an egalitarian relationship characterized by therapist self-disclosure, greater informality and nurturance, and patient advocacy. Third, feminist interventions focus on different etiological factors that play a role in the development of body image disturbance. Primary among proposed etiologies is the role of sexual abuse in the development of body image disturbances.

1.4. CONCLUSIONS

There is a growing appreciation of the complexities inherent in body image *(13)*. Concerns about body image range from a normal desire to look attractive, body

---

*To obtain a copy of the program, please contact the Centre for Health Promotion Studies and Department of Agricultural, Food, and Nutritional Sciences, University of Alberta, 5-10 University Extension Centre, 8303-112 St, Edmonton, AB T6G 2T4, Canada. Tel:(780) 492-9415; Fax: (780) 492-9579.

dissatisfaction, to a pathological concern with thinness or perfection *(3)*. Furthermore, the formation of body image occurs within a social and cultural context. Ideas of beauty and standards of attractiveness vary across ethnic groups, gender, sexual orientation, age, and culture *(61,62)*.

1.5. SCENARIO WITH QUESTIONS AND ANSWERS

1.5.1. Scenario

You are family care physician practicing in a small upscale urban community. Recently, it has come to your attention that two or three of your female patients have brought their adolescent daughters in for their annual physical and have expressed concern about their daughters delayed menarche. The women state that their daughters, aged 14–15, have not begun cycling, even though many of their classmates have been cycling for 2 or 3 years. Even though the time of menarche varies among young girls, this delayed menarche among this group of young girls seems to be more than simply coincidental: These young girls all belong to the local gymnastic club and compete on the same team in local meets. Upon further physical assessment, you find out that their weight is less than that expected for their age and height, and their secondary sexual characteristics also seem to be delayed.

1.5.2. Questions

1. What do you think your next step should be?
2. What are hypothetical causes for the delayed menarche given this scenario?
3. What are some considerations when choosing appropriate body image assessment instruments?

1.5.3. Plausible Answers

1. Although there is not one right answer, there are prudent and reasonable courses of action to take. However, all of the courses should begin with an overall clinical assessment of the causes for the delayed menarche. The course of action would then depend on the results of the assessment.
2. Given this scenario, hypothetical causes could be body image concerns resulting in inadequate caloric intake to compensate for their energy needs. This energy imbalance could then result in delayed menarche.
3. The psychometric properties of the scales should be considered, the assessment instruments should have an internal consistency rating, and test-retest reliability rating of at least 0.70.

REFERENCES

1. Sarwer D, Grossbart T, Didie E. Beauty and society. *Semin Cutan Med Surg* 2003;22(2):79–92.
2. Thompson JK, Heinberg LJ, Altabe M, et al. *Exacting Beauty: Theory, Assessment and Treatment of Body Image Disturbances*. Washington, DC: American Psychological Association, 1999:19–82.
3. Weinshenker N. Adolescence and body image. *School Nurse News* 2002;19(3):12–16.
4. Padgett J, Biro FM. Different shapes in different cultures: body dissatisfaction, overweight, and obesity in African American and Caucasian females. *J Pediatr Adolesc Gynecol* 2003;16(3):349–354.
5. Thompson JK, Smolak L. *Body Image, Eating Disorders, and Obesity in Youth*. Washington, DC: American Psychological Association, 2001.

6. Purzinsky T. Enhancing quality of life in medical populations: a vision of body image assessment and rehabilitation as standards of care. *Body Image* 2004;1:78–81.
7. Stice E, Shaw HE. Role of body dissatisfaction in the onset and maintenance of eating pathology: a synthesis of research findings. *J Psychosom Res* 2002;53(5):985–993.
8. Andrist L. Media images, body dissatisfaction, and disordered eating in adolescent women. *MCN Am J Matern Child Nurs* 2003;28(2):119–123.
9. Abraham SF. Dieting, body weight, body image and self-esteem in young women: doctor's dilemmas. *Med J Aust* 2003;178(12):607–611.
10. Potts NLW. Body image dissatisfaction, self-esteem, and sex-role identity in midlife women. Unpublished doctoral dissertation, Texas Women's University, Denton, TX, 1993:88–137.
11. Tiggemann M. Body image across the adult life span: stability and change. *Body Image* 2005;1(1):29–41.
12. Veron-Guidry S, Williamson DA. Development of a body image assessment procedure for children and preadolescents. *Int J Eat Disord* 1996;20(3):287–293.
13. Littleton HL, Ollendick T. Negative body image and disordered eating behavior in children and adolescents: what places youth at risk and how can these problems be prevented? *Clin Child Fam Psychol Rev* 2003;6(1):51–66.
14. Reas DL, Grilo CM. Cognitive-behavioral assessment of body image disturbances. *J Psychiatr Pract* 2004;10(5):314–322.
15. Gowers S, Shore A. Development of weight and shape concerns in the aetiology of eating disorders. *Br J Psychiatry* 2001;179:236–242.
16. Littleton HL, Axsom D, Pury C. Development of the body image concern inventory. *Behav Res Ther* 2005;43(2):229–241.
17. Weiss F. Body image, eating disorders, and obesity in youth: assessment, prevention, and treatment. *Am J Psychother* 2004;58(3):362–364.
18. Cash TF, Pruzinsky T. *Body Image: A Handbook of Theory, Research, and Clinical Practice.* New York, NY: Guilford Press, 2002.
19. American Psychiatric Association. *Diagnostic and Statistical Manual of Mental Disorders*, 4th ed. Washington, DC: American Psychiatric Associations, 1994:539–5550, 729.
20. Robert-McComb J. Eating disorders. In: Robert-McComb JJ, ed. *Eating Disorders in Women and Children: Prevention, Stress Management, and Treatment.* Boca Raton, FL: CRC Press, 2001:3–38.
21. Cortese AJ. *Provocateur: Images of Women and Minorities in Advertising.* Lanham, MD: Rowman & Littlefield Publishers, 1999.
22. Field AE, Cheung L, Wolf AM, Herzog DB, Gortmaker SL, Colditz GA. Exposure to the mass media and weight concerns among girls. *Pediatrics* 1999;103:36.
23. Baker D, Sivyer R, Towell T. Body image dissatisfaction and eating attitudes in visually impaired women. *Int J Eat Disord* 1998;24:319–322.
24. McComb JR. The physiological consequences of energy-deficiency for adolescent girls and the promotion of this concept on the web. *Clearing House* 2002;75(6):297–300.
25. Parker S, Nichter M, Nichter M, et al. Body image and weight concerns among African American and White adolescent females: differences that make a difference. *Hum Organ* 1995;54:103–115.
26. Green CL. Ethnic evaluations of advertising: interaction effects of strength of ethnic identification, media placement, and degree of racial composition. *J Advert Res* 1999;28:49–65.
27. Jones DP. Cultural views of the female breast. *ABNF J* 2004;15(1):15–21.
28. Matsumoto M. The unspoken way. In: Spector R, ed. *Cultural Diversity in Health and Illness*, 5th ed. Upper Saddle River, NJ: Prentice Hall, 2000:71–96.
29. Spector R. *Cultural Diversity in Health and Illness*, 5th ed. Upper Saddle River, NJ: Prentice Hall, 2000:71–96.
30. Gill K, Overdorf V. Incentives for exerciser in younger and older women. *J Sport Behav* 1994;17:87–97.
31. Krane V, Waldron J, Michalenok J, et al. Body image concerns in female exercisers and athletes: a feminist cultural studies perspective. *Women Sport Phys Act J* 2001;10(1):17–54.
32. Hill AJ, Franklin JA. Mothers, daughters and dieting: investigating the transmission of weight control. *Br J Clin Psychol* 1998;37:3–13.
33. Shoebridge P, Gowers SG. Parental high concern and adolescent-onset anorexia nervosa. A case-control study to investigate direction of causality. *Br J Psychiatry* 2000;176:132–137.

34. McLaren L, Kuh D. Body dissatisfaction in midlife women. *J Women Aging* 2004;16(1):35–54.
35. McLaren L, Kuh D. Women's body dissatisfaction, social class, and social mobility. *Soc Sci Med* 2004;58:1575–1584.
36. Wardle J, Griffin J. Socioeconomic status and weight control practices in British adults. *J Epidemiol Community Health* 2001;55:185–190.
37. Robinson TN, Chang JY, Haydel KF, et al. Overweight concerns and body dissatisfaction among third grade children: the impacts of ethnicity and socioeconomic status. *J Pediatr* 2001;138(2):181–187.
38. Shapiro S, Newcomb M, Loeb TB. Fear of fat, disregulated-restrained eating, and body-esteem: prevalence and gender differences among eight to ten year old children. *J Clin Child Psychol* 1997;26:358–365.
39. Feldman W, Feldman E, Goodman JT. Culture vs. biology: children's attitude toward thinness and fatness. *Pediatrics* 1998;81:190–194.
40. Davison KK, Earnest MB, Birch LL. Participation in aesthetic sports and girls' weight concerns at ages 5 and 7 years. *Int J Eat Disord* 2002;31(3):312–317.
41. Cooper PJ, Goodyer I. Prevalence and significance of weight and shape concerns in girls aged 11–16 years. *Br J Psychiatry* 1997;171:542–544.
42. Packard P, Krogstrand KS. Half of rural girls aged 8 to 17 report weight concerns and dietary changes, with both more prevalent with increased age. *J Am Diet Assoc* 2002;102(5):672–677.
43. Field AE, Camargo CA, Jr., Taylor CB, et al. Peer, parent, and media influences on the development of weight concerns and frequent dieting among preadolescent and adolescent girls and boys. *Pediatrics* 2001;107(1):54–60.
44. Robb-Todter GA. Women's experience of weight and shape changes during pregnancy. Unpublished doctoral dissertation, University of Virginia, Afton, VA, 1996:117–121.
45. Cooper PJ, Taylor M, Cooper Z, et al. The development and validation of the body shape questionnaire. *Int J Eat Disord* 1987;6:485–494.
46. Traub AC, Orbach J. Psychological studies of body image: an adjustable body distorting mirror. *Arch Gen Psychiatry* 1964;11:53–66.
47. Cash TF, Hrabosky JI. Treatment of body image disturbances. In: Thompson JK, ed. *Handbook of Eating Disorders and Obesity*. New York, NY: Wiley, 2004:515–541.
48. Thompson JK. *Handbook of Eating Disorders and Obesity*. New York, NY: Wiley, 2004.
49. Cash TF, Fleming EC. The impact of body-image experiences: development of the body image quality of life inventory. *Int J Eat Disord* 2002;31:455–460.
50. Reed DL, Thompson JK, Brannick MT, et al. Development and validation of the physical appearance state and trait anxiety scale (PASTAS). *J Anxiety Disord* 1991;5:323–332.
51. Garner D, Olmsted M. *Manual for the Eating Disorder Inventory (EDI)*. Odessa, FL: Psychological Assessment Resources, 1984.
52. Garner D. *Manual for the Eating Disorder Inventory-2 (EDI-2)*. Odessa, FL: Psychological Assessment Resources, 1991.
53. Shisslack CM, Renger R, Sharpe T, et al. Development and evaluation of the McKnight Risk Factor Survey for assessing potential risk and protective factors for disordered eating in preadolescent and adolescent girls. *Int J Eat Disord* 1999;25:195–214.
54. Wooley OW, Roll S. The Color-a-Person Body Dissatisfaction Test: stability, internal consistency, validity, and factor structure. *J Pers Assess* 1991;56:395–413.
55. Shore R, Porter J. Normative and reliability data for 11–18 year olds on the eating disorder inventory. *Int J Eat Disord* 1990;25:201–207.
56. Wood KC, Becker JA, Thompson JK. Body image dissatisfaction in preadolescent children. *J Appl Dev Psychol* 1996;17:85–100.
57. Banasiak SJ, Wertheim EH, Koerner J, et al. Test-retest reliability and internal consistency of a variety of measures of dietary restraint and body concerns in a sample of adolescent girls.*Int J Eat Disord* 2001;29:85–89.
58. Thompson JK. Assessing body image disturbance; measures, methodology, and implementation. In: Thompson JK, ed. *Body Image, Eating Disorders, and Obesity*. Washington, DC: American Psychological Association, 1996:80.
59. Paquette M-C, Leung R, Raine K. Development of a body image program for adult women. *J Nutr Educ Behav* 2002;34:172–174.

60. Cash TF. The treatment of body image disturbances. In: Thompson JK, ed. *Body Image, Eating Disorders, and Obesity*. Washington, DC: American Psychological Association, 1996:83–107.
61. Akan GE, Grilo CM. Sociocultural influences on eating attitudes on behaviors, body image, and psychological functioning: a comparison of African-American, Asia American, and Caucasian college women. *Int J Eat Disord* 2002;32:335–343.
62. Barry DB, Gril CM. Eating and body image disturbances in adolescent psychiatric patients: gender and ethnicity patterns. *Int J Eat Disord* 1996;20:135–141.

2 Reproductive Changes in the Female Lifespan

Reid Norman

CONTENTS

2.1 LEARNING OBJECTIVES
2.2 INTRODUCTION
2.3 RESEARCH FINDINGS
2.4 CONCLUSIONS
2.5 SCENARIO WITH QUESTIONS AND ANSWERS

2.1. LEARNING OBJECTIVES

After completing this chapter, you should have an understanding of the following:

- How a woman's body changes during her life.
- How reproductive hormones change throughout the lifespan of a woman.
- The terminology used to describe changes in reproductive capabilities.
- The impact of menopause on health measures.

2.2. INTRODUCTION

When considering reproductive competence in the human female, the average lifespan can be divided into three phases. The first of these is childhood where, except in rare instances of precocious puberty, the reproductive system is quiescent and secondary sexual characteristics are absent. Puberty is the transition between childhood and sexual maturity, and menarche or first menstruation is an important sign of this maturation. During this second stage of sexual maturity, which in the United States last from about 13 to 51 years of age, menstrual cycles occur about once a month except when interrupted by pregnancy or by synthetic steroids used as birth control. Even when hormonal birth control is used, there is usually an attempt to maintain monthly cyclic menstruation. In modern civilizations, menstrual cycles can be suppressed by stressful life events or by lack of nutritional resources, the latter usually because of dieting or exercise. Most women stop having menstrual cycles in their late forties or early fifties and enter the postmenopausal stage. This transition between the fertile

From: *The Active Female*
Edited by: J. J. Robert-McComb, R. Norman, and M. Zumwalt © Humana Press, Totowa, NJ

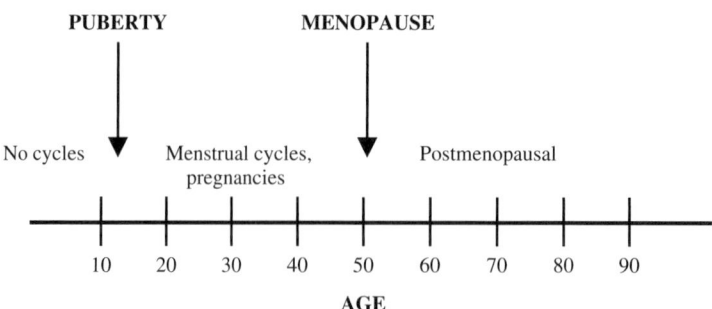

Fig. 2.1. Graphic depiction of the female reproductive capabilities across the lifespan of the average woman.

period and postmenopausal or infertile phase is called menopause (Fig. 2.1). In the past century, the time spent in the postmenopausal state has become extended because the average lifespan has increased from 50 to more than 80 years. The health concerns of postmenopausal women have become a real concern in the past 50 years. The transitions between these three stages, puberty and menopause, bring dramatic changes in levels of the most important female hormones, estrogen and progesterone.

2.3. RESEARCH FINDINGS

2.3.1. Hormonal Changes During Childhood

From shortly after birth until the beginning of sexual maturation, reproductive hormone levels are low and available energy is largely committed to growth and development. Before puberty, boys and girls have similar lean body mass and the same amount of body fat. Body mass index in girls is generally between 15 and 18 and does not change much during the childhood years. Growth rates are similar for boys and girls before the pubertal growth spurt and, in the absence of serious illness or genetic abnormality, 12-year-old girls and boys are the same height and weight on average.

Physicians routinely monitor height, weight, chronological age, bone age, and growth during the previous year to evaluate how development is progressing. Since Title IX legislation, which prohibits sex discrimination in any educational program or activity in institutions from elementary school through college that received federal funding, was passed in 1972, female participation in school athletic programs has increased dramatically. Of the total participants in high school athletics in 1971, < 10% were female. Today, the proportion of male and female athletes is nearly equal. Although exercise is largely beneficial, excessive exercise can utilize energy needed for growth and development and can significantly impact these processes if energy intake is limited by dietary restrictions. This is important because in some competitive sports, and particularly in elite athletes, rigorous training begins before and can potentially influence sexual development.

2.3.2. What Happens at Puberty

Although puberty in human females is generally defined as the process of sexual maturation, and it is certainly that because the changes are primarily driven by the awakening of the ovaries, a more inclusive definition would also encompass the accompanying physiological and behavioral changes that occur during this transition. This process of sexual maturation requires several years, and the ages of 8–14 years are considered the average range of when this process occurs. The defining event, menarche (first menstrual period), occurs at an average of 12.5 years in the United States, but there are differences among ethnic groups and between the United States and other countries. Menarche is a sign that the ovarian cycle is sufficiently functional to support growth and development of the uterine lining. The diagnosis of *primary amenorrhea* is made when menarche does not occur by about age 16. Several years before sexual maturation, increased secretion of steroid hormones from the adrenal gland and ovaries initiate widespread physiologial changes in the body. One of these changes is the adolescent growth spurt. This acceleration in growth lasts for a year or so, then slows, and growth is eventually terminated by fusion of the epiphyseal regions of the long bones where growth occurs and this terminates growth. Estrogen is responsible for epiphyseal fusion in both males and females. During this pubertal transition in females, there is an increase in percentage of body fat, which does not occur in males. Secondary sexual characteristics, such as breast development and genital development, are directed by ovarian estrogens, but axillary (underarm) hair and pubic hair are controlled by androgens from both the ovary and adrenal. The appearance of the female body becomes very different from that of males, and it is primarily the hormones, estrogen and progesterone, from the ovary that drive this change. A major question that has not been answered is why does the secretion of these ovarian steroid hormones increase at this time in life? In other words, what event initiates the process of sexual maturation?

What we do know is that the process of sexual maturation is driven by an increase in the release of a small peptide (a small molecule composed of 10 amino acids) called gonadotropin-releasing hormone (GnRH); what we do not know is why it occurs at this specific time. Recent evidence implicates an obligatory role for a peptide called kisspeptin that directly stimulates GnRH release from the hypothalamus *(1)*. The prevailing hypothesis is that puberty is initiated at some point when the brain is sufficiently mature. When this occurs, pulses of GnRH are released at 1- to 2-h intervals into the pituitary portal system and travel a short distance from the base of the brain (hypothalamus) to the pituitary and stimulate the release of two protein hormones (large hormones also composed of amino acids) called luteinizing hormone (LH) and follicle stimulating hormone (FSH). LH and FSH travel through the circulation to the ovary where they cause growth and development of follicles containing ova (eggs) and at the same time stimulate the secretion of the ovarian hormones (Fig. 2.2). This process begins slowly at first, with LH pulses released only at night. Because the LH (and FSH) levels are not maintained at adult levels throughout the day and night, stimulation of the ovarian follicles is not sufficient to result in ovulation. As puberty progresses, the time when LH pulses are released gradually expands to the daytime, and in the adult, these pulses are observed throughout the day as well as at night. In response to FSH and LH, estrogen release from the stimulated follicles results in

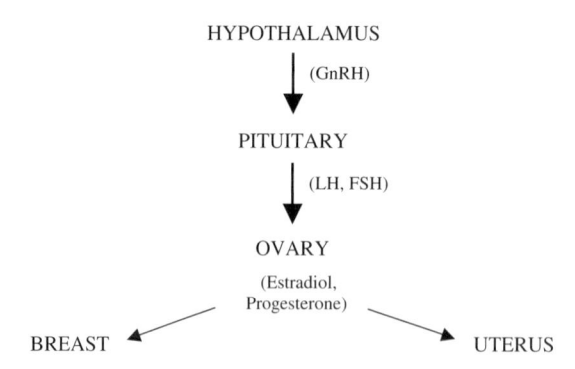

```
                    HYPOTHALAMUS
                         |
                         | (GnRH)
                         ↓
                     PITUITARY
                         |
                         | (LH, FSH)
                         ↓
                       OVARY
                    (Estradiol,
                   Progesterone)
                   ╱            ╲
          BREAST  ←              →  UTERUS
```

Fig. 2.2. This figure shows the primary components of the female reproductive system and the hormones that communicate between the various organs. The hormones produced by each gland are shown in parentheses.

changes in the body including growth of the breasts and hips primarily because of the deposition of fat. Late in the pubertal process, ovulation occurs when the LH and FSH levels are maintained at a level to provide consistent support for the developing follicle. Menarche usually occurs late in the sequence of events defining puberty.

2.3.3. Sexual Maturity: The Reproductive Years

The ability to reproduce is one of the hallmarks of sexual maturity. Sexually mature women who are not taking birth control pills have regular menstrual cycles that average 26–35 days in length and that are (can be) occasionally interrupted by pregnancy and lactation. Menarche signals the beginning of the ability to reproduce and menopause marks the end; the reproductive lifespan lasts nearly 40 years from about 13 years (menarche) to 51 years (menopause). Menstruation (blood and dead cell debris discharged from the uterus through the vagina) occurs at the end of an ovarian cycle, and the lining of the uterus dies and sloughs off indicating that implantation has not occurred. Menstruation is also the beginning of a new cycle when several follicles begin to grow rapidly and by convention the first day of menstrual flow is day 1 of the cycle. This sequence of follicular development, ovulation, and menstruation is repeated at regular intervals, unless interrupted by pregnancy and lactation, until menopause. Many women postpone reproduction with birth control pills or other contraceptive methods for an indefinite period of time to pursue educational or career goals. Even though hormonal birth control pills inhibit follicular development and ovulation, they do not extend the fertile lifespan which peaks in the twenties and declines thereafter.

2.3.4. Menopause: The Climacteric

The average age of menopause (last menstrual period) in the United States is 51, but much like the pubertal transition, menopause is a process that occurs over a period of years. Menopause is recognized when a woman has not had a menstrual period for 12 months. As a woman ages, there is a steady decline in the number of ova (eggs) in her ovaries that can be mustered to develop into follicles with the potential to ovulate. As the number of developing follicles declines, so does the level of estradiol in the circulation. The brain and pituitary, sensing this gradual decline in estrogen, strengthens the signal (LH and FSH) to the ovary to encourage more follicular development and estrogen production. Thus, as a women approaches menopause, there is a gradual increase in circulating LH and FSH levels, eventually reaching postmenopausal levels that remain high because there is no feedback signal (estrogen) from the ovary to control their release. This feedback relationship will be discussed in more detail in Chap. 9. During this time, there is also an adjustment to this new hormonal environment with many psychological and physiological changes, some of which can be unpleasant and disturbing. Symptoms of menopause that most women complain about are vasomotor changes (hot flashes), mood changes, and urogenital problems. Hot flashes are experienced by about 75% of menopausal women and typically last for about 3.8 years *(2)*. Intense heat, sweating, flushing, chills, and clamminess are all symptoms experienced during a hot flash. Once thought to be a figment of the menopausal imagination, hot flashes reflect a real increase in core body temperature and in skin temperature in the digits, cheek, forehead, upper arm, chest, abdomen, back, calf, and thigh *(3)*. It is interesting that menopausal hot flashes occur at the same time pulses of LH are released from the pituitary *(4,5)*. This suggests that the abrupt increases in body temperature are linked to the same central nervous system event that causes the intermittent release of GnRH that stimulates LH release. The current opinion is that estrogen regulates norepinephrine activity in the brain, and since norepinephrine release influences both LH release and body temperature, it is the changes in norepinephrine activity owing to estrogen withdrawal that causes the hot flashes *(3)*. Long-term effects of decreased estrogen levels including increased cardiovascular disease, osteoporosis, and decreased mental function are far more debilitating than the transitional changes that occur at menopause.

2.3.5. Postmenopause: Life Without Estrogen

At the beginning of the twentieth century, the average age at menopause was 50 years, and this age was also the approximate life expectancy for women at that time. Because the life expectancy at the present time for women has increased to over 80 years, most women will live more than a third of their life after menopause, and without estrogen from their ovaries. This extended postmenopausal is a relatively recent phenomenon, and therefore, some of the health-related issues caused by aging are poorly understood and not well documented. Although the increase of 30 years in life expectancy in past 100 years is substantial, the gain in healthy, functional years is less impressive. Many women experience physical and mental impairment in these later years that restricts their social function and isolates them from their friends and family. What is even more disturbing is that because of a variety of environmental influences, many young women have menstrual cycle disturbances that result in the hormonal levels that approximate those seen in menopause. If menstrual cycle disruption is prolonged and particularly if there is amenorrhea, this can result in some of the same consequences at age 30 or 40 that are usually experienced by women in their seventies and eighties.

2.3.5.1. OSTEOPOROSIS

Of all the consequences of aging in women, osteoporosis is the most debilitating and affects the most women. The risk of a lumbar or hip fracture, particularly after the age of

65, approaches 50% in white women. There are effective treatments for this condition including hormone replacement therapy (HRT) and bisphosphonates. HRT, specifically estrogen therapy, reduces bone turnover and improves calcium homeostasis. However, there are drawbacks to HRT, and the risks of breast and uterine cancer in individuals with a family history of these diseases must be considered when decisions regarding the treatment of osteoporosis are made.

2.3.5.2. ALZHEIMER'S DISEASE

A significant percentage of older women have some form of dementia (deterioration of cognitive function) and estrogen may protect against this deterioration *(6)*. Although, compared with previous studies, the recent results of the Women's Health Initiative (WHI) suggest that there is an increased risk of ischemic stroke with estrogen (Premarin) either with or without progesterone (Prempro), this trial used HRT on older menopausal women with obesity as a complicating factor *(7–10)*. Well-controlled studies with the native estrogen, estradiol-17β are needed before a rational, effective treatment regimen for menopausal/postmenopausal women can be safely proscribed. There are studies suggesting that estrogen replacement therapy, if begun at menopause and continued for a few years, is effective in reducing both the risk for osteoporosis and dementia.

2.3.5.3. CORONARY ARTERY DISEASE AND STROKE

The overwhelming evidence from observational studies indicates that estrogen has a protective effect against coronary artery disease *(11)*. This effect of estrogen appears to be limited to prevention of cardiovascular disease and does not ameliorate the progression of coronary disease that is established *(12)*. However, two recent trials, Heart and Estrogen/Progestin Replacement Study (HERS) and WHI, have brought these observational data into question. These two large trials with 2763 women (HERS) and 16,608 women (WHI) found no net benefit (HERS) or an increased risk (WHI) of coronary artery disease with HRT *(13)*.

2.3.5.4. BREAST AND ENDOMETRIAL CANCER

One of the main concerns in women who take HRT for menopausal symptoms is the increased risk of breast and endometrial cancer. The analysis of some 50 studies clearly indicates an increased risk for breast cancer in women taking estrogen alone *(14)*. The risk is increased substantially when women are on combined treatment of estrogen and progesterone *(15)*. Conversely, progesterone has a protective effect against the increased incidence of endometrial cancer in postmenopausal women taking estrogen therapy alone *(16)*.

2.4. CONCLUSIONS

During the lifespan of women, there are dramatic and life-changing transitions associated with the beginning and cessation of reproductive functions. These transitions, puberty and menopause, result in dramatic changes in the anatomy, physiology, and cognitive function of females and are caused by fluctuating levels of estrogen and progesterone. Because the life expectancy for women is now approaching 80 years and menopause occurs at about 50 years of age, the average female will live approximately 30 years after her ovaries have ceased to produce estrogen. This has serious physical and mental health implications.

2.5. SCENARIO WITH QUESTIONS AND ANSWERS

2.5.1. Scenario

One of the female students on your cross-country track team is 16 years old and has not experienced menarche. She is a good athlete and exercises regularly, but not excessively. All of the other girls her age have started their periods. She denies being sexually active.

2.5.2. Questions

1. Should this concern you as a coach responsible, to some extent for the health of your athletes?
2. What, if anything, would you recommend that the girl do?

2.5.3. Plausible Answers

1. When a young woman has not experienced menarche by the age of 16 there is some cause for concern. This situation is called primary amenorrhea and may be due to some medical or developmental problem.
2. There are several reasons to seek medical advice in this situation.

   a. Even though the individual has denied sexual activity, she may be pregnant, and at this age, prenatal care is critical.
   b. There may be cognitive or behavioral problems causing this condition that are not evident, but need medical attention.
   c. There are developmental abnormalities that result in amenorrhea. Some of these are serious and this situation should definitely be investigated.

REFERENCES

1. Messager S, Chatzidaki EE, Ma D, Hendrick AG, Zahn D, Dixon J, Thresher RR, Malinge I, Lomet D, Carlton MBL, Colledge WH, Caraty A, Aparicio SAJR. Kisspeptin directly stimulates gonadotropin-releasing hormone release via G protein-coupled receptor 54. *Proc Natl Acad Sci USA* 2005;102:1761–1766.
2. Avis NE, Crawford SL, McKinlay SM. Psychosocial, behavioral, and health factors related to menopause symptomatology. *Womens Health* 1997;3:103–120.
3. Freedman RR. Pathophysiology and treatment of menopausal hot flashes. *Semin Reprod Med* 2005;23:117–125.
4. Caspar RF, Yen SSC, Wilkes MM. Menopausal flushes: a neuroendocrine link with pulsatile luteinizing hormone secretions. *Science* 1979;205:823–825.
5. Tataryn IV, Meldrum DR, Lu KH, Frumar AM, Judd HL. LH, FSH, and skin temperature during menopausal hot flush. *J Clin Endocrinol Metab* 1979;49:152–154.
6. Wise PM. Estrogens and cerebrovascular stroke: what do animal models teach us? *Ann N Y Acad Sci* 2005;1052:225–232.
7. Rossouw JE, Anderson GL, Prentice RL, LaCroix AZ, Kooperberg C, Stefanick ML, Jackson RD, Beresford SAA, Howard BV, Johnson KC, Kotchen JM, Ockene J. Risks and benefits of estrogen plus progestin in healthy postmenopausal women: principal results from the Women's Health Initiative randomized controlled trial. *JAMA* 2002;288:321–333.

8. Rapp SR, Espeland MA, Shumaker SA, Henderson VW, Brunner RL, Manson JE, Gass MLS, Stefanick ML, Land DS, Hays J, Johnson KC, Coker LH, Dailey M, Bowen D. Effect of estrogen plus progesterone on global cognitive function in postmenopausal women. *JAMA* 2003;289:2663–2672.
9. Shumaker SA, Legault C, Rapport SR, Thal LJ, Wallace RB, Ockene JD, Hendrix SL, Jones BN, Anlouise RA, Jackson RD, Kotchen JM, Wassertheil-Smoller S, Wactawski-Wende J. Estrogen plus progestin and the incidence of dementia and mild cognitive impairment in postmenopausal women. *JAMA* 2003;29:2651–2662.
10. Anderson GL, Limacher M, Assaf AR, Bassford T, Beresford SA, Black H, Bonds D, Brunner R. Effects of conjugated equine estrogen in postmenopausal women with hysterectomy: the Women's Health Initiative randomized controlled trial. *JAMA* 2004;291:1701–1712.
11. Barrett-Conner E. Hormone replacement therapy. *BMJ* 1998;317:457–461.
12. Herrington DM, Reboussin DM, Brosnihan KB, Sharp PC, Shumaker SA, Snyder TE, Furberg CD, Kowalchuk GJ, Stuckey TD, Rogers WJ, Givens DH, Waters D. Effects of estrogen replacement on the progression of coronary-artery atherosclerosis. *N Engl J Med* 2000;343:522–529.
13. Hulley SB, Grady D. The WHI estrogen-alone trial—do things look better? *JAMA* 2004;291:1769–1771.
14. Collaborative Group on Hormonal Factors in Breast Cancer. Breast cancer and hormone replacement therapy: collaborative reanalysis of data from 51 epidemiological studies of 52,705 women with breast cancer and 108,411 women without breast cancer. *Lancet* 1997;350:1047–1059.
15. Willet WC, Colditz G, Stampfer M. Postmenopausal estrogen-opposed, unopposed, or none of the above. *JAMA* 2000;283:534–535.
16. Grady D, Rubin SM, Petitti DE. Hormone therapy to prevent disease and prolong life in postmenopausal women. *Ann Intern Med* 1992;117:1016–1037.

3 Considerations of Sex Differences in Musculoskeletal Anatomy

Phillip S. Sizer and C. Roger James

CONTENTS

3.1 LEARNING OBJECTIVES
3.2 INTRODUCTION
3.3 RESEARCH FINDINGS
3.4 SUMMARY
3.5 CONCLUSIONS

3.1. LEARNING OBJECTIVES

After completing this chapter, you should have an understanding of the following:

- Sexual dimorphism and how it applies to humans.
- Sex differences in general morphology.
- Sex differences in skeletal geometry.
- Sex differences in collagenous, cartilage, and bone tissue.
- Sex differences in the upper extremity anatomy and mechanics.
- Sex differences in the lower extremity anatomy and mechanics.
- Sex differences in the spine anatomy and mechanics.

3.2. INTRODUCTION

A woman's musculoskeletal anatomy is grossly similar, yet individually distinctive from a man's musculoskeletal anatomy. Structural differences exist between the sexes, and these differences are due to both environmental and genetic factors. Sex differences in musculoskeletal anatomy can be described in terms of sexual dimorphism, which refers to physical differences in secondary sexual characteristics between male and female individuals of the same species resulting from sexual maturation *(1)*. Sexual dimorphism is present in many species of birds, spiders, insects, and mammals, among others *(1)*. For example, male pheasants are larger and more brightly colored than female pheasants, some female spiders are larger than their male counterparts, and only male deer grow antlers, just to name a few *(1)*. However, with a few exceptions

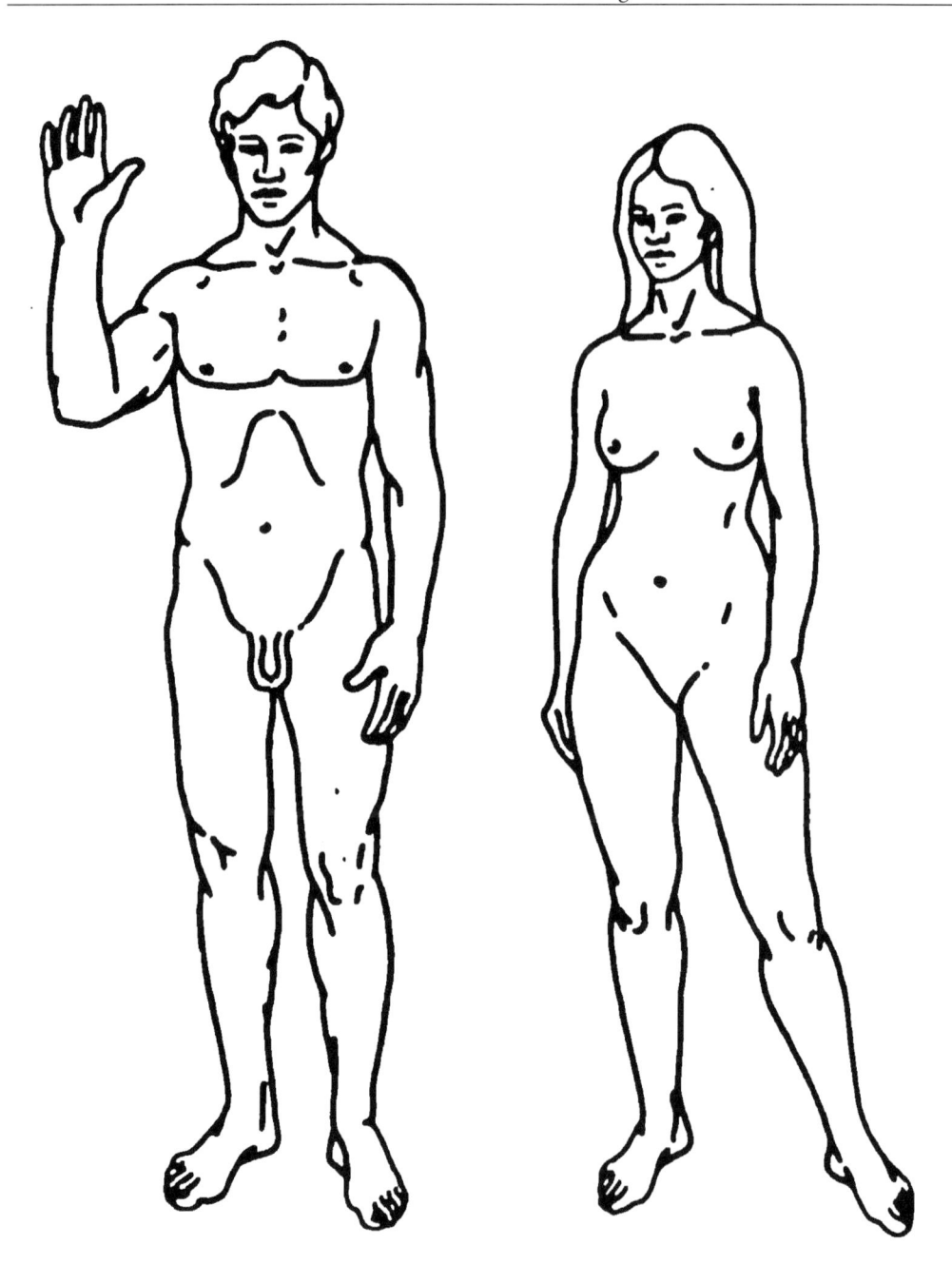

Fig. 3.1. Symbolic representation of men and women as depicted on the plaque of the Pioneer 10 spacecraft in 1972. *Source*: NASA. (www.nasa.gov/centers/ames/images/content/72418main_plaque.jpg). Adapted with permission.

(e.g., facial hair), sexual dimorphism in humans is more subtle compared to other species *(1)*. Yet, most individuals recognize that men and women exhibit different physical characteristics that include differences in body height, weight, shape, size, and alignment of the extremities (e.g., pelvis width, body mass distribution, and ligament and tendon laxity) *(2,3)*. Some of these differences in body structure are widely recognized and ingrained in cultural beliefs and stereotypes. For example, an artist's rendition of a representative man and woman was used to depict the sexes of the human species on the plaque of the Pioneer 10 spacecraft (Fig. 3.1) where the differences in gross structure are evident.

The typical differences in physical characteristics of the sexes are further exemplified by population data. Data from standard growth charts *(4)* demonstrate typical sexual dimorphic differences, but the division between men and women is usually < 1 SD and is age-dependent *(1)*. For example, according to the clinical growth charts provided by the Centers for Disease Control (CDC), boys and girls at the 50th percentile are approximately the same height (usually within 1–2 cm) until puberty *(4)*. However, after puberty (age 20 years), men are an average of approximately 14 cm taller than women *(4)*. Similar relationships are documented for body mass, with a relatively small difference observed prior to puberty and an approximately 12 kg difference (men greater than women) at the age of 20 *(4)*.

Although sexual dimorphism is apparent in general body characteristics, other sexual dimorphic traits are less obvious. Reports of sex differences in skeletal and soft tissue components are prevalent in the literature, and these differences not only explain the differences in general appearance but also may influence movement patterns, injury risk, and the development and progression of selected musculoskeletal pathologies. Consideration of the sex differences in musculoskeletal anatomy is important for both the general public and health care professionals in order to provide a basis for understanding normal and abnormal conditions that may exist. Moreover, a thorough appreciation that men and women have differences in musculoskeletal anatomy may suggest that they have distinctive health care needs. Therefore, the purposes of this chapter are to (1) examine sex differences in the anatomy of selected musculoskeletal components and (2) explore selected regional considerations in female functional pathoanatomy, which are pertinent to women's health issues.

3.3. RESEARCH FINDINGS

3.3.1. Sex Differences in the Anatomy of Selected Musculoskeletal Components

3.3.1.1. SEX DIFFERENCES IN SKELETAL GEOMETRY

There are several differences in skeletal geometry between men and women. In the sports medicine literature, sex differences in musculoskeletal anatomy, including skeletal geometry, have been reported in context with common injuries that occur in active women and have mostly focused on lower extremity characteristics. For example, differences in size, shape, structure, or alignment of the pelvis, femur, tibia, tarsals, and toes all have been reported *(2,5–9)*. Sex differences in the pelvis include a larger inlet and outlet *(2)*, greater interacetabular distance *(6)*, and a greater hip width normalized to femur length *(3,5)* in women compared to men. Sex differences in the femur include increased femoral anteversion *(3)* and a narrower femoral intercondylar notch width *(7,10)* in women. However, other authors have reported no differences in notch width between sexes *(11,12)*. Greater genu recurvatum *(2,9)*, more lateral patellar alignment *(2)*, greater tibial torsion *(9)*, and more bunions and deformities of the lesser toes *(2)* in women have been reported as well. Because of the mechanical linkage and interaction among structures of the lower extremity, skeletal differences in one or more interacting structures may result in differences in overall lower extremity alignment. For example, the quadriceps or Q angle, is formed by the intersection of a line connecting the anterior superior iliac spine and the midpoint of the patella with a line connecting the midpoint of the patella and the tibial tuberosity. In the literature, it is generally reported that Q angle is larger in women than in men *(5,8)* and is a function of the structural and alignment characteristics of the involved bones (e.g., pelvis width, patella position, and tibial torsion). Additionally, large deviations in Q angle have been suggested to contribute to some knee and foot pathologies *(2,13)*, although reports are equivocal.

Further evidence supporting the existence of sex differences in skeletal geometry comes from the areas of forensic anthropology and archeology. Scientists from these areas have used knowledge of skeletal sexual dimorphism to determine the sex of deceased individuals from their skeletal remains. A plethora of literature exists that describes the skeletal geometric characteristics commonly used and the ability of these characteristics to predict sex. The humerus *(14)*, pelvis *(15,16)*, femur *(14,17–22)*, tibia *(23–25)*, talus *(26)*, and calcaneus *(26–28)* all have been used for this purpose. The ability to discriminate sex based on one or more skeletal geometric characteristics varies somewhat by different bones as indicated by the percentage of individuals accurately categorized as male or female in the respective studies cited: tibia (96%) *(24)*, femur (95%) *(18)*, calcaneus (92%) *(27)*, pelvis (88%) *(16)*, and talus (81%) *(26)*. Furthermore, different parameters from the same bone appear to be better discriminators than other parameters *(15,16,18,21–23,28)*. For example, Mall et al. *(21)* reported that a different percentage of individuals could be grouped correctly using a discriminant analysis when evaluating different characteristics of the femur. In this study, the femoral head transverse diameter (89.6% of cases correctly categorized) was the best discriminator, followed by the head circumference (87.7%), vertical head diameter (86.8%), condylar width (81.4%), maximum midshaft diameter (72.4%), and maximum length (67.7%) *(21)*. Using a combination of variables (midshaft diameter and head circumference), 91.7% of the cases were classified correctly *(21)*.

The presence and magnitude of sex differences in skeletal geometry appears to be dependent on a number of factors including skeletal maturity *(15,29)*, environmental stresses (i.e., loading) *(14,18,20,25)*, and genetics (i.e., race) *(16–20,22,30)*. Consideration of these factors is important in establishing sexually dimorphic traits in skeletal geometry. Nevertheless, ample evidence exists to support the presence of sex differences in skeletal geometry.

3.3.1.2. SEX DIFFERENCES IN MUSCULOSKELETAL TISSUES

In addition to sex differences in skeletal geometry, there are several reported sex differences in musculoskeletal tissues, including collagenous tissues (e.g., tendon, ligament, and skin), cartilage, and bone *(31–34)*. Collagen is the primary protein of connective tissues in mammals *(35)*. It is strong, inextensible and provides much of the strength of tendon, ligament, skin, and cartilage *(35,36)*. Additionally, it is the main protein component of bone *(35,36)*.

In collagenous tissues, collagen molecules align themselves based on stress patterns and provide strength against tensile loads *(36)*. Some of the reported sex differences in collagen include differences in thickness *(31)*, orientation *(31)*, content *(37)*, diameter *(32)*, volume *(33)*, and metabolism *(34)* for the specific tissues examined. Some of these sex differences are associated with fiber strength. Moreover, sex differences in collagen degradation have been observed in subjects as young as 2 years of age *(34)*, although other sex differences may not appear until after several decades of life *(33)*. Many disorders of collagenous tissues (e.g., lupus erythematosus, scleroderma, rheumatoid arthritis, dermatomyositis, and Sjögren's syndrome) have been associated with sex differences. Etiological factors associated with collagenous tissue diseases are thought to exist at many levels (e.g., genetic, cellular, organ, age, behavioral, and environmental), but sex hormones are thought to influence the onset and course of these disorders *(38)*.

Sex differences in articular cartilage are reported in the literature *(39–42)* and have been associated with differences between men and women in the onset of osteoarthritis *(40,42,43)*. Epidemiological evidence suggests that women are more likely (1.5–4 times greater risk) to develop osteoarthritis than men *(40)*. Sex differences in articular cartilage morphology have been reported in children (age 9–18), persisting throughout adulthood, and increase during the postmenopausal years *(39–41,44)*. Reported sex differences in articular cartilage morphology include greater cartilage volume, thickness, and surface area in male compared to female subjects *(39–42)*. These differences appear to be partially related to other characteristics such as age, body mass index, bone area, physical activity, and the specific articulation involved *(40,44)*. However, research suggests that sex differences remain at some cartilage sites even after adjustment has been made for these other factors *(39,40)*. Furthermore, the sex differences in cartilage morphology have been associated with a faster cartilage tissue accrual rate in boys compared to girls (i.e., more cartilage tissue early in life) *(40)* and greater cartilage tissue degradation in older women compared to older men (i.e., greater loss of cartilage later in life) *(45)*.

The presence of sex hormone receptors in cartilage tissue is thought to be an indicator that sex hormones influence these accrual and degradation processes *(42,43,45,46)*. Evidence for the role of sex hormones in cartilage morphology and metabolism has been demonstrated in both animal *(42,46)* and human *(45)* models. In mice, sex hormones have been shown to influence inflammation-induced cartilage degradation through modulation of cytokine production and release in granulomatous tissue *(46)*. Additionally, male rats have demonstrated higher levels of proteoglycan and collagen, lower glycosaminoglycan loss, and greater proteoglycan synthesis than female rats in vitro *(42)*. Furthermore, cartilage from female rats was shown to have greater susceptibility to degradation when implanted into female rats compared to male cartilage implanted in female rats or both male and female cartilage implanted into male rats *(42)*. In humans, urinary markers of cartilage degradation have provided evidence that cartilage loss is greater in women than in men *(45)*. Additionally, cartilage degradation was shown to be greater in postmenopausal women compared to age-matched premenopausal women and less in postmenopausal women undergoing hormone replacement therapy compared to postmenopausal women not undergoing the therapy *(45)*. Therefore, sex differences in cartilage morphology and metabolism that exist early in life appear to increase with advancing age and might be explained by differences in sex hormones.

Sex differences in bone tissue are also reported in the literature *(47–50)* and have been associated with differences between men and women in affecting the risk of fracture *(51–53)*. Sex differences in bone tissue vary by skeletal site *(29,54)*, but are reflected by differences in both morphological *(29,48,51,53–57)* and remodeling *(47,49, 50,54,57,58)* characteristics, particularly in osteoporotic individuals *(36,54,59,60)*. Sex differences in bone tissue are present at an early age *(29,47,48,55)*, persist throughout adulthood *(36,49–51,57,58)*, and diverge further in older age *(36,49,50,52–56,58)*. Bone mass accrues during childhood and adolescence and peaks at about the age of 30 years in both men and women *(36)*. A lower peak bone mass has been associated with a greater risk for osteoporosis in later life *(47)*; therefore, sex differences in the development of bone tissue during youth might partially explain some of the bone tissue differences between adult men and women. Several factors influence bone mass accrual, including nutrition (e.g., calcium and vitamin D), physical activity, lifestyle behaviors (e.g., smoking), genetic factors (race and sex), and hormonal factors (e.g., estrogen) *(59,61)*.

In children, serum markers of bone turnover have been shown to change significantly with pubertal age in both boys and girls *(47)*. In girls, these markers were shown to peak at midpuberty and decrease thereafter; in boys, the markers continued to increase through late puberty *(47)*. Moreover, the serum markers of bone turnover were overall higher in boys than girls even after adjustment for age, body weight, and pubertal stage *(47)*. These sex differences in bone turnover likely influence differences in peak bone mass *(47)*, volumetric bone mineral density *(48)*, selected measures of bone area *(29)*, cortical thickness *(29)*, and ultimately compressive *(48)* and bending *(55)* strength. However, some authors have suggested that sex differences in some of these characteristics can be explained by differences in anthropometric dimensions (e.g., height and body total lean mass) *(55)*.

In adults, there are several reported sex differences in bone tissue. Differences between men and women have been reported in bone mineral density (men greater than women) *(57)*, peak bone mass (men greater than women) *(36)*, cortical thickness (men greater than women) *(51)*, age-dependent hormonal responsiveness of osteoblasts (less responsiveness in older cells from women compared to older cells from men) *(49)*, bone turnover (men greater than women) *(57)*, and bone strength (men greater than women) *(36,51)*. In older adults, sex differences in bone mineral density *(52,54)*, cross-sectional area *(52)*, cortical thickness *(52)*, bone width *(52)*, and strength *(36,52,56)* remain substantial, with some differences (i.e., bone mineral density and strength) further diverging compared to younger adult values *(36,52)*. Furthermore, many of these sex differences do not disappear after adjusting for anthropometric factors such as height and weight *(52)*.

The role of sex hormones in the loss of bone mass in older women has been explored widely in the literature. It is well known that a decrease in estrogen production following menopause is the primary contributing factor to the accelerated loss of bone mass in older women compared to older men *(59,60)*. However, older men also lose bone mass and are at greater risk for hip and vertebral fractures compared to their younger counterparts *(50)*. Although men do not experience a physiological event comparable to menopause and therefore do not experience a substantial decline in total serum testosterone or estrogen, some evidence suggests that a decline in the bioavailable estrogen (nonsex hormone-binding globulin-bound) might explain the loss of bone mass in both women and men *(50)*.

3.3.2. Selected Regional Considerations in Female Functional Pathoanatomy

3.3.2.1. UPPER EXTREMITY

3.3.2.1.1. Shoulder Sex-based anatomical differences of the shoulder complex are tissue-specific and pathology-specific. For example, Pandley et al. *(62)* observed significant differences between men and women in the distribution of the articular branch of the axillary artery, which may influence decisions made during shoulder surgery. Differences have been associated with the incidence of external impingement of the shoulder. Although no sex-based differences have been found in the role of the coracoid process in subscapularis impingement *(63)*, differences in the role of the acromion process with external impingement have been observed. Historically, Bigliani et al. *(64)* classified the shape of the acromion process. Type I processes are flat, Type II are mildly hooked, and Type III are drastically hooked. Bigliani et al. *(64)* suggested that differences could lend to the incidence of impingement and any subsequent rotator cuff tearing. Later, Bigliani et al. *(65)* reported that 78% of all full thickness rotator cuff tears were associated with a Type III acromion. More recent investigators have observed increased incidence of full thickness rotator cuff tears in women versus men *(66)*. Selected investigators have suggested that acromion differences are acquired, resulting from altered tension loads imposed by the coracoacromial ligament and deltoid *(67)*. Getz et al. *(68)* observed that Type III acromia were more common in female patients and discovered that Type II acromia were related to posterior capsule adaptive shortening of the glenohumeral joint.

Although external impingement has been associated with sex and age, the relationship between age, sex, and incidence of acromion type is controversial *(68–70)*. More recent investigators have suggested that although inferior acromion changes increased with age, they were not different between sexes *(71)*. However, investigators have observed limited sex-based differences in the acromiohumeral interval with the shoulder at rest, where women exhibited a reduced space compared to men *(72)*.

Although men appear to experience more frequent frank anterior dislocations of the glenohumeral joint *(73)*, women appear to be predisposed to glenohumeral laxity *(74)*. This disparity appears to be related to the notion that not all joint instability results in dislocation, where grades I and II instability represents increased motion and possible humeral head perching on the anterior labrum versus the frank dislocation of the head in grade III *(75)*. Although glenoid fossa inclination appears to influence instability incidence *(76)*, no sex differences in this architectural feature have been noted *(77)*. Thus, the woman's predisposition appears to be more related to increased anterior capsular redundancy and resultant hypermobility, along with decreased joint stiffness *(74)*.

Finally, women between the ages of 40 and 60 are more predisposed to developing secondary idiopathic capsulitis *(78)*. This condition is associated with increased thickening of the anterior superior joint capsule at the coracohumeral ligament *(79)*, along with a noninflammatory synovial reaction in the proximity of the subscapularis tendon *(80)*. These changes demonstrate active fibroblastic proliferation accompanied by tissue transformation into a smooth muscle-like phenotype that is similar to Dupuytren's disease *(81)*.

3.3.2.1.2. Elbow Women appear to be at greater risk for developing tennis elbow because of tendinosis that emerges from mesenchymal changes in the collagenous tendon structures *(82)*. This condition, typically lasting greater than 12 months in duration, is more likely noninflammatory in nature *(83)* and affects one of four possible different regions of the tendinous insertions at the lateral elbow. The tendons that are at risk are specifically located about the lateral epicondyle of the distal humerus. The extensor carpi radialis longus (ECRL) originates on the distal one-third of anterior supracondylar ridge, possesses almost no tendon at the origin, and demonstrates an immediate transition into muscle. Extensor carpi radialis brevis (ECRB) starts from a ($5\,mm \times 5\,mm^2$) area on the superior surface of the lateral epicondyle (10% of origin) and collagen/fascial layers of intracompartmental septa that share fascia with the extensor digitorum communis (EDC) coursing to second and third metacarpal (MC) (especially fascia associated with third MC). Thus, resistive wrist dorsal extension and resistive dorsal extension of the second and third metacarpals may be suggestive of tendonopathy at either EDC or ECRB. The ECRB tendon is juxtaposed between the muscle bellies of ECRL and EDC. This common test outcome merits palpatory discrimination between the two regions for a differential diagnosis. The ECRB can exhibit tendonopathy at its origin, along the tendon between ECRL and EDC or at its musculotendinous junction more distally. Finally, EDC can be found at the anterior surface of lateral epicondyle. If involved in a lateral tendopathy, then resistive dorsal extension of the second through fifth metacarpophalangeal (MCP) joints will differentiate this lesion from a lesion of ECRB. This condition seldom occurs in isolation, but is typically discovered in combination with affliction to ECRB *(84)*.

The woman's predisposition for lateral elbow tendinosis is increased when her estrogen decreases, especially after premature hysterectomy (at less than 35 years of age) and/or lowered estrogen levels from other causes *(82)*. Although inflammatory tendinitis involves a chemically mediated inflammation because of tendon injury *(82)*, the tendinosis to which women are more predisposed produces a nonchemically mediated degenerative change associated with long-term tendon stress *(85)*, resulting in a condition that could persist longer than 12–14 months. This process produces tendon necrosis that manifests as a moth-eaten appearance in the tendon *(86)*. As a consequence, the tendon becomes friable, along with possible bony exostosis at the epicondyle *(82)*. This chronic condition can be accompanied by: an imbalance between vasodilatory and vasoconstrictive innervation *(86)*; substance P, and calcitonin generelated polypeptide (CGRP) proliferation in the vicinity of the affected tendon *(86)*; accompanied by a high concentration of glutamate in the surrounding tissue *(87)*.

Women appear to be more predisposed to ulnar nerve lesions at the medial elbow *(88)*. The medial elbow exhibits three different predilection sites for ulnar nerve entrapment. The nerve first courses underneath the arcade of Struthers just dorsal to the medial intermuscular septum. Distally, the nerve courses through the cubital tunnel, whose boundaries are the medial collateral ligament complex (ceiling), medial epicondyle (medial wall), olecranon process (lateral wall), and cubital tunnel retinaculum (floor). In selected individuals, the retinaculum is dorsally bordered by the anconeus epitrochlearis muscle that is innervated by the radial nerve and is activated simultaneous with the triceps, lending to possible entrapment symptoms during resisted elbow extension. The retinaculum that courses from the medial epicondyle to the olecranon tightens with passive elbow flexion, creating increased nerve entrapment symptoms at end-range passive flexion. Finally, the nerve must course under the retinaculum between the two heads of the flexor carpi ulnaris. As a consequence, entrapment symptoms may increase during resistive wrist flexion–ulnar deviation.

3.3.2.1.3. Wrist and hand Selected few afflictions of the wrist and hand appear to differ between men and women. Tendon pathologies, including tenosynovitis and tendinosis, appear to be more frequent in women *(89)*, but data related to the pathoanatomical and physiological influences on these differences have not been explored. Similarly, carpal tunnel syndrome (CTS) is more common *(90,91)* and prolonged *(92)* in women. However, multiple factors have been explored that may contribute to this difference. The etiology of CTS is multi-factorial, resulting from anatomical, biomechanical, pathophysiological, neuropathological, and psychosocial influences. Anatomic factors, such as tunnel architecture and volume *(93)*, lumbrical anatomy *(94)*, and the shape of the hamate hook *(95)*, have been associated with CTS. Specific anatomical and anthropometric measures appear to influence a woman's greater predisposition to CTS *(96,97)*. Although carpal bone size and scaling do not appear to differ between men and women *(98)*, hand-length ratios, space indices at the wrist, and digital features appear to differ between the sexes *(96)*. Along with these, differences in body mass index appear to predispose women to CTS *(96)*.

The onset and progression of CTS appear to be related to an increase in intra-tunnel pressure *(99)*. Different factors appear to increase this pressure, including tunnel narrowing associated with wrist movements *(100,101)*, carpal instability *(102)*, increased muscle force production *(103)*, and trauma that produces perineural edema and fibrosis *(104)*. Woman may be more susceptible to these influences versus men, because of reduced available space for the median nerve within the tunnel. The median nerve appears to increase in cross-sectional diameter with sustained repetitive hand movement in women compared to men *(97)*, thus compromising the tunnel and potentially increasing pressure within the tunnel in context with the previously discussed factors.

The individual suffering from CTS may experience sensory or motor changes, including paresthesias and/or true numbness that reflect deficits in neurophysiological function. Women have demonstrated greater deficits in neurophysiological function of the median nerve when compared to men *(105,106)*. However, controversy exists over the value of neurophysiological testing for the diagnosis of CTS *(107)*. Orthodromic median sensory latency is typically prolonged with CTS patients *(108)*, and median nerve motor amplitudes are decreased in patients with CTS *(106)*. Yet, Glowacki et al. *(109)* discovered a poor relationship between electrodiagnostic test outcomes versus final symptom profiles. Differences between the median and ulnar motor latencies appear to be important for the diagnosis of CTS *(105)*. Padula *(110)* found that the difference between the median and ulnar motor latencies were greater in patients suffering from CTS versus controls.

The presence of autonomic disturbances appears to be associated with the woman's increased predisposition to CTS *(111)*. This disturbance in neural function could be related to local sympathetic fiber and/or brachial plexus irritation associated with a double crush phenomenon, which has been observed in as many as 40% of all patients suffering from CTS *(112)*. As result, a vasoconstrictive event could lead to decreased perineural microvascular flow and increased protein leakage from the vascular supply that produce epineural and perineural edema *(104)*, as well as increased endoneural pressure and ischemia *(113)*.

3.3.2.2. LOWER EXTREMITY

3.3.2.2.1. Hip joint Women are at greater risk for both stress fracture *(114)* and macrotraumatic frank fracture at the hip *(115)* especially in the femoral neck *(114)*. This predilection appears to be influenced by differences in architecture about the hip and pelvis *(116)*. Acetabular depth and femoral head width appear to be reduced in women versus men *(117)*. The coxadiaphyseal (CD) angle has been reported to be wider in men versus women in selected races, thus potentially predisposing women to higher incidence in stress reactions *(118)*. Women appear to have decreased femoral neck strength versus men, as evidenced by decreased femoral neck cross-sectional moment of inertia *(119,120)*. Compressive stress, defined as the stress in the femoral neck at its weakest cross-section arising from a standardized fall, is higher in women *(120)*. These features interact with women's altered estrogen associated with menstrual irregularities *(121)* and/or menopause *(115)*, thus enhancing women's fracture predisposition.

The outer margin of the hip acetabulum is completely lined with the cartilaginous labrum that serves to enlarge the articular surfaces *(122,123)*. The labrum enhances the articular seal, fluid pressurization load support, and joint lubrication of the hip joint *(124)*. The labrum possesses a variety of sensory endings important to proprioception and nociception *(125)*. The labrum is vascularized in a fashion similar to the meniscus of the knee, where the outer margins are well vascularized and the inner margin is lacking *(126)*. The labrum is at risk for vertical traumatic, as well horizontal degenerative, tears *(123,127)*. The propensity for tears is increased by the deficiencies in the mechanical properties of the labral tissue, especially in women. Labra obtained from male patients have stronger tensile stress than those from female patients *(128)*. Moreover, labral degenerative changes may influence those same mechanical properties *(129,130)*.

Labral tears appear to occur more frequently in the superior region of the structure, because of decreased mechanical properties accompanied by increased demand *(129,131)*. The superior region of the labrum appears to be less well vascularized, lending to the susceptibility of that region to traumatic and degenerative tears *(129,131)*. One significant mechanical contribution to this loading demand is the impact of the femoral neck against this region during full flexion of the hip *(130)*. Femoral neck architecture appears to differ between men and women, where increased thickening and decreased CD angulation of the neck in older women predispose them to anterior acetabular labral trauma, especially when the hip is positioned in full flexion *(132)*.

3.3.2.2.2. Knee complex Little evidence is available to describe sex-based differences in the patellofemoral complex of the knee. One might explain differences in terms of cartilage volume, where sex explains 33–42% of the variation in knee cartilage volumes and women demonstrate decreased cartilage volume versus men *(39)*. However, with T2 magnetic resonance imaging (MRI) examination of young, healthy volunteers, sex-based differences could not be seen for the magnitude and spatial dependency of cartilage *(133)*.

Investigators have attempted to describe sex-based differences in terms of patellofemoral contact areas at various positions of knee flexion. Csintalan et al. *(134)* observed larger contact areas in male posterior patellar surfaces with the knee flexed to 30°. In addition, they observed a greater change in the female's contact pressures in response to varying vastus medialis activity with the knee positioned at 0°, 30°, and 60° flexion. Although no differences were seen by Besier et al. *(135)* with the knee in full extension, they observed larger contact areas in male patellofemoral joints with the knee flexed to 30° and 60°. However, the contact areas were not different when the data were normalized by patellar dimensions of height and width.

More striking is the relationship between sex and knee ligament injury. Injury to the anterior cruciate ligament (ACL) can be a devastating event, and a woman's increased risk for this injury over male counterparts is well documented *(136,137)*. It has been reported that 70% of all ACL injuries are a result of a noncontact mechanism *(136,138)*, where girls and women appear to tear their ACL two to eight times more frequently than men *(139)*.

Because the reason for this is unclear, investigators have explored many possible causes including anatomical differences, hormonal differences, and mechanical differences. One of the classic anatomical factors attributed to sex-based differences in ACL injury is the width of the femoral intercondylar notch, or Grant's notch. The intercondylar notch is found in the roof of the space between the femoral condyles, lending a point where the ACL could crimp or tear during forced rotational noncontact injury *(137)*. The female knee was once thought to possess a smaller notch versus men, lending them to greater vulnerability for traumatic tears *(10,12)*. However, other investigators have suggested that increased female risk was based on differences in the ratio between the notch width and width of the femoral condyles (notch width index) *(140)*.

The role of the notch width has remained controversial. Charlton et al. *(141)* have suggested that the narrower notch width in the female knee simply reflects the smaller diameter ACL within the notch, which still must constrain the relative same loads and stresses as the male ACL. This difference in diameter, along with an increase in creep deformation under sustained loading *(142)*, subsequently renders the ligament to greater injury potential in female athletes. Murshed et al. *(143)* found no differences in notch width characteristics between the sexes, and Ireland et al. *(11)* suggested that any individual with a smaller notch width is at higher risk for injury, regardless of sex.

Static knee postural and alignment characteristics have been considered to be factors that could contribute to woman's greater risk for ACL injury *(144)*. The Q angle is a clinical measure used to determine the position of the knee in the frontal plane *(144)*. Livingston and Mandigo *(145)* compared Q angles between male versus female lower extremities and found no significant sex or right-to-left lower limb differences. Conversely, Tillman et al. *(13)* reported that women exceeded men in quadriceps angle (Q angle) and thigh-foot (TF) angle *(146)*. Yet, the TF angle, which is a measurement of tibia external rotation (toeing out), is not clearly linked to ACL injury *(147)*.

Biomechanical features have been linked to sex-based ACL injury predisposition *(147)*. Kinematically, differences in knee flexion angle at contact (< 30°), tibial rotation in the coronal plane, and frontal plane motion have all been implicated *(148)*. Investigators have associated reduced contact and peak knee and hip flexion during selected load-bearing functional activities with female ACL injury *(149,150)*. Similarly, investigators have observed decreased peak hip abduction in women when cutting *(149,151)*.

Female athletes have exhibited increased valgus motion in the frontal plane during a landing or cutting maneuver, which may serve as a factor in female ACL injury predisposition *(152)*. Numerous investigators have observed this behavior *(149,150, 153)*, along with an increased variability in the valgus motion during the landing and/or cutting sequence *(149)*. Hewett et al. *(154)* reported that these excessive behaviors could be reduced through appropriate jump training. Yet, sex-based sagittal *(153)* and frontal *(155)* plane movement differences have been disputed, where expected differences did not emerge, and the authors suggest that other factors are at play in producing the increased injury risk for the female ACL. Finally, authors have reported increased coronal plane excursion for the hip and knee in women versus men during drop-landing activities *(156)*, producing increased internal rotation during those activities.

Increased joint laxity and anterior tibial translation is associated with noncontact ACL injury *(147,157)*. Trimble et al. *(147)* reported that sex and excessive subtalar joint pronation are the only predictors of knee joint laxity. Women exhibit increased anterior knee joint surface translation during extension *(158)*. This is accompanied by reduced protective hamstring activity during that translation *(139,158)* that renders the female ACL less protected when exposed to anterior shear forces *(159)*. In a similar fashion, the female's ACL injury predisposition may be related to the excessive subtalar joint pronation in the female ankle *(147,149,160)*, which appears to promote the previously discussed increase in tibial internal rotation *(160)* and tibial anterior translation *(147)*. This behavior does not appear to relate to genu recurvatum and the TF angle *(147)*. Still other authors have not observed the sex-based differences in subtalar pronation *(13)*.

Differences in kinetic behaviors when cutting or landing have been attributed to the female ACL injury risk *(161)*. Female athletes have been found to exhibit reduced peak knee flexor moments *(162)* and increased peak knee valgus moments *(150,162,163)* during cutting tasks. Similarly, Kernozek et al. *(150)* found increased greater peak vertical and posterior force than men during landing. These altered kinetic behaviors are accompanied by reduced leg stiffness during rapid load-bearing *(164)*, which could translate into a reduced ability to diffuse stress from the ACL *(165)*.

Stress management appears to be related to appropriate cocontractive behavior between the quadriceps and hamstrings during cutting and landing. Thus, differences in muscle activation, timing, coordination, and force production may serve as a contributing factor to the female ACL injury predisposition *(166,167)*. The woman's difference in muscle activity may begin early, as girls developmentally increase the quadriceps strength disproportionately more than the hamstring strength *(168)*. Although da Fonseca et al. *(169)* questioned the role of sex in cocontractive disturbances, several authors have suggested that female athletes exhibit greater quadriceps versus hamstring activity during landing and cutting *(170–172)*. Increased soleus *(170)* and gastrocnemius *(171)* activity may contribute to the woman's recruitment differences, although decreased hamstring activity may reduce the woman's ability to decelerate and control tibial extension, internal rotation, and anterior translational shearing *(158,173)*.

As a consequence of the woman's differences in kinematic, kinetic, and neuromuscular control strategies, the ACL potentially sustains greater loads. Unfortunately, the female ACL demonstrates different mechanical behaviors amidst these altered strategies. During passive cyclic loading, the female ACL appears to exhibit greater ligament creep versus the male ligament *(142)*. This difference in creep could compound the previously discussed deleterious mechanical effects, as quadriceps electromyographic activity may increase after ACL creep, although hamstrings coactivation is not likely change *(174)*.

Another factor that appears to interact with the anatomical and biomechanical influences contributing to female ACL injury predisposition revolves around changes in sex hormones. Ovarian sex hormone fluctuations have been related to increased noncontact ACL injury *(159,175)*. Estrogen and progesterone receptors have been identified within the substance of the ACL *(176)*, likely responsible for the relationship between peaks in estrogen levels and increased laxity *(177)*. Exposure to estrogen appears to increase metallomatrix protease activity and decrease fibroblastic activity within the ligament, lending to increased tissue laxity *(177)*. Correspondence between injury risk and menstrual phase is controversial at best, where authors have reported increased ACL injury during the follicular *(159,178)* and luteal *(178)* phases.

3.3.2.2.3. Ankle and foot Women appear to be more predisposed to ankle foot injuries than men in selected populations. Heir *(179)* found that women in military office training were at greater risk for developing Achilles tendonopathy and ankle sprains than their male counterparts. Structural differences have been noted in the female foot, which demonstrate smaller width and length, as well as specific shape *(180)*. However, sex differences are not observed in terms of medial longitudinal arch measurements *(181)* or overall arch height *(182)*.

Women appear to be at greater risk for ankle inversion trauma. Hosea et al. *(183)* found that female athletes are at greater risk for grade I inversion trauma, where there is disruption of the anterior talofibular ligament. The same authors found no sex-based differences in grade II (anterior talofibular and calcaneofibular ligament involvement) or grade III (the same ligaments plus the posterior talofibular ligament). This predisposition is not only related to the sports in which female athletes participate *(184)* but also by selected structural differences in the lower extremity. Female athletes' risk of inversion trauma is increased by increased tibia varum and rearfoot eversion in weight bearing *(184)*.

Other structural differences have been noted that add to female predisposition to ankle foot affliction. Women demonstrate decreased cartilage thickness over the talar dome, which is at risk for developing ostechondritis and necrotic changes *(185)*. Women appear to demonstrate increased obliquity of the first metatarsal base, resulting in increased metatarsus primus varus and potential increased incidence of clinical hallux valgus *(186)*. Women demonstrate increased incidence of hallux rigidus in the first metatarsophalangeal joint, with the vast majority of the subjects demonstrating a flat joint configuration.

Achilles tendonopathy has been attributed to numerous factors, including histochemical, pathomechanical, and neurophysiological influences. In addition, sex has been seen as a factor lending to the development of Achilles tendopathy, possibly interacting with other factors. Marked deficiencies have been noted for tissue histochemical responsiveness in female rabbit tendons *(187)*. Sex-based factors may contribute to differences in tendon response and pathology. For example, a female tendon may experience increased load in response to footwear with hard soles, insufficient rearfoot control, or high heels *(188)*, all which have been associated with increased incidence of tendopathy. However, this sex-based predilection for Achilles tendopathy is controversial, where more recent studies have questioned female predisposition *(189)*. Sex-based differences in Achilles tendon properties and pathology may be related to muscle and tendon strength differences, rather than other sex-specific tissue characteristics *(190)*.

3.3.2.3. Spine

3.3.2.3.1. Cervical Sex-related anatomical differences in the cervical spine can be observed in the vertebral structure *(191,192)*, lending to clinically relevant differences in bony processes and the joints they form, as well as the intravertebral and intervertebral spaces through which important neurovascular components course. For example, investigators have observed sex differences in dimensions of lower cervical vertebral laminae and pedicles. Rezcallah et al. *(191)* found that women demonstrated smaller pedicular widths, lengths, and transverse angles at C3 through C7 when compared to men. Similarly, Xu et al. *(192)* observed smaller laminar height, width, thickness, and angulation in women at levels C2 through C5. The role of sex-based cervical vertebral structure differences is not clear *(193)*. Yet, men appear to have a smaller vertebral canal-to-body anterior–posterior diameter ratio versus women, potentially predisposing them to a smaller canal in proportion to their overall axial skeletal morphism and decreased incidence of cervical myelopathy *(194)*. Conversely, women appear to demonstrate greater spinal canal narrowing after whiplash versus their male counterparts *(195)*, possibly contributing to their increased incidence of whiplash-related disorders *(196)*, as well as the latent clinical sequelae and delayed recovery status post-whiplash *(197,198)*.

Female cervical zygapophyseal (facet) joints may be at greater risk for injury during a whiplash trauma versus their male counterparts. Excessive segmental translation has been shown to be a potential cause of injury. Simulated rear impact using human volunteer subjects showed greater degrees of cervical retraction in women that were unaware at time of rear-end impact *(199)*.

Yoganandan et al. *(200)* found that the facet articular surfaces in female cadavers were less adequately covered by cartilage than similar specimens in men. In addition, these joints exhibited a greater distance from the dorsal-most region of the facet joint to the location where the cartilage began to appear (called a cartilage gap), potentially lending these joints to greater translation during unanticipated loads. Stemper et al. *(201)*

found that female cadaveric specimens exhibited increased compression in the dorsal region of the facet joint during the early phase of whiplash. These properties could predispose female facets to injury in the subchondral bone during normal physiological and traumatic loads, especially when accompanied by endplate perforations and older age.

Female cervical discs may be at greater risk for failure when exposed to aphysiological loads. Truumees et al. *(202)* examined the geometric characteristics and loading response of the cartilaginous endplates found in cadaveric cervical discs. They found that the female sex was associated with significantly lower endplate fracture loads when exposed to compression.

3.3.2.3.2. Thoracic The primary sex-based differences that have been observed in the thoracic spine appear to center around differences in postural alignment. Fundamental sex-based differences have been observed in children and adolescents regarding the extent of the thoracic kyphosis in the sagittal plane. The presence and severity of kyphosis is especially observable in women *(203)*. Thoracic kyphosis changes as children age, where the rate of change is greater in women versus men *(204)*. This change in the kyphosis seems to progress in a fashion similar to the lumbar spine during childhood. However, the relationship between the change in kyphosis and lordosis decreases by the age of 15 in girls, but not in boys *(205)*.

Of greater interest are the sex-based differences in the development and progression of adolescent idiopathic scoliosis (AIS), where the individual develops rotatory 3D deformation in the thoracolumbar spine, especially in the frontal plane. A single thoracic curve is the most common in selected populations, followed by other configurations of single and double curves in the thoracolumbar spine *(206)*. Investigators have reported increased incidence of AIS in female adolescents *(206–208)*, where female prevalence appears to be genetically coded *(209)*. Women appear to be at greater risk for developmental curve progression versus men, especially in the age prior to the onset of menses *(208)*. Similarly, girls with scoliosis generally grow faster than girls without the same condition *(210)*. Yet, while an age older than 15 years, skeletal maturity, postmenarchal status, and a history of spine injury are all associated with the prevalence of back pain in people with AIS, sex, family history of scoliosis, leg length discrepancy, curve magnitude, and spinal alignment were not *(211)*.

Investigators have not only looked at the interactions of physical changes with sex in scoliotic patients but also the role of sex in psychosocial factors associated with the condition. Girls with scoliosis seem to be at greater risk for psychosocial stresses, including feelings about poor body development, troubled peer interactions, and health compromising behaviors *(207)*. Yet, although investigators have evaluated the impact of scoliosis on health-related quality of life (HRQoL), the role of sex in that evaluation is controversial. Although Ugwonali et al. *(212)* discovered that male adolescents scored higher on validated self-report instruments that measured HRQoL, Bunge et al. *(213)* found no effect of sex, curve type, and curve size on a similar battery of measures. Additionally, adolescents undergoing brace-based management did not appear to score on HRQoL instruments differently from age-adjusted norms.

3.3.2.3.3. Lumbar Premenopausal women demonstrate decreased bone density in the lumbar vertebrae versus men *(214)*. Ebbesen et al. *(215)* found that adult women demonstrate lower vertebral bone mass than age-matched men. Additionally, women exhibited decreased compressive load tolerances, accompanied by increased mechanical stress *(215,216)*. However bony differences may not be limited to adults. Gilsanz et al. *(217)* found similar bone mass differences in preadolescent and adolescent girls, who demonstrate lower vertebral bone mass than age-matched boys. However, these investigators reported that these differences were likely related to differences in bone size versus bone density *(216,217)*. Naganathan and Sambrook *(218)* went further to report that volumetric bone density of the third lumbar vertebra did not differ between the sexes, whereas observed differences in areal bone density were likely related to differences in bone size.

An account of sex-based differences in low back pain (LBP) is controversial. However, Korovessis et al. *(219)* has reported a higher incidence of LBP and dorsal pain in female youth, especially in those girls involved in sports. However, the structural etiology of sex-based differences in back symptoms has not been fully explored. Evidence of differences in intervertebral disc structure or function is scarce. Gruber et al. *(220)* found that the female sex was a contribution to the cellular proliferation potential within the annulus fibrosis surrounding the disc nucleus, along with a contribution from increased age, degree of degeneration, and surgical derivation. Investigators have found differences in lumbar zygapophyseal facet size, pedicle facet angle, and facet shape *(221,222)*. These differences appear to be related to a greater incidence of degenerative anterolisthesis at L4 in women versus their male counterparts *(221)*.

Sex differences are observable in the posture and postural control of the lumbar spine. Norton et al. *(223)* found a greater lumbar spine lordosis in women versus men. O'Sullivan et al. *(224)* examined the impact of unstable versus stable sitting surfaces on recruitment and control of the superficial lumbar multifidus, transverse fibers of internal oblique, and iliocostalis lumborum pars thoracis. Although these investigators found no sex-based electromyography (EMG) differences, they did observe that women exhibited greater medial-lateral postural sway versus men on an unstable surface. Although the role of posture and postural control in the development of LBP is inconclusive, future studies could examine the role of these factors on the development of lumbar pathology.

Investigators have observed sex-based differences in lumbar muscle cross-sectional areas and muscle geometry *(225)*. These differences, coupled with differences in trunk motor control strategies, could have an influence on biomechanical behaviors in the lumbar spine. Women exhibit decreased Type II fiber diameter versus men, lending to decreased strength and increased endurance of the lumbar muscle groups *(226)*. Moreover, women appear to experience greater compressive and anterior–posterior shear loading at the lower lumbar spine *(227)*. These loading differences appear to be related to altered coactivation of the muscles surrounding the lumbar segments, where women produce greater flexor antagonistic coactivation than men *(228)*. It is likely that these altered behaviors result in distorted strategies for controlling dynamic load conditions. For example, Granata et al. *(229)* examined the effect of sustained flexion postures on protective paraspinal muscle reflexes. These investigators not only observed a detrimental alteration in the reflexive activity after sustained trunk flexion but also found that women demonstrated greater detriments in the protective reflexive response. Moreover, they appear to have decreased stiffness and increased segmental motion in the lumbar spine versus men *(230)*. These factors together could lend the female lumbar segments to the development of clinical lumbar instability *(231)*.

3.3.2.3.4. Sacroiliac and pelvis Women appear to suffer from pain associated with the sacroiliac joint (SIJ) more frequently than men, most likely associated with anatomical differences and hormonal fluctuations. The incidence of SIJ clinical hypermobility in the joint is greatest between the ages of 18–35 years. However, this prevalence appears to be sex-specific, where the SIJ mobility begins to decline at 35 years in men and 45 years in women. Thus, SIJ-related pain that is associated with clinical instability could persist in women after 45 years, especially when the individual is on estrogen replacement.

Anatomical changes are seen in the SIJ throughout the course of life, and those changes appear to be different between men and women *(232–234)*. By the second decade, differences between the sexes are observable *(235)*. Although the male synovial capsule thickens and the joint architecture observably adapts, the female SIJ soft tissues become more pliable as hormones change with the onset of menses.

Although the sacral vertebrae start ossifying in the third decade, the mobility of the female SIJ continues to increase, producing a ratio of mobility of approximately 5 to 1 as compared to men. Pregnancy can increase mobility of the SIJ 2.5 times, increasing the movement disparity between women and men *(235)*. Movement persists in the female SIJ through the fourth and fifth decades, whereas the male SIJ demonstrates a further decline in movement in the same time frame *(236,237)*. Although complementary ridges and depressions form on the iliac and sacral cartilages and the synovial membranes thicken in both men and women, the men appear to be more prone to the development of periarticular osteophytes and sacroiliac bridging, further lending the male SIJ to decreased movement *(235,238–240)*.

The external contours of the SIJ articular surfaces are generally a C-shape in men and an L-shape in women, lending the female joint to greater translation during selected situations such as pregnancy and delivery. The joint surfaces at the S1 level are the largest compared to the smaller surfaces at S2 and S3. Each SIJ surface is approximately 17.5 cm^2 in surface area, well suited for absorption and transfer of large forces *(241)*. The SIJ itself is found deep within the sacrum and ilium. The iliac cartilage is thin (0.5 mm), bluish, dull, and rough, compared to the sacral cartilage that is thick (3 mm), white, shiny, and smooth *(238)*. The iliac cartilage is the same relative thickness in both sexes, in contrast to the sacral cartilage which is thicker in women *(242)*.

Women present with SIJ-related pain in the last trimester of pregnancy *(243,244)*, in response to increased relaxin that changes the stiffness of the elaborate ligament system and produces a hypermobile state in the joint *(245,246)*. The ligamentous system of the SIJ enhances stability by increasing the friction in the SIJ and contributes to a self-locking mechanism *(247–249)*. In addition, the system offers proprioceptive feedback in response to activity in a rich plexus of articular receptors.

The SIJ ligament system can be divided into four different layers, where the most superficial layer being the thoracolumbar fascia to which numerous muscles attach and impose control, including the latissimus dorsi, gluteus maximus, transverse abdominis, and serratus posterior inferior *(249)*. The next layer associated with the SIJ includes the sacrospinous and the sacrotuberous ligaments that constrain sacral nutation and control movement of the pubic symphysis on the anterior side of the pelvic ring *(250,251)*.

The long dorsal sacroiliac ligament (also know as the longissimus ligament) courses from the posterior superior iliac spine (PSIS) to the inferior lateral sacrum outside the coccyx. The ligament is approximately 2 cm wide and 6 cm long. It is the only ligament that maximally tightens during counter-nutation, lending it to strain and clinical symptoms during a woman's third trimester of pregnancy after the baby descends *(252)*. The iliolumbar ligament constrains both SIJ movements and movement of the lower lumbar segments with respect to the sacrum *(253,254)*. Along with the dorsal SIJ ligaments, the self-locking mechanism is further enhanced through the constraints imposed by the deep interosseus ligaments, especially during nutation *(247–249)*. The stiffness of these ligaments decreases under the influence of hormonal changes in the final stages of pregnancy, so that the birthing process can be enhanced. Moreover, the SIJ tends to counter-nutate during these stages, where the sacral base tips posterior and opens the pelvic inlet for fetal decent. Counter-nutation reduces ligament constraint and promotes joint hypermobility that can contribute to postpartum pelvic pain *(245,255,256)*, which can persist several years after delivery *(257)*.

3.4. SUMMARY

Sexual dimorphism in the human musculoskeletal system is apparent, but more subtle than the sex differences often observed in other species. Some musculoskeletal sex differences in humans are present at an early age, although others tend to appear later in life, especially at puberty and menopause. Sex differences in gross skeletal geometry and specific tissue characteristics are common. For example, women are generally shorter, have less body mass, and have a different general morphological appearance than men. Women tend to have different characteristics of specific bones and bony features than men, which have been explained by both genetic and environmental factors. In the pelvis, women tend to have a larger inlet and outlet, greater interacetabular distance, and greater hip width normalized to femur length. In the femur, women tend to have greater femoral anteversion and narrower intercondylar notch width. Additionally, there are several differences in specific characteristics of the femur, such as head diameter and circumference that are relatively strong predictors of sex. In the knee, tibia, and foot, women tend to exhibit greater genu recurvatum, greater quadriceps angle, more lateral patellar alignment, greater tibial torsion, and more bunions and deformities of the toes. Moreover women and men appear to have several differences in collagenous, cartilage, and bone tissues, which may predispose women to certain pathologies such as osteoarthritis and osteoporosis later in life. In some collagenous tissues, there are sex differences in collagen thickness, orientation, content, diameter, volume, and metabolism. In cartilage, women tend to have less cartilage volume, thickness, and surface area at some sites. Prepubescent girls (generally under the age of 13 years) tend to have slower cartilage accrual rate, and postmenopausal women (generally over the age of 50 years) tend to have greater cartilage degradation than their male counterparts at either age. In bone tissue, women tend to have a slower accrual rate in youth and less peak bone mass and slower bone turnover in adulthood compared to men. Finally women tend to have less volumetric bone mineral density, less bone area, less cortical thickness, and less compressive and bending strength at some bony sites compared to men, even after correction for anthropometric differences like height and weight.

Sexual dimorphism can manifest itself in specific differences in each joint system throughout the body, possibly resulting in differences in clinical pathology and symptomology. Although differences in acromion morhpism have been attributed to sex-based differences in clinical impingement at the shoulder, the role of those differences remains controversial. More trustworthy are the female glenohumeral capsular responses that appear to lend to sex-based differences in the incidence of joint laxity and/or secondary idiopathic capsulitis. Hormonal differences appear to lend to tissue changes related to the woman's higher incidence of tendinosis in the lateral elbow tendon structures, while the female predisposition for increased incidence of CTS looks as if it is related to differences in architectural differences in the wrist and hand especially found around the carpal tunnel. Similarly, architectural differences are at the root of the female predilection for fracture responses at the hip joint, while tissue biomechanical differences accompany architectural distinctions in contributing to the female incidence of acetabular labral tears.

The woman's increased risk for ACL injury has received special attention in the literature, which has suggested that several factors are responsible for this incidence. Anatomical, hormonal, mechanical, and neurophysiological differences have all been examined, and multiple mechanisms have been observed. While the femoral intercondylar notch has been examined, its role remains controversial. Similarly, the role of static measures including Q angle and TF angle has remained questionable, while differences in joint movement at both the knee and hip during cutting and landing have been deemed partially responsible for sex-based differences in ACL injury. Joint laxity and tibial translation behaviors appear to contribute to this disparity, along with altered joint motion responses in the subtalar joint. Moreover, locomotor control strategy differences are exhibited by female athletes, lending to their heightened predisposition. Finally, the woman's hormonal fluctuations affect not only changes in ACL architecture but also the biomechanical response of the tissue to stress.

Selected sex-based differences have been observed in the structural and mechanical features of the ankle and foot. These differences appear to contribute to the woman's increased risk for both ankle sprains and Achilles tendonopathy, especially in those who are athletically inclined. The role of cervical spine architectural differences in female musculoskeletal health is unclear. However, the woman's cervical structures that include the facets and intervertebral disc appear to respond more poorly to macrotrauma, such as whiplash. The preadolescent female is more susceptible to developing thoracolumbar postural changes that include excessive kyphosis, lordosis, and/or scoliosis. These differences are not only influenced by physical differences in the vertebrae, articular structures, intervertebral disc, and/or muscle but are apparently influenced by psychomotor control and psychobehavioral differences as well. Finally, SIJ differences between men and women are not only influenced by architectural disparities but additionally by the influence that hormones have on the integrity of the complex capsuloligamentous structures that surround the joint mechanism.

3.5. CONCLUSIONS

The purposes of this chapter were to (1) examine sex differences in the anatomy of selected musculoskeletal components and (2) explore selected regional considerations in female functional pathoanatomy that are pertinent to women's health issues. There exists a plethora of literature that documents sex differences in general body characteristics, skeletal geometry, musculoskeletal tissue characteristics, and joint-specific functional anatomy and pathomechanics. In conclusion, while the musculoskeletal anatomy of men and women is grossly similar, important differences exist that may influence the way in which the general public views and health care professionals respond to women's musculoskeletal health issues.

REFERENCES

1. Anonymous. Sexual dimorphism. *Answers com* 2006 January 12; Available at: http://www.answers.com/sexual%20dimorphism.
2. Smith FW, Smith PA. Musculoskeletal differences between males and females. *Sports Med Arthrosc* 2002;10(1):98–100.
3. Chmielewski T, Ferber R. Rehabilitation considerations for the female athlete. In: Andrews JR, Harrelson GL, Wilk KE, editors. *Physical Rehabilitation of the Injured Athlete*. 3rd ed. Philadelphia, PA: Saunders; 2004. p. 315–28.
4. Anonymous. Clinical growth charts. *CDC gov* 2006 January 12; Available at: http://www.cdc.gov/nchs/about/major/nhanes/growthcharts/clinical_charts.htm.
5. Horton MG, Hall TL. Quadriceps femoris muscle angle: normal values and relationships with gender and selected skeletal measures. *Phys Ther* 1989 November;69(11):897–901.
6. Kersnic B, Iglic A, Kralj-Iglic V, et al. Determination of the femoral and pelvic geometrical parameters that are important for the hip joint contact stress: differences between female and male. *Pflugers Arch* 1996;431(6 Suppl 2):R207–8.
7. Souryal TO, Freeman TR. Intercondylar notch size and anterior cruciate ligament injuries in athletes. A prospective study. *Am J Sports Med* 1993 July;21(4):535–9.
8. Woodland LH, Francis RS. Parameters and comparisons of the quadriceps angle of college-aged men and women in the supine and standing positions. *Am J Sports Med* 1992 March;20(2):208–11.
9. Yoshioka Y, Siu DW, Scudamore RA, et al. Tibial anatomy and functional axes. *J Orthop Res* 1989;7(1):132–7.
10. Shelbourne KD, Davis TJ, Klootwyk TE. The relationship between intercondylar notch width of the femur and the incidence of anterior cruciate ligament tears. A prospective study. *Am J Sports Med* 1998 May;26(3):402–8.
11. Ireland ML, Ballantyne BT, Little K, et al. A radiographic analysis of the relationship between the size and shape of the intercondylar notch and anterior cruciate ligament injury. *Knee Surg Sports Traumatol Arthrosc* 2001 July;9(4):200–5.
12. LaPrade RF, Burnett QM. Femoral intercondylar notch stenosis and correlation to anterior cruciate ligament injuries. A prospective study. *Am J Sports Med* 1994 March;22(2):198–202.
13. Tillman MD, Bauer JA, Cauraugh JH, et al. Differences in lower extremity alignment between males and females. Potential predisposing factors for knee injury. *J Sports Med Phys Fitness* 2005 September;45(3):355–9.
14. Gualdi-Russo E. Study on long bones: variation in angular traits with sex, age, and laterality. *Anthropol Anz* 1998 December;56(4):289–99.
15. LaVelle M. Natural selection and developmental sexual variation in the human pelvis. *Am J Phys Anthropol* 1995 September;98(1):59–72.
16. Patriquin ML, Loth SR, Steyn M. Sexually dimorphic pelvic morphology in South African whites and blacks. *Homo* 2003;53(3):255–62.
17. Igbigbi PS, Msamati BC, Ng'Ambi TM. Intercondylar shelf angle in adult black Malawian subjects. *Clin Anat* 2001 July;14(4):254–7.
18. Iscan MY, Shihai D. Sexual dimorphism in the Chinese femur. *Forensic Sci Int* 1995 June 30;74(1–2):79–87.
19. King CA, Iscan MY, Loth SR. Metric and comparative analysis of sexual dimorphism in the Thai femur. *J Forensic Sci* 1998 September;43(5):954–8.
20. Macho GA. Is sexual dimorphism in the femur a "population specific phenomenon"? *Z Morphol Anthropol* 1990;78(2):229–42.

21. Mall G, Graw M, Gehring K, et al. Determination of sex from femora. *Forensic Sci Int* 2000 September 11;113(1–3):315–21.
22. Purkait R, Chandra H. A study of sexual variation in Indian femur. *Forensic Sci Int* 2004 November 10;146(1):25–33.
23. Iscan MY, Miller-Shaivitz P. Determination of sex from the tibia. *Am J Phys Anthropol* 1984 May;64(1):53–7.
24. Iscan MY, Yoshino M, Kato S. Sex determination from the tibia: standards for contemporary Japan. *J Forensic Sci* 1994 May;39(3):785–92.
25. Ruff CB, Hayes WC. Cross-sectional geometry of Pecos Pueblo femora and tibiae – a biomechanical investigation: II. Sex, age, side differences. *Am J Phys Anthropol* 1983 March;60(3):383–400.
26. Steele DG. The estimation of sex on the basis of the talus and calcaneus. *Am J Phys Anthropol* 1976 November;45(3 Pt. 2):581–8.
27. Bidmos MA, Asala SA. Discriminant function sexing of the calcaneus of the South African whites. *J Forensic Sci* 2003 November;48(6):1213–8.
28. Riepert T, Drechsler T, Schild H, et al. Estimation of sex on the basis of radiographs of the calcaneus. *Forensic Sci Int* 1996 February 9;77(3):133–40.
29. Hogler W, Blimkie CJ, Cowell CT, et al. A comparison of bone geometry and cortical density at the mid-femur between prepuberty and young adulthood using magnetic resonance imaging. *Bone* 2003 November;33(5):771–8.
30. Ashizawa K, Kumakura C, Kusumoto A, et al. Relative foot size and shape to general body size in Javanese, Filipinas and Japanese with special reference to habitual footwear types. *Ann Hum Biol* 1997 March;24(2):117–29.
31. Axer H, von Keyserlingk DG, Prescher A. Collagen fibers in linea alba and rectus sheaths. *J Surg Res* 2001 April;96(2):239–45.
32. Tzaphlidou M. Diameter distributions of collagenous tissues in relation to sex. A quantitative ultrastructural study. *Micron* 2001 April;32(3):333–6.
33. Vitellaro-Zuccarello L, Cappelletti S, Dal PR, et al. Stereological analysis of collagen and elastic fibers in the normal human dermis: variability with age, sex, and body region. *Anat Rec* 1994 February;238(2):153–62.
34. Zanze M, Souberbielle JC, Kindermans C, et al. Procollagen propeptide and pyridinium cross-links as markers of type I collagen turnover: sex- and age-related changes in healthy children. *J Clin Endocrinol Metab* 1997 September;82(9):2971–7.
35. Anonymous. Collagen. *Wikipedia org* 2006; Available at: http://en.wikipedia.org/wiki/Collagen.
36. Nordin M, Frankel VH. *Basic Biomechanics of the Musculoskeletal System.* 3rd ed. Philadelphia, PA: Lippincott Williams & Wilkins; 2001.
37. Osakabe T, Hayashi M, Hasegawa K, et al. Age- and gender-related changes in ligament components. *Ann Clin Biochem* 2001 September;38(Pt. 5):527–32.
38. Tamir E, Brenner S. Gender differences in collagen diseases. *Skinmed* 2003 March;2(2):113–7.
39. Ding C, Cicuttini F, Scott F, et al. Sex differences in knee cartilage volume in adults: role of body and bone size, age and physical activity. *Rheumatology (Oxford)* 2003 November;42(11):1317–23.
40. Jones G, Glisson M, Hynes K, et al. Sex and site differences in cartilage development: a possible explanation for variations in knee osteoarthritis in later life. *Arthritis Rheum* 2000 November;43(11):2543–9.
41. Lanyon P, Muir K, Doherty S, et al. Age and sex differences in hip joint space among asymptomatic subjects without structural change: implications for epidemiologic studies. *Arthritis Rheum* 2003 April;48(4):1041–6.
42. Larbre JP, Da Silva JA, Moore AR, et al. Cartilage contribution to gender differences in joint disease progression. A study with rat articular cartilage. *Clin Exp Rheumatol* 1994 July;12(4):401–8.
43. Cicuttini FM, Wluka A, Bailey M, et al. Factors affecting knee cartilage volume in healthy men. *Rheumatology (Oxford)* 2003 February;42(2):258–62.
44. Faber SC, Eckstein F, Lukasz S, et al. Gender differences in knee joint cartilage thickness, volume and articular surface areas: assessment with quantitative three-dimensional MR imaging. *Skeletal Radiol* 2001 March;30(3):144–50.
45. Mouritzen U, Christgau S, Lehmann HJ, et al. Cartilage turnover assessed with a newly developed assay measuring collagen type II degradation products: influence of age, sex, menopause, hormone replacement therapy, and body mass index. *Ann Rheum Dis* 2003 April;62(4):332–6.

46. Da Silva JA, Larbre JP, Seed MP, et al. Sex differences in inflammation induced cartilage damage in rodents. The influence of sex steroids. *J Rheumatol* 1994 February;21(2):330–7.
47. Fares JE, Choucair M, Nabulsi M, et al. Effect of gender, puberty, and vitamin D status on biochemical markers of bone remodedeling. *Bone* 2003 August;33(2):242–7.
48. Nuckley DJ, Eck MP, Carter JW, et al. Spinal maturation affects vertebral compressive mechanics and vBMD with sex dependence. *Bone* 2004 September;35(3):720–8.
49. Katzburg S, Lieberherr M, Ornoy A, et al. Isolation and hormonal responsiveness of primary cultures of human bone-derived cells: gender and age differences. *Bone* 1999 December;25(6):667–73.
50. Khosla S, Melton LJ, III, Atkinson EJ, et al. Relationship of serum sex steroid levels and bone turnover markers with bone mineral density in men and women: a key role for bioavailable estrogen. *J Clin Endocrinol Metab* 1998 July;83(7):2266–74.
51. Beck TJ, Ruff CB, Shaffer RA, et al. Stress fracture in military recruits: gender differences in muscle and bone susceptibility factors. *Bone* 2000 September;27(3):437–44.
52. Kaptoge S, Dalzell N, Loveridge N, et al. Effects of gender, anthropometric variables, and aging on the evolution of hip strength in men and women aged over 65. *Bone* 2003 May;32(5):561–70.
53. Schuit SC, van der KM, Weel AE, et al. Fracture incidence and association with bone mineral density in elderly men and women: the Rotterdam Study. *Bone* 2004 January;34(1):195–202.
54. Krall EA, Dawson-Hughes B, Hirst K, et al. Bone mineral density and biochemical markers of bone turnover in healthy elderly men and women. *J Gerontol A Biol Sci Med Sci* 1997 March;52(2):M61–7.
55. Forwood MR, Bailey DA, Beck TJ, et al. Sexual dimorphism of the femoral neck during the adolescent growth spurt: a structural analysis. *Bone* 2004 October;35(4):973–81.
56. Mosekilde L. Sex differences in age-related loss of vertebral trabecular bone mass and structure-biomechanical consequences. *Bone* 1989;10(6):425–32.
57. Henry YM, Eastell R. Ethnic and gender differences in bone mineral density and bone turnover in young adults: effect of bone size. *Osteoporos Int* 2000;11(6):512–7.
58. Minisola S, Dionisi S, Pacitti MT, et al. Gender differences in serum markers of bone resorption in healthy subjects and patients with disorders affecting bone. *Osteoporos Int* 2002;13(2):171–5.
59. Smith EL, Smith KA, Gilligan C. Exercise, fitness, osteoarthritis, and osteoporosis. In: Bouchard C, Shephard RJ, Stephens T, Sutton JR, McPherson BD, editors. *Exercise, Fitness, and Health: A Consensus of Current Knowledge*. Champaign, IL: Human Kinetics; 1990. p. 517–28.
60. Harrison JE, Chow R. Discussion: exercise, fitness, osteoarthritis, and osteoporosis. In: Bouchard C, Shephard RJ, Stephens T, Sutton JR, McPherson BD, editors. *Exercise, Fitness, and Health: A Consensus of Current Knowledge*. Champaign, IL: Human Kinetics; 1990. p. 529–32.
61. Anonymous. Osteoporosis: peak bone mass in women. *Osteo org* 2005 March; Available at: http://www.niams.nih.gov/bone/hi/bone-mass.htm.
62. Pandley SK, Shamal S, Kuman S, et al. Articular branch of the axillary artery and its clinical implication. *Nepal Med Coll J* 2003;5(2):61–3.
63. Radas CBPHG. The coracoid impingement of the subscapularis tendon: a cadaver study. *J Shoulder Elbow Surg* 2004;13(2):154–9.
64. Bigliani LU, Morrison DS, April E. The morphology of the acromion and its relationship to rotator cuff tears. *Orthop Trans* 1986;10:228.
65. Bigliani LU, Ticker JB, Flatow EL, et al. The relationship of acromial architecture to rotator cuff disease. *Clin Sports Med* 1991 October;10(4):823–38.
66. Berbig R, Weishaupt D, Prim J, et al. Primary anterior shoulder dislocation and rotator cuff tears. *J Shoulder Elbow Surg* 1999 May;8(3):220–5.
67. Speer KP, Osbahr DC, Montella BJ, et al. Acromial morphotype in the young asymptomatic athletic shoulder. *J Shoulder Elbow Surg* 2001 September;10(5):434–7.
68. Getz JD, Recht MP, Piraino DW, et al. Acromial morphology: relation to sex, age, symmetry, and subacromial enthesophytes. *Radiology* 1996 June;199(3):737–42.
69. Gill TJ, McIrvin E, Kocher MS, et al. The relative importance of acromial morphology and age with respect to rotator cuff pathology. *J Shoulder Elbow Surg* 2002 July;11(4):327–30.
70. Wang JC, Shapiro MS. Changes in acromial morphology with age. *J Shoulder Elbow Surg* 1997 January;6(1):55–9.
71. Mahakkanukrauh P, Surin P. Prevalence of osteophytes associated with the acromion and acromio-clavicular joint. *Clin Anat* 2003 November;16(6):506–10.

72. Graichen H, Bonel H, Stammberger T, et al. Sex-specific differences of subacromial space width during abduction, with and without muscular activity, and correlation with anthropometric variables. *J Shoulder Elbow Surg* 2001 March;10(2):129–35.
73. Gill TJ, Zarins B. Open repairs for the treatment of anterior shoulder instability. *Am J Sports Med* 2003 January;31(1):142–53.
74. Borsa PA, Sauers EL, Herling DE. Patterns of glenohumeral joint laxity and stiffness in healthy men and women. *Med Sci Sports Exerc* 2000 October;32(10):1685–90.
75. Hawkins RJ, Mohtadi NG. Controversy in anterior shoulder instability. *Clin Orthop* 1991 November;(272):152–61.
76. Kronberg M, Brostrom LA. Humeral head retroversion in patients with unstable humeroscapular joints. *Clin Orthop* 1990 November;(260):207–11.
77. Churchill RS, Brems JJ, Kotschi H. Glenoid size, inclination, and version: an anatomic study. *J Shoulder Elbow Surg* 2001 July;10(4):327–32.
78. Arkkila PE, Kantola IM, Viikari JS, et al. Shoulder capsulitis in type I and II diabetic patients: association with diabetic complications and related diseases. *Ann Rheum Dis* 1996 December;55(12):907–14.
79. Omari A, Bunker TD. Open surgical release for frozen shoulder: surgical findings and results of the release. *J Shoulder Elbow Surg* 2001 July;10(4):353–7.
80. Mengiardi B, Pfirrmann CW, Gerber C, et al. Frozen shoulder: MR arthrographic findings. *Radiology* 2004 November;233(2):486–92.
81. Hutchinson JW, Tierney GM, Parsons SL, et al. Dupuytren's disease and frozen shoulder induced by treatment with a matrix metalloproteinase inhibitor. *J Bone Joint Surg Br* 1998 September;80(5):907–8.
82. Nirschl RP. Tennis elbow tendinosis: pathoanatomy, nonsurgical and surgical management. In: Gordon SLBSJFLJ, editor. *Repetitive Motion Disorders of the Upper Extremity*. Rosemont, IL: American Academy of Orthopaedic Surgeons; 1995. p. 467–78.
83. Svernlov B, Adolfsson L. Non-operative treatment regime including eccentric training for lateral humeral epicondylalgia. *Scand J Med Sci Sports* 2001 December;11(6):328–34.
84. Winkel D, Matthijs O, Phelps V. Part 2: The knee. *Diagnosis and Treatment of the Lower Extremities*. Gaithersburg, MD: Spen Publishers, Inc.; 1997.
85. Solveborn SA. Radial epicondylalgia ('tennis elbow'): treatment with stretching or forearm band. A prospective study with long-term follow-up including range-of-motion measurements. *Scand J Med Sci Sports* 1997 August;7(4):229–37.
86. Ljung BO, Lieber RL, Friden J. Wrist extensor muscle pathology in lateral epicondylitis. *J Hand Surg [Br]* 1999 April;24(2):177–83.
87. Alfredson H, Ljung BO, Thorsen K, et al. In vivo investigation of ECRB tendons with microdialysis technique – no signs of inflammation but high amounts of glutamate in tennis elbow. *Acta Orthop Scand* 2000 October;71(5):475–9.
88. Richardson JK, Green DF, Jamieson SC, et al. Gender, body mass and age as risk factors for ulnar mononeuropathy at the elbow. *Muscle Nerve* 2001 April;24(4):551–4.
89. Tanaka S, Petersen M, Cameron L. Prevalence and risk factors of tendinitis and related disorders of the distal upper extremity among U.S. workers: comparison to carpal tunnel syndrome. *Am J Ind Med* 2001 March;39(3):328–35.
90. Moghtaderi A, Izadi S, Sharafadinzadeh N. An evaluation of gender, body mass index, wrist circumference and wrist ratio as independent risk factors for carpal tunnel syndrome. *Acta Neurol Scand* 2005;112(6):375–9.
91. McDiarmid M, Oliver M, Ruser J, et al. Male and female rate differences in carpal tunnel syndrome injuries: personal attributes or job tasks? *Environ Res* 2000 May;83(1):23–32.
92. Mondelli M, Aprile I, Ballerini M, et al. Sex differences in carpal tunnel syndrome: comparison of surgical and non-surgical populations. *Eur J Neurol* 2005 December;12(12):976–83.
93. Pierre-Jerome C, Bekkelund SI, Nordstrom R. Quantitative MRI analysis of anatomic dimensions of the carpal tunnel in women. *Surg Radiol Anat* 1997;19(1):31–4.
94. Siegel DB, Kuzma G, Eakins D. Anatomic investigation of the role of the lumbrical muscles in carpal tunnel syndrome. *J Hand Surg [Am]* 1995 September;20(5):860–3.
95. Richards RS, Bennett JD. Abnormalities of the hook of the hamate in patients with carpal tunnel syndrome. *Ann Plast Surg* 1997 July;39(1):44–6.

96. Boz C, Ozmenoglu M, Altunayoglu V, et al. Individual risk factors for carpal tunnel syndrome: an evaluation of body mass index, wrist index and hand anthropometric measurements. *Clin Neurol Neurosurg* 2004 September;106(4):294–9.
97. Massy-Westropp N, Grimmer K, Bain G. The effect of a standard activity on the size of the median nerve as determined by ultrasound visualization. *J Hand Surg [Am]* 2001 July;26(4):649–54.
98. Crisco JJ, Coburn JC, Moore DC, et al. Carpal bone size and scaling in men versus in women. *J Hand Surg [Am]* 2005 January;30(1):35–42.
99. Hamanaka I, Okutsu I, Shimizu K, et al. Evaluation of carpal canal pressure in carpal tunnel syndrome. *J Hand Surg [Am]* 1995 September;20(5):848–54.
100. Ham SJ, Kolkman WF, Heeres J, et al. Changes in the carpal tunnel due to action of the flexor tendons: visualization with magnetic resonance imaging. *J Hand Surg [Am]* 1996 November;21(6):997–1003.
101. Rempel D, Keir PJ, Smutz WP, et al. Effects of static fingertip loading on carpal tunnel pressure. *J Orthop Res* 1997 May;15(3):422–6.
102. Chen WS. Median-nerve neuropathy associated with chronic anterior dislocation of the lunate. *J Bone Joint Surg Am* 1995 December;77(12):1853–7.
103. Seradge H, Jia YC, Owens W. In vivo measurement of carpal tunnel pressure in the functioning hand. *J Hand Surg [Am]* 1995 September;20(5):855–9.
104. Gelberman RH, Hergenroeder PT, Hargens AR, et al. The carpal tunnel syndrome. A study of carpal canal pressures. *J Bone Joint Surg Am* 1981 March;63(3):380–3.
105. Buschbacher RM. Mixed nerve conduction studies of the median and ulnar nerves. *Am J Phys Med Rehabil* 1999 November;78(6 Suppl):S69–S74.
106. Vennix MJ, Hirsh DD, Chiou-Tan FY, et al. Predicting acute denervation in carpal tunnel syndrome. *Arch Phys Med Rehabil* 1998 March;79(3):306–12.
107. Tetro AM, Evanoff BA, Hollstien SB, et al. A new provocative test for carpal tunnel syndrome. Assessment of wrist flexion and nerve compression. *J Bone Joint Surg Br* 1998 May;80(3):493–8.
108. Vogt T. Median-ulnar motor latency difference in the diagnosis of CTS. *Z Fur Elektroenzephalographie Diagnosis Verwandte Gebiete* 1995;26:141–5.
109. Glowacki KA, Breen CJ, Sachar K, et al. Electrodiagnostic testing and carpal tunnel release outcome. *J Hand Surg [Am]* 1996 January;21(1):117–21.
110. Padula L. A useful electrophysiologic parameter for diagnosis of CTS. *Muscle Nerve* 1995;19:48–53.
111. Verghese J, Galanopoulou AS, Herskovitz S. Autonomic dysfunction in idiopathic carpal tunnel syndrome. *Muscle Nerve* 2000 August;23(8):1209–13.
112. Golovchinsky V. Frequency of ulnar-to-median nerve anastomosis revisited. *Electromyogr Clin Neurophysiol* 1995 March;35(2):67–8.
113. Szabo RM, Gelberman RH. The pathophysiology of nerve entrapment syndromes. *J Hand Surg [Am]* 1987 September;12(5 Pt. 2):880–4.
114. Kiuru MJ, Pihlajamaki HK, Ahovuo JA. Fatigue stress injuries of the pelvic bones and proximal femur: evaluation with MR imaging. *Eur Radiol* 2003 March;13(3):605–11.
115. Seeman E. The structural and biomechanical basis of the gain and loss of bone strength in women and men. *Endocrinol Metab Clin North Am* 2003 March;32(1):25–38.
116. Genser-Strobl B, Sora MC. Potential of P40 plastination for morphometric hip measurements. *Surg Radiol Anat* 2005 April;27(2):147–51.
117. Wang SC, Brede C, Lange D, et al. Gender differences in hip anatomy: possible implications for injury tolerance in frontal collisions. *Annu Proc Assoc Adv Automot Med* 2004;48:287–301.
118. Igbigbi PS. Collo-diaphysial angle of the femur in East African subjects. *Clin Anat* 2003 September;16(5):416–9.
119. Beck TJ, Ruff CB, Scott WW, Jr., et al. Sex differences in geometry of the femoral neck with aging: a structural analysis of bone mineral data. *Calcif Tissue Int* 1992 January;50(1):24–9.
120. Crabtree N, Lunt M, Holt G, et al. Hip geometry, bone mineral distribution, and bone strength in European men and women: the EPOS study. *Bone* 2000 July;27(1):151–9.
121. Korpelainen R, Orava S, Karpakka J, et al. Risk factors for recurrent stress fractures in athletes. *Am J Sports Med* 2001 May;29(3):304–10.
122. Horii M, Kubo T, Inoue S, et al. Coverage of the femoral head by the acetabular labrum in dysplastic hips: quantitative analysis with radial MR imaging. *Acta Orthop Scand* 2003 June;74(3):287–92.

123. Seldes RM, Tan V, Hunt J, et al. Anatomy, histologic features, and vascularity of the adult acetabular labrum. *Clin Orthop Relat Res* 2001 January;(382):232–40.
124. Ferguson SJ, Bryant JT, Ganz R, et al. An in vitro investigation of the acetabular labral seal in hip joint mechanics. *J Biomech* 2003 February;36(2):171–8.
125. Narvani AA, Tsiridis E, Tai CC, et al. Acetabular labrum and its tears. *Br J Sports Med* 2003 June;37(3):207–11.
126. Kelly BT, Shapiro GS, Digiovanni CW, et al. Vascularity of the hip labrum: a cadaveric investigation. *Arthroscopy* 2005 January;21(1):3–11.
127. Stiris MG. [Magnetic resonance arthrography of the hip joint in patients with suspected rupture of labrum acetabulare]. *Tidsskr Nor Laegeforen* 2001 February 28;121(6):698–700.
128. Ishiko T, Naito M, Moriyama S. Tensile properties of the human acetabular labrum-the first report. *J Orthop Res* 2005 November;23(6):1448–53.
129. McCarthy JC, Lee JA. Acetabular dysplasia: a paradigm of arthroscopic examination of chondral injuries. *Clin Orthop Relat Res* 2002 December;(405):122–8.
130. Leunig M, Sledge JB, Gill TJ, et al. Traumatic labral avulsion from the stable rim: a constant pathology in displaced transverse acetabular fractures. *Arch Orthop Trauma Surg* 2003 October;123(8):392–5.
131. Mintz DN, Hooper T, Connell D, et al. Magnetic resonance imaging of the hip: detection of labral and chondral abnormalities using noncontrast imaging. *Arthroscopy* 2005 April;21(4):385–93.
132. Ito K, Minka MA, Leunig M, et al. Femoroacetabular impingement and the cam-effect. A MRI-based quantitative anatomical study of the femoral head-neck offset. *J Bone Joint Surg Br* 2001 March;83(2):171–6.
133. Mosher TJ, Collins CM, Smith HE, et al. Effect of gender on in vivo cartilage magnetic resonance imaging T2 mapping. *J Magn Reson Imaging* 2004 March;19(3):323–8.
134. Csintalan RP, Schulz MM, Woo J, et al. Gender differences in patellofemoral joint biomechanics. *Clin Orthop Relat Res* 2002 September;(402):260–9.
135. Besier TF, Draper CE, Gold GE, et al. Patellofemoral joint contact area increases with knee flexion and weight-bearing. *J Orthop Res* 2005 March;23(2):345–50.
136. Arendt E, Dick R. Knee injury patterns among men and women in collegiate basketball and soccer. NCAA data and review of literature. *Am J Sports Med* 1995 November;23(6):694–701.
137. Fayad LM, Parellada JA, Parker L, et al. MR imaging of anterior cruciate ligament tears: is there a gender gap? *Skeletal Radiol* 2003 November;32(11):639–46.
138. Boden BP, Dean GS, Feagin JA, et al. Mechanisms of anterior cruciate ligament injury. *Orthopedics* 2000 June;23(6):573–8.
139. Rozzi SL, Lephart SM, Gear WS, et al. Knee joint laxity and neuromuscular characteristics of male and female soccer and basketball players. *Am J Sports Med* 1999 May;27(3):312–9.
140. Muneta T, Takakuda K, Yamamoto H. Intercondylar notch width and its relation to the configuration and cross-sectional area of the anterior cruciate ligament. A cadaveric knee study. *Am J Sports Med* 1997 January;25(1):69–72.
141. Charlton WP, St John TA, Ciccotti MG, et al. Differences in femoral notch anatomy between men and women: a magnetic resonance imaging study. *Am J Sports Med* 2002 May;30(3):329–33.
142. Sbriccoli P, Solomonow M, Zhou BH, et al. Neuromuscular response to cyclic loading of the anterior cruciate ligament. *Am J Sports Med* 2005 April;33(4):543–51.
143. Murshed KA, Cicekcibasi AE, Karabacakoglu A, et al. Distal femur morphometry: a gender and bilateral comparative study using magnetic resonance imaging. *Surg Radiol Anat* 2005 April;27(2):108–12.
144. Loudon JK, Jenkins W, Loudon KL. The relationship between static posture and ACL injury in female athletes. *J Orthop Sports Phys Ther* 1996 August;24(2):91–7.
145. Livingston LA, Mandigo JL. Bilateral Q angle asymmetry and anterior knee pain syndrome. *Clin Biomech (Bristol, Avon)* 1999;14(1):7–13.
146. Stuberg W, Temme J, Kaplan P, et al. Measurement of tibial torsion and thigh-foot angle using goniometry and computed tomography. *Clin Orthop Relat Res* 1991 November;(272):208–12.
147. Trimble MH, Bishop MD, Buckley BD, et al. The relationship between clinical measurements of lower extremity posture and tibial translation. *Clin Biomech (Bristol, Avon)* 2002 May;17(4):286–90.
148. McNair PJ, Marshall RN. Landing characteristics in subjects with normal and anterior cruciate ligament deficient knee joints. *Arch Phys Med Rehabil* 1994 May;75(5):584–9.

149. McLean SG, Lipfert SW, van den Bogert AJ. Effect of gender and defensive opponent on the biomechanics of sidestep cutting. *Med Sci Sports Exerc* 2004 June;36(6):1008–16.
150. Kernozek TW, Torry MR, Van Hoof H, et al. Gender differences in frontal and sagittal plane biomechanics during drop landings. *Med Sci Sports Exerc* 2005 June;37(6):1003–12.
151. Pollard CD, Davis IM, Hamill J. Influence of gender on hip and knee mechanics during a randomly cued cutting maneuver. *Clin Biomech (Bristol, Avon)* 2004 December;19(10):1022–31.
152. Hewett TE, Lindenfeld TN, Riccobene JV, et al. The effect of neuromuscular training on the incidence of knee injury in female athletes. A prospective study. *Am J Sports Med* 1999 November;27(6):699–706.
153. Ford KR, Myer GD, Toms HE, et al. Gender differences in the kinematics of unanticipated cutting in young athletes. *Med Sci Sports Exerc* 2005 January;37(1):124–9.
154. Hewett TE, Stroupe AL, Nance TA, et al. Plyometric training in female athletes. Decreased impact forces and increased hamstring torques. *Am J Sports Med* 1996 November;24(6):765–73.
155. Barber-Westin SD, Noyes FR, Galloway M. Jump-land characteristics and muscle strength development in young athletes: a gender comparison of 1140 athletes 9 to 17 years of age. *Am J Sports Med* 2006 March;34(3):375–84.
156. Ford KR, Myer GD, Smith RL, et al. A comparison of dynamic coronal plane excursion between matched male and female athletes when performing single leg landings. *Clin Biomech (Bristol, Avon)* 2006 January;21(1):33–40.
157. Ramesh R, Von Arx O, Azzopardi T, et al. The risk of anterior cruciate ligament rupture with generalised joint laxity. *J Bone Joint Surg Br* 2005 June;87(6):800–3.
158. Hollman JH, Deusinger RH, Van Dillen LR, et al. Gender differences in surface rolling and gliding kinematics of the knee. *Clin Orthop Relat Res* 2003 August;(413):208–21.
159. Wojtys EM, Ashton-Miller JA, Huston LJ. A gender-related difference in the contribution of the knee musculature to sagittal-plane shear stiffness in subjects with similar knee laxity. *J Bone Joint Surg Am* 2002 January;84-A(1):10–6.
160. Beckett ME, Massie DL, Bowers KD, et al. Incidence of hyperpronation in the ACL injured knee: a clinical perspective. *J Athl Train* 1992;27(1):58–62.
161. Chappell JD, Yu B, Kirkendall DT, et al. A comparison of knee kinetics between male and female recreational athletes in stop-jump tasks. *Am J Sports Med* 2002 March;30(2):261–7.
162. Sigward SM, Powers CM. The influence of gender on knee kinematics, kinetics and muscle activation patterns during side-step cutting. *Clin Biomech (Bristol, Avon)* 2006 January;21(1):41–8.
163. McLean SG, Huang X, van den Bogert AJ. Association between lower extremity posture at contact and peak knee valgus moment during sidestepping: implications for ACL injury. *Clin Biomech (Bristol, Avon)* 2005 October;20(8):863–70.
164. Granata KP, Padua DA, Wilson SE. Gender differences in active musculoskeletal stiffness. Part II. Quantification of leg stiffness during functional hopping tasks. *J Electromyogr Kinesiol* 2002 April;12(2):127–35.
165. Decker MJ, Torry MR, Wyland DJ, et al. Gender differences in lower extremity kinematics, kinetics and energy absorption during landing. *Clin Biomech (Bristol, Avon)* 2003 August;18(7):662–9.
166. Besier TF, Lloyd DG, Ackland TR. Muscle activation strategies at the knee during running and cutting maneuvers. *Med Sci Sports Exerc* 2003 January;35(1):119–27.
167. Fagenbaum R, Darling WG. Jump landing strategies in male and female college athletes and the implications of such strategies for anterior cruciate ligament injury. *Am J Sports Med* 2003 March;31(2):233–40.
168. Ahmad CS, Clark AM, Heilmann N, et al. Effect of gender and maturity on quadriceps-to-hamstring strength ratio and anterior cruciate ligament laxity. *Am J Sports Med* 2006 March;34(3):370–4.
169. da Fonseca ST, Vaz DV, de Aquino CF, et al. Muscular co-contraction during walking and landing from a jump: comparison between genders and influence of activity level. *J Electromyogr Kinesiol* 2006 June;16(3):273–80.
170. Padua DA, Carcia CR, Arnold BL, et al. Gender differences in leg stiffness and stiffness recruitment strategy during two-legged hopping. *J Mot Behav* 2005 March;37(2):111–25.
171. Hurd WJ, Chmielewski TL, Snyder-Mackler L. Perturbation-enhanced neuromuscular training alters muscle activity in female athletes. *Knee Surg Sports Traumatol Arthrosc* 2006 January;14(1):60–9.
172. Myer GD, Ford KR, Hewett TE. The effects of gender on quadriceps muscle activation strategies during a maneuver that mimics a high ACL injury risk position. *J Electromyogr Kinesiol* 2005 April;15(2):181–9.

173. Wojtys EM, Huston LJ, Schock HJ, et al. Gender differences in muscular protection of the knee in torsion in size-matched athletes. *J Bone Joint Surg Am* 2003 May;85-A(5):782–9.
174. Chu D, LeBlanc R, D'Ambrosia P, et al. Neuromuscular disorder in response to anterior cruciate ligament creep. *Clin Biomech (Bristol, Avon)* 2003 March;18(3):222–30.
175. Slauterbeck JR, Fuzie SF, Smith MP, et al. The menstrual cycle, sex hormones, and anterior cruciate ligament injury. *J Athl Train* 2002 September;37(3):275–8.
176. Liu SH, al Shaikh R, Panossian V, et al. Primary immunolocalization of estrogen and progesterone target cells in the human anterior cruciate ligament. *J Orthop Res* 1996 July;14(4):526–33.
177. Slauterbeck JR, Hardy DM. Sex hormones and knee ligament injuries in female athletes. *Am J Med Sci* 2001 October;322(4):196–9.
178. Deie M, Sakamaki Y, Sumen Y, et al. Anterior knee laxity in young women varies with their menstrual cycle. *Int Orthop* 2002;26(3):154–6.
179. Heir T. Musculoskeletal injuries in officer training: one-year follow-up. *Mil Med* 1998 April;163(4):229–33.
180. Wunderlich RE, Cavanagh PR. Gender differences in adult foot shape: implications for shoe design. *Med Sci Sports Exerc* 2001 April;33(4):605–11.
181. Gilmour JC, Burns Y. The measurement of the medial longitudinal arch in children. *Foot Ankle Int* 2001 June;22(6):493–8.
182. Michelson JD, Durant DM, McFarland E. The injury risk associated with pes planus in athletes. *Foot Ankle Int* 2002 July;23(7):629–33.
183. Hosea TM, Carey CC, Harrer MF. The gender issue: epidemiology of ankle injuries in athletes who participate in basketball. *Clin Orthop Relat Res* 2000 March;(372):45–9.
184. Beynnon BD, Renstrom PA, Alosa DM, et al. Ankle ligament injury risk factors: a prospective study of college athletes. *J Orthop Res* 2001 March;19(2):213–20.
185. Sugimoto K, Takakura Y, Tohno Y, et al. Cartilage thickness of the talar dome. *Arthroscopy* 2005 April;21(4):401–4.
186. Hyer CF, Philbin TM, Berlet GC, et al. The obliquity of the first metatarsal base. *Foot Ankle Int* 2004 October;25(10):728–32.
187. Hart DA, Kydd A, Reno C. Gender and pregnancy affect neuropeptide responses of the rabbit Achilles tendon. *Clin Orthop Relat Res* 1999 August;(365):237–46.
188. Wang YT, Pascoe DD, Kim CK, et al. Force patterns of heel strike and toe off on different heel heights in normal walking. *Foot Ankle Int* 2001 June;22(6):486–92.
189. Haims AH, Schweitzer ME, Patel RS, et al. MR imaging of the Achilles tendon: overlap of findings in symptomatic and asymptomatic individuals. *Skeletal Radiol* 2000 November;29(11):640–5.
190. Muraoka T, Muramatsu T, Fukunaga T, et al. Elastic properties of human Achilles tendon are correlated to muscle strength. *J Appl Physiol* 2005 August;99(2):665–9.
191. Rezcallah AT, Xu R, Ebraheim NA, et al. Axial computed tomography of the pedicle in the lower cervical spine. *Am J Orthop* 2001 January;30(1):59–61.
192. Xu R, Burgar A, Ebraheim NA, et al. The quantitative anatomy of the laminas of the spine. *Spine* 1999 January 15;24(2):107–13.
193. Lim JK, Wong HK. Variation of the cervical spinal Torg ratio with gender and ethnicity. *Spine J* 2004 July;4(4):396–401.
194. Hukuda S, Kojima Y. Sex discrepancy in the canal/body ratio of the cervical spine implicating the prevalence of cervical myelopathy in men. *Spine* 2002 February 1;27(3):250–3.
195. Pettersson K, Karrholm J, Toolanen G, et al. Decreased width of the spinal canal in patients with chronic symptoms after whiplash injury. *Spine* 1995 August 1;20(15):1664–7.
196. Versteegen GJ, Kingma J, Meijler WJ, et al. Neck sprain after motor vehicle accidents in drivers and passengers. *Eur Spine J* 2000 December;9(6):547–52.
197. Suissa S. Risk factors of poor prognosis after whiplash injury. *Pain Res Manag* 2003;8(2):69–75.
198. Hendriks EJ, Scholten-Peeters GG, van der Windt DA, et al. Prognostic factors for poor recovery in acute whiplash patients. *Pain* 2005 April;114(3):408–16.
199. Siegmund GP, Sanderson DJ, Myers BS, et al. Awareness affects the response of human subjects exposed to a single whiplash-like perturbation. *Spine* 2003 April 1;28(7):671–9.
200. Yoganandan N, Knowles SA, Maiman DJ, et al. Anatomic study of the morphology of human cervical facet joint. *Spine* 2003 October 15;28(20):2317–23.
201. Stemper BD, Yoganandan N, Pintar FA. Gender- and region-dependent local facet joint kinematics in rear impact: implications in whiplash injury. *Spine* 2004 August 15;29(16):1764–71.

202. Truumees E, Demetropoulos CK, Yang KH, et al. Failure of human cervical endplates: a cadaveric experimental model. *Spine* 2003 October 1;28(19):2204–8.
203. Ryan SD, Fried LP. The impact of kyphosis on daily functioning. *J Am Geriatr Soc* 1997 December;45(12):1479–86.
204. Fon GT, Pitt MJ, Thies AC, Jr. Thoracic kyphosis: range in normal subjects. *AJR Am J Roentgenol* 1980 May;134(5):979–83.
205. Widhe T. Spine: posture, mobility and pain. A longitudinal study from childhood to adolescence. *Eur Spine J* 2001 April;10(2):118–23.
206. Chiu YL, Huang TJ, Hsu RW. Curve patterns and etiologies of scoliosis: analysis in a university hospital clinic in Taiwan. *Changgeng Yi Xue Za Zhi* 1998 December;21(4):421–8.
207. Payne WK, III, Ogilvie JW, Resnick MD, et al. Does scoliosis have a psychological impact and does gender make a difference? *Spine* 1997 June 15;22(12):1380–4.
208. Soucacos PN, Zacharis K, Soultanis K, et al. Risk factors for idiopathic scoliosis: review of a 6-year prospective study. *Orthopedics* 2000 August;23(8):833–8.
209. Axenovich TI, Zaidman AM, Zorkoltseva IV, et al. Segregation analysis of idiopathic scoliosis: demonstration of a major gene effect. *Am J Med Genet* 1999 October 8;86(4):389–94.
210. Loncar-Dusek M, Pecina M, Prebeg Z. A longitudinal study of growth velocity and development of secondary gender characteristics versus onset of idiopathic scoliosis. *Clin Orthop Relat Res* 1991 September;(270):278–82.
211. Ramirez N, Johnston CE, Browne RH. The prevalence of back pain in children who have idiopathic scoliosis. *J Bone Joint Surg Am* 1997 March;79(3):364–8.
212. Ugwonali OF, Lomas G, Choe JC, et al. Effect of bracing on the quality of life of adolescents with idiopathic scoliosis. *Spine J* 2004 May;4(3):254–60.
213. Bunge EM, Juttmann RE, de Kleuver M, et al. Health-related quality of life in patients with adolescent idiopathic scoliosis after treatment: short-term effects after brace or surgical treatment. *Eur Spine J* 2007;16:83–9.
214. Cheng WC, Yang RS, Huey-Jen HS, et al. Effects of gender and age differences on the distribution of bone content in the third lumbar vertebra. *Spine* 2001 April 15;26(8):964–8.
215. Ebbesen EN, Thomsen JS, Beck-Nielsen H, et al. Age- and gender-related differences in vertebral bone mass, density, and strength. *J Bone Miner Res* 1999 August;14(8):1394–403.
216. Gilsanz V, Boechat MI, Gilsanz R, et al. Gender differences in vertebral sizes in adults: biomechanical implications. *Radiology* 1994 March;190(3):678–82.
217. Gilsanz V, Boechat MI, Roe TF, et al. Gender differences in vertebral body sizes in children and adolescents. *Radiology* 1994 March;190(3):673–7.
218. Naganathan V, Sambrook P. Gender differences in volumetric bone density: a study of opposite-sex twins. *Osteoporos Int* 2003 July;14(7):564–9.
219. Korovessis P, Koureas G, Papazisis Z. Correlation between backpack weight and way of carrying, sagittal and frontal spinal curvatures, athletic activity, and dorsal and low back pain in schoolchildren and adolescents. *J Spinal Disord Tech* 2004 February;17(1):33–40.
220. Gruber HE, Norton HJ, Leslie K, et al. Clinical and demographic prognostic indicators for human disc cell proliferation in vitro: pilot study. *Spine* 2001 November 1;26(21):2323–7.
221. Iguchi T, Wakami T, Kurihara A, et al. Lumbar multilevel degenerative spondylolisthesis: radiological evaluation and factors related to anterolisthesis and retrolisthesis. *J Spinal Disord Tech* 2002 April;15(2):93–9.
222. Masharawi Y, Rothschild B, Salame K, et al. Facet tropism and interfacet shape in the thoracolumbar vertebrae: characterization and biomechanical interpretation. *Spine* 2005 June 1;30(11):E281–E292.
223. Norton BJ, Sahrmann SA, Van Dillen FL. Differences in measurements of lumbar curvature related to gender and low back pain. *J Orthop Sports Phys Ther* 2004 September;34(9):524–34.
224. O'Sullivan P, Dankaerts W, Burnett A, et al. Lumbopelvic kinematics and trunk muscle activity during sitting on stable and unstable surfaces. *J Orthop Sports Phys Ther* 2006 January;36(1):19–25.
225. Marras WS, Jorgensen MJ, Granata KP, et al. Female and male trunk geometry: size and prediction of the spine loading trunk muscles derived from MRI. *Clin Biomech (Bristol, Avon)* 2001 January;16(1):38–46.
226. Ng JK, Richardson CA, Kippers V, et al. Relationship between muscle fiber composition and functional capacity of back muscles in healthy subjects and patients with back pain. *J Orthop Sports Phys Ther* 1998 June;27(6):389–402.

227. Marras WS, Davis KG, Jorgensen M. Spine loading as a function of gender. *Spine* 2002 November 15;27(22):2514–20.
228. Granata KP, Orishimo KF. Response of trunk muscle coactivation to changes in spinal stability. *J Biomech* 2001 September;34(9):1117–23.
229. Granata KP, Rogers E, Moorhouse K. Effects of static flexion-relaxation on paraspinal reflex behavior. *Clin Biomech (Bristol, Avon)* 2005 January;20(1):16–24.
230. Brown MD, Holmes DC, Heiner AD, et al. Intraoperative measurement of lumbar spine motion segment stiffness. *Spine* 2002 May 1;27(9):954–8.
231. Cook C, Brismee JM, Sizer PS, Jr. Subjective and objective descriptors of clinical lumbar spine instability: a Delphi study. *Man Ther* 2006 February;11(1):11–21.
232. Bowen V, Cassidy JD. Macroscopic and microscopic anatomy of the sacroiliac joint from embryonic life until the eighth decade. *Spine* 1981 November;6(6):620–8.
233. Lin WY, Wang SJ. Influence of age and gender on quantitative sacroiliac joint scintigraphy. *J Nucl Med* 1998 July;39(7):1269–72.
234. Faflia CP, Prassopoulos PK, Daskalogiannaki ME, et al. Variation in the appearance of the normal sacroiliac joint on pelvic CT. *Clin Radiol* 1998 October;53(10):742–6.
235. Brooke R. The sacro-iliac joint. *J Anat* 1924;58:299–305.
236. Brunner C, Kissling R, Jacob HA. The effects of morphology and histopathologic findings on the mobility of the sacroiliac joint. *Spine* 1991 September;16(9):1111–7.
237. Bellamy N, Park W, Rooney PJ. What do we know about the sacroiliac joint? *Semin Arthritis Rheum* 1983 February;12(3):282–313.
238. Sashin D. A critical analysis of the anatomy and the pathologic changes of the sacroiliac joints. *J Bone Joint Surg* 1930;12:891–910.
239. Vleeming A, van Wingerden JP, Snijders CJ, et al. Mobility in the SI-joints in the elderly: a kinematic and roentgenologic study. *Clin Biomech (Bristol, Avon)* 1992;7:170–8.
240. Dar G, Peleg S, Masharawi Y, et al. Sacroiliac joint bridging: demographical and anatomical aspects. *Spine* 2005 August 1;30(15):E429–32.
241. Dreyfuss P, Dryer S, Griffin J, et al. Positive sacroiliac screening tests in asymptomatic adults. *Spine* 1994 May 15;19(10):1138–43.
242. Salsabili N, Valojerdy MR, Hogg DA. Variations in thickness of articular cartilage in the human sacroiliac joint. *Clin Anat* 1995;8(6):388–90.
243. Kristiansson P, Svardsudd K. Discriminatory power of tests applied in back pain during pregnancy. *Spine* 1996 October 15;21(20):2337–43.
244. Snijders CJ, Seroo JM, Snijder JGT, et al. Change in form of the spine as a consequence of pregnancy. *Digest of the 11th Int Conf Med Biol Eng*. Ottawa, ON: Conference Committee; 1976. p. 670–1.
245. Mens JM, Vleeming A, Stoeckart R, et al. Understanding peripartum pelvic pain. Implications of a patient survey. *Spine* 1996 June 1;21(11):1363–9.
246. Rathmell JP, Viscomi CM, Bernstein IM. Managing pain during pregnancy and lactation. In: Raj P, editor. *Practical Management of Pain* 3rd ed. St. Loui, MO: Mosby, Inc; 2000. p. 196–211.
247. Gerlach UJ, Lierse W. Functional construction of the sacroiliac ligamentous apparatus. *Acta Anat (Basel)* 1992;144(2):97–102.
248. Lee D. Biomechanics of the lumbo-pelvic-hip complex. *An Approach to the Examination and Treatment of the Lumbo-Pelvic-Hip Region*. New York, NY: Churchill Livingstone; 1999. p. 43–72.
249. Vleeming A, Pool-Goudzwaard AL, Stoeckart, et al. The posterior layer of the thoracolumbar fascia. Its function in load transfer from spine to legs. *Spine* 1995 April 1;20(7):753–8.
250. Kapandji IA. *The Physiology of the Joints*. Edinburgh, London and New York, NY: Churchill Livingstone; 1974.
251. Moore KL, Dalley AF. The lower limb. In Clinical Oriented Anatomy (4th Ed.). Philadelphia PA: Lippincott Williams & Wilkins, 1999. p. 550.
252. Vleeming A, Pool-Goudzwaard AL, Hammudoghlu D, et al. The function of the long dorsal sacroiliac ligament: its implication for understanding low back pain. *Spine* 1996 March 1;21(5):556–62.
253. Luk KD, Ho HC, Leong JC. The iliolumbar ligament. A study of its anatomy, development and clinical significance. *J Bone Joint Surg Br* 1986 March;68(2):197–200.
254. Pool-Goudzwaard AL, Kleinrensink GJ, Snijders CJ, et al. The sacroiliac part of the iliolumbar ligament. *J Anat* 2001 October;199(Pt. 4):457–63.

255. Ostgaard HC. Assessment and treatment of low back pain in working pregnant women. *Semin Perinatol* 1996 February;20(1):61–9.
256. Damen L, Buyruk HM, Guler-Uysal F, et al. The prognostic value of asymmetric laxity of the sacroiliac joints in pregnancy-related pelvic pain. *Spine* 2002 December 15;27(24):2820–4.
257. Noren L, Ostgaard S, Johansson G, et al. Lumbar back and posterior pelvic pain during pregnancy: a 3-year follow-up. *Eur Spine J* 2002 June;11(3):267–71.

II Preoccupation with Body Image Issues and Disordered Eating Issues in the Active Female

4 Body Image and Eating Disturbances in Children and Adolescents

Marilyn Massey-Stokes

Contents

4.1 Learning Objectives
4.2 Introduction
4.3 Research Findings
4.4 Positive Youth Development
4.5 Future Direction
4.6 Summary and Conclusions
4.7 Scenario with Questions and Answers

4.1. LEARNING OBJECTIVES

After completing this chapter, you should have an understanding of the following:

- Body image disturbances (BIDs) and their prevalence among children and adolescents.
- Eating disturbances and their prevalence among children and adolescents.
- The connection between eating disorders (EDs) and other risk behaviors and psychological disorders.
- Risk and protective factors for body image and eating disturbances.
- Promoting healthy body image and preventing eating disturbances among youth.
- The application of Social Development Strategy (SDS) in the health promotion and prevention process.

4.2. INTRODUCTION

The Centers for Disease Control and Prevention (CDC) has identified six categories of health-risk behaviors that contribute to the leading causes of morbidity and mortality among youth and adults—alcohol and other drug use, injury and violence, tobacco use, sexual behaviors, physical inactivity, and unhealthy dietary behaviors. These interrelated risk behaviors usually begin in childhood, persist into adulthood, and are preventable. Not only are these behaviors linked to health problems, they also contribute to numerous educational and social problems that negatively impact our

From: *The Active Female*
Edited by: J. J. Robert-McComb, R. Norman, and M. Zumwalt © Humana Press, Totowa, NJ

nation *(1)*. The CDCs Youth Risk Behavior Surveillance System (YRBSS) monitors these six categories of health-risk behaviors among youth through national, state, and local data collection. Data collected during October 2004–January 2006 revealed the following selected information about dietary-related behaviors among US students in grades 9–12: 45.6% of students were trying to lose weight during the 30 days preceding the survey; 12.3% of students had gone without eating for 24 hours or more to lose weight or to keep from gaining weight during the 30 days preceding the survey; 6.3% of students had taken diet pills, powders, or liquids without a physician's advice to lose weight or to keep from gaining weight; and 4.5% of students had vomited or taken laxatives to lose weight or to keep from gaining weight during the 30 days preceding the survey *(2)*.

BIDs, eating concerns, and weight issues clearly present major challenges for children and youth. Other data reveal that among American high school students, 30% of girls and 16% of boys, engage in disordered eating behaviors, including bingeing, vomiting, fasting, laxative and diet pill use, and compulsive exercise *(3)*. Furthermore, childhood obesity in the United States has increased at an alarming rate, with approximately 9 million American children classified as obese *(4)*. Research has indicated that American children are dissatisfied with their body shapes or weights *(5–7)*, even children as young as 6 years *(8,9)*. Up to 70% of normal-weight adolescent girls report feeling fat and engaging in unhealthy eating practices for weight loss purposes *(10)*. Studies also have shown that children as young as 8 and 9 years engage in dieting practices *(11–14)*. Moreover, girls who frequently diet are 12 times as likely to engage in binge eating as girls who do not diet *(15)*.

Enhancing health and quality of life for all children and adolescents is a desired outcome of health promotion and prevention programs. Health is not a static condition, but a dynamic interplay among the dimensions of physical, mental, emotional, social, and spiritual health. Consistent with this view of health, promoting healthy body image and preventing eating disturbances among children and adolescents should focus on the "whole child." For example, undernutrition, which often occurs with eating disturbances, can have detrimental effects not only on children's physical health (e.g., growth retardation and delayed maturation) *(16)*, but also on their cognitive development *(17)*. Undernourished students often experience irritability, decreased concentration, nausea, headache, lack of energy, and increased susceptibility to illness *(17)*. Therefore, it is very difficult for these youth to focus their energy on mastering important developmental tasks, such as succeeding in academics and developing social-emotional skills and a positive identity.

There is a broad evidence base that encompasses numerous facets of body image and eating disturbances that negatively impact the whole child. Researchers and clinicians now recognize the triad of body image difficulties, EDs, and obesity as an interrelated set of body weight and shape disturbances that cause substantial problems for children and adolescents *(18)*. In addition, many researchers have viewed eating problems on a continuum, beginning with body dissatisfaction and weight concerns and ending with clinical EDs *(19,20)*. In this chapter, the term "eating disorders" refers to anorexia nervosa and bulimia nervosa. The terms "eating disturbances" and "disordered eating" are broader and refer to a range of unhealthy diet-related behaviors such as obsession with body weight and shape, excessive restrictive eating, skipping meals, laxative and diet pill use, cycles of binge eating and dieting, self-induced vomiting, and excessive exercise with the sole purpose of "purging" calories obtained from dietary intake. The purpose of this chapter is to present an overview of some of the key findings concerning body image and eating disturbances in children and adolescents and discuss viable avenues for promoting healthy body image and preventing eating disturbances among this population.

4.3. RESEARCH FINDINGS

4.3.1. Body Image Disturbances in Children and Adolescents

The term *body image* is the subjective depiction of physical appearance *(21)* and is comprised of behavioral, perceptual, cognitive, and affective experiences *(22)*. Numerous studies have shown a connection between BID and low self-esteem, psychosocial distress, and early-onset depression *(7)*. In addition, the relationship between body image dissatisfaction and BID has been strongly linked to EDs such as anorexia nervosa and bulimia nervosa *(7)*. As mentioned earlier, body dissatisfaction has been seen in children as young as 6 years. As age increases, ideal body size generally becomes progressively thinner *(23,24)*. Therefore, body image and BID in children and adolescents command attention in both research and practice *(7,25)*.

Numerous research studies show that high numbers of Caucasian American children experience body dissatisfaction *(7)*. However, there is evidence that body dissatisfaction may be increasing among girls in minority ethnic groups. For example, Neumark-Sztainer and colleagues *(26)* found high levels of dieting and disordered eating among all ethnic groups. They also discovered that although dieting was more prevalent among adolescents from higher socioeconomic families, disordered eating was widespread in adolescents from lower socioeconomic families as well. Other research has shown that African-American adolescent females have reported being more comfortable with their bodies than those among other ethnic groups, but they also have exhibited a high drive for thinness *(5)*. In addition, there are indications that Mexican-American female adolescents desire to be thinner *(27)*.

There also are developmental trends in body image and weight concerns, and these trends vary by gender and across ethnic groups. For example, research suggests that body dissatisfaction and weight concerns increase with age, particularly among females *(7,28–30)*. These trends are important because there is evidence that body dissatisfaction in young girls can lead to eating problems *(7,31,32)* and early-onset depression *(33,34)* later.

In terms of assessing body image, most researchers focus on two separate components of BID—perceptual body-size distortion and the affective (attitudinal) aspect. Perceptual body-size distortion is comprised of inaccurate perceptions of one's body size (e.g., individuals with EDs often overestimate their actual body size). The affective element relates to dissatisfaction with one's body size, shape, or some other aspect of physical appearance *(25)*. Although most studies have focused on the distortion component, greater consistency has been found by using attitudinal measures. Moreover, these two components appear to function independently *(35)*. Therefore, body image is considered multidimensional, and the assessment of BID requires a variety of methods and techniques *(36)*. There are a variety of assessment instruments, most of which have been developed with adult samples *(37)*. Banasiak and colleagues *(37)* examined reliability in numerous assessments that measure dietary restraint and body concerns. Their findings suggest that many of the instruments developed on adults can be used with middle adolescent girls (9th grade) when proper steps have been taken to ensure that girls understand the terminology used in the instruments. For example, glossaries can be developed for assessment instruments to improve their reliability for use with adolescents. There also is a need for reliable and valid measures of BID and ED symptoms in preadolescents. To help fill the gap in terms of measuring body image in children and adolescents, Veron-Guidry and Williamson *(38)* conducted a study in which they extended the Body Image Assessment (BIA) procedure for adults *(39)* to children and adolescents, developed norms, and evaluated reliability and validity of the adapted BIA procedure. Their data supported the validity of the BIA-C (children) and BIA-P (preadolescents) procedures and confirmed results from the adult BIA procedure *(39)*. Generally speaking, instruments used to measure body image and BID in children and adolescents should have sound psychometric properties (e.g., a test-retest reliability of a least 0.70) and be evidence-based. In addition, researchers have supported the use of video distortion methods, which have been successfully implemented with children as young as 5 and 6 years, as well as custom computer software for measuring body size estimations *(25)*. (For a thorough discussion of assessing body image and BID in children and adolescents, *see* Gardner *(25)*.)

Although the research literature concerning body image in children and adolescents has provided an enlightening knowledge base, Smolak *(40)* points out that there are salient research questions that still need to be addressed, such as (1) how body dissatisfaction varies at different ages for different genders and across different ethnic and socioeconomic groups; (2) the need to develop more accurate measurements of body image, particularly in young children and adolescents from various ethnic and socioeconomic groups; and (3) developmental trends in body image development. In addition, more research needs to target body image importance (overestimated views about body shape and weight) and its role in the development of body dissatisfaction, weight loss strategies, and disordered eating *(41,42)*. Likewise, there is a need to investigate whether childhood body dissatisfaction, high body mass index, and eating disturbances are risk factors for later development of EDs, obesity, or depression *(6)*.

4.3.2. Eating Disorders in Children and Adolescents

Clinically diagnosable EDs *(43)* are atypical among prepubescent children *(7,44)*; yet, they rank as the third most common chronic illness among adolescent females, with an incidence of up to 5% *(45–47)*. EDs are related to other risk behaviors, such as tobacco use, alcohol and other drug abuse, sexual activity, and suicide attempts *(43)*. EDs often lead to multiple negative outcomes that affect the whole child, ranging from a preoccupation with eating that can significantly hinder healthy growth and development to deleterious medical complications such as severe malnutrition, osteoporosis, acute psychiatric emergencies, heart and other organ damage, and even death *(48)*. In fact, death rates from EDs are among the highest for any mental illness *(49)*. More specifically, the mortality rate among those with anorexia nervosa is approximately 12 times higher than the death rate among US females aged 15–24 from all causes of death *(50)*.

The average age of onset for anorexia nervosa is 14–18 years and late adolescence or early adulthood for bulimia nervosa *(43)*. Younger children may have significant problems related to body image, eating, and weight management that do not meet the diagnostic criteria for an ED, but can increase risk for developing an ED later. The American Academy of Pediatrics *(51)* and Society for Adolescent Medicine *(48)* assert that an ED can still exist in the absence of established diagnostic criteria. Patients who do not fully meet the DSM-IV®-TR criteria for anorexia or bulimia, but experience the same medical and psychological consequences of these disorders, usually are diagnosed with an "ED not otherwise specified" (EDNOS) *(52)*. The majority of adolescents in ED treatment centers meet the EDNOS criteria *(48)*. Warning signs include obsessive thinking about food, weight, shape, or exercise; unhealthy weight management practices; and failure to maintain a healthy body weight/composition for gender and age. Similarly, along the spectrum of eating disturbances, researchers often use the terms "partial syndrome" or "sub-clinical" to indicate similar characteristics. Although most children may not meet DSM-IV®-TR diagnostic criteria for an ED, those who exhibit symptoms of EDNOS or a sub-clinical eating disturbance can still suffer considerable social and educational impairment that may require clinical intervention *(18)*. Even in the absence of clinical intervention, obsessive inner dialogues about body weight and shape and the consequences of disordered eating significantly interfere with the developmental needs and resilience of children and adolescents *(53,54)*.

EDs often coexist with other psychological disorders, such as depression, obsessive-compulsive disorder, anxiety, bipolar disorder, substance abuse (including prescription drugs), and personality disorders (e.g., borderline personality) *(43,48,55)*. In some cases, the ED is a secondary symptom to an underlying psychological disorder; and in other cases, the psychological disorder may be secondary to the ED. Adding to this complexity, a young person with an ED may present other self-destructive behaviors, such as self-injury (e.g., cutting) *(56)*.

EDs are comprised of a complex array of biopsychosocial issues that should be addressed by a multidisciplinary team of medical, nutritional, mental health, and nursing professionals who have expertise in child–adolescent health and are experienced in treating BID and EDs *(48)*. Although early interdisciplinary treatment increases the likelihood of successful recovery, there are numerous barriers to this type of health care, including time, cost, and inadequate insurance benefits. In addition, patients and their families often exhibit ambivalence or resistance to the treatment process *(48)*. Failure to detect an ED in its early stages can exacerbate the illness and make it much more difficult to treat *(48,51)*.

4.3.3. Risk and Protective Factors for Body Image and Eating Disturbances

Body image and eating disturbances are extremely complicated and appear to result from a complex interplay among a myriad of risk factors and deficient protective factors. Generally speaking, risk factors are those conditions that increase the likelihood that an individual will develop an eating problem; and protective factors are those conditions that mitigate the risk (i.e., decrease the chances that disordered eating will occur). Numerous research studies have identified multiple risk and protective factors for body image and eating disturbances that can be divided into four primary categories: biological, individual, familial, and sociocultural. The following discussion is not inclusive, but represents an overview of some of the key evidence-based risk and protective factors within these categories.

Biological risk factors include genetic predisposition to EDs, mood disorders *(57)*, neurochemical (e.g., serotonin) imbalances *(58,59)*, and early puberty *(60–62)*. Individual risk factors encompass negative body image *(36)* and body dissatisfaction *(63)*, temperament (e.g., negative emotionality) *(64–66)*, personality characteristics (e.g., perfectionism) *(67)*, low self-esteem *(14,62,68)*, inadequate coping skills *(69)*, substance abuse *(70)*, and overweight/obesity *(71–73)*.

Familial risk factors include EDs in first-degree biological relatives *(43,69,74,75)*, maladaptive parental behaviors and dysfunctional family relations *(36,61,70,76–79)*, family pressure to adhere to the thin ideal *(80)*, alcohol misuse *(81)*, and physical or sexual abuse *(79,82,83)*, although this latter aspect is controversial. It is important to point out that familial influences are often difficult to separate from broader sociocultural influences because most (if not all) families are influenced to some degree by current societal standards *(84)*. With that in mind, other sociocultural risk factors include societal glamorization of the thin ideal *(85)*, media exposure promoting thinness *(86,87)*, and peer influences promoting dieting and adherence to the thin standard *(80,88)*. It is important to note that some of these risk factors were identified through cross-sectional studies; therefore, they cannot completely meet the definition of "risk factor" until they are found to be significant in longitudinal studies *(69,89)*. Moreover, there may be critical developmental periods during which exposure to risk factors may have greater influence on the development of eating disturbances (e.g., puberty) *(70)*.

In contrast, there is little research about protective factors and how they may buffer individuals against developing eating disturbances and clinical EDs *(69,70)*. The primary protective factor that has received the most empirical support is positive family relationships *(34)*. Protective factors that emerged from a qualitative research study with high school girls were self-acceptance, family acceptance, positive peer influences, and knowledge about the hazards of dieting *(90)*. Furthermore, individual protective factors that have been suggested include self-directedness and assertiveness *(91)*, the ability to effectively cope with life stressors *(91,92)*, high self-esteem *(92)*, and a genetic predisposition for slimness *(91,93)*. In addition to positive relationships with parents *(34,61,76)*, other familial protective factors that have been considered include living in a family that does not overemphasize body weight and physical attributes *(91,93)*, living in a family where parents do not misuse alcohol *(94)*, and social support from the family *(95)*.

Sociocultural factors that may be protective are cultural messages that embrace different body shapes and sizes *(91)*, participation in sports that do not emphasize thinness for successful performance *(68,91,96)*, close relationships with friends who do not overstress body weight *(91,97)*, and social support from peers *(34,92,95)*.

4.3.4. Primary Prevention

Primary prevention focuses on keeping body image and eating disturbances from developing among children and adolescents (i.e., stopping the problem before it starts). Because of the varied tasks that comprise healthy development across childhood and adolescence, prevention programs must not only be age and developmentally appropriate, but also address the relevant skills and challenges for each stage of development

(98). Because body dissatisfaction and inappropriate weight management practices start early and can lead to more serious body image and eating problems later, prevention should begin at an early age (e.g., elementary school years) *(7)*.

Health education promoting healthy body image along with healthy eating and physical activity is a principal tool for primary prevention. Health education can be implemented both formally and informally in a variety of settings, including homes, schools, health care facilities, and the wider community. Primary prevention is effective when it incorporates strategies that emphasize multifarious components, including, but not limited to, (1) development of positive self-esteem and healthy body image; (2) development of essential life skills, including social-emotional and effective coping skills; (3) provision of experiences that encourage the development of self-efficacy; (4) skills training for lifelong balanced nutrition and physical activity; (5) opportunities to develop media literacy skills and learn how to challenge sociocultural myths and attitudes regarding body shape and size; and (6) positive youth development.

4.3.4.1. INDIVIDUAL

A crucial aspect of prevention that targets the individual is the development of self-esteem, which has emerged as a significant predictor of eating problems in numerous research studies *(69,99,100)*. Branden *(100)* defines self-esteem as possessing two components—self-respect and self-efficacy. Among the hallmarks of self-respect are an individual's assurance of his/her value and basic right to experience a fulfilling life, plus comfort in appropriately asserting thoughts, wants, and needs *(100)*. A caring adult can foster a child's self-respect through increasing the individual's sense of positive uniqueness (i.e., characteristics that set her apart from others in a way that brings honor and a healthy sense of pride). Self-respect does not arise from physical characteristics, but rather from internal attributes. A child also needs to feel significant and a distinct sense of belonging. This sense of significance and belonging should begin in the family, but also can be fostered through other settings such as the school and other community-based groups. For example, a school-based prevention program called Everybody's Different *(101)* focuses on an individual's uniqueness. Initial findings revealed that boys and girls and those students who were overweight or at high risk of developing body image or eating problems experienced significantly improved body image compared to controls. In addition, improvement in body satisfaction of high-risk female and male students was still statistically significant 12 months after the program *(99)*.

Self-efficacy is a person's belief or confidence that she/he can successfully accomplish a task *(100)*. A sense of competence (power) is foundational for sound mental and emotional health and preventative against the development of negative body image and eating problems. For example, Troop and Treasure *(102)* found that women who reported experiencing more feelings of helplessness and lower levels of mastery in childhood were more likely to develop EDs in adulthood. One avenue for instilling a sense of competence in children and adolescents is to help them develop health literacy, which is comprised of these four competencies: effective communication; self-directed learning; critical thinking and problem solving; and responsible, productive citizenship *(103)*. Health literate individuals are more likely to be resilient and make healthy, life-affirming choices. In terms of preventing body dissatisfaction and eating problems, health literacy competencies are imperative. For example, critical thinking and problem solving are integral skills youth need to effectively counter the barrage of unrealistic, "thin" sociocultural messages they frequently encounter from a variety of sources.

Self-efficacy also can be promoted through goal setting, particularly short-term goals, so that children and youth can take "baby steps" of progress. In turn, they can experience multiple increments of success, thereby developing confidence in their abilities. In addition, repeated opportunities for rehearsal of essential life skills help youth increase a sense of mastery. In turn, this sense of competence empowers them to persist in spite of setbacks and press forward to learn new skills. Positive social support is another critical element for fostering self-efficacy that can be derived from the family, peers, and caring adults and mentors (e.g., school personnel, health care providers, and other members of the community). These significant relationships provide opportunities for children to cultivate resilience and emulate positive role models and mentors.

4.3.4.2. FAMILY

As primary caregivers in the early years of life, parents are considered the main teachers and socializing agents for a child's interaction with the larger environment *(104)*. Families are a major health and social influence in the lives of children and adolescents; therefore, strong family involvement is essential to health promotion and prevention. There are numerous strategies that the family can employ to promote positive body image and healthy eating and physical activity habits in children and youth. First of all, positive, nurturing relationships are integral to family health. Parents/caregivers should make it an utmost priority to invest time in building strong, loving, and supportive relationships with their children. In addition, parents/caregivers should be intentional in building self-esteem and fostering resilience in children, starting at a very early age. Parents also should encourage and model essential life skills such as clear communication, problem solving, decision making, and stress management. Even preschool-aged children can be taught about the value of eating healthy foods and being physically active for health and wellness, the importance of respecting different body types, and how to effectively communicate feelings and needs. All children, regardless of their weight and size, should feel that they are unconditionally loved and accepted by the family. If a family suspects that a child is engaging in restrictive eating and/or other maladaptive behaviors, they should seek help from a qualified professional. (For a more comprehensive list of suggestions, *see* Levine *(105)*.)

Studies have shown that family involvement plays a key role in a variety of school-based health promotion efforts targeting children and youth, including cardiovascular health promotion *(106–108)*, fruit and vegetable consumption *(109)*, and alcohol prevention *(110)*. The family also can play a powerful role in preventing negative body image and eating problems in children and adolescents *(6,9,104,111)*; yet, few school-based prevention programs have included a family involvement component *(104)*. School-based prevention initiatives can include a wide array of family involvement activities, such as home-based activities that the family completes together, family fun nights at the school, parent education workshops, and experiential "role play" activities for skill development. Furthermore, families and schools can partner to foster resilience in youth by forming caring, nurturing relationships, communicating high expectations, and providing them with meaningful opportunities to participate in pro-social activities *(112)*.

4.3.4.3. SOCIOCULTURAL GROUPS

Progress is lagging in terms of altering cultural norms concerning thinness, body image, and weight management practices. Experts contend that sociocultural changes must occur in order to experience a decline in the growing numbers of children and adolescents with body image problems and disordered eating *(51,113)*. Sociocultural influences encompass schools, peers, health care providers, media, and the larger society, all of which interact dynamically in the prevention of body image and eating disturbances.

4.3.4.3.1. Schools Findings from the National Longitudinal Study on Adolescent Health revealed that of all the factors that influence adolescent health-risk behavior, the most critical are the family and schools *(114)*. Schools are logical venues for primary prevention programs because of the sheer number of children they serve, and upper elementary appears to a viable age with which to target interventions. Enhanced knowledge, critical-thinking and problem-solving skills, and realistic perceptions may offset unhealthy influences concerning body shape, weight, and eating practices that are likely to be well established by the middle-school years *(115)*.

Of the 42 school-based prevention programs that Levine and Smolak *(113)* located and reviewed, 10 had been developed for elementary-aged children. Results of these programs varied, ranging from those revealing minimal change to those demonstrating increases in knowledge and body satisfaction. However, in some cases, at least for a significant minority of the girls, program content may have increased or created negative attitudes about their bodies. An upper-elementary prevention curriculum that appears to hold promise is *Healthy Body Image: Teaching Kids to Eat and Love Their Bodies Too!* *(116)*. An uncontrolled pilot study with 222 boys and girls in grades 4–6 revealed positive changes in knowledge of program content, acceptance of diverse body weights and shapes, positive body esteem, and rejection of the thin ideal *(117)*. Later, a controlled study with 415 boys and girls measured pretest and posttest changes related to (1) body image; (2) knowledge about the biology of size, shape, and restricted hunger or dieting; (3) body size prejudice; (4) media awareness; (5) self-image; and (6) lifestyle behavior. Students who completed the curriculum showed significant or noteworthy improvements in most areas compared to a control group not exposed to the curriculum *(118)*. Another prevention program entitled *Eating Smart, Eating for Me* is designed for 4th and 5th grade students and centers around five goals encompassing healthy body image, healthy eating and physical activity, and media literacy. A controlled evaluation of this 10-lesson curriculum revealed an increase in student knowledge about nutrition and decreased negative attitudes toward overweight people. However, attitudes like body esteem and behaviors did not significantly change *(119)*. A 2-year follow-up study revealed that when compared to new controls, adolescents who participated in the prevention program 2 years earlier were more knowledgeable, engaged in fewer unhealthy weight management behaviors, and had higher body esteem. The latter difference was particularly significant for girls *(120)*.

In terms of evaluated school-based programs that have been developed for middle schools and high schools, most have had only a modest effect on knowledge with little impact on attitudes, behavior, or weight change. This partly could be because of the short-term nature of these interventions. Programs may be more effective if they are designed for the stages of change for the participants; are based on the lived experiences of the participants; include relational components; coincide with developmental issues; and incorporate active, experiential, and peer-mediated formats *(34,111,113,121)*.

There seems to be some disagreement as to whether universal programs or programs targeting high-risk groups have the best likelihood for achieving successful outcomes. Research has demonstrated that it is feasible to deliver different interventions by providing a universal prevention program to all students while targeting those at risk *(122)*. In either case, school-based prevention programs should be a part of an overall coordinated school health program (CSHP) that promotes the health and well-being of students through eight interrelated components—health education; healthy school environment; school health services; school food services; counseling, psychological, and social services; physical education; staff health promotion; and family and community involvement *(123)*. For example, a comprehensive program for preventing body image and eating disturbances can include (1) evidence-based curricula for preventing body image and eating problems; (2) a school environment that cultivates caring, supportive relationships; (3) school food services that reinforce healthy eating; (4) counseling services for all students, particularly those at high risk, that include peer support groups and referral networks; (5) daily, quality physical education that emphasizes the importance of staying physically active for health and vitality; (6) ongoing prevention training for staff; and (7) outreach activities that form partnerships among school personnel, families, and the broader community. Furthermore, individuals who are responsible for implementing prevention programs should be stakeholders in the process. For example, engaging teachers is paramount to implementing a successful prevention curriculum *(124)*. This argument can be extended to include all of those involved in a CSHP—administrators, teachers, school nurses, school counselors/psychologists, school social workers, school food service workers and other school staff, students, families, and the larger community. When each of these individuals embraces the importance of health promotion and takes ownership of prevention efforts that can change the school milieu, there is a strong possibility for a reduction in the incidence of body image and eating disturbances among youth.

4.3.4.3.2. Peers Peer groups often are perceived as risk factors for harmful health behaviors because of their role as strong sociocultural influences in the lives of children and youth. However, peer groups that model pro-social values actually can serve as crucial protective factors. A qualitative study of high school girls found that positive peer influences (e.g., being dissuaded from dieting or purging) protected girls from excessive weight concerns that can lead to eating disturbances *(90)*. Relationships with peers who value health and wellness and practice life-affirming behaviors are powerful and absolutely vital to the prevention of body image and eating disturbances in children and adolescents. Healthy peers do not place overemphasis on body weight and size and value themselves and others for who they are on the inside (e.g., their spirit, character, talents, and gifts). They also can challenge each other's unrealistic thoughts and self-defeating actions concerning body weight and size. In addition, peer mentoring programs and peer support groups can make significant contributions to the prevention process. Peer inclusion that fosters a positive sense of belonging is a fundamental need of children and adolescents; and young people should have multiple opportunities to cultivate positive, healthy relationships with peers who model pro-social values and health-enhancing behaviors. These relationships can be formed with peers at school, during after-school programs and extracurricular activities, and in faith-based youth groups and other community organizations.

4.3.4.3.3. Health care providers Because of the potentially irreversible effects of EDs on all aspects of child and adolescent health, the high mortality rate of EDs, and evidence that early treatment is vital for improved outcomes, the role of health care professionals in recognizing and preventing EDs is paramount *(48,51)*. Many families look to health care providers as primary sources of credible health information, which places health care professionals in a powerful leadership position to make a significant difference in health promotion and prevention efforts. There are a number of ways that health care providers can proactively reduce the risk of body image and eating problems among children and adolescents. For example, office visits provide a ripe opportunity for communicating important information to children and their parents, including strategies for healthy eating and physical activity to enhance health and quality of life. It is important to note that many families are seeking the advice and care of physicians and/or dieticians regarding childhood obesity. It is imperative that health professionals be cognizant of the delicate balance when counseling children and their families about obesity and its risks. It is essential not to overemphasize dieting and to help children and adolescents build self-esteem while addressing weight issues *(51)*. Health care professionals also can help both parents and children understand what to expect during various developmental stages, particularly puberty. Additionally, health care providers can contribute to primary prevention through school-based and community-based interventions that deliver screenings and education and through participating in advocacy efforts aimed at changing cultural norms that children and adolescents experience *(51)*. The BodyWise *(125)* information packet designed for health care providers offers useful tips and strategies to facilitate the prevention process.

4.3.4.3.4. Media Children and adolescents must be armed with knowledge and life skills to combat the extreme focus on appearance that pervades our society. A very influential medium for communicating society's thin ideal is the media, which has assumed a ubiquitous role in homes across America. According to a Kaiser Family Foundation study, "Young people today live 'media-saturated' lives, spending an average of nearly $6\frac{1}{2}$ (6:21) hours a day with media" *(126)*. Media literacy education is a valuable weapon against media messages that communicate unrealistic body images. Media literacy has been defined as "the ability to access, analyze, evaluate, and produce communication in a variety of forms" *(127)*. Media literacy education can help girls enhance their sense of self-acceptance and empowerment regarding media portrayal of female bodies and teach them how to effectively counter messages that promote unrealistic body images and unhealthy eating *(128,129)*. Prevention programs like GO GIRLS!™ *(130)* can build self-esteem, enhance body image, increase resilience, and empower participants to be strong media advocates. Most (if not all) media literacy education for the prevention of EDs has targeted high school and college-aged females, which leaves an alarming gap in terms of reaching prepubescent and younger girls. Even young children are highly impressionable to negative health messages, including ones about body weight and size. Therefore, media literacy education should target different age groups of children and adolescents and hold relevance in terms of the types of media each group uses.

The proliferation of Internet websites that advocate or encourage youth to practice disordered eating warrants additional attention *(48)*. These pro-anorexia (pro-ana) and pro-bulimia (pro-mia) websites promote disordered eating as a lifestyle and offer dangerous and negative reinforcement for youth who are practicing disordered eating or thinking about doing so. Disturbingly, these sites far outnumber that of professional or recovery sites *(131)*, which calls for the need to increase awareness among parents, teachers, health care providers, and other caring adults and mentors.

4.3.4.3.5. Society In order for prevention of body image and eating disturbances among children and adolescents to be successful, there must be inside-out changes at the grassroots level that lead to broader sociocultural changes. These changes should

Conceptual Building Blocks	Foundation	Desired Outcome	Goal
Developmental change is inevitable. Normal changes of puberty include weight gain and temporary out-of-proportion growth; fat does not by itself define "overweight." Genetics and other internal weight regulators strictly limit the degree to which shape, weight & Body Mass Index can be manipulated through healthy means. Restricted or restrained hunger (dieting) results in predictable consequences that are *counterproductive* to weight loss and interfere with normal hunger regulation.	Recognize and respect basic biology; understand what *cannot* be controlled about size, shape and hunger.	**Accept the innate body: "This is the body I was born to have."**	**Healthy Body Image**

Balance attention to *many* aspects of identity. Looks are only one part. Consistently satisfy hunger with *enough* varied, wholesome food in a stable, predictable manner. Limit sedentary choices to promote a physically active lifestyle at all ages. Choose role models that reflect a realistic standard.	Emphasize what *can* be influenced or chosen.	**Enjoy eating for health, energy, and hunger satisfaction.** **Create a physically active lifestyle for fitness, endurance, fun, relaxation and stress relief.**	**Prevention of Unhealthy and Disordered Eating**
Promote historical perspective on today's cultural attitudes related to body image. Teach critical thinking about media messages that influence body image. Support others in resisting unhealthy norms about weight, dieting, low nutrient food choices, eating for entertainment, and sedentary entertainment.	Develop social and cultural resiliency.	**Develop autonomy, self esteem, confidence, and the ability for critical thinking.**	

Fig. 4.1. The Model for Healthy Body Image©. (*Source*: Developed by Kathy J. Kater, LICSW. ©1998 – www.BodyImageHealth.org. May be photocopied for educational purposes.)

include how females are portrayed in the media; expectations regarding gender roles; acceptance of a wide range of body weights and shapes; and increased opportunities for healthy eating and physical activity, particularly in low-income neighborhoods *(132)*. In addition, Kater *(133)* addresses four "toxic" sociocultural myths that need to be overcome in order to help children and adolescents develop healthy body image: valuing image over substance (e.g., marketing extremely thin models as if they were normal); denial of biological diversity of body sizes and shapes; denial of the harmful effects of restrictive eating for weight loss purposes; and discounting the value of eating well and being physically fit. The Model for Healthy Body Image (MHBI) *(134)* was designed to combat these myths by helping children learn to value health and wellness and resist sociocultural pressures that promote negative body image and unhealthy behaviors (Fig. 4.1). This model is addressed in an elementary prevention curriculum *(116)* so that children can develop healthy body image before they hit the transitional period of the middle school years.

Other credible prevention strategies and programs that address various sociocultural influences can be located online. In addition to BodyWise *(125)*, the National Eating Disorders Association (http://www.edap.org) provides an array of prevention information and tips.

4.4. POSITIVE YOUTH DEVELOPMENT

Numerous researchers and practitioners have recognized the importance of positive youth development in promoting health-enhancing behaviors and preventing health-risk behaviors among youth. In a comprehensive study of positive youth development programs in the United States, Catalano and colleagues *(135)* found that various youth development approaches can result in positive youth behavior outcomes (e.g., enhanced relational skills and quality of peer and adult relationships, improved problem solving, and self-efficacy) and prevention of youth risk behaviors (e.g., drug and alcohol use, smoking, school misbehavior, and risky sexual behavior). Data from longitudinal studies in the period from the 1970s to the 1990s have identified similar risk and protective factors in individuals, peer groups, families, schools, and neighborhoods that can accurately predict diverse youth problem behaviors *(136–141)*. However, exposure to ever-increasing numbers of protective factors has been found to prevent problem behaviors in spite of the presence of multiple risks *(142,143)*.

As a forerunner in the research and implementation of positive youth development, the Social Development Research Group *(144)* emphasizes the importance of a comprehensive approach to preventing youth problem behaviors by addressing both risk and protective factors. The framework for guiding the positive youth development process is SDS, which is part of a comprehensive model of behavioral development—The Social Development Model *(138)*. SDS organizes protective factors into a framework for promoting positive youth development, despite the presence of risk. Additionally, SDS focuses on the outcome of health-enhancing behaviors through exposing children and adolescents to two critical protective factors: (1) pro-social bonding to family, school, and peers and (2) healthy beliefs and clear standards for behavioral norms. The mechanisms that help create these protective factors are opportunities for meaningful participation in productive pro-social roles, life skills to facilitate participation in these roles, consistent systems of recognition and positive reinforcement for pro-social involvement, and individual characteristics (Fig. 4.2).

Positive youth development programs usually strive to achieve one or more of the following objectives: foster resilience; promote healthy bonding; promote social,

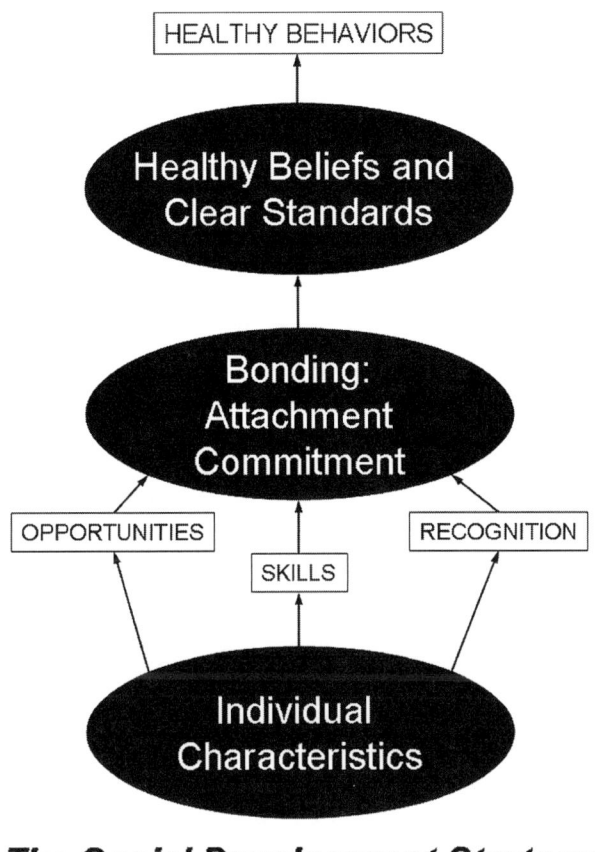

Fig. 4.2. The Social Development Strategy. (*Source*: The Social Development Research Group, School of Social Work, University of Washington [http://depts.washington.edu/sdrg/index.html]. Used with permission.)

emotional, cognitive, behavioral, and moral competencies; foster self-determination; foster self-efficacy; foster clear and positive identity; cultivate spirituality; foster belief in the future; provide recognition for positive behavior; provide opportunities for pro-social involvement; and foster pro-social norms *(135)*. Although positive youth development programs can target youth in one particular setting, the majority of successful programs capitalize on the resources of multiple settings (family, schools, and community) *(135)*.

Positive youth development that is guided by SDS can be considered a viable framework for promoting healthy body image and preventing eating disturbances among children and adolescents. In order to develop positive body image and healthy eating behaviors, all children and adolescents must be enveloped in environments that consistently provide the protective factors depicted in Fig. 4.2. Youth who live with multiple risk factors for developing body image and eating problems can benefit from all of these steps and need to be even more fully immersed in protective environments. For positive youth development to be successful, families and communities must identify healthy beliefs and clear standards for healthy body image and eating behaviors. Then they must foster those healthy beliefs and communicate the standards in multiple arenas of a young person's life—at home and school, in peer groups, within youth-serving and faith-based organizations, and in all segments of the larger community. Children and adolescents must have numerous opportunities to cultivate strong, relational bonds with those who embrace and model positive body image along with other health-promoting behaviors, such as healthy eating and physical activity. These protective bonds can be created through providing youth meaningful opportunities for participation/involvement in a variety of pro-social activities (e.g., advocacy efforts aimed at changing cultural norms promoting the thin ideal). In addition, children and adolescents need repetitive opportunities to develop crucial cognitive, social-emotional, and behavioral skills to help them successfully accomplish these activities as well as important developmental tasks. They also need to be consistently recognized and affirmed for their pro-social involvement at all levels. Finally, recognizing and nurturing individual strengths can enhance children's resilience and provide a foundation for external protective factors to optimally function.

4.5. FUTURE DIRECTION

There are several prevention questions that deserve further study. For example, which protective factors should be targeted and enhanced to reduce the incidence of body image and eating disturbances among children and adolescents from different age and ethnic groups? Which interventions are more effective—universal or targeted ones? How long should interventions last, and what are the most cost-effective approaches? *(111)* How can high-risk youth be reached, and what interventions are most effective with this group? *(48,111)* Future studies also need to directly compare promising prevention programs in randomized trials, particularly in ecologically valid settings where large numbers of children and adolescents can be reached (e.g., schools) *(145)*. There is an additional need to further explore the pathogenesis of early-onset EDs and enhance the current diagnostic system to address the "unique spectrum of early-onset EDs and the development of effective treatments for adolescent eating disorders" (p. 500) *(48)*. Finally, research is needed to determine whether positive youth development programs that employ SDS can produce positive outcomes in terms of decreasing body image and eating disturbances among children and youth.

4.6. SUMMARY AND CONCLUSIONS

Body image and eating disturbances are highly complex and wield a harsh blow to our nation in terms of health care costs, diminished quality of life, and tragic loss of life. Youth, parents, school personnel, health care providers, and entire communities can be effective change agents for the prevention of body image and eating disturbances among children and adolescents. Health promotion and prevention efforts need to be multifaceted and encompass the whole child by addressing physical, mental, emotional, social, and spiritual health. Furthermore, these efforts should target interests, skills, and challenges that are distinctive for particular stages of development *(104)*. No individual should underestimate his or her ability to make a positive difference in the prevention of body image and eating disturbances in children and adolescents. Even small changes have the capacity to produce a powerful ripple effect that can transform society. When individuals, families, and communities unite and engage in proactive endeavors to promote health and quality of life for children and adolescents, they create a legacy of tremendous impact and enduring value *(146)*.

4.7. SCENARIO WITH QUESTIONS AND ANSWERS

4.7.1. Scenario

You are a health care provider who has received a telephone call from a school nurse about her concerns regarding several girls (ages 11–14) who are struggling with negative body image and engaging in restrictive eating and unhealthy exercise behaviors. The nurse asks you for guidance/assistance in dealing with this issue.

4.7.2. Questions

1. What knowledge and skills base would be useful for the girls to possess regarding body image, nutrition, and physical activity?
2. What ideas can you discuss with the nurse regarding ED prevention programs in the school setting?

4.7.3. Plausible Answers

1. Information about the normal developmental changes that accompany puberty; media literacy education to help them acquire the skills to effectively counter the unrealistic thin ideal that pervades our society; knowledge and skills for eating well and staying physically active to promote health and quality of life (as opposed to achieving a specific body weight or size); health literacy; stress management and other coping skills; and competencies for positive youth development.
2. Consider how to infuse prevention into an existing CSHP; consider using an evidence-based prevention curricula that is developmentally appropriate for that age group; investigate the feasibility of setting up peer-mentoring programs; and brainstorm ways that the family, school, and community can partner in planning and implementing positive youth development programs targeting the prevention of BID and EDs.

REFERENCES

1. Centers for Disease Control and Prevention. Healthy Youth! Atlanta, GA, 2006. (Accessed May 10, 2006 at http://www.cdc.gov/HealthyYouth/healthtopics/index.htm).
2. Centers for Disease Control and Prevention. Youth Risk Behavior Surveillance—United States, 2005. MMWR 2006;55(SS-5):1–108. (Accessed June 16, 2006 at http://www.cdc.gov/mmwr/PDF/SS/SS5505.pdf).
3. Austin B, Ziyadeh N, Leliher A, Zachary A, Forman S. Screening high school students for eating disorders: results of a national initiative. *J Adolesc Health* 2001;28:96.
4. Institute of Medicine. *Preventing Childhood Obesity: Health in the Balance*. Washington, DC: Institute of Medicine, 2005.
5. Dounchis JZ, Hayden HA, Wilfley DE. Obesity, body image, and eating disorders in ethnically diverse children and adolescents (pp. 67–101), in *Body Image, Eating Disorders, and Obesity in Youth*, Thompson JK, Smolak L, Eds. Washington, DC: American Psychological Association, 2001.
6. Schur, EA, Sanders, M, Steiner H. Body dissatisfaction and dieting in young children. *Int J Eat Disord* 2000;27:74–82.

7. Smolak L, Levine MP. Body image in children (pp. 41–66), in *Body Image, Eating Disorders, and Obesity in Youth*, Thompson JK, Smolak L, Eds. Washington, DC: American Psychological Association, 2001.
8. Davison K, Markey C, Birch L. Eitology of body dissatisfaction and weight concerns among 5-year-old girls. *Appetite* 2000;35:143–151.
9. Smolak L, Levine MP. Toward an empirical basis for primary prevention of eating problems with elementary school children. *Eat Disord* 1994;2:293–307.
10. Ferron C. Body image in adolescence in cross-cultural research. *Adolescence* 1997;32:735–745.
11. Gustafson-Larson AM, Terry, RD. Weight-related behaviors and concerns of fourth-grade children. *J Am Diet Assoc* 1992; 92(7):818–822.
12. Maloney M, McGuire J, Daniels, S, Specker B. Dieting behavior and eating attitudes in children. *Pediatrics* 1989;84(3):482–489.
13. Pierce J, Wardle J. Cause-and-effect beliefs and self-esteem of overweight children. *J Child Psychol Psychiatry* 1997;38(6):645–650.
14. Thomas K, Ricciardelli LA, Williams RJ. Gender traits and self-concept as indicators of problem eating and body dissatisfaction among children. *Sex Roles* 2000;43:441–458.
15. Neumark-Sztainer D. *I'm, Like, SO Fat!* New York: Guildford Press, 2005.
16. Netemeyer SB, Williamson DA. Assessment of eating disturbance in children and adolescents with eating disorders and obesity (pp. 215–233), in *Body Image, Eating Disorders, and Obesity in Youth*, Thompson JK, Smolak L, Eds. Washington, DC: American Psychological Association, 2001.
17. Tufts University School of Nutrition Science and Policy. Statement on the Link Between Nutrition and Cognitive Development in Children. Boston: Center on Hunger, Poverty and Nutrition Policy, 1998. (Accessed June 15, 2006 at http://www.centeronhunger.org/cognitive.html).
18. Thompson JK, Smolak L. Introduction: Body image, eating disorders, and obesity in youth—The future is now (pp. 1–18), in *Body Image, Eating Disorders, and Obesity in Youth*, Thompson, JK, Smolak L, Eds. Washington, DC: American Psychological Association, 2001.
19. Shisslak C, Crago M, Estes L. The spectrum of eating disturbances. *Int J Eat Disord* 1993;18:209–219.
20. Stice E, Killen J, Hayward C, Taylor C. Support for the continuity hypothesis of bulimic pathology. *J Consult Clin Psychol* 1998;66:784–790.
21. Fisher S. The evolution of psychological concepts about the body (pp. 3–20), in *Body Images: Development, Deviance, and Change*, Cash TF, Pruzinsky T, Eds. New York, NY: Guildford Press, 1990.
22. Thompson JK, Heinberg LJ, Altabe M, Tantleff-Dunn S. Exacting beauty: theory, assessment and treatment of body image disturbance. Washington, DC: American Psychological Association, 1999.
23. Cohn L, Adler N, Irwin C, Millstein S, Kegeles S, Stone G. Body-figure preferences in male and female adolescents. *J Abnorm Psychol* 1987;96:276–279.
24. Gardner RM, Stark K, Jackson N, Friedman B. Development and validation of two new body-image assessment scales. *Percept Mot Skills* 1999;89:981–993.
25. Gardner RM. Assessment of body image disturbance in children and adolescents (pp. 193–213), in *Body Image, Eating Disorders, and Obesity in Youth*, Thompson JK, Smolak L, Eds. Washington, DC: American Psychological Association, 2001.
26. Neumark-Sztainer D, Story M, Falkner NH, Teuhring T, Resnick MD. Sociodemographic and personal characteristics of adolescents engaged in weight loss and weight/muscle gain behaviors: who is doing what? *Prev Med* 1999;28:40–50.
27. Joiner G, Kashubeck S. Acculturation, body image, self-esteem, and eating-disorder symptomatology in adolescent Mexican American women. *Psychol Women Q* 1996;20(3):419–435.
28. Gardner R, Sorter R, Friedman B. Developmental changes in children's body images. *J Soc Behav Pers* 1997;12:1019–1036.
29. Littleton HL, Ollendick T. Negative body image and disordered eating behavior in children and adolescents: what places youth at risk and how can these problems be prevented? *Clin Child Fam Psychol Rev* 2003;6(1):51–66.
30. Mellin L, Irwin C, Scully S. Prevalence of disordered eating in girls: a survey of middle-class children. *J Am Diet Assoc* 1992;92:851–853.
31. Attie I, Brooks-Gunn J. Development of eating problems in adolescent girls: a longitudinal study. *Dev Psychol* 1989;25:70–79.

32. Killen JD, Taylor CB, Hayward C, Wilson DE, Haydel KF, Hammer, LD, et al. Pursuit of thinness and onset of eating disorder symptoms in a community sample of adolescent girls: a three-year prospective analysis. *Int J Eat Disord* 1994;16:227–238.
33. Kaplan SL, Busner J, Pollack S. Perceived weight, actual weight and depressive symptoms in a general adolescent sample. *Int J Eat Disord* 1988;7:107–113.
34. McCabe M, Marwitt SJ. Depressive symptomatology, perceptions of attractiveness, and body image in children. *J Child Psychol Psychiatry* 1993:34:1117–1124.
35. Garner DM, Garfinkel PE. Body image in anorexia nervosa: measurement, theory and clinical implications. *Int J Psychiatry Med* 1981;11:263–284.
36. Keeton WP, Cash TF, Brown TA. Body image or body images?: comparative, multidimensional assessment among college students. *J Pers Assess* 1990;54:213–230.
37. Banasiak SJ, Wertheim EH, Koerner J, Voudouris NJ. Test-retest reliability and internal consistency of a variety of measures of dietary restraint and body concerns in a sample of adolescent girls. *Int J Eat Disord* 2001;29:85–89.
38. Veron-Guidry S, Williamson DA. Development of a body image assessment procedure for children and preadolescents. *Int J Eat Disord* 1996;20:287–293.
39. Williamson DA, Davis CJ, Bennett SM, Goreczny AJ, Gleaves, DH. Development of a simple procedure for body image assessment. *Behav Assess* 1989;11:433–446.
40. Smolak L. Body image in children and adolescents: where do we go from here? *Body Image* 2004;1:15–28.
41. Cooper PJ, Fairburn CG. Confusion over the core psychopathology of bulimia nervosa. *Int J Eat Disord* 1993;13:385–389.
42. Ricciardelli LA, McCabe MP. Sociocultural and individual influences on muscle gain and weight loss strategies among adolescent boys and girls. *Psychol Sch* 2003;40(2):209–224.
43. American Psychiatric Association. Diagnostic and statistical manual of mental disorders. Washington, DC: American Psychiatric Association, 2000.
44. Thompson JK, Smolak L. Introduction: body image, eating disorders, and obesity in youth – the future is now (pp. 1–18), in *Body Image, Eating Disorders, and Obesity in Youth*, Thompson JK, Smolak L, Eds. Washington, DC: American Psychological Association, 2001.
45. Croll J, Neumark-Sztainer D, Story M, Ireland M. Prevalence and risk and protective factors related to disordered eating behaviors among adolescents: relationship to gender and ethnicity. *J Adolesc Health* 2002;31:166–175.
46. Fisher M, Golden NH, Katzman DK, Kreipe RE, Rees J, Schebendach J, et al. Eating disorders in adolescents: a background paper. *J Adolesc Health* 1995;16:420–437.
47. Leichner P. Disordered eating attitudes among Canadian teenagers. *Can Med Assoc J* 2002;166: 707–708.
48. Society for Adolescent Medicine. Eating disorders in adolescents: position paper of the Society for Adolescent Medicine. *J Adolesc Health* 2003;33:496–503.
49. U.S. Department of Health and Human Services – Office on Women's Health. Eating Disorders 2002. (Accessed May 18, 2006 at http://www.4woman.gov).
50. Sullivan PF. Mortality in anorexia nervosa. *Am J Psychiatry* 1995;152(7):1073–1074.
51. American Academy of Pediatrics. Policy statement. Identifying and treating eating disorders. *Pediatrics* 2003;111:204–211.
52. American Psychiatric Association, Work Group on Eating Disorders. Practice guidelines for the treatment of patients with eating disorders (revision). *Am J Psychiatry* 2000;157 (1 Suppl):1–39.
53. Cash TF. The psychology of physical appearance: aesthetics, attributes, and images (pp. 51–79), in *Body Images: Development, Deviance, and Change*, Cash TF, Pruzinsky T, Eds. New York, NY: Guilford, 1990.
54. Powers PS. Initial assessment and early treatment options for anorexia nervosa and bulimia nervosa. *Psychiatr Clin North Am* 1996;19:639–655.
55. Jordan J, Joyce PR, Carter FA, Horn J, McIntosh, VVW, Luty SE, et al. Anxiety and psychoactive substance use disorder comorbidity in anorexia nervosa or depression. *Int J Eat Disord* 2003;34: 211–219.
56. Something Fishy – Website on Eating Disorders. Associated mental health conditions and addictions. (Accessed May 19, 2006 at http://www.something-fishy.org/isf/mentalhealth.php).
57. Nolen-Hoeksema S. *Abnormal Psychology* (2nd ed). Boston: McGraw Hill, 2001.

58. Kaye W, Strober M. Neurobiology of eating disorders, in *Neurobiological Foundations of Mental Illness*, Charney D, Nestler E, Bunney W, Eds. New York, NY: Oxford, 1999.
59. Strober M. Family-genetic studies of eating disorders. *J Clin Psychol* 1991;52:9–12.
60. Graber JA, Brooks-Gunn J, Praikoff RL, Warren MP. Prediction of eating problems: an 8-year study of adolescent girls. *Dev Psychol* 1994;30:823–834.
61. Swarr AE, Richards MH. Longitudinal effects of adolescent girls' pubertal development, perceptions of pubertal timing, and parental relations on eating problems. *Dev Psychol* 1996;32:636–646.
62. Williams JM, Currie C. Self-esteem and physical development in early adolescence: pubertal timing and body image. *J Early Adolesc* 2000;20:129–149.
63. Cattarin JA, Thompson JK. A three-year longitudinal study of body image, eating disturbance, and general psychological functioning in adolescent females. *Eat Disord* 1994;2:114–125.
64. Leon GR, Fulkerson JA, Perry CL, Keel PK, Klump KL. Three- to four-year prospective evaluation of personality and behavioral risk factors for later disordered eating in adolescent girls and boys. *J Abnorm Psychol* 1995;28:181–196.
65. Stice E, Killen, JD, Hayward C, Taylor CB. Age of onset for binge eating and purging during late adolescence: a 4-year survival analysis. *J Abnorm Psychol* 1998;107:671–675.
66. Martin GC, Wertheim EH, Prior M, Smart D, Sanson A, Oberklaid F. A longitudinal study of the role of childhood temperament in the later development of eating concerns. *Int J Eat Disord* 2000;27:150–162.
67. Tobin-Richards MH, Boxer AM, Petersen AC, The psychological significance of pubertal change: sex differences in perceptions of self during early adolescence (pp. 127–154), in *Girls at Puberty: Biological and Psychosocial Perspectives*, J Brooks-Gunn, Petersen AC, Eds. New York, NY: Plenum, 1983.
68. Tiggemann M. The impact of adolescent girls' life concerns and leisure activities on body dissatisfaction, disordered eating, and self-esteem. *J Genet Psychol* 2001;162:133–142.
69. Shisslak CM, Crago M. Risk and protective factors in the development of eating disorders (pp. 103–125), in *Body Image, Eating Disorders, and Obesity In Youth*, Thompson JK, Smolak L, Eds. Washington, DC: American Psychological Association, 2001.
70. Steiner H, Kwan W, Shaffer TG, Walker S, Miller S, Sagar A, et al. Risk and protective factors for juvenile eating disorders. *Eur Child Adolesc Psychiatry* (Suppl 1) 2003;12:38–46.
71. Coughlin JW, Heinberg LJ, Marinilli A, Guarda AS. Body image dissatisfaction in children: prevalence and parental influence. *Healthy Weight J* 2003;17(4):56–59.
72. Keel PK, Fulkerson JA, Leon GR. Disordered eating precursors in pre- and early adolescent girls and boys. *J Youth Adolesc* 1997;26:203–216.
73. Fairburn CG, Welch S, Doll HA, Davies BA, O'Connor ME. Risk factors for bulimia nervosa: a community-based case-control study. *Arch Gen Psychiatry* 1997;54:509–517.
74. Agras S, Hammer L, McNicholas F. A prospective study of the influence of eating-disordered mothers on their children. *Int J Eat Disord* 1999; 25(3): 253–262.
75. Park RJ, Senior R, Stein A. The offspring of mothers with eating disorders. *Eur Child Adolesc Psychiatry* (Suppl 1) 2003;12:110–119.
76. Byely L, Archibald AB, Graber J, Brooks-Gunn J. A prospective study of familial and social influences on girls' body image and dieting. *Int J Eat Disord* 2000;28:155–164.
77. Humphrey LL. Family relationships (pp. 263–282), in *Psychobiology and Treatment of Anorexia Nervosa and Bulimia Nervosa*, Halmi KA, Ed. Washington, DC: American Psychiatric Press, 1994.
78. Minuchin S, Rosman BL, Baker L. *Psychosomatic Families: Anorexia Nervosa in Context*. Cambridge, MA: Harvard University Press, 1978.
79. Neumark-Sztainer D, Story M, Hannan, PJ, Beuhring T, Resnick MD. Disordered eating among adolescents: association with sexual/physical abuse and other familial psychosocial factors. *Int J Eat Disord* 2000;28:249–258.
80. Sands ER, Wardle J. Internalization of ideal body shapes in 9–12-year-old girls. *Int J Eat Disord* 2003;33(2):193–204.
81. Chandy JM, Harris L, Blum RW, Resnick MD. A three-year longitudinal study of body image, eating disturbance, and general psychological functioning in adolescent females. *Eat Disord*, 1994;2: 114–124.
82. Connors ME. Relationship of sexual abuse to body image and eating problems (pp. 149–170), in *Body Image, Eating Disorders, and Obesity in Youth*, Thompson JK, Smolak L, Eds. Washington, DC: American Psychological Association, 2001.

83. Wonderlich S, Crosby R, Mitchell J, Thompson K, Redlin J, Demuth G, et al. Pathways mediating sexual abuse and eating disturbance in children. *Int J Eat Disord* 2001;29:270–279.
84. Steinberg AB, Phares, V. Family functioning, body image, and eating disturbances (pp. 127–147), in *Body Image, Eating Disorders, and Obesity in Youth*, Thompson JK, Smolak L, Eds. Washington, DC: American Psychological Association, 2001.
85. Wiseman MA, Gray JJ, Mosimann JE, Ahrens AH. Cultural expectations of thinness in women: an update. *Int J Eat Disord* 1992;11:85–89.
86. Gowers SG, Shore A. Development of weight and shape concerns in the aetiology of eating disorders. *Br J Psychiatry* 2001;179:236–242.
87. Groesz LM, Levine MP, Murnen SK. The effect of experimental presentation of thin media images on body satisfaction: a meta-analytic review. *Int J Eat Disord* 2002;31:1–16.
88. Lieberman M, Gauvin L, Bukowski WM, White DR. Interpersonal influence and disordered eating behaviors in adolescent girls: the role of peer modeling, social reinforcement, and body-related teasing. *Eat Behav* 2001;2:215–236.
89. Burt B. Definitions of risk. Report for the Consensus Development Conference on Diagnosis and Management of Dental Caries Throughout Life, March 26–28, 2001. (Accessed May 18, 2006 at http://www.nidcr.nih.gov).
90. Wertheim EH, Paxton SJ, Schutz HK, Muir SL. Why do adolescent girls watch their weight? An interview study examining sociocultural pressures to be thin. *J Psychosom Res* 1997;42:345–355.
91. Rodin J, Streigel-Moore, RH, Silberstein LR. Vulnerability and resilience in the age of eating disorders: risk and protective factors for bulimia nervosa (pp. 361–383), in *Risk and Protective Factors in the Development of Psychopathology*, Rolf J, Masten AS, Cicchetti D, Neuchterlein KH, Weintraub S, Eds. Cambridge: Cambridge University, 1990.
92. Striegel-Moore RH, Cachelin FM. Body image concerns and disordered eating in adolescent girls: risk and protective factors (pp. 85–108), in *Beyond Appearance: A New Look at Adolescent Girls*, Johnson NG, Roberts MC, Eds. Washington, DC: American Psychological Association, 1999.
93. Connors MA. Developmental vulnerabilities for eating disorders (pp. 285–310), in *The Developmental Psychopathology of Eating Disorders*, Smolak L, Levine MP, Striegel-Moore R, Eds. Mahwah, NJ: Erlbaum, 1996.
94. Chandy JM, Harris L, Blum RW, Resnick MD. Female adolescents of alcohol misusers: disordered eating features. *Int J Eat Disord* 1995;17:283–289.
95. Berndt TJ, Hestenes SL. The developmental course of social support: Family and peers (pp. 77–106), in *The Developmental Psychopathology of Eating Disorders*, Smolak L, Levine MP, Striegel-Moore R, Eds. Mahwah, NJ: Erlbaum, 1996.
96. Smolak L, Murnen S, Ruble A. Female athletes and eating problems: a meta-analysis. *Int J Eat Disord* 2000;27:371–380.
97. Gerner BG, Wilson PH. The relationship between friendship factors and adolescent girls' body image concern, body dissatisfaction, and restrained eating. *Int J Eat Disord* 2005;37:313–320.
98. Compas BE. Promoting positive mental health during adolescence (pp. 159–179), in *Promoting the Health of Adolescents*, Millstein SG, Petersen AC, Nightingale EO, Eds. New York, NY: Oxford, 1993.
99. O'Dea JA. Evidence for a self-esteem approach in the prevention of body image and eating problems among children and adolescents. *Eat Disord* 2004;12:225–239.
100. Branden N. *The Six Pillars of Self-Esteem.* New York, NY: Bantam, 1994.
101. O'Dea JA. *Everybody's Different: A Self-Esteem Program for Young Adolescents.* Sydney: University of Sydney, 1995.
102. Troop NA, Treasure JL. Setting the scene for eating disorders, II. Childhood helplessness and mastery. *Psychol Med* 1997;27:531–538.
103. Joint Committee on National Health Education Standards. *Achieving Health Literacy: An Investment in the Future.* Atlanta, GA: American Cancer Society, 1995.
104. Graber JA, Brooks-Gunn J. Prevention of eating problems and disorders: including parents. *Eat Disord* 1996;4:348.
105. Levine P. Prevention guidelines & strategies for everyone: 50 says to lose the "3 Ds" – dieting, drive for thinness, and body dissatisfaction. National Eating Disorders Assocation, Seattle, WA, 2002. (Accessed May 23, 2006 at http://www.nationaleatingdisorders.org).
106. Hopper CA, Monoz KD, Gruber M, Nguyen KP. The effects of a family fitness program on the physical activity and nutrition behaviors of third-grade children. *Res Q Exerc Sport* 2005;76: 130–139.

107. Nader PR, Sellers DE, Johnson CC, Perry CL, Stone EJ, Cook KC, et al. The effect of adult participation in a school-based family intervention to improve children's diet and physical activity: the Child and Adolescent Trial for Cardiovascular Health. *Prev Med* 1996;25:455–464.
108. Sallis JF, McKenzie TL. Physical education's role in public health. *Res Q Exerc Sport* 1991;62:124–137.
109. Domel S, Baranowski T. Development and evaluation of a school intervention to increase fruit and vegetable consumption among 4th and 5th grade students. *J Nutr Educ* 1993;25:345–349.
110. Perry CL, Williams CL, Veblen-Morenson S., Toomey TL, Komro KA, Anstine PS, et al. Project Northland: outcomes of a community-wide alcohol use prevention program during early adolescence. *Am J Public Health* 1996;86: 956–965.
111. Taylor CB, Altman T. Priorities in prevention research for eating disorders. *Psychopharmacol Bull* 1997;33(3):413–417.
112. Benard B. Fostering resilience in children. ERIC Digest, EDO-PS-95-9, August 1995. Clearinghouse on Elementary and Early Childhood Education University of Illinois at Urbana-Champaign Children's Research Center, Champaign, IL. (Accessed June 28, 2006 at http://resilnet.uiuc.edu/library/benard95.html).
113. Levine MP, Smolak L. Primary prevention of body image disturbances and disordered eating in childhood and early adolescence (pp 237–260), in *Body Image, Eating Disorders, and Obesity in Youth*, Thompson JK, Smolak L, Eds. Washington, DC: American Psychological Association, 2001.
114. Resnick MD, Bearman PS, Blum RW, Bauman KE, Harris KM, Jones J, et al. Protecting adolescents from harm. Findings from the National Longitudinal Study on Adolescent Health. *JAMA* 1997;278:823–832.
115. Thelen M, Powell A, Lawrence C, Kuhnert, M. Eating and body image concerns among children. *J Clin Child Psychol* 1992;21:41–46.
116. Kater K. Healthy body image: teaching kids to eat and love their bodies too! Curriculum available from the National Eating Disorders Association, 603 Stewart St., Suite 803, Seattle, WA 98101, http://www.nationaleatingdisorders.org, 1998.
117. Kater KJ, Rohwer J, Levine MP. An elementary school project for developing healthy body image and reducing risk factors for unhealthy and disordered eating. *Eat Disord* 2000;8:3–16.
118. Kater KJ, Rohwer J, Londre K. Evaluation of an upper elementary school program to prevent body image, eating and weight concerns. *J Sch Health* 2002;72(5):199–204.
119. Smolak L, Levine MP, Schermer F. A controlled evaluation of an elementary school primary prevention program for eating problems. *J Psychosom Res* 1998;44:339–353.
120. Smolak L, Levine MP. A two-year follow-up of a primary prevention program for negative body image and unhealthy weight regulation. *Eat Disord* 2001;9:313–325.
121. Piran N. Prevention: can early lessons lead to a delineation of an alternative model? A critical look at prevention with schoolchildren. *Eat Disord* 1995; 3:28–36.
122. Abascal L, Brown JB, Winzelberg AJ, Dev, P, Taylor CB. Combining universal and targeted prevention for school-based eating disorder programs. *Int J Eat Disord* 2003;35:1–9.
123. Centers for Disease Control and Prevention. Healthy youth! Coordinated school health program. Atlanta, GA, 2006. (Accessed June 1, 2006 at http://www.cdc.gov/HealthyYouth/CSHP).
124. Smolak L, Harris B, Levine MP, Shisslak, CM. Teachers: the forgotten influence on the success of prevention programs. *Eat Disord* 2001;9:261–265.
125. U.S. Department of Health and Human Services – Office on Women's Health. BodyWise handbook. Washington, DC, April 2005. (Accessed June 15, 2006 at http://4woman.gov/bodyimage/bodywise/bp/BodyWise.pdf.cfm).
126. Rideout V, Roberts DF, Foehr, UG. Executive summary: generation M – media in the lives of 8–18 year-olds. Kaiser Family Foundation 2005. (Accessed May 24, 2006 at http://www.kff.org).
127. Kaiser Family Foundation. Media literacy, Fall 2003. (Accessed May 24, 2006 at http://www.kff.org).
128. Irving L, DuPen J, Berel S. A media literacy program for high school females. *Eat Disord* 1998;6:119–131.
129. Piran N, Levine M, Irving L. GO GIRLS! Media literacy, activism, and advocacy project. *Healthy Weight J* Nov/Dec 2000:89–90.
130. The National Eating Disorders Association. GO GIRLS!™ A high school prevention program and curriculum. Available from the National Eating Disorders Association, 603 Stewart St., Suite 803, Seattle, WA 98101 (http://www.nationaleatingdisorders.org).

131. Chelsey EB, Alberts JD, Klein JD, Kreipe RE. Pro or con? Anorexia nervosa and the internet. *J Adolesc Health* 2003;32:123–124.
132. Neumark-Sztainer D. School-based programs for preventing eating disturbances. *J Sch Health* 1996;66:64–71.
133. Kater K. Promotion of healthy weight: helping without harming. Connect® 2006; May–June: 22–25.
134. BodyImageHealth.org. Model for healthy body image, 1998. (Accessed June 22, 2006 at www.BodyImageHealth.org).
135. Catalano RF, Berglund ML, Ryan JAM, Lonczak HS, Hawkins JD. Positive youth development in the United States: research findings on evaluations of positive youth development programs. *Prev Treat* 2002;5(1):15[Article A].
136. Hawkins JD, Arthur M, Catalano RF. Preventing substance abuse (pp. 343–427), in *Building a Safer Society: Strategic Approaches to Crime Prevention, Crime and Justice: A Review of Research*, Tonry M, Farrington D, Eds. Chicago, IL: University of Chicago, 1995.
137. Lonczak HS, Huang B, Catalano RF, Hawkins JD, Hill, KG, Abbott RD, et al. The social predictors of adolescent alcohol misuse: a test of the Social Development Model. *J Stud Alcohol* 2001;62:179–189.
138. Catalano RF, Hawkins, JD. The social development model: a theory of antisocial behavior (pp. 149–197), in *Delinquency and Crime: Current Theories*, Hawkins JD, Ed. New York: Cambridge University, 1996.
139. Lonczak HS, Abbott RD, Hawkins JD, Kosterman R, Catalano RF. Effects of the Seattle Social Development Project on sexual behavior, pregnancy, birth, and sexually transmitted disease outcomes by age 21 years. *Arch Pediatr Adolesc Med* 2002;156:438–447.
140. Huang B, Kosterman R, Catalano RF, Hawkins JD, Abbott RD. Modeling mediation in the etiology of violent behavior in adolescence: a test of the Social Development Model. *Criminology* 2001;39:75–107.
141. Hawkins FD, Guo J, Hill K, Battin-Pearson S, Abbott R. Long-term effects of the Seattle Social Development Intervention on school bonding trajectories. *Appl Dev Sci* 2001;5:225–236.
142. Hawkins JD, Catalano RF, Miller JY. Risk and protective factors for alcohol and other drug problems in adolescence and early adulthood: implications for substance abuse prevention. *Psychol Bull* 1992;112:64–105.
143. Institute of Medicine (IOM), Committee on Prevention of Mental Disorders. *Reducing Risks for Mental Disorders: Frontiers for Preventive Intervention Research*, Mrazek PJ, Haggerty RJ, Eds. Washington, DC: National Academy Press, 1994.
144. Social Development Research Group, School of Social Work, University of Washington, Seattle, 2006. (Accessed June 16, 2006 at http://depts.washington.edu/sdrg/index.html).
145. Stice E, Shaw, H. Eating disorder prevention programs: a meta-analytic review. *Psychol Bull* 2004;130:206–227.
146. Massey-Stokes MS. The role of parents, school personnel, and the community in the primary prevention of eating disorders in children (pp. 239–259), in *Eating Disorders in Women & Children: Prevention, Stress Management and Treatment*, Robert-McComb J, Ed. Boca Raton, FL: CRC, 2000.

5 The Female Athlete Triad: Disordered Eating, Amenorrhea, and Osteoporosis

Jacalyn J. Robert-McComb

CONTENTS
- 5.1 LEARNING OBJECTIVES
- 5.2 INTRODUCTION
- 5.3 RESEARCH FINDINGS
- 5.4 CLOSING REMARKS
- 5.5 SUMMARY AND CONCLUSIONS
- 5.6 SCENARIO WITH QUESTIONS AND ANSWERS

5.1. LEARNING OBJECTIVES

After completing this chapter, you should have an understanding of the following:

- The American College of Sports Medicine (ACSM) position stand on the female athlete triad.
- The difference between disordered eating and eating disorders.
- The progressive nature of menstrual disturbances in athletes.
- Osteopenia and osteoporosis.
- The interrelatedness of disordered eating, amenorrhea, and osteoporosis.
- Athletes at greatest risk for developing signs and symptoms associated with the triad.
- The growing health concern of the triad for allied health professionals.
- The general consensus regarding reproductive dysfunction in the female athlete.

5.2. INTRODUCTION

In 1992, the term Female Athlete Triad was coined to describe three distinct, but frequently interrelated disorders found in the female athletic population *(1,2)*. These disorders are disordered eating, amenorrhea, and osteoporosis. Disordered eating refers to a wide spectrum of harmful and often ineffective eating behaviors used in attempts to achieve a perceived ideal body weight. Although the definitions of amenorrhea are somewhat arbitrary, amenorrhea can be described as primary or secondary. Primary

From: *The Active Female*
Edited by: J. J. Robert-McComb, R. Norman, and M. Zumwalt © Humana Press, Totowa, NJ

amenorrhea (delayed menarche) is the absence of menstruation by age 16 in a girl with secondary sex characteristics *(3)*. Secondary amenorrhea is the absence of three or more consecutive menstrual cycles after menarche or rather after cycles have been established *(3–5)*.

The terms used by clinicians to describe amenorrhea caused by exercise or stress are functional (a functional problem, i.e., altered hormonal patterns, rather than an anatomical problem) or hypothalamic amenorrhea, and these are used widely in the clinical and scientific literature. Adaptive responses to psychological, physical, or nutritional stressors are similar in humans and nonhuman primates and include activation of the hypothalamic-pituitary-adrenal (HPA) axis and, sometimes, suppression of the hypothalamic-pituitary-gonadal (HPG) axis also referred to as the hypothalamic-pituitary-ovarian (HPO) axis, thus the term hypothalamic amenorrhea. This stress-induced suppression of the HPG axis leading to chronic anovulation and amenorrhea is therefore called functional hypothalamic anovulation or functional hypothalamic amenorrhea and is reversible.

Osteoporosis is a disease characterized by low bone mass and microarchitectural deterioration of bone tissue leading to bone fragility. The principle cause of premenopausal osteoporosis in active young women is decreased ovarian hormone production and hypoestrogenemia as a result of hypothalamic amenorrhea *(6–9)*.

The diagnostic criteria for low bone mass as defined by the World Heath Organization *(10)* is as follows: (1) normal—bone mineral density (BMD) that is no more than 1 SD below the mean of young adults; (2) ostopenia—BMD between 1 and 2.5 SD below the mean of young adults; (3) osteoporosis—BMD more than 2.5 SD below the mean of young adults; and (4) severe osteoporosis—BMD more than 2.5 SD below the mean of young adults plus one or more fragility fractures.

In 1992, the ACSM Task Force on Women's Issues held a conference on the Female Athlete Triad *(11)*. In 1997, ACSM published its Position Stand on the Female Athlete Triad. An excerpt taken from ACSM Position Stand on the Female Athlete Triad is as follows *(2)*:

1. *The Female Athlete Triad is a serious syndrome consisting of disordered eating, amenorrhea, and osteoporosis. The components of the Triad are interrelated in etiology, pathogenesis, and consequences. Because of the recent definition of the Triad, prevalence studies have not yet been completed. However, it occurs not only in elite athletes but also in physically active girls and women participating in a wide range of physical activities. The Triad can result in declining physical performance, as well as medical and psychological morbidity's and mortality.*
2. *Internal and external pressures placed on girls and women to achieve or maintain unrealistically low body weight underlies the development of these disorders.*
3. *The Triad is often denied, not recognized, and under reported. Sports medicine professionals need to be aware of the interrelated pathogenesis and the varied presentation of components of the Triad. They should be able to recognize, diagnose, and treat or refer women with any one component of the Triad.*
4. *Women with one component of the Triad should be screened for the other components. Screening for the Triad can be done at the time of the preparticipation examination and during clinical evaluation of the following: menstrual change, disordered eating patterns, weight change, cardiac arrhythmias including bradycardia, depression, or stress fracture.*

> 5. *All sports medicine professionals, including coaches and trainers, should learn about preventing and recognizing the symptoms and risks of the Triad. All individuals working with active girls and women should participate in athletic training that is medically and psychologically sound. They should avoid pressuring girls and women about losing weight. They should know basic nutrition information and have referral sources for nutritional counseling and medical and mental health evaluation.*
> 6. *Parents should avoid pressuring their daughters to diet and loose weight. Parents should be educated about the warning signs of the Triad and initiate medical care for their daughters if signs are present.*
> 7. *Sports medicine professionals, athletic administrators, and officials of sport governing bodies share a responsibility to prevent, recognize, and treat the Triad. The sport governing bodies should work toward offering opportunities for educational programs for coaches to educate them and to lead them toward professional certification. They should work toward developing programs to monitor coaches and others to ensure safe training practices.*
> 8. *Physically active girls and women should be educated about proper nutrition, safe training practices, and the warning signs and risks of the Triad. They should be referred for medical evaluation at the first sign of any of the components of the Triad.*
> 9. *Further research is needed into the prevalence, causes, prevention, treatment, and sequelae of the Triad.*

Although any athlete may suffer from the disorders of the Female Athlete Triad, girls and women who participate in sports that place a premium on appearance and thinness are especially susceptible *(12)*. According to the Consensus Statement of the United States Olympic Committee and ACSM, girls and women who participate in the following categories of sports are most susceptible *(13)*.

1. Sports where performance is subjectively judged (i.e., dance, figure skating, diving, gymnastics, and aerobics).
2. Endurance sports (i.e., distance running, cycling, and cross-country skiing).
3. Sports where athletes wear revealing clothing (i.e., volleyball, swimming, diving, cross-country skiing, track and field, and cheerleading).
4. Sports with weight categories (i.e., wrestling, weightlifting, rowing, and some martial arts).
5. Sports where a prepubescent body is emphasized (i.e., figure skating, gymnastics, and diving).

Additional stressors that contribute to disorders of the triad in young athletes are the natural biological changes that occur in puberty or the increase in sex-specific fat during puberty. These young athletes must not only cope with these biological changes but must also conform to the pressures to stay thin for increased sport performance. Disordered eating is often an unhealthy attempt at coping with these stressors in their lives.

Early recognition and awareness of athletes most at risk and the early signs of an eating disorder is essential when developing medical protocols for the triad of disorders. It is important for allied health professionals to be able to recognize disordered eating patterns before these subclinical disorders progress to a clinical diagnosis.

5.3. RESEARCH FINDINGS

5.3.1. Disordered Eating and Eating Disorders

You might say that there is a continuum ranging from healthy eating patterns, disordered eating, and eating disorders. The term disordered eating includes a spectrum of abnormal eating behaviors that range from mild restricting behaviors and occasional binging and purging to those that meet the diagnostic criteria for Eating Disorders in the Diagnostic and Statistical Manual of Mental Disorders-DSM-IV-TR *(14)*. The primary types of eating disorders are anorexia nervosa (AN) and bulimia nervosa (BN). The diagnostic criteria for these disorders are outlined in Tables 5.1 and 5.2 *(14)*. However, many times, there is vacillation between the behaviors associated with AN and BN.

Individuals with AN may move back and forth between the two types of subgroups in AN as there is a similarity in characteristics between these two subgroups. The restricting subtype accomplishes weight loss through dieting, fasting, and excessive exercise. The bulimic subtype purge after binge eating or even after the consumption of a small amount of food through self-induced vomiting or the misuse of laxatives, diuretics, or enemas.

BN (Table 5.2) is similar to AN in terms of a distorted body image and a drive for thinness, yet the individual's body weight is typically normal throughout the illness *(15)*. The incidence of BN seems to be higher than AN; yet, because of the

Table 5.1
DSM-IV- TR Diagnostic Criteria for Anorexia Nervosa

A. Refusal to maintain body weight at or above a minimally normal weight for age and height (e.g., weight loss leading to maintenance of body weight less than 85% of that expected; or failure to make expected weight gain during period of growth, leading to body weight less than 85% of that expected).
B. Intense fear of gaining weight or becoming fat, even though underweight.
C. Disturbance in the way in which one's body weight or shape is experienced, undue influence of body weight or shape on self-evaluation, or denial of the seriousness of the current low body weight.
D. In postmenarcheal females, amenorrhea, i.e., the absence of at least three consecutive menstrual cycles. (A woman is considered to have amenorrhea if her periods occur only following hormone, e.g., estrogen, administration.)

Specific type:

Restricting type: during the current episode of anorexia nervosa, the person has not regularly engaged in binge-eating or purging behavior (i.e., self-induced vomiting or the misuse of laxatives, diuretics, or enemas)

Binge-eating/purging type: during the current episode of anorexia nervosa, the person has regularly engaged in binge eating or purging behavior (i.e., self-induced vomiting or the misuse of laxatives, diuretics, or enemas)

Note: In order to determine the values for criteria A (< 85% of ideal body weight) as listed by the DSM-IV-TR, the professional can use a simple formula to roughly determine ideal weight for a young woman of average bone size. Begin with 100 pounds and to add 4 pounds for every inch over five feet in height. For a very slight bone size, add only 3 pounds for every inch over five feet.
Source: American Psychiatric Association *(14)*. With permission.

Table 5.2
DSM-IV-TR Diagnostic Criteria for Bulimia Nervosa

A. Recurrent episodes of binge eating. An episode of binge eating is characterized by both of the following:

   (1) Eating in a discrete period of time (e.g., within any 2-h period), an amount of food that is definitely larger than most people would eat during a similar period of time and under similar circumstances
   (2) A sense of lack of control over eating during the episode (e.g., a feeling that one cannot stop eating or control what or how much one is eating)

B. Recurrent inappropriate compensatory behavior in order to prevent weight gain, such as self-induced vomiting; misuse of laxatives, diuretics, enemas, or other medications; fasting; or excessive exercise.
C. The binge eating and inappropriate behaviors both occur, on average, at least twice a week for 3 months.
D. Self-evaluation is unduly influenced by body shape and weight.
E. The disturbance does not occur during episodes of anorexia nervosa.

Specific type:

Purging type: during the current episode of bulimia nervosa, the person has regularly engaged in self-induced vomiting or the misuse of laxatives, diuretics, or enemas

Nonpurging type: during the current episode of bulimia nervosa, the person has used other inappropriate compensatory behaviors, such as fasting or excessive exercise, but has not engaged in self-induced vomiting or the misuse of laxatives, diuretics, or enemas

Source: American Psychiatric Association *(14)*. With permission.

secretiveness of individuals with this disorder, it is difficult to estimate the exact number of incidences *(1)*. Screening tools to help identify these disorders are discussed in Chap. 7 *(16–18)*.

Recently, two additional categories of eating disorders have been added or proposed—these disorders are termed Eating Disorder Not Otherwise Specified (EDNOS) and Binge Eating Disorder (BED). The Diagnostic Criteria for EDNOS and BED can be found in the Diagnostic and Statistical Manual of Mental Disorders *(14)*. EDNOS is a diagnosis for eating disorders that meet some but not all of the specific criteria for AN or BN. For example, all of the criteria for AN are met except that the individual has regular menses or all of the criteria for BN are met except that the binge eating and inappropriate compensatory mechanisms occur at a frequency of less than twice a week or for a duration of less than 3 months. BED is a diagnosis that has been proposed for possible inclusion as an official category in the DSM-IV-TR *(14)*. This disorder now falls under the eating disorders not otherwise specified category in the DSM-IV. This disorder is listed as a potential category that warrants future study. The term "compulsive overeating" has also been used to identify binge eating in the absence of self-induced vomiting or purgative abuse. The term "compulsive" may be misleading and obscures the fact that overeating may be explained in most instances by a long history of excessive dieting.

5.3.2. The Progressive Nature of Menstrual Disturbances in Athletes

As discussed in the Sect. 5.2, classification of menstrual cycle status is somewhat arbitrary and the profile of menstrual dysfunction varies considerably. There seems to be general agreement on the definition of primary amenorrhea in the exercise literature (acyclic at age 16 with secondary sex characteristics). However, secondary amenorrhea has been defined as the absence of at least three to six menstrual cycles in the position stand by ACSM *(2)*; yet, the International Olympic Committee defines secondary amenorrhea as having one or fewer menstrual cycles a year *(19)*. Athletes with oligomenorrhea have been defined as having cycles that occur at intervals longer than 35 days *(4,19,20)*. Yet, Sabatini *(21)* defines oligomenorrhea as an increase in cycle length to 45–90 days, with fewer than six menses a year. Because of these inconsistencies in definitions, lack of reporting, and group bias selection, there is a large range (3.4–66%) in the reported incidences of amenorrhea in athletes *(15)*. However, it seems that the prevalence of amenorrhea is larger in athletes participating in sports where leanness is advantageous than in nonathletes. There also seems to be a progressive nature of menstrual disturbances in athletes.

Even though the profile of menstrual dysfunction varies considerably in athletes, there is a progression that develops ranging from the least severe form of menstrual dysfunction (luteal phase deficiency) to amenorrhea with no follicular development, no ovulation, no menstrual bleeding, and a permanent hypoestrogenic state *(1)*. During the progression of menstrual dysfunction, there is an increased risk of endometrial cancer because the follicle starts to develop, but the process ceases before ovulation creating an environment of unopposed estrogen *(22)*. The progressive nature of menstrual disturbances in athletic women resembles the pattern depicted below. Stages 1–3 are usually asymptomatic but may present as infertility *(23)*.

1. Regular cycles with a shortened luteal phase—progesterone production stops early.
2. Regular cycles with inadequate progesterone production.
3. Regular cycles with failure to develop and release an egg (ovulation).
4. Irregular cycles but still ovulating.
5. Irregular cycles and anovulation.
6. Absence of menses and anovulation.

5.3.3. Differentiating Osteoporosis from Osteopenia: Should Osteopenia be used as a Criteria for the Triad Rather than Osteoporosis?

In an article published in the British Journal of Sports Medicine *(24)*, information was presented which calls attention to the actual prevalence of osteoporosis in female athletes. The authors suggested that perhaps, the triad of disorders should be disordered eating, amenonrrhea, and ostopenia rather than osteoporosis. Osteoporosis does occur in athletes; however, it is much rarer than ostopenia. Pettersson et al. *(25)* found that only 10% of amenorrheic distance runners aged 21 ± 3 had osteoporosis as defined by the World Health Organization (BMD more than 2.5 SD below the mean of young adults); yet, 50% had ostopenia (BMD between 1 and 2.5 SD below the mean of young adults). Rutherford *(26)* found that 13% of amenorrheic triatheletes and distance runners aged 29.5 ± 7.5 had osteoporosis; yet, 40% had ostopenia. Kaufman et al. *(27)* found that in a group of elite ballet dancers ($n = 21$), ostopenia was found in the spine and arm only in the dancers who had a history of amenorrhea.

This does not negate the importance of low bone mass in female athletes. The condition of osteopenia is important in an athletic population because the risk of developing osteoporosis is greater than in individuals with normal bone mass. Recognition of the prevalence of osteopenia in athletes has important implications because the keystones for management of osteopenia are physical activity and adequate nutrition, not pharmacotherapy, as advocated for osteoporosis.

5.3.4. The Interrelatedness of Disordered Eating, Amenorrhea, and Osteoporosis

One disorder leads to another: disordered eating → amenorrhea → osteoporosis. Chronic energy deprivation directly affects the HPG axis by decreasing the amplitude and frequency of the pulsatile release of gonadotropic releasing hormone (GnRH) produced by the arcuate nucleus of the hypothalamus. The decreased release of GnRH causes decreased release of lutenizing hormone (LH) and follicle stimulating hormone by the pituitary. Without the LH surge at midcycle, ovulation cannot take place, estrogen and progesterone production decreases because of lack of ovarian stimulation, and menses either occurs irregularly or not at all *(28)*. Suppression of reproductive axis affects bone resorption and remodeling because of hypoestrogenism, ultimately leading to low bone mass.

Women deposit most of their bone mass between 11 and 14 years of age and reach peak bone mass between 25 and 30 years of age *(28)*. Adolescent peak bone mass in girls is achieved in the final stages of pubertal progression *(29)* with minimal increases in BMD occurring 2 years after menarche *(30)*. Peak bone mass is determined by estrogen status, diet, exercise, body weight, gender, and genetic influences *(31)*. Adequate levels of sex steroids during adolescence is necessary to achieve peak bone mineralization *(32)*. Low levels of sex steroids affect bone mass to a greater degree during pubertal development *(33)*. Low estrogen levels seen in women with hypothalamic amenorrhea lead to increased bone turnover and bone resorption, causing reductions in trabecular and cortical bone *(34)*.

Weight-bearing exercise is known to increase bone mass by inducing mineralization of bone but is maturity dependent. There is a critical period for bone response to weight-bearing exercise during early puberty and premenarchal years *(35)*. Not achieving maximum bone density during this critical time in adolescence may be a factor in osteoporosis *(36)*. Additionally, optimum levels of factors that enhance bone formation such as estrogen, testosterone, growth hormone, and insulin-like growth factor 1 (IGF-I) in premenarchal years improve the effect of exercise and mechanical loading on bone turnover and mineralization *(35)*.

Research has shown that in athletic amenorrhea, estrogen replacement alone did not normalize low bone density *(30,37)*. Therefore, mechanisms other than estrogen deficiency may also account for low BMD in women whose amenorrhea is associated with caloric imbalance, nutritional inadequacy, and intense exercise. Young women who consistently diet may slow their metabolic rates that could affect bone metabolism through decreased leptin levels *(27)*. Leptin receptors have been reported to be found in bone and may be important to osteoblastic function *(27,38)*. Thus, dieting may indirectly affect bone metabolism through a mechanism involving a leptin pathway *(27)*.

A depressed metabolic rate, associated with lower BMD and lowered leptin levels, has been found in women ballet dancers with a history of amenorrhea *(27)*.

5.3.5. Athletes at Greatest Risk for Developing Signs and Symptoms Associated with the Triad

It has been reported that the prevalence of disordered eating and eating disorders is higher in athletes than nonathletes, and particularly in athletes competing in sports that emphasize leanness or a low body weight *(39,40)*. It has also been reported that amenorrhea is more prevalent in the athletic population (3–66%) than in the general population (2–5%) *(5)*. Although osteoporosis has been reported in amenorrheic triathletes and distance runners (10–13%), osteopenia is much more prevalent (40–50%) *(25,26)*.

However, Torstveit and Sundgot-Borgen *(41)* conducted a study in Norway in which elite athletes ($n = 699$) and nonathlete controls ($n = 607$) were divided into leanness and nonleanness groups. The purpose of their study was to look at the risk factors for the triad components and not the endpoints of the triad per se. Using their criteria to define risk of the triad (scales on eating disorder inventories, BMI < 18.5, self-reported menstrual dysfunction, and stress fractures), they found that more than 6 of 10 females were classified at risk of the female triad, with small differences between normal active females and elite athletes. They also found that a higher percentage of both athletes competing in leanness sports and nonathlete controls were classified as at risk of the triad compared with athletes competing in nonleanness sports. So in this instance, athletes competing in nonleanness sports would be less at risk than the general active population. They also found that a higher percentage of athletes compared to controls reported the occurrence of menstrual dysfunction as defined in their study. However, when dividing the athletes into leanness and nonleanness groups, athletes competing in the nonleanness group reported the same amount of menstrual dysfunction as controls. They concluded that the high number of women classified at risk of the triad both in the sports where leanness is emphasized and in the nonathlete controls may actually reflect the present situation in young females today. Society's focus on the female body—on thinness and low weight—may lead to dissatisfaction with one's own body, further leading to dieting and unhealthy eating behaviors and subsequent loss of menstruation and bone mass.

5.3.6. The Growing Health Concern of the Triad for Allied Health Professionals

There is a general agreement among researchers and leading sport organizations that disordered eating and menstrual dysfunction is a health issue for many women competing in sports focusing on leanness and/or low body weight *(2,13,41,42)*. For example, in a study of collegiate gymnasts, 62% displayed some type of disordered eating. Twenty-six percent of these athletes vomited on a daily basis, 24% used diet pills, 12% fasted, and 75% were told by their coach that they weighed too much *(43)*.

Many colleges and high schools do not use a medical history form that asks questions that might help determine if athletes are at risk of developing the signs and symptoms associated with disordered eating, amenorrhea, and osteoporosis. Therefore, it is up to allied health professionals such as athletic trainers, school nurses, team physicians, physical therapists, nutritionists, and exercise physiologists to implement such a screening device. The history and physical form that is recommended by ACSM includes questions about menstrual history, body image, recent weight loss, and a history of fractures. The menstrual history should include the age at menarche, the frequency and duration of periods, and the use of hormonal therapy. The nutritional history should include a 24-h recall of food intake, the usual number of meals and snacks, and a list of forbidden foods. The body weight history should include the highest and lowest weights since menarche and the athlete's satisfaction with her present weight. Athletes should be questioned to find out if they are preoccupied with their weight (Appendices 7–8). Even though it is suggested that athletes should be asked if they have ever tried to control their weight by vomiting or using laxatives, diet pills, or diuretics *(44)*, there may be a better way to read between the lines as denial of this problem is common.

Orthopedic nurses are in a perfect position to screen young female athletes regarding the risk factors associated with the triad of disorders because orthopedic nurses are often called on to provide preseason physicals *(44)*. Physical therapists, exercise physiologists, and athletic trainers typically can spend more time with athletes than other clinicians involved in their care and can serve as a first-line defense in the detection of the triad of disorders *(45)*.

5.4. CLOSING REMARKS

Many factors have been postulated to cause the disturbances in the reproductive axis in female athletes. Low body weight or body fat, poor nutrition, disordered eating behavior, psychological stress, energy imbalance, initiating training too soon, and excessive exercise have been cited in the research literature. It seems that it is a combination of many of these stressors that results in a hormonal imbalance in the athlete which ultimately affects not only the reproductive axis but also bone metabolism.

Hypothalamic amenorrhea has been associated with long-term dieting and restricted energy intake resulting in an energy imbalance. Overall, the athlete is energy-deprived and this triggers suppression of the HPO axis also referred to in the literature as HPG axis. Athletes may not compensate enough in their diet to accommodate for increased energy expenditures brought about by intensive training. This may be intentional or unintentional but the end result is the same, an energy-deprived host with a lowered metabolic rate and thermal response to food *(28,46)*.

Excessive exercise or psychological stress unaltered by sufficient caloric intake or coping mechanisms also contributes to hypothalamic amenorrhea through the HPA *(28)*. Stressful events result in increased secretion of corticotropin-releasing hormone (CRH) in both animals and women *(47,48)*. CRH release slows the pulsatile release of GnRH in animal studies, thereby decreasing LH levels *(47)*. These endocrine changes result in a chronic hypoestrogenic state that adversely affects both the reproductive system and the skeletal system.

In addition, initiation of intense exercise prior to the complete maturation of the HPO axis is thought to place adolescence at an increased risk of developing amenorrhea *(8)*.

Preadolescents experiencing an energy drain secondary to high metabolic demands and caloric restriction are particularly vulnerable to menstrual abnormalities.

5.5. SUMMARY AND CONCLUSIONS

In the period from 1970s to 1990s, participation by girls and women in organized athletics has increased dramatically *(44)*. President Nixon signed Title IX into law in 1972, which required that all school districts receiving federal funding provide equal opportunities for men and women. This increase in women participating in sports and exercise, especially where leanness is paramount, has brought about new health concerns for active females. The triad of disorders seen in female athletes are interrelated and represent imbalances nutritionally, hormonally, and psychologically. However, the disorders associated with the triad can be prevented and are not a result of exercise or sports performance alone. There are many positive benefits to participation in sports and exercise, and most would argue that the benefits far outweigh the risks.

5.6. SCENARIO WITH QUESTIONS AND ANSWERS

This scenario has been adapted from an intervention example taken from the article "The Female Athlete: An Emerging Role for Physical Therapy" *(45)*.

5.6.1. Scenario

Rachel is a 22-year-old, female, elite, national caliber cross-country and long-distance track athlete. She is 5'4" tall and weighs 107 lb. She intends to compete in the next Olympic trials. She has a history of stress fractures both on the right side on the fifth metatarsal and midtibial shaft. She reports that she has to decrease her training volume for a couple of days and thereafter can push through her pain. On her menstrual history questionnaire, she reported cycles every 27 days for an extended time interval, followed by cycles at intervals every 20 days for an extended period of time, and more recently her cycles are occurring less frequently than every 35 days, however, the interval is not consistent. Caloric intake is approximately 1000 kcal/day. A DEXA scan of the lumbar spine revealed her age adjusted T-score to be 2 SD below the mean for young adults.

5.6.2. Questions

1. Based on her menstrual history questionnaire what would you suspect about her menstrual history progression?
2. How would you classify her bone density?
3. Does her caloric intake seem reasonable?

5.6.3. Plausible Answers

1. It would seem that her cycles were normal (every 27 days) initially. Possibly, the luteal phase of her cycle shortened (< -22 days) which was followed by irregular cycles. Cycles that occur at intervals longer than 35 days be classified as oligomenorrheic if this pattern continued.

2. She would be classified as having ostopenia (BMD between 1 and 2.5 SD below the mean of young adults).
3. No it does not; caloric intake seems inadequate in order to achieve her sports performance goals.

REFERENCES

1. Lebrun C, Rumball J. Female athlete triad. *Sports Med Arthrosc* 2002;10:23–32.
2. Otis C, Drinkwater B, Johnson M, et al. ACSM position stand on the female athlete triad. *Med Sci Sports Exerc* 1997;29(5):i–ix.
3. Shangold MM, Rebar RW, Wentz AC, et al. Evaluation and management of menstrual dysfunction in athletes. *JAMA* 1990;263:1665–1669.
4. Loucks AB, Horvath SM. Athletic amenorrhea: a review. *Med Sci Sports Exerc* 1985;17(1):56–72.
5. Otis CL. Exercise associated amenorrhea. *Clin Sports Med* 1992;11(3):351–362.
6. Drinkwater BL, Bruemner B, Chesnut CH. Menstrual history as a determinant of current bone density in young athletes. *JAMA* 1990;263(4):545–548.
7. Drinkwater BL, Nilson K, Chesnut CH, et al. Bone mineral content of amenorrheic and eumenorrheic athletes. *N Engl J Med* 1984;311(5):277–280.
8. Marcus R, Cann C, Madvig P, et al. Menstrual function and bone mass in elite woman distance runners. *Ann Intern Med* 1985;102:158–163.
9. Rigotti NA, Nussbaum SR, Herzog DB, et al. Osteoporosis in women with anorexia nervosa. *N Engl J Med* 1984;311:1601–1606.
10. Kanis JA, Melton J, Christiansen C, et al. The diagnosis of osteoporosis. *J Bone Miner Res* 1994;9:1137–1141.
11. Yeager KK, Agostini R, Nattiv A, et al. The female athlete triad. *Med Sci Sports Exerc* 1993;25: 775–777.
12. Otis CL. Too slim, amenorrheic, fracture-prone: the female athlete triad. *ACSM's Health Fit Journal* 1998;2(1):20–25.
13. United States Olympic Committee. The female athlete triad: a responsible approach for coaches. *Olympic Coach* 1999;9(4):6–9.
14. American Psychiatric Association. Diagnostic and Statistical Manual of Mental Disorders: DSM-IV-TR. 4th ed., text revision. Washington, DC: American Psychiatric Association, 2000:583–595, 785–788.
15. Putukian M. The female athlete triad. *Clin Sports Med* 1998;17:67–5696.
16. Garner DM, Garfinkel PE. The eating attitudes test: an index of the symptoms of anorexia nervosa. *Psychol Med* 1979;9:273–279.
17. Henderson M, Freeman CP. A self-rating scale for bulimia the 'bite'. *Br J Psychiatry* 1987;150:18–24.
18. Thelan M. BULLIT-R. *J Consult Clin Psychol* 1984:52;863–872.
19. Marshall LA. Clinical evaluation of amenorrhea in active and athletic women. *Clin Sports Med* 1994;13:371–387.
20. Loucks AB. Effects of exercise training on the menstrual cycle: existence and mechanisms. *Med Sci Sports Exerc* 1990;22(3):275–279.
21. Sabatini, S. The female athlete triad. *Am J Med Sci* 2001;322:193–195.
22. Rose MZ, Lee TC, Maffulli N, et al. Special gynecological problems of the young female athlete. In: Maffulli N et al., eds. *Sports Medicine for Specific Ages and Abilities*. London: Churchill Livingstone, 2001:139–147.
23. Atlanta Reproductive Health Centre. Menstrual disturbances in female athletes. Atlanta, Georgia, 2003. (Accessed July 19, 2005 at http://www.ivf.com/amenath.html).
24. Khaun KM, Lui-Ambrose T, Sran MM, et al. New criteria for the female athlete triad. *Br J Sports Med* 2002;36:10–13.
25. Pettersson U, Stalnacke B, Ahlenius G, et al. Low bone mass density at multiple skeletal sites, including the appendicular skeleton in amenorrheic runners. *Calcif Tissue Int* 1999;64:117–125.
26. Rutherford OM. Spine and total body bone mineral density in amenorrheic endurance athletes. *J Appl Physiol* 1993;74:2904–2908.

27. Kaufman BA, Warren M, Dominguez JE, et al. Bone density and amenorrhea in ballet dancers are related to a decreased resting metabolic rate and lower leptin levels. *J Clin Endocrinol Metab* 2005;87(6):2777–2783.
28. Lo B, Hebert C, McClean A. The female athlete triad no pain, no gain? *Clin Pediatr (Phila)* 2003;42:573–580.
29. Gibson JH, Mitchell A, Reeve J, et al. Treatment of reduced bone mineral density in athletic amenorrhea: a pilot study. *Osteoporos Int* 1999;10:284–289.
30. Hergenroeder AC. Bone mineralization hypothalamic amenorrhea and sex steroid therapy in female adolescents and young adults. *J Pediatr* 1995;126:683–689.
31. Cummings D, Cummings C. Estrogen replacement therapy and female athletes. *Sports Med* 2001;31:1025–1031.
32. Biller BM, Coughlin JF, Saxe V. Osteopenia in women with hypothalamic amenorrhea: a prospective study. *Obstet Gynecol* 1991;78:996–1001.
33. Golden NH, Lanzkowsky L, Schebendach J, et al. The effects of estrogen-progestyin treatment on bone mineral density in anorexia nervosa. *J Pediatr Adolesc Gynecol* 2002;15:135–143.
34. Miller KIK, Klibanski A. Amenorrheic bone loss. *J Clin Endocrinol Metab* 1999;84:1775–1783.
35. MacKelvie KJ, Khan KM, McKay. Is there a critical period for bone response to weight-bearing exercise in children and adolescents? A systematic review. *Br J Sports Med* 2002;36:250–257.
36. Eliakim A, Beyth Y. Exercise training, menstrual irregularities and bone development in children and adolescents. *J Pediatr Adolesc Gynecol* 2003;16:201–206.
37. Warren MP, Fox RP, DeRogatis AJ, et al. Osteopenia in hypothalamic amenorrhea: a 3-year longitudinal study. Program of the 76th Annual Meeting of the Endocrine Society, Anaheim, CA, 1994, p 1044 (Abstract).
38. Thomas T, Gori F, Burguera B, et al. Leptin acts on human marrow stromal precursor cells to enhance osteoblast differentiation and to inhibit adipocyte differentiation: a potential mechanism for increased bone mass in obesity. Program of the 80th Annual Meeting of the Endocrine Society, New Orleans, LA, 1998, p 90 (Abstract).
39. Byrne S, McLean N. Elite athletes: effects of the pressure to be thin. *J Sci Med Sport* 2002;5:80–94.
40. Sundgot-Borgen J, Torstveit MK. Prevalence of eating disorders in elite athletes is higher than in the general population. *Clin J Sport Med* 2004;14:25–32.
41. Torstveit MK, Sundgot-Borgen J. The female athlete triad: are elite athletes at increased risk? *Med Sci Sports Exerc* 2005;37(2):184–193.
42. Beals KA, Manore MM. Disorders of the female athlete triad among collegiate athletes. *Int J Sport Nutr Exerc Metab* 2002;12:281–293.
43. Calabrese LH. Nutritional and medical aspects of gymnastics. *Clin Sports Med* 1985;4:23–37.
44. Donaldson ML. The female athlete triad: a growing health concern. *Orthop Nurs* 2003;22(5):322–323.
45. Papanek P. The female athlete triad: an emerging role for physical therapy. *J Orthop Sports Phys Ther* 2003;33:594–614.
46. Kazis K, Iglesias E. The female athlete triad. *Adolesc Med* 2003;14(1):87–95.
47. Norman RL, McClone J, Smith CJ. Restraint inhibits luteinizing hormone secretion in the follicular phase of the menstrual cycle in rhesus macques. *Biol Reprod* 1994;50:1;16–26.
48. Warren MP, Stiehl AL. Exercise and female adolescents: effects on the reproductive and skeletal systems. *JAMA* 1999;54:115–120.

6 Disordered Eating in Active Middle-Aged Women

Jacalyn J. Robert-McComb

CONTENTS

6.1 LEARNING OBJECTIVES
6.2 INTRODUCTION
6.3 RESEARCH FINDINGS
6.4 CONCLUSIONS
6.5 SCENARIO WITH QUESTIONS AND ANSWERS

6.1. LEARNING OBJECTIVES

After completing this chapter, you should have an understanding of the following:

- The risk factors for developing disordered eating in middle-aged women.
- Life stressors in midlife that may predispose women to disordered eating.
- Thirty Something and Beyond (TSAB) treatment programs.

6.2. INTRODUCTION

Eating disorders are among the most common psychiatric problem that affects young women *(1)*. However, research suggests that more middle-aged women are seeking treatment for eating disorders *(2)*. An article published in Primary Psychiatry entitled "Eating Disorders in Middle and Late Life: A Neglected Problem" calls attention to this problem *(3)*.

It is not surprising that physically active women of all ages are many times the victims of an eating disorder. Environmental risk factors for developing an eating disorder range from participation in activities that promote thinness, such as ballet dancing at a young age, to participating in endurance sports, such as triathlons, for the older recreational athlete *(4,5)*. However, it is not the participation in the sport per se that causes the eating disorder but a multitude of environmental and interpersonal factors. There is general agreement that the benefits of sport participation or activity for women of all ages far outweigh the risks of participation in sport and activity if

From: *The Active Female*
Edited by: J. J. Robert-McComb, R. Norman, and M. Zumwalt © Humana Press, Totowa, NJ

adequate nutritional needs are met. The salient point is that even "seemingly" healthy physically active women cannot escape the barrage of environmental influences that may lay the foundation for an eating disorder.

Unfortunately, the diagnosis of an eating disorder is difficult, and more than one-half of all cases go undetected. Up to one-third of women with type 1 diabetes may have an eating disorder, and these women are especially at high risk of microvascular and metabolic complications (6). The difficulty in detecting an eating disorder in active middle-aged woman is particularly challenging because this disorder would not normally be expected or screened for by a physician.

6.3. RESEARCH FINDINGS

6.3.1. Risk Factors for Developing Disordered Eating in Middle-Aged Women

Zerbe (3) asserts that women in midlife face different issues that would precipitate an eating disorder than women entering adolescence and early adulthood. However, regardless of whether an eating disorder has an early or later onset, research strongly suggests that there are certain environmental stressors and personality traits that lay the foundation for the development of an eating disorder. Issues such as participation in sports where leanness is emphasized or serves as an advantage in performance, the central role of beauty in femininity, the media's emphasis on appearance, physical changes during the lifespan, and personality traits have all been implicated in the development of eating disorders in children and adolescence (7).

Personality traits that may contribute to eating disorders are low self-esteem, difficulty in expressing negative emotions, difficulty resolving conflict, being a perfectionist, and anxiety avoidance (1). These are characteristics seen in the older women as well as in children and adolescents; therefore, eating disorders, even though more prevalent in children and adolescents, are not age dependent (8,9).

However, age may play a role in the etiology of eating disorders at both ends of the spectrum. Lewis and Cachelin (9) found that in a nonclinical sample of 250 aged women, the women who endorsed more disordered eating also reported a greater fear of aging. However, this fear may be because of an underlying personality trait, such as perfectionism or social insecurity, as not all middle-aged women report that physical changes associated with aging contribute to eating disorder symptoms. Rather, the middle-aged women (2) have reported clinically elevated scores on perfectionism, ineffectiveness, and social insecurity as measured by the Eating Disorder Inventory. Furthermore, these scores were not significantly different between younger and older women (2).

Several studies report that middle-aged women are just as preoccupied with their physical appearance as adolescent and young adult women (8,9). Allaz et al. (8) found that in a nonclinical sample of 1053 women above the age of 30, 71% expressed the desire to loose weight despite being of normal weight. In fact, women often describe their main goal of exercise as weight loss and toning their muscles to enhance body shape and attractiveness rather than on the health benefits of exercise (10,11).

6.3.2. Life Stressors in Midlife that may Predispose Females to Disordered Eating

Anecdotal descriptions of eating disorders in middle-aged women have suggested that loss is an important trigger *(3)*. Death of a loved one, divorce, a loss of youthfulness, changing appearance, physical deterioration, and the empty nest syndrome have all been implicated as the precipitant of an eating disorder in middle-aged women. Many middle-aged women who suffer from eating disorders have also been in abusive relationships. Others are coping with significant medical issues such as breast cancer and the long-term effects of osteoporosis. However, many women who report eating disorder symptoms in midlife also report eating disorder symptoms in adolescence and early adulthood *(2)*.

6.3.3. Thirty Something and Beyond Treatment Programs

The Renfrew Center specializes in the treatment of eating disorders and related mental health issues (http://www.renfrewcenter.com). It is a Joint Commission on Accreditation of Healthcare Organizations (JCAHO) accredited organization that established its roots in 1985. The treatment programs offered at the Renfrew Center include traditional psychotherapy, psychiatry, and nutritional counseling as well as art therapy, dance movement therapy, and psychodrama. Through the history of offering treatment programs for women with eating disorders, some observations were made that motivated the performance improvement team to research and develop a separate program for women more than the age of 35, called TSAB.

Examples of issues addressed for TSAB treatment program include domestic violence, balancing relationships, needs and recovery, self-care, and finding a voice *(12)*. The women also attend a series entitled Lifespan Challenges which is composed of four groups: (1) Wellness, (2) Nutrition, (3) Exercise and Self-Nurturing, and (4) Reclaiming Your Energy. There are also art therapy classes, relationship groups, and grief and loss support groups for the women to participate in.

The TSAB treatment program offered at the Renfrew Center seems to have a positive effect on midlife women who enter this program as evidenced by their pre-post test scores drive for thinness, bulimia, and body dissatisfaction from the Eating Disorder Inventory *(2)*.* Also the Renfrew Center Foundation provides treatment access for women and girls who might otherwise not be able to afford treatment (http://www.renfrew.org).

6.4. CONCLUSIONS

Physically active midlife women, even though not given the attention that the young female athlete is concerning the triad of disorders (disordered eating, menstrual dysfunction, and osteoporosis), are still subject to these disorders. These women may be competing in sports where thinness is seen as an obvious function of training, such as in endurance sports, and the disorder goes unnoticed. However, the number

---

*For more information on treatment programs offered by the Renfrew Center for middle-aged women who may be suffering from an eating disorder, you may call 1-800-RENFREW or visit them on the web (http://www.renfrewcenter.com).

of middle-aged women seeking treatment for eating disorders is on the rise *(2,3)*. Even women aged 61–92 identified weight as their greatest concern when asked what bothered them about their bodies *(13)*. Other studies find comparable unhealthy levels of dieting and disordered eating in elderly age groups as in the young *(14)*.

Many of these women have had signs and symptoms of disordered eating since their early teens; however, they have managed to survive somehow. However, they cannot escape the debilitating consequences of disordered eating and energy imbalance that results in osteoporosis if gone undetected. Yet, half of all cases go undetected *(15)*.

6.5. SCENARIO WITH QUESTIONS AND ANSWERS

6.5.1. Scenario

You have been Eileen's primary care physician for 30 years. In fact, you have treated her whole family. Eileen has always been active and at the age of 59 still walks regularly and attends aerobics classes at the local fitness center. Eileen is 5'4" and has weighed between 130 and 135 lb for the last 10 years. It has been a year since her regular physical exam. You noticed that when she walked in for her yearly exam she looked more distraught than usual. Looking at her medical history, you see that she has lost 22 lb and now weighs 113 lb. You questioned her about her eating patterns but she said that they have not changed in the last year. You also sent her in for a series of laboratory tests to rule out disease states that could cause this unusual weight loss but the results came back negative.

6.5.2. Questions

1. What type of questions might be appropriate to identify a possible eating disorder in middle-aged women?
2. What are risk factors that may predispose Eileen to an eating disorder in midlife?
3. What are treatment options that may be available to her at her age?

6.5.3. Plausible Answers

1. You could begin by asking her about any major changes in her life recently.
2. Risk factors for middle-aged women include but are not limited to death of a loved one, divorce, a loss of youthfulness, changing appearance, physical deterioration, and coping with a debilitating illness.
3. Although there are many treatment options available, you would encourage her to seek counseling with a psychiatrist or psychologist who specializes in eating disorders. If she is interested in group therapy or residential treatment, it would be best for her to identify treatment centers that have programs for older women, as many of the issues that these women have to deal with are different from younger women.

REFERENCES

1. Kreipe RE, Birndorf SA. Eating disorders in adolescents and young adults. *Med Clin North Am* 2000;84:1027–1049.
2. Forman M. A descriptive overview of middle-aged women with eating disorders. Perspective: A Professional Journal of the Renfrew Centre Foundation 2004;Summer:1–4.
3. Zerbe K. Eating disorders in middle and late life: a neglected problem. *Prim Psychiatry* 2003;10:80–82.

4. Torstveit MK, Sundgot-Borgen J. The female athlete triad: are elite athletes at increased risk? *Med Sci Sports Exerc* 2005;37(2):184–193.
5. Mehler PS. Diagnosis and care of patients with anorexia nervosa in primary care settings. *Ann Intern Med* 2001:134:1048–1059.
6. Walsh JM, Wheat ME, Freund K. Detection, evaluation, and treatment of eating disorders; the role of the primary care physician. *J Gen Intern Med* 2000;15:577–590.
7. Striegel-Moore R, Marcus, M. Eating disorders in women: current issues and debates. In: Stanton AL, Gallant SJ, eds. *The Psychology of Women's Health: Progress and Challenges in Research and Application.* Washington, DC: American Psychological Association, 1995:445–487.
8. Allaz A, Bernstein M, Rouget P, et al. Body weight preoccupation in middle age and ageing women: a general population survey. *Int J Eat Disord* 1998;23:287–294.
9. Lewis D, Cachelin F. Body image, body dissatisfaction, and eating attitudes in midlife and elderly women. *Eating Disorders* 2001;9:29–39.
10. Gill K, Overdorf V. Incentives for exercise in younger and older women. *J Sport Behav* 1994;17:87–97.
11. Krane V, Waldron J, Michalenok J, et al. Body image concerns in female exercisers and athletes: a feminist cultural studies perspective. *Women Sport Phys Act J* 2001;10(1):17–54.
12. Grishkat H. Thirty-something and beyond: residential treatment programming for midlife eating disorders. *Perspective* 2004;Summer:7–9.
13. Clarke LH. Older women's perception of ideal body weights: the tension between health and appearance motivations for weight loss. *Ageing Soc* 2002;22:751–753.
14. Hetherington MM, Burnett L. Ageing and the pursuit of slimness: dietary restraint and weight satisfaction in elderly women. *Br J Clin Psychol* 1994;33:391–400.
15. Becker AE, Grinspoon SK, Klibanski A, et al. Eating disorders. *N Engl J Med* 2000;84:1027–1049.

7 Eating Disorder and Menstrual Dysfunction Screening Tools for the Allied Health Professional

Jacalyn J. Robert-McComb

CONTENTS

7.1 LEARNING OBJECTIVES
7.2 INTRODUCTION
7.3 RESEARCH FINDINGS
7.4 CONCLUDING REMARKS
7.5 SCENARIO WITH QUESTIONS AND ANSWERS

7.1. LEARNING OBJECTIVES

After completing this chapter, you should have an understanding of the following:

- The primary care provider's role in the detection of an eating disorder.
- Commonly used screening tools for eating disorders.
- Screening tools specifically designed for the female athlete.
- Credible self-assessment and educational tools on the Web for eating disorders.

7.2. INTRODUCTION

The annual National Eating Disorders Screening Program (http://www.mentalhealthscreening.org/events/nedsp/index.aspx) is sponsored by a number of professionally recognized organizations, including the American Psychiatric Association, American Psychological Association, American Academy of Pediatrics, and National Collegiate Athletic Association (NCAA). The Screening Program incorporates educational materials, a short screening questionnaire, and a post-questionnaire follow-up by a knowledgeable eating disorders counselor. However, there are no national screening guidelines for eating disorders *(1–3)*.

Target groups for screening should include young women with low body mass index (BMI) compared with age norms, patients consulting with weight concerns who are not overweight, women with menstrual disturbances or amenorrhea, patients

From: *The Active Female*
Edited by: J. J. Robert-McComb, R. Norman, and M. Zumwalt © Humana Press, Totowa, NJ

with gastrointestinal symptoms, patients with physical signs of starvation or repeated vomiting, and children with poor growth *(4)*.

7.3. RESEARCH FINDINGS

7.3.1. The Primary Care Provider's Role in the Detection of an Eating Disorder

The family physician's office is an ideal setting to identify eating disorders and initiate treatment in a timely fashion *(5)*. Research has shown that early diagnosis with intervention and diagnosis at an earlier age correlated with improved outcomes *(6)*. The role of the physician in the screening of eating disorders is vital and cannot be overemphasized. Many times the primary care physician serves as the first line of defense in the diagnosis. However, following diagnosis, the role of the primary care physician is predominantly that of medical monitor providing surveillance to manage the medical complications.

Normally, a good medical history is the most powerful tool to diagnose an eating disorder. Physical findings such as low BMI, amenorrhea, bradycardia, gastrointestinal disturbances, skin changes, and laboratory studies can help diagnose an eating disorder *(5)*. However, during the early course of an eating disorder, physical examination and laboratory findings may be normal. The Office on Women's Health, Department of Health and Human Services, suggests that the following physical and behavioral/emotional characteristics as listed in Table 7.1 be used during the routine screening of adolescent and preadolescent patients for the detection of an eating disorder. In addition, it may be necessary to ask the patient questions about weight and dieting.

Table 7.1
A primary care provider reference tool for screening of adolescent and pre-adolescent patients to detect eating disorders

Physical	*Behavioral/emotional*
• Primary or secondary amenorrhea • Weight loss greater than 10% • Failure to gain expected weight during the adolescent growth spurt • Overweight • Lanugo hair • Hypothermia • Dry hair or skin, dehydration • Weight fluctuation in a short period of time • Bloating and abdominal discomfort • Damaged teeth • Enlargement of lymph or salivary glands	• Recurrent or excessive dieting when not overweight • Eating in secret • Eating large quantities of food in a short period of time • Excessive concerns about perceived body image that are incongruous with actual weight • Compulsive or overly rigid exercising • Depression • Use of self-inducted vomiting, laxatives, starvation, diuretics, or other extreme measures to loose weight

Source: From Office on Women's Health. Screening for Eating Disorders-Tips for Health Care Providers (httpFACSMhttp://www.womenshealth.gov/BodyImage/bodywise/hp/HCP-eattips.pdf). With permission.

Anstine and Grinenko *(7)* suggest that a positive response to any of the following questions warrants further evaluation.

- How many diets have you been on in the past year?
- Do you think you should be dieting?
- Are you dissatisfied with your body size?
- Does your weight affect the way you think about yourself?

Although more direct questions can be asked to identify an eating disorder, the nature of the illness is secretive. It is likely that the more embarrassing symptoms of an eating disorder, such as vomiting, will not be readily disclosed by the individual. The intensity of the questioning has to be balanced with the relationship between the individual and health care provider and the readiness for the individual to disclose their illness. Therefore, the allied health professional has to sometimes read between the lines and look for physical and behavioral characteristics that may signify an eating disorder. Examples of more probing questions that can be used to identify signs and symptoms of an eating disorder can be found in Table 7.2.

Table 7.2
Questions to ask the patient and/or the parents if an eating disorder is suspected

---

Questions to ask the patient
  Weight history
    Has there been any change in your weight?
    What's the most you ever weighted and when? The least and when?
    Are you terrified about being overweight?
  History of dieting
    Have you ever tried to loose weight?
    What kinds of diets have you tried?
    Have you ever tried to loose weight or control your weight by vomiting, taking diet pills or laxatives, or excessive dieting?
    Are you currently dieting or trying to loose weight?
  Emotions and eating
    Do you feel extremely guilty after eating?
    Do you feel that food controls your life?
    Have you gone on eating binges where you feel you may not be able to stop?
  Current eating/exercise habits
    Are you satisfied with the way you eat?
    What did you eat yesterday?
    Have you gone on eating binges where you feel you could not stop?
    How much do you exercise in a typical week?
    Have you ever fainted?
  Attitudes about weight and shape
    How do you feel about the shape and size of your body?
    What do you think your ideal weight should be?

(Continued)

Table 7.2
(Continued)

Menstrual history
   Are your menstrual periods regular?
   When was your last menstrual period?
Questions to ask parents
   Does your child make negative remarks about his or her body?
   Have you noticed any changes in his or her food-related habits? If so, what?
   Are you concerned about your child's weight? Eating habits? Exercise habits?

Source: From Office on Women's Health. Screening for Eating Disorders-Tips for Health Care Providers (httpFACSMhttp://www.womenshealth.gov/BodyImage/bodywise/hp/HCP-eattips.pdf) and from the National Eating Disorders Screening Program (http://www.mentalhealthscreening.org/events/nedsp/index.aspx). With permission.

7.3.2. Commonly Used Screening Tools for Eating Disorders

7.3.2.1. SCOFF QUESTIONNAIRE

If an eating disorder is suspected, one of the most practical screening tools to use in the primary care setting *(8)* is the SCOFF questionnaire (Appendix 4). Because of its 12.5% false-positive rate, this test is not sufficiently accurate for diagnosing eating disorders, but it is an appropriate screening tool.

7.3.2.2. THE EATING ATTITUDES TEST

One of the most widely used standardized measure of symptoms and concerns of eating disorders is The Eating Attitudes Test (EAT-26) by Garner et al. *(9)* (see Appendix 5 or view http://river-centre.org/orderaet26.html). It has been found to be relatively valid and reliable. However, the EAT-26 does not discriminate between individuals with bulimia nervosa or anorexia nervosa and does not yield a specific diagnosis for an eating disorder. It should be used as a two-stage screening process in which those who score at or above a cutoff score of 20 should be interviewed by qualified mental health professional to make a formal diagnosis.*

7.3.2.3. BULIT-R

Another screening tool used widely in the research literature is the BULIT-R by Thelen et al. *(10)*. This instrument has been shown to be a reliable and valid measure for identifying individuals who may suffer from bulimia nervosa in both a clinical and nonclinical population. The internal consistency is very high, and the test-retest stability over a 2-month period is very stable *(11)*. The BULIT-R screening instrument can be found in Appendix 6 along with the scoring instructions.†

---

*Garner et al. do not charge for permission to use the EAT-26 screening instrument in research or clinical work as long as their 1982 publication is cited *(9)*.

†While there is generally no charge for the use of the BULIT-R, Dr. Thelen requests that permission to use the instrument be obtained through him (ThelenM@missouri.edu).

7.3.3. Screening Tools Specifically Designed for the Female Athlete

According to the American Academy of Pediatrics on Sports Medicine and Fitness, many times the physicians who care for female athletes are unaware of their unique medical concerns *(12)*. A joint publication by leading medical societies outlined specific orthopedic and medical concerns for female athletes *(13)*. Among these are disordered eating, amenorrhea, and osteoporosis/osteopenia; when interrelated, these disorders are more commonly known as the Female Athlete Triad. At the high school and collegiate level, the preparticipation physical examination (PPE) offers an opportunity to screen for components of the Female Athlete Triad. The medical history form is considered the most important aspect of the PPE *(14)*.

It has been suggested that a separate section, consisting of both a gynecological and nutritional component be directed at the female athlete *(15)*. Straightforward questions concerning body image and disordered eating may lead to underreporting *(16)*. Therefore, the questionnaire should address specific areas of concern without causing the athlete to minimize her condition *(17)*. Johnson *(17)* proposed that the questionnaire should contain questions about the athlete's menstrual history, questions about food intake within the past 24 h as well as foods that the athlete may refuse to eat, and questions about satisfaction with weight and attempts at controlling weight. An example of a supplemental medical history questionnaire that may be added to the PPE for female athletes can be found in Appendix 7. There are presently no standardized instruments used in the PPE for female athletes. Results of a survey to NCAA Division 1 programs indicated that there is a pressing need for more standardized eating disorder and menstrual dysfunction screening, prevention, and treatment programs among NCCA Division I schools *(18)*.

There are some instruments in the research literature that have been developed to identify eating pathology in female athletes; however, the widespread use of these instruments in the athletic environment is more of an exception rather than the rule. Female athletes may differ from other patients with classic eating disorder in both their psychological makeup and observed manifestations. Behaviors vary and probably fit better within the classification of Eating Disorder Not Otherwise Specified *(19)*. McNulty et al. *(3)* developed a screening tool to identify eating pathology in female athletes. The Female Athlete Screening Tool (FAST) has a high internal consistency (Cronbach's $\alpha = 0.87$). Athletes with eating disorders scored higher on the FAST as compared to athletes without eating pathology and nonathletes with eating disorders ($p < 0.001$), demonstrating discriminant validity. The FAST can be found in Appendix 8 along with the scoring instructions. By using this screening tool, professionals can help identify athletes who may need additional psychological screening for an eating disorder and also need additional help with their aberrant eating patterns.

7.3.4. Credible Self-Assessment and Educational Tools on the Web for Eating Disorders

Education about eating disorders and self-knowledge assessment tools are readily available for girls and women of all ages on the Web; however, it is important that the information be accurate and the site is credible. The opportunities for self-knowledge on the Web are particularly suitable for those individuals who are not quite comfortable with disclosing the fact that they may have an eating disorder and to an authorative figure, yet suspect that they may have the early signs and symptoms of this illness. Table 7.3 lists a number of credible websites that individuals of varying ages may view for additional information about eating disorders. Table 7.3 does not include all of the available resources, as there are approximately 723,000 sites for screening for eating disorders and 10,100,000 for kids and eating disorders.

Table 7.3
Credible educational and screening tools on the Web for eating disorders

Target audience	Kids (KidsHealth also has separate areas for teens and parents)
Web site name	Kids and Eating Disorders
Internet address	http://kidshealth.org/kid/health_problems/learning_problem/eatdisorder.html
Sponsor	The Nemours Foundation's Center for Children's Health Media
Resources	KidsHealth is the largest and most visited site on the Web providing doctor-approved health information about children from before birth through adolescence.
Target audience	Female Athletes
Web site name	BODYSENSE: A Positive Body Image Initiative for Athletes
Internet address	http://www.bodysense.ca/aboutus/indexe.html
Sponsor	Canadian Centre for Ethics in Sport
Resources	Dedicated to the promotion of healthy sport settings for athletes with the ultimate goal of preventing eating disorders
Target audience	College Students
Web site name	College Response: National Eating Disorder Screening Program (NEDSP)
Internet address	http://www.mentalhealthscreening.org/college/eating.aspx
Sponsor	Screening for Mental Health
Resources	The College Response National Eating Disorder Screening Program (NEDSP) is designed to educate and screen college students for eating disorders, and to connect at-risk students with the resources they need.
Note:	Colleges and universities must register to participate, encourage your college to do so today. All results are confidential.
Target audience	Adults
Web site name	Dr. John Grohol's Psych Central
Internet address	http://psychcentral.com/disorders/eating_disorders/
Sponsor	Copyright 1992–2005 John M. Grohol, Psy.D., is a specialist in online mental health services
Resources	Site provides education and screening tools for eating disorders (short 5-item quiz adapted from the SCOFF and the EAT-26) as well as information and screening for other mental illnesses
Target audience	Women
Web site name	Women's Health Matters
Internet address	http://www.womenshealthmatters.ca/index.cfm
Sponsor	Sunnybrook and Women's Health Sciences Centre
Resources	Provides reliable and evidence-based information on women's health
Note:	also view http://www.healthymeasures.ca/HM_Eng_Home.html

7.4. CONCLUDING REMARKS

Although research is still in its infancy, studies show that there is considerable genetic propensity for developing an eating disorder *(20–22)*. Kendler *(21)* proposed that eating disorders have a genetic component that requires an environmental factor for gene expression (i.e., participation in sports where lean body mass is emphasized, not having the opportunity to learn normal eating behavior from strong role models, sexual abuse, etc.). Sports and fitness activities where extreme leanness is emphasized may serve as a stimulus for the development of eating pathology and the related adverse health consequences. Unfortunately, we do not know what the genetic characteristics are that make a person likely to develop an eating disorder. We do know that being female and being adolescent are biologically relevant.

There is also a common element of attempted weight loss in patients with eating disorders and a preoccupation with weight, shape and eating, and a concomitant fear of loss of control over their bodies *(23)*. However, simply being underweight, experiencing a negative-energy balance, or experiencing amenorrhea do not necessarily signify an eating disorder. Amenorrhea can occur at a high, normal, or low body weight or BMI. Signs and symptoms are synergistic; adolescents who exhibit signs and symptoms of eating disorders also show poor impulse control, depression, anxiety, perfectionism, confusion about feelings, and personality problems. Common comorbid diagnosis for patients with bulimia nervosa are posttraumatic stress disorder, borderline or other personality disorder, depression, and anxiety disorders *(20–24)*.

Abraham *(23)* has proposed that we simplify the criteria for an eating disorder for physicians and allied health professionals for detection and screening purposes. She refers to anorexia nervosa as a negative-energy control disorder and proposes the following criteria for screening purposes: (1) maintenance of a negative-energy balance; (2) negative effect of behavior and disordered thoughts on physical, psychological, and social health; and (3) no medical or major psychiatric illness accounting for the negative-energy status. She refers to bulimia as an episodic positive-energy control disorder and proposes the following criteria for screening purposes: (1) episodic positive-energy balance that is not appropriate for overall energy balance; (2) negative effect of behavior and disordered thoughts on physical, psychological, and social health; and (3) no medical or major psychiatric illness accounting for the negative-energy status. The features that she proposes for these disorders center around a fear of loss of control over eating or weight, body, or exercise and a preoccupation with thoughts of eating, exercise, body weight, or shape. In closing, the family care physician can play an important role in diagnosing an eating disorder by recognizing the possibility of an eating disorder based on the medical history questionnaire, asking appropriately targeted questions, and using practical screening tools.

7.5. SCENARIO WITH QUESTIONS AND ANSWERS

7.5.1. Scenario

Candice is a 17-year-old cross-country runner. You have been her family physician since birth. Her parents have scheduled her annual physical in late October this year. Her last physical was approximately 11 months ago in early December the previous year. She is 65 inches tall and now weighs 100 lb. Looking at her chart, you realize that she has not gained weight since her last physical. She earnestly began her high school cross-country training this year in June and competition started in late August and will finish in the month of November. She is very serious about her training and competition.

You ask your office staff to type age, height, and weight in the KidsHealth BMI Calculator (http://www.kidshealth.org/kid/exercise/weight/bmi.html). You discover that her BMI is 16.6 which puts her below the 5th percentile which indicates that she is underweight for her height and age. Her periods are irregular and she reports about 5 periods a year.

7.5.2. Questions

1. Would Candice be considered to be in an environment that would predispose her to an eating disorder?
2. What are some questions that you should ask Candice?
3. What are some logical explanations for her weight hypothetically excluding the possibility of an eating disorder?

7.5.3. Plausible Answers

1. Yes, she is in a sport where leanness enhances performance.
2. Are you currently dieting or trying to lose weight?
   What did you eat yesterday?
   Do you feel extremely guilty after eating?
   How much do you exercise in a typical week?
   What do you think your ideal weight should be?
3. If Candice does not seem preoccupied with her weight and body image, perhaps, she is simply training intensively and not taking in enough calories unintentionally. However, if she is unwilling to increase her caloric intake to maintain a balanced nutritional energy state following consultation with a nutritionist or other authoritative figure, disordered eating should be suspected.

REFERENCES

1. Favaro A, Santonastaso P. Construction and validation of a new screening questionnaires for eating disorders: the inventory for the screening of eating disorders. *Epidemiol Psichiatr Soc* 2000;(1):26–35.
2. Perry L, Morgan J, Reid F, et al. Screening for symptoms of eating disorders: reliability of the SCOFF screening tool with written compared to oral delivery. *Int J Eat Disord* 2002;32(4):466–472.
3. McNulty KY, Adams CH, Anderson JM, et al. Development and validation of a screening tool to identify eating disorders in female athletes. *J Am Diet Assoc* 2001;101(8):886–892.
4. National Collaborating Centre for Mental Health. Eating disorders. Core interventions in the treatment and management of anorexia nervosa, bulimia nervosa and related disorders. Leicester (UK): British Psychological Society; 2004. 260 p. [408 references] (Accessed August 8, 2005 at http://www.guideline.gov/summary/summary.aspx?doc_id=5066&nbr=3550&string=eati).
5. Pritts S, Susman J. Diagnosis of eating disorders in family care. *Am Fam Physician* 2003;67(2):297–304.
6. Herzog DB, Nussbaum KM, Marmor AK. Comorbidity and outcome in eating disorders. *Psychiatr Clin North Am* 1996;19:843–859.
7. Anstine D, Grinenko D. Rapid screening for disordered eating in college-aged females in the primary care setting. *J Adolesc Health* 2000;26(5):338–342.
8. Morgan JF, Reid F, Lacey JH. The SCOFF questionnaire: assessment of a new screening tool for eating disorders. *BMJ* 1999;319:1467–1468.

9. Garner DM, Olmsted MP, Bohr Y, et al. The eating attitudes test: psychometric features and clinical correlates. *Psychol Med* 1982;9:273–279.
10. Thelen MH, Mintz LB, Wal JSV. The Bulimia Test—Revised: validation with DSM-IV Criteria for Bulimia Nervosa. *Psychol Assess* 1996;8:219–221.
11. Thelen MH, Farmer J, Wonderlich S, et al. A revision of the Bulimia Test: the BULIT-R. *Psychol Assess* 1991;3:119–124.
12. American Academy of Pediatrics Committee on Sports Medicine and Fitness. Medical concerns in the female athlete. *Pediatrics* 2000;106:610–613.
13. Herring SA, Bergfeld JA, Boyajiian-O'Neill LA, et al. Female athlete issues for the team physician: a consensus statement. *Med Sci Sports Exerc* 2003;35:1785–1793.
14. Myers A, Sickles T. Preparticipation sports examination. *Prim Care* 1998;25:225–236.
15. Tanner SM. Preparticipation examination targeted for the female athlete. *Clin Sports Med* 1994;13:337–353.
16. Sanborn CF, Hosea M, Siemers BJ, et al. Disordered eating and the female athlete triad. *Clin Sports Med* 2000;19:199–213.
17. Johnson, M. Tailoring the preparticipation exam to female athletes. *Phys Sportsmed* 1992;20(7):61–72.
18. Beals KA. Eating disorder and menstrual dysfunction screening, education and treatment programs: survey results from NCAA Division I schools. *Phys Sportsmed* 2003;31:33–38.
19. Rumball JS, Leburn CM. Use of the preparticipation physical examination (PPE) form to screen for the "female athlete triad" in Canadian Intercollegiate Athletics Union (CIAU) universities [abstract]. *Clin Sports Med* 2000;11:291–292.
20. Klump KL, Kaye WH, Strober M. The evolving genetic foundations of eating disorders. *Psychiatr Clin North Am* 2001;24:215–225.
21. Kendler KS. Twin studies of psychiatric illness: an update. *Arch Gen Psychiatry* 2001; 58:1005–1014.
22. Wade T, Martin NG, Tiggerman M. Genetic and environmental risk factors for the weight and shape concerns characteristic of bulimia nervosa. *Psychol Med* 1998;28:761–771.
23. Abraham S. Dieting, body weight, body image and self-esteem in young women: doctors' dilemmas. *Med J Aust* 2003;178(12):607–611.
24. Kendler KS, MacLean C, Neale M, et al. The genetic epidemiology of bulimia nervosa. *Am J Psychiatry* 1991;148:1627–1637.

8 Education and Intervention Programs for Disordered Eating in the Active Female

Jacalyn J. Robert-McComb

CONTENTS

    8.1   LEARNING OBJECTIVES
    8.2   INTRODUCTION
    8.3   RESEARCH FINDINGS
    8.4   SUMMARY
    8.5   SCENARIO WITH QUESTIONS AND ANSWERS

8.1. LEARNING OBJECTIVES

After completing this chapter, you should have an understanding of the following:

- The importance of self-esteem in the prevention and treatment of disordered eating.
- The Transtheoretical Model (TTM) used to explain the process of health behavior change.
- Athletes@Risk® prevention program.
- General treatment principles.
- Medical criteria for hospitalization.
- Eating disorder organizations and resources.

8.2. INTRODUCTION

Low self-esteem, accompanied with perfectionism, is a well-recognized trait of those with disordered eating and could be a precipitating factor in the development of eating disorders *(1–2)*. The terms self-concept, self-worth, and self-image are often used interchangeably with self-esteem and are all based on self-perception. Self-esteem can defined as the extent to which a person feels positive about himself or herself *(3)*. Often, self-esteem is described as dichotomous, with a person possessing either high or low self-esteem. People with low self-esteem feel positive about themselves when they encounter affirmative experiences. Conversely, when they face negative experiences, they are disapproving of themselves. People with high self-esteem embrace and benefit

From: *The Active Female*
Edited by: J. J. Robert-McComb, R. Norman, and M. Zumwalt © Humana Press, Totowa, NJ

from positive experiences and have developed strategies to mollify negative feedback. In short, they have learned how to "offset" negative experiences *(4–5)*.

Effective prevention and intervention programs must recognize the importance of enhanced self-esteem and embrace the development of this concept in their programs. Education to prevent disordered eating can only be effective if the individual understands and accepts herself, even her limitations. Many times, physically active women have a heightened awareness of the body and its limitations *(5)*. These limitations may contribute to low self-esteem (loosing a race, finishing last, etc.). Rosenberg *(6)* describes self-esteem as consisting of three major components: (1) social identities (how an individual defines him or herself in society); (2) personal dispositions (perceptions of traits, preferences, and response tendencies); and (3) physical characteristics (height, weight, body fat distribution, attractiveness, etc.). Although there are many models for self-concept, all models recognize that the development of positive self-esteem is multidimensional and an individual's perception of self can be affected by the social, emotional, and physical involvement in sport and exercise.

8.3. RESEARCH FINDINGS

8.3.1. The Transtheoretical Model Used to Explain the Process of Health Behavior Change

Numerous theoretical frameworks have been proposed to explain and predict the process of health behavior change. One frequently used is the TTM developed by Prochaska and DiClemente as a model of intentional behavior change *(7)*. The TTM assumes that individuals vary in motivation and readiness to change their behavior, and as well, realistically acknowledges that relapse is normal under situations that involve such significant behavior change. Four related concepts considered central to health behavior change are included within this model: stage of change, self-efficacy, decisional balance, and the processes of change.

The five stages of change are precontemplation (PC, no intention to change health behaviors within the next 6 months); contemplation (C, seriously considering behavioral change within the next 6 months); preparation (P, still lack commitment to change, but investigating the possibility of change within the next 30 days); action (A, actively modifying problematic behavior within the last 6 months); and maintenance (M, self-control of the behavior established more than 6 months ago).

In the PC stage, information needs to be provided about the behavioral and potential medical problems associated with the behavior. In the C stage, health professionals must help individuals assess the pros and cons of the behavior change so that they will make a commitment to change. Health professionals must encourage initial small steps to initiate change, no one can force someone to change, and individuals must begin to place greater significance on the benefits of behavior change in the P stage. The A stage occurs when benefits outweigh the costs. Initiating a new health behavior change is bound to be fraught with relapse, and health professionals should act to reinforce an individual's self-confidence along with their decision to change. The M stage occurs when individuals are able to continue the new behavior. Conceptually, progression through these stages during attempts at behavioral change is expected to be linked to differences in self-efficacy, decisional balance, and the processes of change *(8)*.

Self-efficacy and decisional balance represent beliefs about behavior that are common to many social cognition models. Specifically, self-efficacy refers to an individual's confidence in his or her ability to perform a specific behavior, which is expected to increase as an individual moves through the stages *(9)*. Indeed, self-efficacy for health behavior change reliably predicted stages of change; precontemplators and contemplators had the lowest efficacy although those in the maintenance stage exhibited the highest efficacy *(10)*.

Decisional balance relates to the pros (benefits) and cons (costs) of the behavior, for example, eating gives me more energy, yet it might make me fat. Individuals who change their behaviors have positive decisional balance because the positive beliefs about the behavior outweigh the negative ones. Additionally, the pros increase although the cons decrease across the stages of change.

Last, the TTM also includes processes of change and specifically define a process of change as a "type of activity that is initiated or experienced by an individual in modifying affect, behavior, cognition or relationships" *(7)*. Health professionals can assist the process of change and the maintenance of the new behavior by providing follow-up support *(11)*.

8.3.2. Athletes @ Risk® Program

Targeted educational programs for female athletes can be effective in any of the stages of change. The Athletes@Risk® program is a preventative educational program for female athletes in both recreational and competitive sport who are at risk for developing disordered eating, amenorrhea, and osteoporosis (Table 8.1). The program has been designed for health professionals who interact with female athletes who are at risk for developing the Female Athlete Triad and who may be prone to anxiety disorders, who have frequent muscular injuries, low self-esteem, and problems coping with lifestyle stressors.*

8.3.3. General Treatment Principles

Treating eating disorders is both a science and an art *(12)*. Even the treatment goals for anorexia nervosa (AN) and bulimia nervosa (BN) are well defined (Table 8.2), the method of achieving these goals is less certain *(13)*.

Achieving and maintaining a normal weight and nutritional status is essential for recovery for both AN and BN. Perfectionist attitudes, low self-esteem, unrelenting pursuit of thinness, intolerance of mood fluctuations, and poor coping skills are problematic for both AN and BN. Recovery is unlikely without a fundamental change in these attitudes.

---

*For further information about the program you can access http://www.sunnybrookand womens.on.ca/programs/wcacc/clinicsandservices/sportcare or you may call Sport C.A.R.E. at Sunnybrook & Women's College Health Sciences Centre at (416) 323-7527 or 1-800-363-9353 (North America), or by fax at (416) 323-6153.

Table 8.1
Overview of the components in the Athletes@Risk® program developed by Dr. Julia Alleyne at the University of Toronto

Session no. 1:
  Understanding the health consequences
  The Female Athlete Triad
  Prevention, recognition, and treatment
  The continuum between disordered eating and an eating disorder
  Prevention and physical consequences
  Exploring osteoporosis in the younger women
  Identification and decreasing the risk
  Menstruation and the female athlete
    Interactive tools:
      Are you at risk?
      Commonly asked questions
      Word search
Session no. 2: my body, my sport
  Body image and self-esteem
  The genetics of body shape
  Body image and shape through history and culture
  The truth about body composition testing
  Body image and injury: dealing with the scars
  Healthy self-esteem strategies for teens
    Interactive tools:
      Body contour rating
      Determining body esteem
      Body image diary
      Developing self-talk: being positive and realistic
      Athletes response to injury
      Simple relaxation techniques
Session no. 3: getting strong, getting fit
  Stretch and strength: keeping balanced
  How do we get stronger
  Women and strength training
  The seasons of strength training
  Additional benefits of getting stronger
  Getting fit
  Aerobic, anaerobic, and interval training
  Performance aids: what is safe and what is not?
  Overtraining: signs and symptoms
  Injury prevention and treatment principles
  Female-specific issues in physical activity
    Interactive tools:
      Strength training on the go
      Heart smart learning
      Performance aids challenge
      Injury management game
      Female-specific issues crossword
Session no. 4: Food as fuel
  Nutrition: the basics
  Using the food guide for good nutrition

Getting enough fuel for activities
How much is enough?
What happens if needs are not met?
Eating for performance: before, during, and after
Why do we eat?
  Interactive tools:
    Healthy choices at fast food restaurants
    Food and water log
    Analyze a food label
    Nutrition jeopardy
    Olympic quiz
Session no. 5: Life skills and wellness
  Stress management
  Coping with stress
  Perception and control
  Self-esteem and stress management
  Developing healthy sexuality
  Harassment and abuse
  Boundary setting
  Physical, social, and sexual boundaries
  Assertiveness training
    Interactive tools:
      Are you stressed?
      Simple relaxation techniques
      Are you non-assertive?
      Developing assertive behavior

Note: Athletes@Risk® is a licensed preventive education program. For more information, access http://www.sunnybrookandwomens.on.ca/programs/wcacc/clinicsandservices/sportcare.

Research has shown that the therapist should function in the therapeutic encounter as a parent, teacher, guide, and coach and that the personality of the therapist is a major therapeutic element in the treatment of patients with AN *(13)*. The therapist must make every effort to engage the family, especially if treating patients with AN under 18 years of age *(14)*. Family therapy has shown to be very effective in younger patients with AN of shorter duration (less than 2 years).

Although most patients with BN can be treated in an outpatient setting, the first issue a clinician must decide with AN is the treatment setting. Most clinicians will recommend inpatient treatment for a patient who weighs less than 75% of average weight, has severe metabolic disturbances, if feeling suicidal, or has failed to improve after a period of outpatient or partial program treatment *(15)*. Table 8.3 lists the medical criteria for hospitalization established by the American Psychiatric Association *(16)*.

Traditionally, inpatient treatment is continued until a patient reaches a reasonable healthy body weight. Discussing a target weight is one of the most important initial tasks of weight restoration. Currently, most clinicians use a body mass index of $18.5 \, kg/m^2$ as the minimal healthy weight for a patient older than 16 years *(13)*.

The treatment of BN is usually conducted in an outpatient setting. Cognitive behavioral therapy (CBT) is the treatment of choice and is effective whether conducted

Table 8.2
Treatment goals for anorexia nervosa and bulimia nervosa

Anorexia nervosa	Bulimia nervosa
Restore weight and improve eating habits	Identify the factors and processes that maintain the binge-purge-starve cycle
Change dysfunctional attitudes	
Treat concomitant medical complications	Help the individual identify strategies to overcome the disturbed eating pattern
Work with the family	Change dysfunctional thoughts
Pharmacotherapy to treat depression, moderate obsessive perfectionism, to target certain neurotransmitters, or to treat concomitant complications	Build coping skills
	Note: Cognitive behavioral therapy (CBT) has been found to be most effective. However, self-help (SH), using a written manual based on the principles of CBT, has gained wide appeal as the patient may use it with or without the guidance of a therapist
Prevent relapse (most difficult task)	

Table 8.3
Medical criteria for hospitalization

	Adults	Children and Adults
Heart rate	< 40 bpm	In 40s
Blood pressure	< 90/60 mmHg	< 80/60 mmHg; orthostatic blood pressure changes (> 20 bpm increase in heart rate or > 10–20 mmHg drop in blood pressure)
Glucose	< 60 mg/dL	< 60 mg/dL
Electrolytes	Potassium < 3 meq/l; Electrolyte imbalance; dehydration	Hypokalemia or hypophosphatemia, dehydration
Temperature	< 97.0 °F	< 97.0 °F
Hepatic, renal, or cardiovascular	Abnormal laboratory profiles, ECG abnormalities	Abnormal laboratory profiles, ECG abnormalities
Weight	< 75% of healthy body weight	Acute weight decline with food refusal even if not < 75% of healthy body weight

Motivation to recover	Very poor to poor; preoccupied with egosyntonic thoughts	Very poor to poor; preoccupied with egosyntonic thoughts
Comorbid disorders	Any existing psychiatric disorder that would require hospitalization	Any existing psychiatric disorder that would require hospitalization
Purging behavior (laxatives and diuretics)	Needs supervision during and after all meals and in bathrooms	Needs supervision during and after all meals and in bathrooms
Ability to care for self; ability to control Exercise	Complete role impairment; structure required to keep patient from Compulsive exercising	Complete role impairment; structure required to keep patient from Compulsive exercising
Environmental stress	Severe family conflict, unable to provide structured treatment in home, lives alone without an adequate support system	Severe family conflict, unable to provide structured treatment in home, lives alone without an adequate support system

Source: American Psychiatric Association (16). With permission.

individually (17) or in a group setting (18). The most widely used CBT is the version developed by Fairburn and is implemented over a period of 18 weeks (19). In the first stage, behavioral techniques are used to replace binge eating with a stable pattern of regular eating. In the second stage the goal is to eliminate dieting, the focus is on the thoughts, beliefs, and values that reinforce dieting. The third stage is focused on the maintenance of these new healthy behaviors and thought patterns.

Self-help (SH), using a written-based manual based on the principles of CBT, has been shown to be effective for BN patients and is more accessible than CBT (20). The patient may use it without any guidance or with the help of a therapist. Under guidance, the program usually consists of seven sessions and is conducted over a period of 12 weeks. The most widely used SH manual, written by Fairburn, provides a step-by-step discussion of implementing the program (21).

8.3.4. Eating Disorder Organizations and Resources

Table 8.4 lists Web resources for the treatment of eating disorders that are targeted for physicians.

Appendix 9 lists additional educational resources for the prevention and treatment of eating disorders and other related mental illnesses.[†]

[†]For more information about the National Mental Health Association or additional resources, please call 1-800-969-NMHA (6642) or visit their website at http://www.nmha.org/infoctr/index.cfm.

Table 8.4
Resources for physicians for the treatment of eating disorders

The Academy for Eating Disorders (AED)
3728 Old McLean Village Dr,
McLean, VA 22101.
Phone: (703) 556-9222
Fax: (703) 556-8729
Web site: http://www.aedweb.org
E-mail: aed@degnon.org
Back issues of the AED newsletter are available in the Newsletters area of the site. A useful tool for educating patients and their families can be accessed by clicking on the "For the Public", link on the home page. Physicians can search for professionals with experience in treating those with eating disorders in the Members Only section of the site. The member directory can be searched by practice parameters, discipline, geographic area, or last name.

American Psychiatric Association (APA)
1400 K St, NW,
Washington, DC 20005.
Phone: (202) 682-6000
Fax: (202) 682-6850
Web site: http://www.psych.org/psychpract/treatg/pg/pracguide.cfm
E-mail: apa@psych.org
This site features tools such as Practice Guidelines and a Quick Reference Guide for Eating Disorders. From their homepage you can access these tools by using the site's web search engine or by clicking on the Psychiatric Practice link which will direct you to http://www.psych.org/psychpract/treatg/pg/pracguide.cfm.

Internet Mental Health (IMH)
601 W Broadway Suite 902,
Vancouver, BC,
Canada, V5Z4C2.
Phone: (604) 876-2254
Fax: (604) 876-4929
Web site: http://www.mentalhealth.com
This site features descriptions of anorexia and bulimia, as well as the latest research on those disorders. A news area, called the "Magazine," is also available, as are links to other Internet eating disorders resources, including Websites and a discussion group.

National Association of Anorexia and Associated Disorders (ANAD)
PO Box 7, Highland Park, IL 60035.
Hotline: (847) 831-3438
Fax: (847) 433-4632
Website: http://www.anad.org
E-mail: anad20@aol.com
A page with eating disorders information appropriate for distribution to patients including the results of a study run by ANAD can be accessed through the "Information" link on the home page. Customized packets containing additional material can be ordered by clicking on the designated home page links. The site also contains an area dedicated to legislative news relevant to

mental, health and eating disorders, and an area that discusses insurance fraud. A list of referrals and support groups is available by calling the hotline.

National Eating Disorders Association [Formerly Eating Disorders, Awareness, and Prevention (EDAP) and American Anorexia Bulimia Association (AABA)]
603 Stewart St. Suite 803,
Seattle, WA 98101.
Phone: (206) 382 3587 or
(800) 931-2237 (toll-free information and referral hotline)
Fax: (206) 829-8501
Web: http://www.edap.org
E-mail: info@NationalEatingDisorders.org
An order form for patient education materials that can be printed out and faxed to EDAP is accessible through the "Materials" link. The ED area has fact sheets on different topics relevant to eating disorders that can be printed out and given to patients. Subjects include general information, how to help a friend, and men and eating disorders. A resource list is also available that features a reading list, video resources, links to related sites, and additional resources on subjects such as body image and dieting. Physicians can obtain referrals by calling the toll-free referral hotline.

Something Fishy
Website: http://www.something-fishy.org
News, descriptions of eating disorders, and a treatment finder are all offered on this extensive site. Online support through chat rooms, message boards, and American Online Instant Messaging (AIM) is also available. Sections on dangers associated with eating disorders, helping loved ones, recovery, and cultural issues provide useful information not always seen on other Web sites. "Doctors and Patients" is a section that provides a practical discussion of medications as well as blood and lab tests. An area entitled "Tips for Doctors" dispenses functional advice about what patients fear most. Their concerns, for example, include not being taken seriously by a physician or that the physician will notify their parents. The additional resources area lists organizations, other Web sites, and hotline numbers; recommends written material; and provides links to research.

Source: From Patient Care. Identifying and managing eating disorders. November 30, 2001 (www.patientcareonline.com/patcare/). With permission.

8.4. SUMMARY

Practice guidelines for the treatment of patients with eating disorders have been developed by psychiatrists who are in active clinical practice and are available on the Web at http://www.psych.org/psych_pract/treatg/pg/eating_revisebook_index.cfm?pf=y. These guidelines were approved by the American Psychiatric Association in 1999 and published in 2000 *(16)*. These guidelines are not intended to serve as a standard of medical care but rather provide recommendations for treating patients with eating disorders.

Theoretical perspectives differ in the treatment of eating disorders, and the interplay among the many treatment variables is very complex and not well understood. However, regardless of theoretical perspectives, it seems that the personality of the health professional and the therapeutic relationship developed between the health professional and patient is an important element in recovery *(13)*.

8.5. SCENARIO WITH QUESTIONS AND ANSWERS

8.5.1. Scenario

You are a team physician for a women's high school cross-country team. You have noted that many of the women report menstrual disturbances in their health history questionnaire. In light of the TTM of intentional behavior change developed by Prochaska and DiClemente *(7)*, you have decided to propose that the school counselor adopts the Athletes@Risk Program developed by Dr. Julia Alleyne at the University of Toronto.

8.5.2. Questions

1. Why would adoption of this program be a wise decision for women who may have menstrual disturbances?
2. Would you hypothesize that the female athletes would be in the same stage of change and would respond uniformly to the presentation of the materials?

8.5.3. Plausible Answers

1. Menstrual disturbances, particularly in young female athletes, can be caused by an energy imbalance and/or by psychological stress related to high performance. The objectives of the program is to educate female athletes about (1) the health consequences of the triad, (2) healthy eating habits, (3) positive self-esteem and body image, (4) safe training practices, and (5) stress management. These educational objectives would at least provide the foundation for the decision to make a behavior change based on informative health information.
2. No, conceptually, progression through these stages during attempts at behavioral change is expected to be linked to differences in self-efficacy, decisional balance, and the processes of change. However, information needs to be provided about the behavioral and potential medical problems associated with the behavior in the first stage of behavior change. The process of change would be highly individualized for each athlete.

REFERENCES

1. Silverstone PH. Is chronic low self-esteem the cause of eating disorders? *Med Hypotheses* 1992;39:311–315.
2. Thompson RA, Sherman RT. *Helping Athletes with Eating Disorders*. Champaign, IL: Human Kinetics, 1993.
3. Sonstroem RJ. Exercise and self-esteem. In: Terjung RL, ed. *Exercise and Sports Science Reviews*, vol. 12. Lexington, MA: The Collamore Press, 1984:123–155.
4. Brown JD, Mankowski TA. Self-esteem, mood, and self-evaluations: changes in mood and the way you see you. *J Pers Soc Psychol* 1993;64(3):421–430.
5. Lindeman A. Self-esteem: its application to eating disorders and athletes. Int J Sport Nutr 1994;4:237–252.
6. Rosenberg M. *Society and the Adolescent Self-Image*. Princeton, NJ: Princeton University Press, 1965.

7. Prochaska J, DiClemente C. Toward a comprehensive model of change. In: Miller WE, Heather N, eds. *Treating Addictive Behaviors*. London: Plenum Press, 1998:3–27.
8. Prochaska J, Marcus B. The transtheoretical model: applications to exercise. In: Dishman RK, ed. *Advances in Exercise Adherence*. Champaign, IL: Human Kinetics, 1994:161–180.
9. Marcus B, Selby V, Niaura R, et al. Self-efficacy and the stages of exercise behavior change. *Res Q Exerc Sport* 1992;63:63–66.
10. Marcus B, Owen N. Motivational readiness; self-efficacy and decision making for exercise. *J Appl Soc Psychol* 1992;22:3–16.
11. Bass M, Turner L, Hunt S. Counseling female athletes: application of the stages of change model to avoid disordered eating, amenorrhea, and osteoporosis. *Psychol Rep* 2001;88:1153–1160.
12. Lucas AR. *Demystifying Anorexia Nervosa*. New York, NY: Oxford University Press, 2004.
13. Hsu LK. Eating disorders: practical interventions. *Womens Health* 2004;59:113–124.
14. Lock J. Treating adolescents with eating disorders in the family context. *Child Adolesc Psychiatr Clin N Am* 2002;11:331–342.
15. American Psychiatric Association. Practice guidelines for eating disorders (revision). *Am J Psychiatry* 2000;57:Suppl:1–39.
16. American Psychiatric Association. *Practice Guidelines for the Treatment of Patients with Eating Disorders*, 2nd edition. Washington, DC: American Psychiatric Press; 2000;57:5–56.
17. Agras WS, Walsh BT, Fairburn CG, et al. A multicenter comparison of CBT and IPT for patients with BN. *Arch Gen Psychiatry* 2000;57:459–466.
18. Chen E. Touyz SW, Beumont PJ, et al. Comparison of group and individual CBT for patients with BN. *Int J Eat Disord* 2003;33:241–254.
19. Fairburn CG, Marcus MD, Wilson GT. CBT for bulimia nervosa: a treatment manual. In: Fairburn CG, Wilson GT, eds. *Binge Eating*. New York, NY: Guilford Press, 1993:361–404.
20. Kaye WH, Nagata T, Weltzin TE, et al. Double blind placebo controlled administrates of fluoxetine in restrictive type AN. *Br J Psychiatry* 2000;57:459–466.
21. Fairburn CG. *Overcoming Binge Eating*. New York, NY: Guilford Press, 1995.

III Reproductive Health

9 The Human Menstrual Cycle

Reid Norman

CONTENTS

9.1 LEARNING OBJECTIVES
9.2 INTRODUCTION
9.3 RESEARCH FINDINGS
9.4 CONCLUSIONS
9.5 SCENARIO WITH QUESTIONS AND ANSWERS

9.1. LEARNING OBJECTIVES

After completing this chapter, you should have an understanding of the following:

- The hormonal changes that occur throughout a menstrual cycle.
- Reproductive hormones and where they are produced.
- How secretion of pituitary and ovarian hormones is regulated.
- How birth control pills work.

9.2. INTRODUCTION

Menstruation is caused by a complex series of hormonal events that lead to the growth and demise of the lining of the uterus and the eventual sloughing off of the uterine lining and discharge through the vagina (menstruation). Ultimately, the menstrual cycle is a reflection of cyclic changes in secretion of steroid hormones (estrogen and progesterone) from the ovary. The ovarian cycle in women and other female primates is coordinated by a series of chemical signals between the brain, pituitary, and ovary. The most conspicuous sign that occurs as a result of these events is menstruation or the monthly (usually) shedding of blood and cellular debris through the vagina. Generally, regular menstruation is a sign of regular ovulation, and menstrual cycles that are prolonged or not regular suggest that ovarian cycles are disrupted. Not only does the hypothalamic-pituitary-ovarian axis need to function normally, but malfunction of other endocrine organs such as the thyroid and adrenal can also result in menstrual cycle disturbances.

The environment within the body (internal milieu) is under tight homeostatic regulation and, in the absence of genetic defects or disease, is kept exceedingly

From: *The Active Female*
Edited by: J. J. Robert-McComb, R. Norman, and M. Zumwalt © Humana Press, Totowa, NJ

constant. It is this homeostatic regulation that allows us to live in a widely fluctuating external environment with relative ease. However, there are environmental factors that can negatively impact reproduction, the foremost of these being availability of food *(1)*, and reduced energy availability (nutritional stress) is thought to be the main factor in menstrual irregularity *(2,3)*. Along with nutritional stress, psychological and physical stress may also influence reproductive cycles. In times of social stress, such as war, menstrual cycle disturbances may approach 50% of sexually mature women *(4)*.

9.3. RESEARCH FINDINGS

9.3.1. How is the Brain Involved in Reproduction?

At the transition between childhood and adulthood, the initial signal that activates (or reawakens) the reproductive system at puberty is thought to originate in the brain. It is not clear what event that actually initiates the process, but it is the increased secretion of gonadotropin-releasing hormone (GnRH, also known as luteinizing hormone releasing hormone or LHRH) from a few thousand neurons in the hypothalamus that begins the process of sexual maturation. GnRH is a small peptide (10 amino acids in length) produced in neurons and released from the base of the brain in pulses into a set of specialized vessels, the pituitary portal system, and travels a very short distance to the pituitary gland. These pulses of GnRH occur at about hourly intervals during the first half of the menstrual cycle and cause the release of LH and follicle stimulating hormone (FSH) from the pituitary. We cannot reliably measure GnRH in the peripheral circulation, but there is a very high coincidence of LH pulses in the circulation that follow GnRH pulses in the portal system *(5,6)*. Therefore, because LH secretion is a faithful indicator of GnRH release, we can analyze GnRH secretion, in part, by measuring LH in the peripheral circulation. Even so, it has been difficult to determine whether the amplitude of GnRH secretion changes throughout the cycle, but there is general agreement that the frequency of GnRH release is highest in the follicular phase and slows down in the luteal phase when progesterone is present.

9.3.2. Pituitary Hormones: Luteinizing Hormone and Follicle Stimulating Hormone

LH and FSH are secreted from the cell type called the gonadotroph in the pituitary. This gonadotroph cell can secrete either FSH or LH or both. FSH and LH are protein hormones composed of amino acids and sugars. They are released from the pituitary and travel through the circulatory system where they bind to specific receptors on the cell surface of specific cells in the ovary in females and in the testis in males. The secretion of FSH is regulated by GnRH, estrogen, progesterone, activin, and inhibin. Activin stimulates and inhibin suppresses FSH release. Activin and inhibin are protein hormones secreted by the ovary. Secretion of LH is regulated primarily by GnRH, estrogen, and progesterone. Both FSH and LH can easily be measured in the peripheral circulation, and after puberty, the circulating concentration of these hormones is a useful diagnostic tool for the clinician.

9.3.3. The Ovaries Release Steroids and Ova

The ovaries are located in the pelvis, a long distance from the pituitary. LH and FSH are released from the pituitary and travel through the circulatory system, and although most cells in the body are exposed to these hormones, they only bind to those cells that have the specific LH and FSH receptors on their surface. How these hormones interact with cells in the ovary is an intricate and fascinating story, but too detailed to describe here. However, the main points regarding the cyclic changes in the ovary are important for understanding the menstrual cycle. At the beginning of the cycle (day 1 of menstruation), FSH is elevated and the increase in this hormone stimulates several follicles (8–10) to begin their final stages of growth and development for ovulation. After a few days, one of these follicles becomes the dominant follicle and is destined to be the one that will ovulate or release the mature ova into the Fallopian tube. As the dominant follicle grows, cells within that structure called granulosa cells release estrogen into the circulation, and estrogen levels increase throughout the first 14 days of the cycle (follicular phase). Estrogen causes growth of the lining of the uterus (endometrium) in preparation for possible implantation of the developing embryo if fertilization occurs. High levels of estrogen are produced when the follicle is ready to ovulate, and this is the signal to the brain and pituitary to release LH to cause ovulation. When the ovum is released from the follicle, the granulosa cells within the follicle transform into luteal cells under the influence of LH. The luteal (yellow) cells primarily secrete progesterone, the dominant hormone during the luteal phase. Progesterone from the corpus luteum causes additional changes in the uterus in preparation for implantation of the embryo and is a crucial hormone for the maintenance of pregnancy.

9.3.4. The Uterus and Fallopian Tubes

The uterus (womb) is located in the pelvis and is the site of implantation of the fertilized egg after it develops into a blastocyst. The uterus is composed of an outer muscular layer (myometrium) and a lining (endometrium) that undergoes dramatic changes in response to estrogen and progesterone during the menstrual cycle. At the end of the menstrual cycle, the endometrium degenerates and is shed through the vagina as the menstrual flow. As the new cycle begins and the follicle begins to secrete estrogen, the endometrium begins to proliferate and thicken. Hence, the follicular phase of the cycle is also referred to as the proliferative phase. After ovulation, the endometrium becomes secretory in the final preparation for implantation and pregnancy. The luteal phase is sometimes called the secretory phase. If fertilization and implantation do not occur, progesterone and estrogen secretion diminishes and the endometrium degenerates and is shed. A new crop of follicles begins to develop and the cycle begins again.

9.3.5. Feedback Relationships

The hypothalamus, pituitary, and ovary all release chemical messengers (reproductive hormones) and communicate with each other with these hormones. The control of the ovarian cycle is complicated, but a few key points regarding the feedback relationships between the pituitary and ovarian hormones are important for understanding how the cycles are maintained. As we stated before, at the beginning of the

```
                    HYPOTHALAMUS  ◄─────────┐
                         │                   │
                       (GnRH)                │
                         │                   │
                         ▼              negative
           ┌──────►  PITUITARY  ◄────     and
           │             │              positive
           │         (LH, FSH)          feedback
      negative          │                   │
      feedback          ▼                   │
           │          OVARY                 │
           │        (Estradiol,             │
           │        Progesterone) ──────────┘
           └──────── (Inhibin)
```

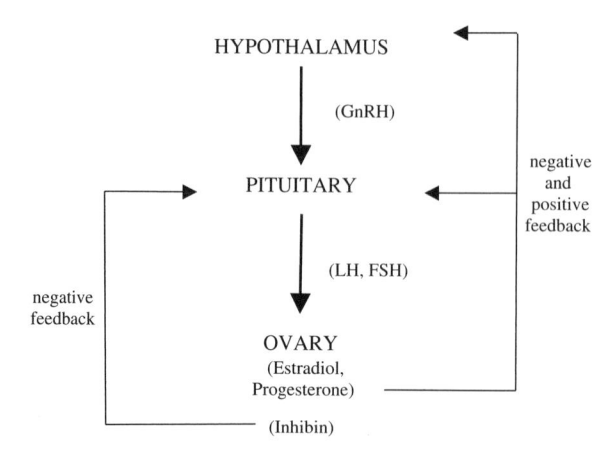

Fig. 9.1. This figure shows the feedback relationships between the ovarian hormones and the hypothalamic-pituitary axis. Steroid (estrogen and progesterone) and protein (inhibin) hormones produced by the ovary regulate the secretion of gonadotropin-releasing hormone (GnRH) from the hypothalamus and luteinizing hormone (LH) and follicle stimulating hormone (FSH) from the pituitary to provide the appropriate stimulation for follicular development and ovulation. These feedback relationships are the basis for cyclic ovarian function.

cycle, a few follicles begin to develop under the stimulation of FSH. FSH is elevated at the beginning of the cycle because estrogen, progesterone, and inhibin levels are low. These hormones from the ovary control FSH secretion, and if the circulating levels of these hormones are low, FSH rises. As the ovarian follicle begins to mature, estrogen and inhibin levels gradually increase in the circulation and maintains the release of both FSH and LH from the pituitary at a level that will continue the process of follicular maturation. This relationship where LH and FSH are kept in check by the ovarian hormones they stimulate is called negative feedback (Fig. 9.1). Around day 12 of the cycle when the follicle matures and is ready to release the ovum, it produces high levels of estrogen which increase blood levels of estrogen sufficient to trigger the midcycle or ovulatory release of LH. This ovulatory surge of LH causes the follicle to rupture and release the ovum. This release of LH by high levels of estrogen is called positive feedback and only occurs at midcycle when estrogen levels have exceeded a threshold level for several hours. After ovulation, the corpus luteum (yellow body) secretes progesterone, estrogen, and inhibin, and these hormones keep LH and FSH secretion in check. The corpus luteum has about a 2-week lifespan, and unless a pregnancy is established, the corpus luteum will die. At the end of the cycle, estrogen and progesterone fall to low levels, and this will release the negative feedback inhibition, FSH levels in the blood increase, and recruit a new group of follicles to begin a new cycle (Fig. 9.2 for hormonal changes throughout the cycle). These negative and positive feedback signals keep the reproductive hormones in balance and are the basis by which the ovarian cycle is controlled.

9.3.6. Pregnancy and Lactation

When a woman becomes pregnant, menstrual cycles cease because elevated levels of estrogen and progesterone suppress pituitary LH and FSH release so that new

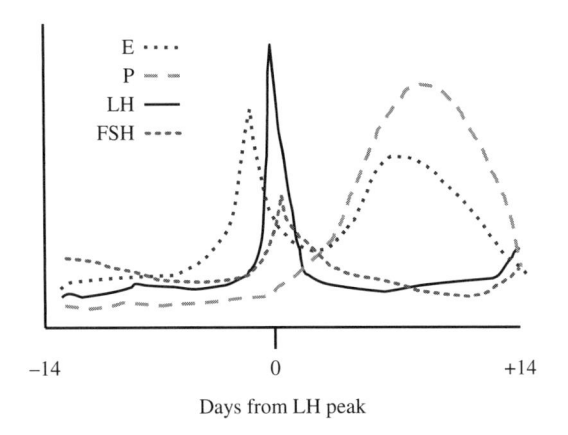

Fig. 9.2. Changing levels of pituitary and ovarian hormones throughout the menstrual cycle plotted in reference to the day of the ovulatory LH peak. E, estradiol; FSH, follicle stimulating hormone; LH, luteinizing hormon; and P, progesterone.

follicles do not mature. Levels of progesterone and estrogen are initially supported by a hormone from the embryo called human chorionic gonadotropin or hCG. This is the hormone that is measured in pregnancy tests. If, after a woman delivers, she nurses her child, the hormone prolactin is elevated and can also suppress follicular development as well. This is called lactational amenorrhea and is a common experience, particularly when the nutritional needs of the child are not supplemented, and nursing is the only source of energy. In primitive societies where children were nursed for several years, lactational amenorrhea served as an effective method of birth control *(7)*. However, it is not a reliable form of birth control and should not be used as such.

Regardless of self-reported sexual activity, pregnancy should always be considered when investigating the cause of amenorrhea in a woman of childbearing age because it is the most common cause of that condition.

9.3.7. Hormonal Birth Control

There are several different birth control methods that use different strategies for controlling fertility. One of these methods, hormonal birth control or oral contraceptive, uses synthetic steroid hormones and the inherent negative feedback system to suppress gonadotropin (LH and FSH) release. These hormonal methods inhibit growth and development of follicles so that ovulation and pregnancy do not occur. Birth control pills contain either a progestin (a compound that acts like progesterone) alone or a combination of an estrogen and a progestin. Progestin-alone pills do not block ovulation but affect the lining of the uterus so that there is a hostile environment for implantation and also retard sperm penetration into the uterus. The steroids in the combination pill suppress LH and FSH release and prevent follicular development and ovulation. The steroid regimen in the combination pill more closely mimics the changes in estrogen and progesterone that occur in the menstrual cycle than does the progestin-alone pill. These combination pills can be given so that women who want to have regular menstrual cycle can do so. Modern birth control pills have a very low risk of side effects and are very effective.

9.4. CONCLUSIONS

Regular menstrual cycles are the result of predictable cyclic changes in reproductive hormones produced by the pituitary and ovary. A few neurons at the base of the brain in the hypothalamus provide the ongoing stimulus, GnRH pulses, for the synthesis and secretion of LH and FSH which stimulate the ovary to produce the ovarian steroids, estrogen and progesterone, and the small peptides inhibin and activin. It is the feedback relationships between the pituitary and ovarian hormones that provide the basis for the cyclic function of the reproductive system. If fertilization occurs, there is the possibility of a pregnancy that interrupts the cycle for the duration of the pregnancy. Nursing or lactation can also suppress the ovarian cycles, but should not be relied on for birth control. The negative feedback relationship between the ovarian steroids and pituitary gonadotropins is the basis by which hormonal birth control works. Synthetic steroid hormones that can be administered orally suppress the secretion of LH and FSH and thus interrupt the growth and development of ovarian follicles.

9.5. SCENARIO WITH QUESTIONS AND ANSWERS

9.5.1. Scenario

You are the head coach for a college women's track team. One of the most dedicated young women in your college track team has, on her own initiative, recently increased her training schedule for the conference finals that will occur in about a month. In the past year, she has shown consistent improvement in her performance and has a good chance of taking first place in two of the events in which she competes. If she does perform well, your team has a good chance to win the conference title. This title would be a first for the college and would be a significant achievement in your career. Recently, this young woman confided to your assistant coach that she has stopped having menstrual periods.

9.5.2. Questions

1. Should you be concerned about this revelation?
2. Should you encourage the athlete to seek medical advice regarding this development or wait until after the conference finals?

9.5.3. Plausible Answers

1. Yes is the answer to both questions. You should always be concerned about changes in menstrual cycles in athletic women. It could be a result of increased intensity of exercise, but it could also indicate that the woman is pregnant. When a woman experiences changes in her menstrual cycle, she should see her doctor.
2. As the most common cause of amenorrhea in young women is pregnancy, the woman in question should be encouraged to see her physician.

REFERENCES

1. Bronson FH. *Mammalian Reproductive Biology*. Chicago: University of Chicago Press, 1989.
2. Loucks AB, Thuma JR. Luteinizing hormone pulsatility is disrupted at a threshold of energy availability in regularly menstruating women. *J Clin Endocrinol Metab* 2003;88:297–311.

3. Loucks AB, Redman LM. The effect of stress on menstrual function. *Trends Endocrinol Metab* 2004;15:466–471.
4. Whitacre FE, Barrera B. War amenorrhea. *JAMA* 1944;124:399–403.
5. Levine JE, Pau KY, Ramirez VD, Jackson, GL. Simultaneous measurement of luteinizing hormone-releasing hormone and luteinizing hormone release in unanesthetized, ovariectomized sheep. *Endocrinology* 1982;111:1449–1455.
6. Clarke IJ, Cummins JT. The temporal relationship between gonadotropin releasing hormone (GnRH) and luteinizing hormone (LH) secretion in ovariectomized ewe. *Endocrinology* 1982;111:1737–1739.
7. Short RV. Lactational infertility in family planning. *Ann Med* 1993;25:175–180.

10 Abnormal Menstrual Cycles

Reid Norman

CONTENTS

10.1 LEARNING OBJECTIVES
10.2 INTRODUCTION
10.3 RESEARCH FINDINGS
10.4 CONCLUSIONS
10.5 SCENARIO WITH QUESTIONS AND ANSWERS

10.1. LEARNING OBJECTIVES

After completing this chapter, you should have an understanding of the following:

- The terminology of normal and abnormal menstrual function.
- The causes of menstrual dysfunction.
- Consequences of menstrual dysfunction in young women.
- Health concerns because of menstrual dysfunction.

10.2. INTRODUCTION

Regular menstrual cycles in young women reflect cyclic ovarian activity and generally are associated with a healthy lifestyle. When a woman experiences a change from regular, cyclic menstrual function to irregular or acyclic function, that change is an indication of either a pregnancy or of some underlying pathology, and the cause for the change in cyclic function should be determined by a physician as soon as possible.

To communicate effectively with your peers and students regarding menstrual cycles, you should know and understand normal and abnormal menstrual function and the terminology describing that function. Although there is not complete agreement on the strict definition of all the terms, the terms described next are a reasonable approximation of these definitions (1). These words describing menstrual function are used widely in the scientific literature, but you should assume that the average lay person will probably not be aware of the meaning of these words. Therefore, to avoid misunderstanding and confusion, you should define what you mean when you use these terms in discussion of menstrual problems.

From: *The Active Female*
Edited by: J. J. Robert-McComb, R. Norman, and M. Zumwalt © Humana Press, Totowa, NJ

10.2.1. Eumenorrhea (Regular)

This refers to menstrual cycles that occur consistently at intervals of 25–38 days. These cycle lengths are generally observed in sexually mature women except during pregnancy and during the pubertal and the perimenopausal (around the time of menopause; this can be several years before the actual cessation of menstrual function) transitions when menstrual cycles are more variable.

10.2.2. Oligomenorrhea (Irregular)

This term refers to infrequent menses or menstrual cycles that occur inconsistently at intervals of 39–90 days.

10.2.3. Amenorrhea (Acyclic)

This is when menstrual cycles occur at intervals of greater than 90 days or when there is the complete absence of menstruation. There are two types of amenorrhea that are defined based on whether the woman has experienced previous menstrual cycles.

10.2.3.1. PRIMARY AMENORRHEA

This is when a young woman has not experienced menarche by age 16. In other words, a girl has primary amenorrhea if she has not had her first period by the age of 16.

10.2.3.2. SECONDARY AMENORRHEA

It is defined as the absence of a menstrual period for six consecutive times or months after menarche has occurred. It is estimated that as high as 5% of the adult women in the United States experience secondary amenorrhea (2). There are a number of adjectives to describe the causes of amenorrhea such as dietary, emotional, jogger, postpartum, and lactational (nursing), but these will be avoided for the most part. However, terms used by clinicians to describe amenorrhea caused by exercise or stress are functional or hypothalamic amenorrhea, and these are used widely in the clinical and scientific literature.

10.3. RESEARCH FINDINGS

10.3.1. Eumenorrhea

In Chap. 9, we have discussed the normal menstrual cycle.

10.3.2. Oligomenorrhea

This term actually means "few menses." Although very light menstruation is sometimes referred to as oligomenorrhea, medical professionals have a more narrow definition, applying this term only to the frequency of menstrual periods. There are many reasons for irregular menstruation, but the most frequent cause is what clinicians refer to as PCOS or polycystic ovary syndrome (3) that affects approximately 6% of women (4). Low energy availability is another cause of oligomenorrhea and amenorrhea (5) and is probably an important consideration for elite high school and college-aged female athletes, particularly in those sports where body image is considered important. Female athletes will often restrict their food intake to maintain a desired body image even though they are expending considerable energy training for their sport. This can result in an energy deficit (low energy availability) sufficient to inhibit menstrual function. Psychological stress is another factor to consider, although it is hard to separate from other causes.

10.3.2.1. PCOS Syndrome

PCOS syndrome was originally described by Stein and Leventhal *(6)*, and its pathophysiology is still poorly understood. This syndrome is associated with irregular menstruation, increased secretion of androgens, and the lack of ovulation (anovulation). The diagnosis of PCOS, as defined by the Rotterdam workshop *(7)*, requires two of the following three criteria: (1) infrequent or lack of ovulation, (2) clinical signs of excess androgen, and (3) polycystic ovaries with ultrasound. The typical symptoms that women who have this syndrome experience are irregular menstruation, infertility, and some evidence of increased androgen secretion or hyperandrogenism. Menstrual irregularity may be difficult to document without charting of the cycle for several months. This can easily be done by keeping track of the days on a calendar when bleeding occurs. Excessive bleeding can also be documented this way. Infertility is a complaint more associated with married women who wish to become pregnant and not generally a concern of single, college-aged women. Physical symptoms of increased androgen secretion in women are acne and/or oily skin and increased hair growth on the face, back, between the breasts, upper arms, upper and lower abdomen, and inner thighs *(8)*. However, a woman's perception of hirsutism (increased facial and body hair) may be altered if other women in the family or in the cultural community are hirsute, and therefore, this may not be of particular concern to her. Another concern regarding this syndrome is that it is associated with a high frequency of insulin resistance and increased risk for type 2 diabetes *(9)*, hypertension, and cardiovascular disease *(10)*. Because early diagnosis could result in amelioration or prevention of the serious consequences and financial burden of this disease *(1)*, it is recommended that women with menstrual cycle disturbances seek medical advice promptly.

10.3.2.2. Low Energy Availability

Low energy availability can result from several different behaviors, and often, there is more than one contributing factor to disrupted cycles resulting from reduced availability of energy. Disrupted menstrual function because of intense exercise, such as that found in elite or even highly competitive female athletes, coupled with restricted food intake is a situation that is not uncommon in high school and college athletes. The total amount of energy available is partitioned by the body (1) for basic metabolic needs necessary for survival, (2) for growth, and (3) for reproduction. If energy intake is restricted below the level needed to support all of these biological processes, those that are necessary for growth and reproduction slow or cease completely. This is an oversimplification of the idea that reproduction in animals is initiated only when there is sufficient energy to support the successful completion of that process *(11)*, but the basic idea is that reproduction in female mammals requires a lot of energy and that natural selection favors those females with the ability to allocate energy resources so that survival is not compromised by unsuccessful attempts to reproduce. To accomplish this task of successfully reproducing and surviving at the same time, animals have developed two capabilities: (1) they have some internal way to measure how much energy they take in and how much energy they are using and (2) they use environmental cues to predict the availability of food. This latter capability is particularly important to mammals that have a long pregnancy. Change in day length is one of the most common environmental factors animals used to predict the availability of food, and most animals give birth in the spring when food will be available for several months. However, in women, as discussed in Chap. 11, psychological stress is probably one of the most important environmental factors that affect reproduction.

10.3.2.3. OTHER CAUSES

Other causes of menstrual cycle disturbances are hypothyroidism, elevated levels of prolactin (hyperprolactinemia), Crohn's disease, eating disorders, and, sometimes, there is no apparent cause (idiopathic).

10.3.3. Amenorrhea

Amenorrhea literally means "the absence of menses." The complete absence of menstrual cycles in a nonpregnant, sexually mature woman is a symptom of something seriously wrong. As the most common cause for the sudden onset of amenorrhea is pregnancy, that possibility should always be ruled out as the first cause. It has been documented that as many as 5% of women of reproductive age at sometime experience amenorrhea *(2,12,13)*. However, girls and women engaged in athletics often welcome the absence of periods and do not consider this a serious medical issue. In addition to some of the aforementioned medical conditions that can cause disruption of menstrual cycles, there can also be serious long-term consequences of suppressed ovarian function.

High levels of prolactin or hyperprolactinemia can also cause amenorrhea. Lactational amenorrhea is common in women who supply most of the nutritional needs of their infant by breast-feeding. Pituitary tumors (prolactinomas) that secrete prolactin also suppress gonadotropin secretion which results in cessation of menstrual cycles. Approximately 50% of the individuals who have these tumors also have galactorrhea *(14)*, or secretion of milk from the breast, which often goes unnoticed. If the tumor is actively growing, it can progressively cause visual deficits, blurred vision, and headaches. If not treated, long-term consequences of hyperprolactinemia include osteoporosis because of suppression of ovarian steroid secretion. Medical management of prolactin-secreting tumors is effective using dopamine agonists that suppress the tumors and optimize prolactin levels.

Anorexia nervosa is probably the most recognizable cause of amenorrhea in young women who are not pregnant. Serious eating disorders, particularly anorexia nervosa and bulimia nervosa, are often associated with menstrual disorders. Both of these conditions have their highest incidence during adolescent years and can lead to death if not treated by appropriate professionals. In young female athletes, amenorrhea is often part of a syndrome called the Female Athlete Triad. This term was first used in the literature associated with sports medicine *(15)* to describe a condition in young female athletes who exhibited disordered eating, loss of bone (osteopenia), and amenorrhea.

Women at greatest risk for this condition are those training in sports where low body weight is desired or required, but it can occur in women involved in a wide range of physical activities. Reversible amenorrhea where there is no physical cause for the condition is called functional amenorrhea. This condition is the result of diminished or suppressed gonadotropin-releasing hormone release from the brain to the extent that there is not sufficient luteinizing hormone and follicle stimulating hormone to stimulate ovarian function (16). Because of low gonadotropin levels, follicles in the ovary do not develop, estrogen levels are low, and the lining of uterus (endometrium) does not proliferate. This endocrine situation is much like the prepubertal condition where menstruation does not occur.

Another possible cause of amenorrhea in young women is ovarian failure or gonadal dysgenesis. This is a rare condition, occurring in about 1 of 10,000 individuals.

10.4. CONCLUSIONS

Menstrual cycles are a reflection of cyclic ovarian function, and irregular or absent cycles may indicate some underlying pathology. Regardless of the cause, prolonged disruption of ovarian function can have serious consequences, even if there is no serious medical condition causing the abnormal cycles.

10.5. SCENARIO WITH QUESTIONS AND ANSWERS

10.5.1. Scenario

One of the athletes in your tennis team has told her trainer that her menstrual cycles recently became irregular and now have completely stopped. She is a highly competitive athlete, but this is the off-season and although she exercises regularly, it is not excessive. She also told her trainer that sometime her bra was moist, even when she was not exercising. The athlete is not underweight and denies sexual activity.

10.5.2. Questions

1. Should you discuss this issue of menstrual cycle disturbances with the athlete?
2. Should you recommend that she seeks medical attention?
3. What do you think is the cause of the amenorrhea?

10.5.3. Plausible Answers

1. You should tell this athlete (and all of your female athletes) that irregular menstruation or the lack of menstruation is a sign that their reproductive system in not functioning as it should and that this condition is potentially dangerous to her health.
2. Any time there is a change in menstrual cycles there is a cause for concern. This individual could be pregnant or have an underlying problem that could be serious and therefore should seek a medical help.
3. As the young woman has stopped menstruating and she is not exercising excessively, this is probably not an energy-related issue. She could have an eating disorder or possibly a prolactin-secreting pituitary tumor. She might have galactorrhea given that sometimes her bra is moist.

REFERENCES

1. Loucks AB, Horvath SM. Athletic amenorrhea: a review. *Med Sci Sports Exerc* 1985;17:56–72.
2. Peterson F, Fries H, Nillius SJ. Epidemiology of secondary amenorrhea. Incidence and prevalence rates. *Am J Obstet Gynecol* 1973;117:80–86.
3. Hull MG. Ovulation failure and induction. *Clin Obstet Gynecol* 1981;8:753–785.
4. Azziz R, Woods KS, Reyna R, et al. The prevalence and features of the polycystic ovary syndrome in an unselected population. *J Clin Endocrinol Metab* 2004;89:2745–2749.
5. Loucks AB, Thuma JR. Luteinizing hormone pulsatility is disrupted at a threshold of energy availability in regularly menstruating women. *J Clin Endocrinol Metab* 2003;88:297–311.
6. Stein I, Leventhal M. Amenorrhea associated with bilateral polycystic ovaries. *Am J Obstet Gynecol* 1935;29:181–191.
7. Rotterdam ESHRE/ ASRM-Sponsored PCOS Consensus Workshop. Revised 2003 consensus on diagnostic criteria and long-term health risks related to polycystic ovary syndrome. *Fertil Steril* 2004;81:19–25.
8. Hillard PJA, Deitch HR. Menstrual disorders in the college age female. *Pediatr Clin North Am* 2005;52:179–197.
9. Ovalle F, Azziz R. Insulin resistance, polycystic ovary syndrome, and type 2 diabetes mellitus. *Fertil Steril* 2002;77:1095–1105.
10. Legro RS. Polycystic ovary syndrome and cardiovascular disease: a premature association? *Endocr Rev* 2002;24:302–312.
11. Bronson FH. *Mammalian Reproductive Biology*. Chicago, IL: The University of Chicago Press, 1989.
12. Bachman GA, Kemmann E. Prevalence of oligomenorrhea and amenorrhea in a college population. *Am J Obstet Gynecol* 1982;144:98–102.
13. Singh KB. Menstrual disorders in college students. *Am J Obstet Gynecol* 1981;140:299–302.
14. Thorner MO, Vance, ML, Laws ER Jr, et al. The anterior pituitary. In: Wilson JD, Foster DW (eds), *Williams Textbook of Endocrinology*. Philidelphia, PA: WB Saunders, 1998; pp 249–340.
15. Yeager KK, Agostine R, Nativ A, et al. The female athlete triad: disordered eating, amenorrhea, osteoporosis. *Med Sci Sports Exerc* 1993;25:775–777.
16. Loucks AB, Vaitukaitis J, Cameron JL. The reproductive system and exercise in women. *Med Sci Sports Exerc* 1992;24(Suppl.):S288–S293.

11 Psychological Stress and Functional Amenorrhea

Reid Norman

CONTENTS

11.1   LEARNING OBJECTIVES
11.2   INTRODUCTION
11.3   RESEARCH FINDINGS
11.4   CONCLUSIONS

11.1. LEARNING OBJECTIVES

After completing this chapter, you should have an understanding of the following:

- The definition of functional hypothalamic amenorrhea (FHA).
- The causes of FHA.
- The importance of seeking medical attention for this condition.
- The long-term medical consequences of low estrogen levels.

11.2. INTRODUCTION

FHA is the absence of menstrual cycles for more than 6 months without any anatomic or pathologic cause for the condition. The term "functional" is used to indicate that the lack of menstrual cycles is because of improper functioning of the hypothalamic-pituitary-ovarian (HPO) axis rather than to an anatomic (organic) problem. In women with amenorrhea, it is important for the health care professional to rule out possible metabolic, neurological, or organic causes of this condition that can cause serious health-related problems if not treated. Common medical problems that can result in amenorrhea are prolactin-secreting pituitary adenomas, thyroid dysfunction, and polycystic ovary syndrome. The underlying deficit in FHA is suppressed gonadotropin-releasing hormone (GnRH) release from the hypothalamus resulting in low gonadotropin (luteinizing hormone [LH] and follicle stimulating hormone [FSH]) and estrogen levels. When estrogen levels are low, the endometrium of the uterus does not develop and there is no periodic menstruation.

From: *The Active Female*
Edited by: J. J. Robert-McComb, R. Norman, and M. Zumwalt © Humana Press, Totowa, NJ

11.3. RESEARCH FINDINGS

Since primitive times, it has been a common belief among women that emotional trauma results in disrupted menstruation and that regular menstrual bleeding is a sign of mental health (1). In 1930s, in the introduction to a manuscript reporting the effects of various damaging agents on the female sex organs, Selye (2) was the first to recognize that mental stress was among those changes in the environment that were well known from human pathology to disturb the sexual cycle in the female. Refeinstein (3) was also among the first to link disrupted menstrual cycles with "overt or latent psychological disturbances." He recognized that psychogenic or functional amenorrhea was a problem associated with the brain even before it was accepted that reproductive function was governed by the hypothalamus and before the existence of a hypothalamic factor that stimulated the release of LH and FSH (GnRH) was conclusively demonstrated. Although there is relationship between the severity of the stress and the proportion of women who develop amenorrhea (4), because of the differences in how individuals cope with and respond to stress, it is difficult to establish a threshold at which psychogenic stress interferes with the normal menstrual cycle (5). It is thought that a variety of stressors including mild energy deprivation induced by dieting, exercise, and psychosocial distress, such as performance pressure, unrealistic goals, and negative attributions, act synergistically to alter neuroendocrine function, resulting in suppression of the GnRH drive (6,7).

Studies in nonhuman primates suggest that an acute psychological stress can inhibit gonadotropin secretion from the pituitary. LH release was suppressed in both male and female monkeys by acute changes in environment (8,9). These studies showed that when both male and female animals were moved from their home cage to a restraining chair, LH release was suppressed and levels of stress hormones, Adrenocorticotrophic Hormone (ACTH) and cortisol, were elevated. As long as an animal remained in the chair, LH was suppressed, but within a few hours after an individual animal was returned to its cage, pulsatile LH secretion resumed and stress hormones returned to normal. This suppressive effect on gonadotropin release is most likely mediated by the endogenous opiates because naloxone, an opioid antagonist, blocked the inhibitory effect of psychological stress on LH release in these animals (8). Therefore, this conclusion would be consistent with the general notion of how stress inhibits reproductive function, i.e., by suppressing the release of GnRH from the hypothalamus and subsequently the function of the pituitary and ovary. In addition to endogenous opiates, other peptides found in the brain and associated with the stress response have been implicated in the suppression of GnRH release and may be involved in disrupted menstrual function. Both corticotropin-releasing hormone (CRH) and vasopressin are released during stress and can inhibit GnRH release. CRH stimulates the release of ACTH from the pituitary which in turn stimulates cortisol secretion by the adrenal. CRH also can directly inhibit the release of GnRH in vitro (10,11) and in vivo (12) from the hypothalamus, and CRH neurons have direct connections with GnRH neurons in the hypothalamus (13). Because opiate antagonists block the inhibitory action of CRH on GnRH and LH release, it is thought that endogenous opiates mediate the action of CRH (14).

Another major cause of FHA is low energy availability. When there is less energy available than is needed for all daily activities, reproduction is suppressed. This has been elegantly demonstrated in both female monkeys and women. Loucks (15,16) and her colleagues have convincingly shown that it is the availability of energy and not the stress of exercising that alters both LH secretion and the diurnal rhythm of leptin in women *(17)*. When female rhesus monkeys were kept on a constant diet and increased their energy expenditure by increasing the time they exercised each day, all of the animals eventually became amenorrheic *(18)*. In animals that then were provided additional calories while keeping their training constant, ovulatory menstrual cycles were reestablished *(19)*. These studies in both human and nonhuman primates convincingly demonstrate that reproductive cycles are dependent on adequate energy availability and are quickly disrupted when energy is limited.

Psychiatric histories of women with FHA have been compared with amenorrheic (from organic causes) and eumenorrheic controls *(20)*. Women with FHA had more dysfunctional attitudes, did not cope as well with ordinary stresses, and showed more interpersonal dependence than did eumenorrheic women. These women also more often had a history of psychiatric mood disorders than did women with normal cycles, but they were not different in this regard from controls with organic amenorrhea. These psychiatric disorders are associated with elevation in hypothalamic-pituitary-adrenal (HPA) activity and when combined with performance anxiety and dietary restriction contribute to significant endocrine dysfunction which may result in FHA.

The long-term negative health consequences of FHA may include increased risk of cardiovascular disease, osteoporosis, depression, other psychiatric conditions, and dementia *(21)*. Although the underlying mechanisms that cause FHA are not well understood, disturbances of either the HPA axis or the hypothalamic-pituitary-thyroid (HPT) axis contribute to the malfunction of the HPO axis *(21)*. Treatment of the symptoms of FHA with hormone replacement therapy addresses the consequences of low estrogen levels but does not correct the underlying endocrine problems. It is suggested that treatment of the behavioral problems with psychological therapy will potentially have a better chance of correcting the underlying endocrine problems and restore cyclic ovarian function *(21)*.

11.4. CONCLUSIONS

The lack of regular, cyclic menstruation in reproductive age women in which there is no organic or anatomic cause is called FHA. The cause of this malfunction in the HPO axis is generally attributed to some sort of environmental stress: psychological, physical, or nutritional. Because FHA by definition is a chronic condition, it can have serious health consequences if not treated. As the underlying causes of the FHA are thought to be dysfunctional attitudes and behaviors that result in stimulation of the HPA axis or suppression of the HPT axis, current opinion is that cognitive or behavioral therapy has the best potential for ameliorating this condition *(21)*.

REFERENCES

1. Kroger, W. and Freed S.L. *Psychosomatic Gynecology*. 1956. The Free Press, Illinois.
2. Selye H. The effect of adaptation to various damaging agents on the female sex organs in the rat. *Endocrinology* 1939;25:615–624.
3. Refeinstein, E.C. Jr. Psychogenic or "hypothalamic" amenorrhea. *Med. Clin. North Am.* 1946;30,1103–1115.
4. Drew, F.L. The epidemiology of secondary amenorrhea. *J. Chronic Dis.* 1961;14,396–407.

5. Ferin, M. Stress and the reproductive cycle. *J. Clin. Endocrinol. Metab.* 1999;84,1768–1774.
6. Marcus, M.D., Loucks, T.L. and Berga, S.L. Psychological correlates of functional hypothalamic amenorrhea. *Fertil. Steril.* 2001;76,310–316.
7. Young, E.A. and Korszun, A. The hypothalamic-pituitary-gonadal axis in mood disorders. *Endocrinol. Metab. Clin. North Am.* 2002;31,63–78.
8. Norman, R.L. and Smith C.J. Restraint inhibits LH and testosterone secretion in intact male rhesus macaques: effects of concurrent naloxone administration. *Neuroendocrinology* 1992;55:405–415.
9. Norman, R.L., McGlone, J.J. and Smith C.J. Restraint inhibits LH secretion in the follicular phase of the menstrual cycle in female rhesus macaques. *Biol. Reprod.* 1994;50:16–26.
10. Gambacciani, M., Yen, S.S.C. and Rasmussen, D. GnRH release from the medial basal hypothalamus: in vitro inhibition by corticotropin-releasing factor. *Neuroendocrinology* 1986;43:533–536.
11. Nikolarakis, K.E., Almeida, O.F.X. and Herz, A. Corticotropin-releasing factor (CRF) inhibits gonadotropin-releasing hormone (GnRH) release from superfused rat hypothalami in vitro. *Brain Res.* 1986;377:388–390.
12. Petraglia, F., Sutton, S., Vale, W. and Plotsky, P. Corticotropin-releasing factor decreases plasma luteinizing hormone levels in female rats by inhibiting gonadotropin-releasing hormone release into hypophyseal portal circulation. *Endocrinology* 1987;120:1083–1088.
13. MacLusky, J.N., Naftolin, F. and Leranth, C. Immunocytochemical evidence for direct synaptic connections between corticotropin-releasing factor (CRF) and gonadotropin-releasing hormone (GnRH) containing neurons in the preoptic area of the rat. *Brain Res.* 1988;439:391–395.
14. Barbarino, A., De Marinis, L., Tofani, A., Della Casa, S., Damico, C., Mancini, A., Corsello, S.M., Sciuto, R. and Barini, A. Corticotropin-releasing hormone inhibition of gonadotropin release and the effect of opioid blockade. *J. Clin. Endocrinol. Metab.* 1989;68:523–528.
15. Loucks, A.B., Verdun, M. and Heath E.M. Low energy availability, not stress of exercise, alters LH pulsatility in exercising women. *J. Appl. Physiol.* 1998;84:37–46.
16. Loucks AB, Thuma JR. Luteinizing hormone pulsatility is disrupted at a threshold of energy availability in regularly menstruating women. *J. Clin. Endocrinol. Metab.* 2003;88:297–311.
17. Hilton, L.K. and Loucks, A.B. Low energy availability, not exercise stress, suppresses the diurnal rhythm of leptin in healthy young women. *Am. J. Physiol. Endocrinol. Metab.* 2000;278:E43–E49.
18. Williams, N.I., Caston-Balderrama, A.L., Helmerich, D.L., Parfitt, D.B., Nosbisch, C. and Cameron, J.L. Longitudinal changes in reproductive hormones and menstrual cyclicity in cynomolgus monkeys during strenuous exercise training: abrupt transition to exercise-induced amenorrhea. *Endocrinology* 2001;142:2381–2389.
19. Williams, N.I., Helmreich D.L., Parfitt, D.B., Caston-Balderrama, A. and Cameron J.L. Evidence for a causal role of low energy availability in the induction of menstrual cycle disturbances during strenuous exercise training. *J. Clin. Endocrinol. Metab.* 2001;86:5184–5193.
20. Giles, D.E and Berga, S.L. Cognitive and psychiatric correlates of functional hypothalamic amenorrhea: a controlled comparison. *Fertil. Steril.* 1993;60,486–492.
21. Berga, S.L and Loucks, T.L. The diagnosis and treatment of stress-induced anovulation. *Minerva Ginecol.* 2005;57,45–54.

12 Effects of the Menstrual Cycle on the Acquisition of Peak Bone Mass

Mimi Zumwalt

CONTENTS

12.1 LEARNING OBJECTIVES
12.2 INTRODUCTION
12.3 RESEARCH FINDINGS
12.4 CONCLUSION
12.5 SCENARIO WITH QUESTIONS AND ANSWERS

12.1. LEARNING OBJECTIVES

After completing this chapter, you should have an understanding of the following:

- The pertinent female reproductive anatomy and physiology of a normal menstrual cycle, including the different phases and role of various hormones.
- The different components of bone along with its function, biochemistry, and metabolism.
- The interaction between different ovarian hormones and various bone components.
- The definition, importance, and effective methods of achieving peak bone mass.
- The results of an abnormal menstrual cycle on attainment of peak bone mass, how to go about assessing/measuring bone density, and ways to address this issue further to help minimize bone loss and maintain bone density.

12.2. INTRODUCTION

As the skeleton provides complete structural support for the entire body, having the maximal amount of strong bones, especially during the growing phase, will serve to protect against osteoporosis (decreased bone mass because of bone loss) later on in life *(1)*. In addition, bone is far from static; in fact, this living tissue is quite dynamic, new bone continually replaces old bone (remodeling and repair) depending on mechanical, physiological, and hormonal stimuli. The latter, systemic hormonal milieu, plays a very important role during puberty and in females is manifested by the menstrual cycle *(2–4)*. It is estimated that about 50% of adult peak bone mass is accumulated during the adolescent growth spurt, with 60–90% of all skeletal bone being laid down during these teenage years *(5–7)*. Once peak bone mass is reached in the mid to late twenties,

From: *The Active Female*
Edited by: J. J. Robert-McComb, R. Norman, and M. Zumwalt © Humana Press, Totowa, NJ

the amount of bone withdrawn from or deposited in the skeleton—predetermined by inherited genes (approximately 75% unmodifiable)—is subsequently influenced by environmental factors (approximately 25% modifiable) *(3,7)*. The amount of bone that is gained during this period equates to the quantity of bone which will be lost during the rest of one's adult life *(7,8)*. As such, measures to maximize and protect the amount of bone obtained during the growing period will serve the body well in activities of daily living and other life endeavors beyond these early years *(9)*.

This chapter will focus primarily on the female menstrual cycle and its influence on peak bone mass achieved during and after adolescence.

12.3. RESEARCH FINDINGS

12.3.1. Anatomy of the Female Sex Organs and Physiology of the Menstrual Cycle

The female reproductive system includes essential (i.e., sex-specific, including the uterus and ovaries) and accessory organs (Fig. 12.1). Pertinent anatomy involved with the menstrual cycle consists of the ovaries. This paired organ has both a reproductive and an endocrine function, as this is where eggs are formed and female sex hormones are being produced *(10,11)*. Three levels of hormones are normally secreted in a feedback loop termed the hypothalamic-pituitary-ovarian (HPO) axis *(12)*: one by the hypothalamus—gonadotropin-releasing hormone (GnRH); another by the anterior pituitary—follicle stimulating hormone (FSH) and lutenizing hormone (LH) in response to GnRH; and the third tier consists of two gonadal steroids, estrogen and progesterone, secreted from the ovaries in response to FSH and LH. These hormones are secreted in different amounts during various phases of the monthly menstrual cycle *(3,11)*. Indeed, the onset of menarche or menses in females signals the transition from childhood to the pubertal, reproductive state, correlating with both body height and bone maturation *(12)*. This pivotal sexual stage can begin as early as 11 years, but on average it occurs between 12 and 13 years of age, about 2 years after the appearance of secondary sex characteristics (breast and pubic hair development) *(6,11)*. If menstruation does

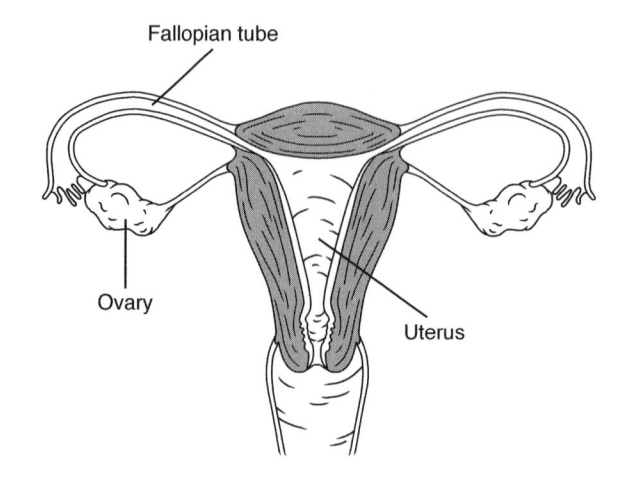

Fig. 12.1. Female reproductive unit.

not occur by age 16, this delay of menarche (primary amenorrhea) may be because of excessive exercise or insufficient intake of appropriate nutrition, among other factors *(13,14)*.

Once regular menses begins, this cycle normally occurs monthly between 20 and 45 days, on average approximately every 4 weeks (28 days), with menstrual flow lasting about 3–5 days *(6,11)*. Typically, three separate phases encompass the normal monthly menstrual cycle: the first or follicular phase (growth of the endometrium or uterine lining stimulated by estrogen), the second or ovulatory phase (ovulation or egg release in response to LH), and the last or luteal phase (transformation of the endometrium from proliferating into secreting type of tissue under progesterone influence) *(13,15)*. Specifically, onset of the follicular/proliferative phase is stimulated by FSH more so than LH; this is marked by the first day of menstrual bleeding. During this time, both progesterone and estrogen levels in the blood are low. Estrogen then increases steadily, peaking just before ovulation (under the influence of LH surge). This rise occurs about the middle of the cycle, or day 14, when one mature egg or follicle is released. The luteal/secretory phase then takes over during the second half of the cycle under the continuing influence of LH; this is when progesterone also increases above estrogen levels, though the latter hormone concentration remains fairly high. These two hormones, estrogen and progesterone, are now being secreted by the corpus luteum which then degenerates, and levels of both hormones also fall concurrently as long as the egg is neither fertilized nor implanted in the uterus, which signifies that pregnancy has not occurred. The cycle then repeats itself by shedding of the endometrium, manifested by menstrual flow, to restart the proliferative phase all over again under FSH influence *(6,11)* (Fig. 12.2).

12.3.2. Effects of Female Sex Hormones on Bone

The skeleton is one of the main target tissues for estrogen. Bone cells contain receptors for this very important ovarian hormone; thus, it is a key regulator of bone turnover, adjusting the bone mass "set point" by keeping in balance the amount being formed with the amount which is then reabsorbed (coupling effect) *(7,12)*. Estrogen acts indirectly through inhibiting bone resorption by osteoclasts while increasing the activity

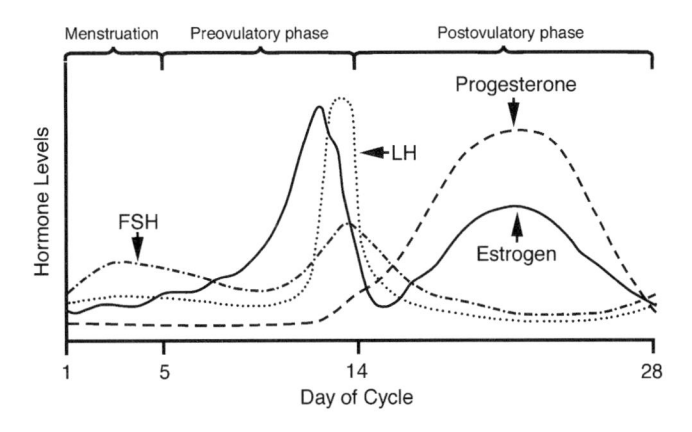

Fig. 12.2. Menstrual cycle. FSH, follicle stimulating hormone; LH, lutenizing hormone.

of bone-forming cells, osteoblasts *(4,11)*. Estrogen also strongly affects longitudinal growth of the diaphyses or shafts of long bones, causing closure of their growth plates at both ends, which in turn influences the final stature/height in females *(11)*. Additionally, having the appropriate blood levels of progesterone is also important for optimal bone health; these two ovarian hormones act in concert to help maximize bone deposition and minimize bone loss *(5,16)*. In other words, the two sex steroids play an essential role in homeostasis of the skeleton by exerting their direct effects through receptors on bone cells and by exerting indirect effects on other systemic hormones regulating calcium balance. In fact, even though mechanical loading, up to a point, will also exert a positive effect on bone formation, the net result is enhanced further through the dual action of estrogen and progesterone, more so than if either one acts on the skeleton singly, i.e., a synergistic relationship exists for both *(17)*.

To summarize, the normal, monthly occurrence of the female menstrual cycle represents an intricate interaction of three organs—hypothalamus, anterior pituitary, and ovaries—resulting in a feedback mechanism whereby several secreted hormones influence each other to effect maturation of an egg (follicle), which is then released (ovulates). Next, the endometrium becomes prepared to receive the mature follicle, and, if no implantation occurs, then sloughing of the uterine lining manifests itself by menstrual bleeding and the cycle repeats. Regulation of this monthly cycle is controlled by LH and FSH from the anterior pituitary gland in response to GnRH from the hypothalamus *(13)*. Disturbances in any portion of this hormonal loop between the reproductive system and these two specific centers in the brain could, in turn, affect the quality/quantity of menstruation and secondarily affect bone deposition, especially during the critical time of initial peak bone mass acquisition of the pubertal/adolescent stage *(14,16,18)*.

12.3.3. Bone Composition, Physiology, Function, and Interaction with Ovarian Hormones

Skeletal tissue is composed of five different bone cell types, along with noncellular substances, organic matrix (35% by weight) and inorganic mineral (65%). The two key types of bone cells are osteoblasts, which are involved in bone formation, and osteoclasts, whose role is bone resorption. Components making up the organic matrix include glycoprotein and protein (90% collagen), bathing in a sea of gelatin-like mucopolysaccharide (ground substance). These collagen fibers represent specific sites where inorganic calcium/phosphate hydroxyapatite crystals are deposited prior to the mineralization process of bone. Ninety-eight to ninety-nine percent of the body's total calcium is sequestered in the skeleton, which serves as a mineral bank and releases just enough into the bloodstream to keep serum calcium levels fairly constant. In other words, as this mineral is extremely tightly regulated, the body will rob calcium from its main reservoir, bone, to maintain adequate blood levels in order to carry out essential, vital tasks such as blood clotting and muscle contraction *(2,4,19)*. Calcium, however, cannot be absorbed and incorporated into bone unless vitamin D is readily available *(2,8)*. This is where ovarian hormones come into play as they possess an important role in the metabolism of calcium and vitamin D. Indeed, estrogen helps to regulate the absorption of this very important inorganic mineral, subsequently contributing to bone formation, as previously stated. In effect, any irregularities in the

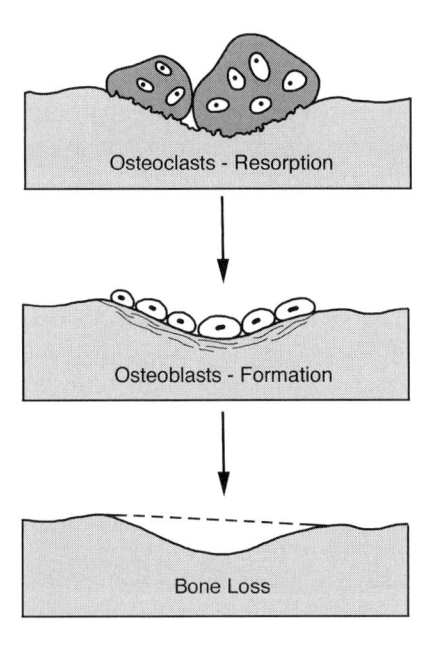

Fig. 12.3. Bone remodeling.

menstrual cycle will negatively affect bone deposition in the long run by negating estrogen's positive influence on calcium balance. Additionally, with less protection from estrogen, osteoclasts will be affected to a greater extent than osteoblasts, resulting in an uncoupling effect, which in turn will ultimately culminate in undesirable bone loss which could manifest itself as early osteoporosis *(4,6,20,21)* (Fig. 12.3). Alarmingly, studies have shown that bone mass in athletes with absent menses is 10–15% lower in the lumbar spine than females with normal monthly cycles, demonstrating that the hypoestrogenic state is associated with a reduction in regional bone density *(7,22)*.

12.3.4. Peak Bone Mass

Peak bone mass signifies the maximal quantity of bone which can be gained primarily during the adolescent years (with the critical window being between 9 and 20 years), while the skeleton is undergoing an accelerated growth in both size and density, i.e., rapid increase in weight and height. This "adolescent growth spurt" occurs 1–2 years preceding the rapid deposition of bone into the skeleton. In females, up to 90% of peak bone mass is accumulated by 18 years of age *(3,7)*. After this prime period, bone can still continue to grow in terms of strength and density, up until about age 25–30, at which time true peak bone mass is reached *(2,3,23)*. However, some researchers have shown that this acquisition may occur as young as during the late teens *(7)*. After attainment of peak bone mass, measures to maintain as much bone density as possible are of paramount importance as physiological bone loss will inevitably occur gradually over time as one ages *(2,7,23,24)*. In fact, before reaching menopause in their fourth or fifth decade, young females lose about 0.3% of their entire skeleton each year after the final acquisition of peak bone mass *(2)*.

The quantity of bone deposited in the skeleton is influenced by both genetic (uncontrollable) and environmental (controllable) factors *(3,7)*. Mechanical loading in the form of weight-bearing/resistance exercise, appropriate and adequate nutrition, i.e., sufficient consumption of key bone building nutrients such as calcium, magnesium, zinc, vitamins C and D, and normal, regular monthly menses all contribute to optimal bone health *(2,5–7,9,21)*. As far as nutritional intake is concerned, an improper diet with a less than optimal amount of calories will adversely affect the pulsatile secretion of LH. As this hypothalamic hormone is dependent on availability of energy, its aberrant serum levels will alter the menstrual cycle and ultimately affect bone balance *(25)*. Thus, a delay in menarche (initiation of menses), dysmenorrhea (irregular menstruation), oligomenorrhea (insufficient number of cycles per year), or amenorrhea (absence of menses) all will interfere with the final attainment of peak bone mass by causing more rapid bone loss. Specifically, women with menstrual disorders will lose as much as 2% of their skeleton per year, which is almost 10-fold more than the usual, natural rate. This is primarily because of a deficiency of estrogen and, if not corrected promptly, ultimately will result in premature osteoporosis *(2,26)*. Such menstrual irregularities/disorders could be caused by being extremely thin, rapid weight loss, excessively intense exercise, and disordered eating, along with associated extreme physical/psychological stress *(6,13,17)*. In other words, exercising excessively can negatively affect both the reproductive and the skeletal systems, altering the body's hormonal milieu and causing a reduction in bone mass *(27)*. One study has found that total vigorous, intense training for more than 8 hours per week could lead to amenorrhea and, subsequently, osteoporosis with its inherent risks *(15)*.

12.3.5. Results of an Abnormal Menstrual Cycle on Peak Bone Mass

Once it is established that menstrual dysfunction is indeed occurring, potential sources should be investigated and ruled out to help slow down and hopefully ultimately arrest the process of accelerated bone loss. It must be understood, however, that the bone which is lost may not be totally recovered, even upon resumption of normal menses. Specifically, when facing the problem of irregular menses occurring in an adolescent female, causes for secondary amenorrhea should be sought such as pregnancy, endocrine disorders, anatomic defects, or tumors of the involved organs, which can interfere with the HPO axis *(7,13,14,17,23)*. After these potential pathological causative factors are excluded, then information regarding eating habits, weight control behavior, and exercise patterns should be obtained. This detailed history is used to evaluate whether the affected young patient is at risk for premature osteoporosis because of the Female Athlete Triad (diagnosis of exclusion)—a constellation of amenorrhea, disordered eating, and osteoporosis *(27)*. Females more at risk for this potentially devastating condition are usually involved in sports requiring an aesthetic/athletic look or those needing to weigh less for better performance (gymnasts, dancers, runners, divers, etc.) *(6,7,25,28)*. In that case, the afflicted female athlete's young, thin, fragile bones are at two to threefold increased risk for stress and/or frank fractures, similar to older, perimenopausal women in their forties and fifties, especially if the duration of absent menses is longer than 6 months *(14,26,28)*. Even those involved in a regular program of exercising, especially under high loads such as gymnastics, may partially offset the osteopenic bone caused by being amenorrheic, but this is not enough if an inadequate amount of calories is being consumed. In fact, research has shown that bone mass in females who are sedentary, eat well, and have normal monthly cycles is actually higher than those athletes who exercise to the point of losing their menses *(2)*. Similarly, another study has shown that increases in circulating FSH levels above 20 mIU/L are linked to progressive bone loss in perimenopausal women, again stressing the importance of maintaining regular menstruation to preserve bone density *(14)*.

Currently, dual energy X-ray absorptiometry (DEXA) is the "gold standard" for measurement of bone mineral density (BMD) to assess bone heath, and thus, it should be used to help diagnose osteoporosis and monitor progress of treatment *(2,7,9,13,21,23,24,28,29)*. This technique uses emission of X-rays at two separate energy levels to distinguish bone from the surrounding soft tissue with very low radiation exposure. The time for scanning with this type of device is brief, and BMD measurements are more accurate (within 1–2%) as compared to other methods *(2,7)*. If the DEXA scan result for BMD is between 1.0 SD and 2.5 SD below the mean for young adults, then bone is considered osteopenic. Even at this level, there exists 2–2.5 times increased risk of spine or hip fractures because of increased fragility of the skeleton. On the other hand, bone is deemed osteoporotic (with associated higher fracture risk) if the BMD is equal to or greater than 2.5 SD from the mean *(2,7,13,24)* (Table 12.1). One study demonstrated that reduced urinary sex steroid hormones are found during the luteal phase of the menstrual cycle in premenopausal women with lower BMD (10th versus 50th–75th percentile), again linking monthly menses to bone mass acquisition *(29)*. Adjunctively, other methods to assess bone metabolism include serum osteocalcin and alkaline phosphatase (for bone formation), and urine collagen breakdown products (telopeptide crosslinks) as indicators of resorption and current rate of bone loss *(2,7,9)*.

The next step in management of osteopenia/osteoporosis in an adolescent female is to correct any offending factor contributing to the rapid bone loss by ensuring adequate and appropriate dietary intake of nutrition, making sure she is engaged in no more than just moderate exercise sessions (5–20% reduction in quantity of training), helping her to change body composition by increasing body weight slowly and gradually (try to gain about 5% or 0.5 to 1 lb per week) and, finally, conferring with her to confirm that she is maintaining the correct goal weight *(6,27,28)*. As far as the disturbance in menses is concerned, these monthly cycles need to be regulated by some sort of hormonal replacement therapy (HRT)—the oral contraceptive pill is a popular form of pharmacological treatment for this purpose *(27)*. At this point, consulting a gynecologic specialist is the next appropriate step to help determine

Table 12.1
Classification of bone mineral density according to the World Health Organization

Normal bone density	< 1 SD below the mean
Osteopenia	1–2.5 SD below the mean
Osteoporosis	> 2.5 SD below the mean
Severe osteoporosis	> 2.5 SD below the mean/fragility fracture(s)

which hormone(s) to prescribe and specific dosages to be given. Moreover, specific lifestyle changes that are recommended include avoidance of inactivity, refraining from smoking cigarettes, and minimizing drinking alcohol *(1)*. As far as dietary intake is concerned, avoid caffeine and excess protein as both of these substances tend to contribute to calcium wasting and, thus, could secondarily affect bone formation. Along the same line, ingesting foodstuff with adequate amounts of calcium and vitamin D to aid in bone building is a good general rule. As far as exercising, make sure the mode is of the weight-bearing kind (loading in the erect position) and low impact type such as walking, rather than excessive high impact activities such as running or jumping, which could be more injurious to bone and eventually cause skeletal failure *(2,3,6–9,13,15,23,28)*.

In conclusion, evaluation and treatment for females with menstrual disturbances include conducting a detailed history and physical exam, along with ordering certain laboratory work. It is also prudent to approach management of the female at risk by compiling a multidisciplinary team, involving an internist to rule out chronic disease, a psychiatrist/psychologist to address associated eating disorders and other psychological issues, a gynecologist to rule out anatomic abnormalities and for prescribing HRT, a dietician for nutritional counseling, and an endocrinologist to address other possible pathology interfering with the hormonal feedback loop *(14,15,25)*. Additionally, it is very important to involve family members, teachers, trainers, other players, and coaches to solicit additional support for the involved patient/athlete, to hopefully help her fully recover within a reasonable period of time *(25,28)*.

12.4. CONCLUSION

The role of the female menstrual cycle in terms of its contribution toward acquisition of peak bone mass has been fairly well elucidated. A regular, well functioning monthly cycle is of paramount importance in bone deposition during the adolescent growth spurt and then in prevention of bone loss after peak bone mass is achieved in the second decade of life. The main hormone involved in bone regulation is estrogen which is secreted by the ovaries, affecting both the osteoclasts and the osteoblasts, keeping these two types of bone cells in balance in terms of their function on skeletal resorption and formation, respectively. Additionally, the organs involved with controlling the menses are the hypothalamus, anterior pituitary, and, of course, the ovaries. Each organ secretes different hormones that interact in a complex, feedback mechanism to regulate the monthly female cycle, in order to maximize bone deposition and minimize bone loss. If menarche is delayed or if menstrual dysfunction occurs or disappears entirely, then the protective mechanism of estrogen on bone is lost; if this is not corrected fairly promptly, it will eventually result in osteopenia or frank osteoporosis, ultimately increasing one's susceptibility to fractures. In fact, osteoporotic prevention starts as early as the initial onset of menarche during adolescence. Additionally, other factors, such as lifestyle habits (nutrition, training, etc.), also play a very essential part in contributing to bone building (formation) or bone loss *(17,24)*. As such, measures taken to ensure that the monthly female menstrual cycle is functioning correctly and optimally are of utmost importance to positively influence the final attainment of peak bone mass *(17)*.

12.5. SCENARIO WITH QUESTIONS AND ANSWERS

12.5.1. Scenario

You are a primary care sports medicine physician in a multispecialty group practice consisting of pediatrics, internal medicine, OB/GYN, and family physicians. For the past few years, you have been the team physician for local high school athletes, including track and field events and other court sports. Lately, the team athletic trainer has brought in a few of the female long-distance runners with similar complaints of vague, aching pain in their feet and legs, especially after a couple of weeks' practice, e.g., running around the track. Upon further questioning, these female athletes also reveal that their menses have been very irregular and/or absent over the last year or so.

12.5.2. Questions

1. What should be your next action in terms of history taking, physical?
2. What are some of the causes of menstrual irregularity/absence in young female athletes?
3. What are the consequences of dysmenorrheal or amenorrhea on bone?
4. What are your recommendations for restoration of menses?
5. How would you go about treating osteopenia/osteoporosis?

12.5.3. Plausible Answers

1. Detailed questions about frequency, duration, and intensity of training should be sought; along with exactly how many monthly menstrual cycles these female athletes are actually missing. Additionally, dietary habits/behaviors should also be questioned, to see whether excessive exercising sessions are occurring and/or whether inappropriate/inadequate nutrition is being consumed. Inspect for signs of overtraining and/or indicators of possible eating disorders. Obtain appropriate laboratory studies including a blood count to rule out anemia (indicator of poor nutrition) and complete metabolic panel to look for electrolyte disturbances, bone indices, and possible endocrine abnormalities. Also, order a urinary analysis to search for markers of bone turnover. A DEXA scan should then be requested to assess bone density, and, finally, radiographs of the involved extremity should be ordered to discover potential stress fractures.
2. First and foremost, pregnancy needs to be considered in an amenorrheic athlete. Next, some sort of disturbance in the hormonal feedback loop between the hypothalamus, pituitary, and ovaries could be another possibility, whether an anatomic anomaly or tumorous growth. This is where subspecialists such as an endocrinologist or gynecologist can be quite helpful in further evaluation for these conditions.
3. Beyond the above considerations, the consequential effect of altered menses on the skeleton is accelerated bone loss, which will eventually result in osteoporosis and its associated consequences such as fragility fractures.
4. After establishing that the osteoporosis is because of excessive training and/or unsound nutrition, involve the athletic trainer and coach; explain to these athletes that treatment of future osteoporosis begins with prevention of bone thinning while their skeleton is developing; that it is extremely important for them to start with good habits about nutrition and training to maximize the amount of bone to be deposited during growth.

5. Do not initiate treatment on your own. Set up consultations with a dietitian to correct the disordered eating, consult a gynecologist to begin hormonal therapy, and, finally, discuss with a psychologist/counselor the possibility of advising the athlete to change her mind set about nutrition and training in general, and her body, specifically.

REFERENCES

1. Elgan C, Dykes AK, Samsioe G. Bone mineral density changes in young women: a two year study. *Gynecol Endocrinol* 2004 October; 19 (4): 169–177.
2. Lane J, Russell L, Khan S. Osteoporosis. *Clin Orthop Relat Res* 2000 March; 372: 139–150.
3. National Institutes of Health Osteoporosis and Related Bone Diseases-National Resource Center. Osteoporosis: Peak Bone Mass in Women 2005 March; URL: http://niams.nih.gov/bone/hi/bone_mass.htm.
4. Resnick D, Manolagas S, Niwayama G. Histogenesis, anatomy, and physiology of bone. *Anatomy and Physiology*. Philadelpia: Saunders, 1989: 16–28.
5. Brown S. Better bones at every age. *Let's Live Magazine* 2000 October.
6. Lebrun C. Menstrual cycle dysfunction. *Am Coll Sports Med* 2000 October; URL: http://medicalreporter.health.org/tmv112000/menstrualcycledysfunction.html.
7. Nichols D, Bonnick S, Sanborn C. Clinics in Sports Medicine. *Athl Woman* 2000 April; 19 (2): 233–249.
8. United States Department of Health and Human Services. Bone Health and Osteoporosis: A Report of the Surgeon General. Chapter 6: Determinants of Bone Health 2004 October; URL: http://www.surgeongeneral.gov /library /bonehealth /chapter_6.html.
9. Lane JM, Nydick M. Osteoporosis: current modes of prevention and treatment. *J Am Acad Orthop Surg* 1999 January; 7(1): 19–31.
10. Altruis Biomedical Network. Anatomy and Physiology. E-Gynecologic.com 2000; URL: http://www.e-gynecologic.com/index.html.
11. Guyton A, Hall J. Female physiology before pregnancy; and the female hormones. In: Schmitt W, Gruliow R, Faber P, Norwitz A, Shaw P, eds. *Textbook of Medical Physiology*, 10th ed. Pennsylvania: Saunders, 2001: 929–943.
12. Berne R, Levy M, Koeppen B, Stanton B, eds. The reproductive glands. *Physiology*, 5th ed. St Louis, MO: Mosby, 920–978.
13. Harmon K. Evaluating and treating exercise-related menstrual irregularities. *Phys Sportsmed* 2002 March; 30 (3).
14. Khan A, Syed Z. Bone densitometry in premenopausal women: synthesis and review. *J Clin Densitom* 2004 January; 7 (1): 85–92.
15. Nelson L. Amenorrhea. eMedicine 2005 May. URL: http://www.emedicine.com/med/topic117.htm.
16. Csermely T, Halvax L, Schmidt E, Zambo K, Vadon G, Szabo I, Szilagyi A. Occurrence of osteopenia among adolescent girls with oligo/amenorrhea. *Gynecol Endocrinol* 2002 April; 16 (2): 99–105.
17. Balasch J. Sex steroids and bone: current perspectives. *Hum Reprod Update* 2003 May–June; 9(3): 207–222.
18. Galuska DA, Sowers MR. Menstrual history and bone density in young women. *J Womens Health Gend Based Med* 1999 June; 8 (5): 647–656.
19. Altruis Biomedical Network. Osteoporosis. E-Gynecologic.com 2000; URL: http://www.e-gynecologic.com/index.html.
20. Thys-Jacobs S. Micronutrients and the premenstrual syndrome: the case for calcium. *J Am Coll Nutr* 2000; 19 (2): 220–227.
21. Templeton K. Secondary osteoporosis. *J Am Acad Orthop Surg* 2005 November; 13 (7): 475–486.
22. Drinkwater BL, Nilson K, Chesnut CH III, Bremner WJ, Shainholtz S, Southworth MB. Bone mineral content of amenorrheic and eumenorrheic athletes. *N Engl J Med* 1984 August; 311 (5): 277–281.
23. Erickson S, Sevier T. Osteoporosis in active women: prevention, diagnosis, and treatment. *Phys Sportsmed* 1997 November; 25 (11).
24. Theodorou S, Theodorou D, Sartoris D. Osteoporosis: a global assessment of clinical and imaging features. *Orthopaedics* 2005 November; 28 (11): 1346–1353.

25. Women's Health Orthopaedic Edition. *Updated Guidelines from the American Academy of Pediatrics.* Assessing health in young female athletes. 2000 November–December; 3 (6): 205–207.
26. Slemenda C, Longcope C, Peacock M, Hui S, Johnston C. Sex steroids, bone mass, and bone loss. *J Clin Invest* 1996 January; 97 (1): 14–21.
27. Warren MP and Stiehl AL. Exercise and female adolescents: effects on the reproductive and skeletal systems. *J Am Med Womens Assoc* 1999 Summer; 54 (3): 115–120.
28. Hobart J, Smucker D. The female athlete triad. *Am Fam Physician* June 2000; 61 (11).
29. Sowers M, Randolph JF, Crutchfield M, Jannausch ML, Shapiro B, Zhang B, La Pietra M. Urinary ovarian and gonadotropin hormone levels in premenopausal women with low bone mass. *J Bone Miner Res* 1998 July; 13 (7): 1191–1202.

IV Prevention and Management of Common Musculoskeletal Injuries in Active Females

13 Prevention and Management of Common Musculoskeletal Injuries in Preadolescent and Adolescent Female Athletes

Mimi Zumwalt

CONTENTS

13.1   LEARNING OBJECTIVES
13.2   INTRODUCTION
13.3   RESEARCH FINDINGS
13.4   CONCLUSIONS
13.5   SCENARIO WITH QUESTIONS AND ANSWERS

13.1. LEARNING OBJECTIVES

After completing this chapter, you should have an understanding of the following:

- The important changes from prepubescence to postpubescence in females as far as anatomy and physiology of the musculoskeletal system is concerned.
- The pertinent differences and similarities between young females and males in terms of anatomy, physiology, and biomechanics of the musculoskeletal system.
- The occurrence of various common musculoskeletal injuries more unique to young female athletes, especially during the adolescent growth spurt.
- Several measures for prevention of athletic injuries from occurring in preadolescent and adolescent females.
- Different methods of orthopedic treatment for musculoskeletal injuries incurred by young female athletes, including specific recommendations for a regular conditioning program once healing is complete in order to enhance their physical fitness profile.

13.2. INTRODUCTION

The initial growing period, especially transitioning from late childhood to adolescence, represents a time of tremendous transformation in not only the physical characteristics, but other aspects of the body as well. Generally speaking, we are talking about

From: *The Active Female*
Edited by: J. J. Robert-McComb, R. Norman, and M. Zumwalt © Humana Press, Totowa, NJ

growth, development, and maturation during these ever so important teenage years. First, a few terms will be defined growth refers to a size increase of the body or part(s) thereof. Development, on the other hand, means the functional alterations occurring along with anatomical/structural growth. Maturation is reached when the body has achieved its full potential in terms of growth and development, i.e., attainment of the adult form/status. Specific bodily units involved in this early stage of life change include the reproductive/sexual, physiological, and musculoskeletal systems. However, children must not be viewed as merely tiny adults, especially while they are undergoing a period of rapid growth during the adolescent stage. Unique musculoskeletal conditions can occur while the body is growing, and they need to be recognized and addressed appropriately. In addition, a with the increased emphasis in youth sports and female involvement with athletic activities, there comes a higher risk/frequency of orthopedic injuries *(1)*. This chapter will focus on several of the more common musculoskeletal concerns of the young female athlete and how best to go about preventing and treating these orthopedic issues.

13.3. RESEARCH FINDINGS

13.3.1. The Important Changes from Prepubescence to Postpubescence in Females as far as Anatomy and Physiology of the Musculoskeletal System is Concerned

After the initial rapid increase in height during the first 2 years of life (when 50% of adult height is attained) and up until the onset of puberty, a female's body grows steadily in height and weight, staying fairly constant during childhood, then both parameters really take off fast in terms of growth velocity, peaking at around the age of 12 or so. In girls, final adult stature is reached between the ages of 16 and 17. It is an established fact that regular exercise, along with an appropriate diet, is essential for proper bone growth in terms of width, density, and strength by mineral (calcium among others) deposition into the skeletal matrix *(1–3)*. Alongside the bony framework providing structural support/protection for the body, musculotendinous and ligamentous attachment to the skeleton helps synergistically by providing dynamic and static restraints while allowing motion/movement of the head, trunk, and limbs, respectively. Muscle mass itself also increases steadily (from hypertrophy or enlarging fiber size) in response to hormonal influence, reaching its peak between the ages of 16 and 20 years in females. As a corollary to the increase in muscle mass, an accompanied gradual rise in muscle strength occurs in conjunction as well and reaches a maximum by 20 years of age in young female adults *(1)*. Before puberty, girls and boys are comparable in terms of muscular strength. Then around the age of 15 or 16, adolescent females are only about three-fourth as strong as pubescent males. This strength difference is more marked in the upper versus the lower extremities *(4)*.

In addition to longitudinal growth and gaining in body size/weight, morphologically the young female body shape is further transformed during the adolescent growth spurt partly in preparation for later childbearing *(1,4)*. Under the influence of estrogen, fat deposition increases which more than doubles the percentage of total body weight present at birth (25% from 10–12%). Both sex-specific (surrounding breasts and hips) and subcutaneous tissue fat accumulate throughout the body of a young woman.

The mechanism of adipose tissue storage, unlike that of muscle, stems from both hypertrophy and hyperplasia (increase in number of fat cells), and the latter process can continue throughout one's lifespan dependent on diet and activity, among other factors. As for motor skills, continued enhancement from the developing neuromuscular and endocrine systems occurs, and then, physical activity starts to plateau out at the time of puberty onset in young females. Girls tend to slow down physically as a whole during adolescence, partly because of greater fat deposition, but more markedly from societal norms, though this trend is currently changing in modern times *(1)*.

13.3.2. The Pertinent Differences and Similarities Between Young Females and Males in Terms of Anatomy, Physiology, and Biomechanics of the Musculoskeletal System

Up to a certain age in both females and males, growth and development parallel each other, but at the onset of puberty, the differing levels of secreted sex hormones, namely the ratio of estrogen to testosterone, starts to separate young girls and boys in terms of maturation of the musculoskeletal system. Specifically speaking, the maximal rate and final growth in both weight and height in boys occurs about 2 years later than in females (14–14.5 years old), is more intense, and lasts longer until approximately the age of 18 years *(1,5)*. Likewise, under the influence of a sudden 10 times increase in testosterone production during puberty, boys markedly gain muscle mass at an accelerated rate, resulting in 40% of total body weight as compared to the 25% present at birth, and muscle mass does not reach its peak quantity until 18–25 years of age. Muscular strength in boys, just as in young girls, also improves along with the increased muscle mass, especially around age 12, but again at a much faster rate, stronger proportion, and longer duration in males, peaking between the age of 20 and 30 *(1,5,6)*. Regarding adipose tissue deposition, boys do not tend to accumulate a large quantity of fat because of much lower estrogen levels. Body fat in males ultimately reaches only 15% of total body weight, which is 10% less than in females. As for the difference in ability, coordination, and learning of motor skills, the key period lies between the ages of 9 and 12 for both sexes; after age 12, the physical performance in males continues to accelerate and far exceeds that of adolescent females during puberty because of more muscle mass/strength and less fat accumulation *(1,5)*. Along the same lines, development of speed also favors adolescent males, which ultimately results in higher fitness levels as compared to girls. In fact, the sprint velocity increases yearly from 5 years of age in both sexes, but maxes out in females between the ages of 13 and 15, yet does not peak in boys until the age of 16. Two phases encompass the phenomenon of speed development; the first begins about age 8 in both girls and boys, most likely attributed to improved coordination aided by the maturing nervous system. The second phase occurs at about age 12 in females, and anywhere from age 12 to 15 in males as a natural progression from the larger body size, muscle mass and, along with that, speed, strength, power, and endurance *(5)*. Morphologically, pubescent boys maintain their body structure, similar to prepubescent girls, again affected by the ratio of testosterone to estrogen. Recent studies have demonstrated, interestingly that neuromuscular control/preference especially in the lower limbs of adolescent females during certain sporting activities differs from that of male athletes, potentially putting these young girls much more at risk for knee injuries *(7–10)*.

13.3.3. The Occurrence of Various Common Musculoskeletal Injuries more Unique to Young Female Athletes Especially During the Adolescent Growth Spurt

As previously outlined, the transition from prepubescence to postpubescence in females brings on a multitude of bodily changes, the majority of which involves the musculoskeletal system. While the skeleton is growing rapidly, muscles, tendons, ligaments, and other connective tissues must also enlarge/elongate to accommodate this bony growth. As a matter of fact, under endocrine and neural influence, locomotion and movement must be coordinated and adapted to the transforming skeletal framework in order to carry out life tasks and other physical endeavors, such as recreational athletics or competitive sports. Consequently, teenage female athletes involved in certain sporting activities are exposed to a higher risk of orthopedic injuries. This is partly because of a marked rise in the number of school-aged females playing sports, from 1 female participating in competition compared to 27 male athletes in 1972 to the ratio of 1 female athlete competing to 3 males 30 years later. The mechanism that incites musculoskeletal injuries in athletic activity is a matter of straightforward basic mathematics. Variables included in impact maneuvers of contact sports are as follows: weight times speed equates to momentum, which is proportional to the force of a collision *(11)*. An injury can also occur by an unusual landing attitude of the lower more so than the upper extremity, i.e., non-linear single or asymmetric limb stance (angled or twisting moment on the involved joints) *(12–17)* (Fig. 13.1). Aside from acute orthopedic injuries occurring by sudden explosive episodes, more gradual, chronic bouts of repeated force over time can also cause damage to the musculoskeletal system, resulting in overuse injuries and can eventually progress to stress fractures *(11,18)*.

Before addressing specific injuries occurring in young female athletes during their growing years, we will review a few pertinent terms and elaborate on several important issues. Physis refers to the growth plate located near the ends of long bones contributing to longitudinal growth. Apophysis is a similar structure but differs in that it is in close proximity to the tendinous attachment on bone (Fig. 13.2). These growing anatomical structures are at more risk of injury when exposed to excessive force. Additionally, bone of the immature skeleton is more porous and not as dense as in the adult as it is more vascular and has less mineral content, which translates to being more flexible but weaker structurally. The offset of this characteristic is that the periosteum or envelope surrounding growing bones is thicker, stronger, and biologically more active, and thus greatly promotes the process of healing at a much faster rate after skeletal trauma in children. Physiological fusion of growth plates begins during the preteen years and is completed by the early twenties, occurring several years earlier in girls than boys. Prior to completion of the fusing process, any physical damage incurred while the body is changing may lead to temporary or even permanent growth disturbance, which could end up being a leg length discrepancy or angular deformity of the affected limb. Another factor contributing to the differing pattern of orthopedic injuries in young athletes is the relative strength of tendons/ligaments versus growth plates. The former structures, including capsules (connective tissue surrounding joints), have sufficient strength and therefore can resist any application of force better than the latter, weaker physis and adjacent cartilage/bone. Consequently, any stress of sufficient magnitude to tear ligaments in adults tends to disrupt the physeal cartilage or bone of younger

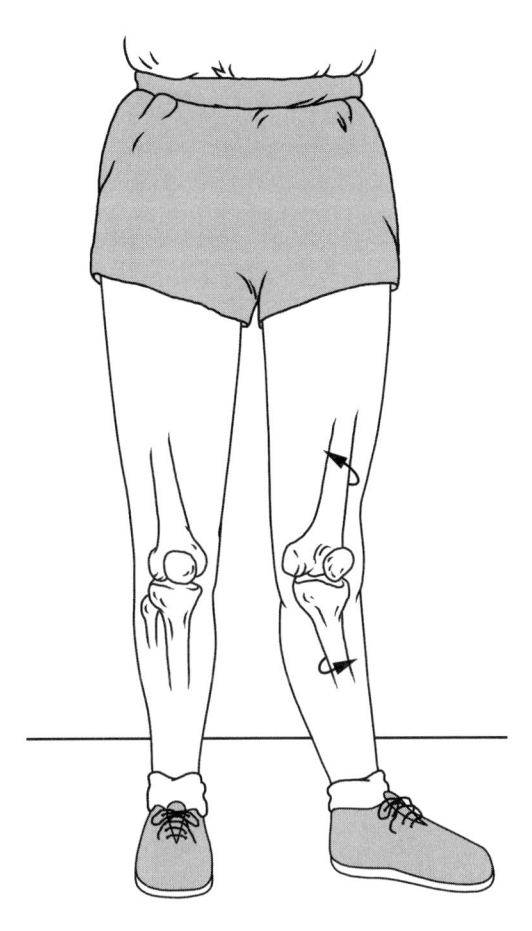

Chapter 13 / Musculoskeletal Injuries in Preadolescent and Adolescent Female Athletes 159

Fig. 13.1. Risky limb landing attitude.

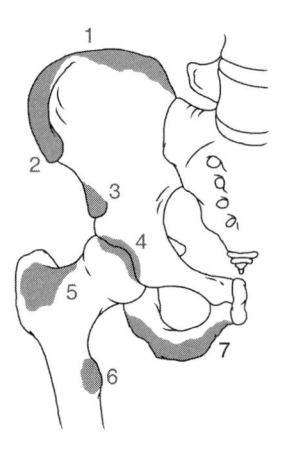

Fig. 13.2. Pelvic apophyses. (1) Iliac crest (2) Anterior superior iliac spine (3) Anterior inferior iliac spine (4) Triradiate cartilage (5) Greater trochanter (6) Lesser trochanter (7) Ischial tuberosity.

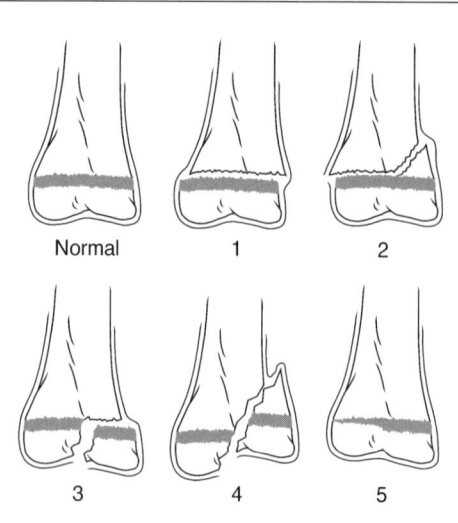

Fig. 13.3. Growth plate injuries. (1) Minor injury through growth plate (2) Injury involves metaphysis of bone (3) Involves epiphysis (4) Both metaphysis and epiphysis involved (5) Crushing type trauma involving growth plate.

children, resulting in growth plate injuries or avulsion fractures (Fig. 13.3). The latter results from sudden violent muscular contractions transmitted through various tendons inserting onto bone. Another source of damage imparted upon the young musculoskeletal system involves chronic, repetitive, and submaximal mechanical load applied over and over to the same area, causing microtrauma which eventually culminates in bursitis of soft tissues, strains of tendons, sprains of ligaments, stress reactions, or even frank fractures of bone! This type of orthopedic injury tends to occur more often in the lower rather than the upper extremity and results primarily from overtraining, i.e., too soon, too frequent, too intense, too much, and/or too fast *(11)*. As for which anatomical sites are more prone to fatigue fractures, these depend on the types of movement performed/required in specific athletic activities. For example, in the lower extremity, volleyball and basketball players are at increased risk of stress injury to the tibia because of repeated jumping; and runners, especially those engaged in training over 20 miles weekly, can damage their femoral neck or fibula from excessive impact. In a similar fashion, because ballerinas have to be "en pointe" while dancing on the tips of their toes, they tend to sustain stress fractures of the second metatarsal more readily. Likewise in the upper extremity, though with much less incidence, ulna stress fractures occur more often in tennis and fast-pitch softball players because of the nature of their racquet striking the ball or mechanism of pitching with their forearms, respectively; along the same lines, the risk of radial fatigue fractures is higher in gymnasts as a result of repetitive load-bearing moves on their wrists. Additionally, although quite rare, swimmers can injure their humeri from repeated overhead loading in water *(18)*. Furthermore, other athletic activities involving repetitive back extension maneuvers, i.e., dance, figure skating, and gymnastics, can potentially place excessive load on the lumbar spine, resulting in a stress fracture of the pars interarticularis (spondylolysis) *(4)* (Fig. 13.4). Additional variables contributing to other overuse type of injuries include environmental factors such as inadequate playing equipment, difficult surface terrain, faulty/worn-out footwear and inappropriate technique, or altered internal

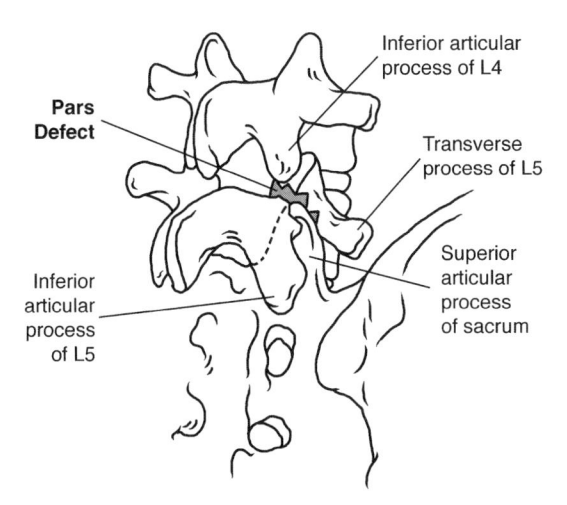

Fig. 13.4. Spondylolysis.

matters/anatomic features inherent within the athlete herself, i.e., systemic ligament laxity or tendinous tightness, and "miserable malalignment," among others *(4,11)*. For instance, with the upper extremity, structural differences in young female athletes place them more at a mechanical disadvantage and, thus, at an increased risk for certain types of overuse injuries. This stems from a shorter humerus compared to total arm length, less muscular strength, and more joint laxity. Consequently, sports involving excessive overhead activity, such as throwing and swimming, can contribute to rotator cuff tendinosis and shoulder subluxation *(4)*.

Regarding the "miserable malalignment syndrome," this entity is quite unique to young female athletes and is a constellation of lower extremity anatomical structures being misaligned associated with primarily quadriceps neuromuscular dominance, placing them at increased risk for knee and especially anterior cruciate ligament (ACL) injuries during various sporting activities *(7,8,10)*. Specifically speaking, we are talking about a widened pelvis, genu valgum, increased internal tibial torsion, and pes planus. The Q angle is accentuated, contributing to patellar maltracking which can cause anterior knee pain *(4)* (Fig. 13.5). Additionally, adolescent female athletes tend to recruit their quadriceps prior to their hamstrings (the reverse is true in males) when they land from a jump, along with having less hip/knee flexion and an elevated valgus knee moment. The resulting attitude/alignment of the lower body places a higher anterior shear force on the tibia which puts more strain on the knee, leading to one of the theories behind a several fold increased incidence of ACL tears in females involved in certain sports, i.e., basketball, soccer, and softball *(19,20)* (Fig. 13.6).

13.3.4. Several Measures for Prevention of Athletic Injuries from Occurring in Preadolescent and Adolescent Females

So what can be done to prevent musculoskeletal injuries from occurring in young female athletes as they go through the difficult growing years of adolescence? Well, we will redefine the different types of orthopedic trauma incurred by these females during various sporting activities. First of all, acute injuries while playing team sports are

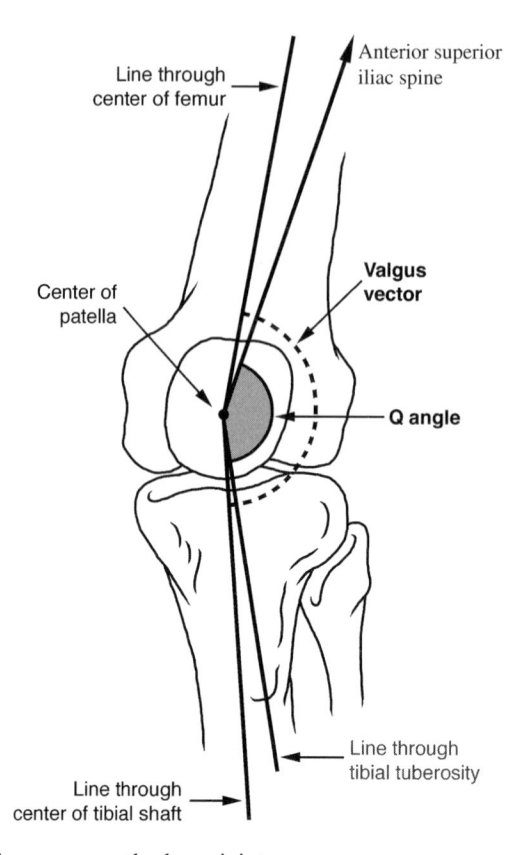

Fig. 13.5. Alignment/forces across the knee joint.

Fig. 13.6. Risky versus safe landing positions.

difficult to control as other players are involved. However, preventive measures should include employing appropriate safety equipment, either worn or used by the athletes themselves, and avoiding extreme surrounding environmental conditions during actual competitions. In terms of the more chronic type of overuse injury, especially of the lower extremity, consider instituting/modifying the six S's: (1) altered Structure or anatomic malalignment, (2) Shoe wear pattern or status, (3) Surface type or irregular topography, (4) Stretching for flexibility, (5) Strengthening of imbalanced muscle groups, and (6) avoiding too much Speed too soon. As an example, for those participating in track and field events, well-fitted and padded/cushioned footwear, along with correct running stride mechanics, is extremely important in the prevention of uneven force distribution and excessive stress transmission through the feet with regard to shoe–ground interaction *(11,21)*. Beyond these external variables, the main factors that can be modified through training are correct technique/skill level, altered parameter of play, and physical condition of the athlete *(9,22,23)*. For example, at the beginning of the regular sports season, athletes should not increase the volume of their workouts more than 10% weekly to avoid overtraining *(11)*.

The other very important variable to consider as far as prevention of musculoskeletal trauma in adolescent female athletes is their individual fitness profile. Studies have shown that preseason (several weeks) workouts, i.e., strength, endurance, conditioning, and plyometrics programs, can increase physical fitness and help in decreasing the risk of knee/ACL injuries *(23–25)*. A previous study has shown that even only 8 weeks of training in preadolescents can result in increased muscle strength of approximately 75% *(6)*. As the period of training is relatively short, these regimens represent the initial physiological adaptations involving the neuromuscular system, incorporating resistance/flexibility exercises, jumping/landing drills, and speed/agility maneuvers; all of these contribute toward improving muscular strength/endurance, anaerobic power, and sprint velocity *(1,26)*. One of the most crucial elements involved with typical conditioning programs is teaching an athlete how to land with the lower limbs in a "safer" position to help protect against potentially devastating ACL injuries, which could easily put her out for the rest of the sports season or even longer *(8,25)* (Fig. 13.6)!

13.3.5. Different Methods of Orthopedic Treatment for Musculoskeletal Injuries Incurred by Young Female Athletes, Including Specific Recommendations for a Regular Conditioning Program once Healing is Complete to Enhance Their Physical Fitness Profile

Once a female athlete has sustained an acute or overuse type of injury to her body, she should strive to follow these principles (PRICE): pain control and Protect the injured part from further harm, relative Rest for the limb (not absolute immobility as the surrounding joints will become stiff) while maintaining gentle range of motion (ROM), apply intermittent Ice for inflammation (20–30 min every 6 h), use Compression wraps for swelling, and Elevate the limb above the heart to limit edema as well. This is essentially the basic first-aid type treatment for any kind of soft tissue trauma. If a fracture occurs, whether because a sudden force or from gradual, repeated stress application, then add the following precaution—no weight bearing on the injured extremity, and if deemed severe enough, proceed toward orthopedic consultation and strict immobilization with external splinting/bracing/casting; if the displacement is of sufficient magnitude, it could ultimately require surgical internal fixation. As for the current training regimen, this must be modified so as not to add further insult to the already injured extremity. Reduction in the levels of play, i.e., magnitude, intensity, duration, and frequency, is a must to allow timely healing. Cross-training with a lower or no impact type of activity may be necessary, such as swimming or stationary bicycling, to keep from losing cardiorespiratory fitness. As far as medications are concerned, over-the-counter nonsteroidal anti-inflammatory drugs may be used, with the caveat that not so much is ingested that pain is taken away as a protective mechanism. Another consideration is the medication's potential toward a delayed healing response because of blunting of the body's natural process of inflammation. Finally, engage in a formal physical therapy rehabilitation program for joint ROM and muscular strength/endurance with progression toward sports-specific exercises after the initial period of recovery. Once functional testing of the affected extremity documents the restoration of objective measures of strength, girth, flexibility, and endurance, among other parameters, demonstrating that an athlete is indeed physically back to preinjury status and again ready to participate in sports, clearance can be granted for gradual return to play *(11)*.

Of utmost importance is the prevention of another injury to the same site, so a maintenance conditioning program should be instituted and continued throughout the sports season, then extended longer beyond that period to help maximize performance and minimize injury *(9,11)*. Several studies have demonstrated the beneficial effects of strength training in children and adolescents in terms of enhancing athletic performance, and there is minimal risk of trauma or growth disturbance as long as these programs are well supervised and submaximal exercises are being performed *(5,6,9,27,28)*. Warming up the involved joints by performing flexibility moves and gentle motions appears to afford some protection from risk of injury because of improved proprioception of surrounding soft tissue restraints, despite another study finding that stretching has a minimal prophylactic protective effect *(29,30)*. Specific guidelines regarding a resistance program for building muscular strength include the following: begin at age 7 or 8; add warm-up and cool-down periods before and after exercise; use own body weight, light medicine balls, or dumbbells; perform workouts that last between 20–30 min or longer, no more than 2–4 days per week (with rest days in between); initially, start with 1 set of 10–15 repetitions two times a week; progress to 1–3 sets of 6–15 repetitions concentrating on 6–8 different exercises, focusing on major muscle groups with mild tension/resistance while learning the correct technique of lifting; finally, sequentially, accelerate training by small increments of weight (2–5 lb) or 5–10% increase in the exercise load weekly once the amount lifted is no longer challenging/fatiguing. For those young athletes who desire to maximize performance in various competitions, they should also practice sports-specific moves/skills as well. Additionally, engaging in any type of aerobic or other endurance-type training is also well advised to enhance overall health in the long term *(5,6,22,23,27,31)*. The only caveat to this principle is that once training stops, regression or detraining occurs after several weeks of inactivity, so the athlete must continue with the conditioning program to retain the highest level of fitness benefits. One word of caution however: always refrain from power-lifting type maneuvers/drills, i.e., single maximal lifts in skeletally immature athletes, in order to avoid the potential of growth plate injury *(5,22,28)*.

13.4. CONCLUSIONS

In summary, in young females, the time span from childhood to adolescence is accompanied by a myriad of bodily changes, some of which resemble the changes experienced by males, while others are totally different. The main components involved with this life transition period of puberty include reproductive, hormonal, nervous, and musculoskeletal systems. The latter undergoes an extreme makeover to involve morphological and neuromuscular alterations. Growth, development, and then eventually maturation into the adult form is faced with an inherent risk, mainly orthopedic trauma from soft tissues (ligament and musculotendinous structures) lagging behind the fast-growing skeletal system (bone). Acute and chronic injuries may occur in preadolescent and adolescent female athletes participating in certain sporting activities. It is important to recognize these unique patterns of musculoskeletal trauma in order to protect the athlete and prevent injury and/or treat the offending problem. When in doubt, or if the injury is severe, do seek appropriate orthopedic consultation. The bottom line lies in helping to maximize gains and minimize risks by involving these young females in a well-supervised maintenance conditioning program. As such, they can compete and perform their best, while adapting to their ever-changing body during their teenage years and even further beyond that into adulthood.

13.5. SCENARIO WITH QUESTIONS AND ANSWERS

13.5.1. Scenario

You are a pediatrician who has just recently completed a primary care sports medicine fellowship and decided to join a multispecialty group practice in a suburban town. One of your partner's physician's assistants (PA) has been involved with taking care of a local youth girls' soccer team. These young female athletes are under tremendous pressure to win as their team will disband if they do not make the playoffs this season; as a result, these female soccer players have been required to "train" extra hard every weekday before and after practice. Their training program consists of running for a couple of miles around the track, then lifting weights for strength and power, and then run again for another mile or so. Evidently, their coach used to work with male football players, and he is incorporating the majority of football drills into the regular sports workout during the competitive season for these young female athletes. As far as he is concerned, the heavier weight moved the better. "Work through the pain!" echoes his motto. As much as these teenage girls want a good chance at the championship trophy, they are beginning to get discouraged, mainly because they cannot seem to get rid of their muscle aches and pains despite resting on the weekends and taking over-the-counter medications. In fact, their performances during the games have also started to deteriorate a bit, which makes their coach exercise them even harder. The PA is coming to you for advice concerning this group of adolescent females.

13.5.2. Questions

1. What do you think is happening to these young female athletes?
2. What questions should you ask of these athletes and how will you go about approaching the coach about his female soccer players' problems?

3. What elements of the history, physical exam, and diagnostic studies should you obtain to help you evaluate and treat these young female athletes?
4. What initial measures should you institute?
5. Who else should you consult for assistance in the management of their musculoskeletal concerns?

13.5.3. Plausible Answers

1. It appears as though the soccer coach is employing training techniques that are more specific to a different sport and group of athletes and do not fit these adolescent female athletes' training needs. Their complaints of pain may be coming from overuse injuries because of working out too much and too frequently.
2. You may want to inquire about other types of athletic activity, along with the specifics of the exercises these females are performing. Then, go talk to their trainer before discussing the situation with their coach. Approach him in a nonthreatening manner, explain the fact that his players are trying as hard as they can, but their bodies are not quite used to the rigors of his method of training yet.
3. Pertinent questions you will want to ask about these athletes' histories should include their previous playing experience, whether it is their first competitive season (rookie) or are they veterans at this game. You also would want to know about their past and present physical fitness profile to determine if indeed they are in good enough shape to undergo rigorous training outside of their soccer workouts. Try to rule out alternative causes of musculoskeletal complaints other than strains and sprains, such as stress reactions or fractures. Perform a thorough physical exam of the involved bones and joints. Obtain radiographs and, if these are negative, then order bone scans and/or magnetic resonance imaging to further delineate details of the orthopedic pathology. Check out laboratory blood work to see if any abnormal indices are present to indicate a metabolic or endocrine source contributing to their problem.
4. As for the initial treatment of these soccer players, institute the PRICE principle after talking things over with their trainer and coach. Discuss decreasing the training volume and intensity, cross-train with other activities so as not to stress the injured extremities, and hopefully maintain physical fitness. Avoid provocative exercise maneuvers while attempting to modify regular practice so that these athletes do not become completely deconditioned in the meantime. Once they are over the acute inflammatory stage, gradually increase their training to tolerance to help get them back into the game.
5. Do hold the affected young females back from any sort of impact activity and seek consultation from an orthopedist or surgical sports specialist if they are not responding to your treatment, especially if you are worried about stress fractures.

REFERENCES

1. Wilmore J. and Costill D. Growth development and the young athlete: Chapter 17. Special Populations in Sport and Exercise. *Physiology of Sport and Exercise: Human Kinetics*, Human Kinetics Champaign, IL, 1994: 401–421.
2. Rogol A., Clark P., and Roemmich J. Growth and pubertal development in children and adolescents: effects of diet and physical activity. *Am J Clin Nutr* 2000; 72: 521S–528S.
3. Schoenau E. and Frost H. M. The "Muscle-Bone Unit" in children and adolescents. *Calcif Tissue Int* 2002 April; 70: 405–407.
4. Timmerman M. Medical problems of adolescent female athletes. *Wis Med J* 1996 June; 95 (6): 351–354.
5. Borms J. The child and exercise: an overview. *J Sports Sci* 1986; 4: 3–20.

6. Bencke J., Damsgaard R., Saekmose A., Jorgensen P., and Klausen K. Anaerobic power and muscle strength characteristics of 11 years old elite and non-elite boys and girls from gymnastics, team handball, tennis and swimming. *Scand J Med Sci Sports* 2002; 12: 171–178.
7. Ahmad C., Clark M., Heilmann N., Schoeb S., Gardner T., and Levine W. Effect of gender and maturity on quadriceps-to-hamstring strength ratio and anterior cruciate ligament laxity. *Am J Sports Med* 2006; 34(3): 370–374.
8. Barber-Westin S., Noyes F., and Galloway M. jump-land characteristics and muscle strength development in young athletes a gender comparison of 1140 athletes 9 to 17 years of age. *Am J Sports Med* 2006; 34 (3): 375–384.
9. Hewett T., Ford K., and Myer G. Anterior cruciate ligament injuries in female athletes. Part 2, A meta-analysis of neuromuscular interventions aimed at injury prevention. *Am J Sports Med* 2006; 34 (3): 490–498.
10. Withrow T., Huston L., Wojtys E., and Ashton-Miller J. The relationship between quadriceps muscle force, knee flexion, and anterior cruciate ligament strain in an in vitro simulated jump landing. *Am J Sports Med* 2006; 34 (2): 269–274.
11. Martin T. and Martin J. Special issues and concerns for the high school and college-aged athletes. *Pediatr Clin North Am* 2002 June; 49 (3); 333–352.
12. Barber-Westin S., Galloway M., Noyes F., Corbett G., and Walsh C. Assessment of lower limb neuromuscular control in prepubescent athletes. *Am J Sports Med* 2005; 33 (12): 1853–1860.
13. Chappell J., Yu B., Kirkendall D., and Garrett W. A comparison of knee kinetics between male and female recreational athletes in stop-jump tasks. *Am J Sports Med* 2002 March–April; 30 (2): 261–267.
14. Colby S., Francisco A., Yu B., Kirkendall D., Finch M., and Garrett W. Electromyographic and kinematic analysis of cutting maneuvers implications for anterior cruciate ligament injury. *Am J Sports Med* 2000; 28 (2): 234–240.
15. Ford K., Myer G., and Hewett T. Valgus knee motion during landing in high school female and male basketball players. *Med Sci Sports Exerc* 2003; 35 (10): 1745–1750.
16. Ford K., Myer G., Toms H., and Hewett T. Gender differences in the kinematics of unanticipated cutting in young athletes. *Med Sci Sports Exerc* 2005; 37 (1): 124–129.
17. Hass C., Schick E., Tillman M., Chow J., Brunt D., and Cauraugh J. Knee biomechanics during landings: comparison of pre- and postpubescent females. *Med Sci Sports Exerc* 2005; 37 (1): 100–107.
18. Verma R. and Sherman O. Athletic stress fractures: Part II. The lower body Part III. The upper body-with a section on the female athlete. *Am J Orthop* 2001 December; 30 (12): 848–860.
19. Hewett T., Zazulak B., Myer G., and Ford K. A review of electromyographic activation levels, timing differences, and increased anterior cruciate ligament injury incidence in female athletes. *Br J Sports Med* 2005; 39:347–350.
20. Powell J. and Barber-Foss K. Sex-related injury patterns among selected high school sports. *Am J Sports Med* 2000; 28 (3): 385–391.
21. Patritti B. Running shoe cushioning impacts foot-ground interface. *Biomechanics* 2004 May 1: 57–67.
22. Benjamin H. and Glow K. Strength Training for Children: Risks Versus Benefits. Illinois Chapter AAP. Illinois Pediatrician Sports Medicine Articles. URL: http://www.illinoisaap.org/sportsarticles.htm.
23. Faigenbaum A. and Kang J. Youth strength training: facts, fallacies and program design considerations. *Am Coll Sports Med* 2005 October/November/December; 15 (4): 5–7.
24. Faigenbaum A. and Chu D. Plyometric training for children and adolescents. *American College of Sports Medicine* 2001 December. URL: http://www.acsm.org/Content/ContentFolders/Publications/CurrentComment/2001/plyometr.pdf
25. Myer G., Ford K., McLean S., and Hewett T. The effects of plyometric versus dynamic stabilization and balance training on lower extremity biomechanics. *Am J Sports Med* 2006; 34 (3): 445–455.
26. Balabinis C., Psarakis C., Moukas M., Vassiliou M., and Behrakis P. Early phase changes by concurrent endurance and strength training. *J Strength Cond Res* 2003 May; 17 (2): 393–401.
27. Bernhardt D., Gomez J., Johnson M.D., Martin T.J., Rowland T.W., Small E., LeBlanc C., Malina R., Krein C., Young J.C., Reed F.E., Anderson S.J., Anderson S.J., Griesemer B.A., Bar-Or O., and Committee on Sports Medicine and Fitness. Strength training by children and adolescents. *Pediatrics* 2001 June; 107 (6): 1470–1472.
28. Guy J. and Micheli L. Strength training for children and adolescents. *J Am Acad Orthop Surg* 2001 January/February; 9 (1): 29–36.

29. Bartlett MJ and Warren P J. Effects of warming up on knee proprioception before sporting activity. *Am J Sports Med* 2002 April; 36 (2): 132–134.
30. Herbert R. and Gabriel M. Effects of stretching before and after exercising on muscle soreness and risk of injury: systematic review. *BMJ* 2002 August; 325: 1–5.
31. Izquierdo M., Hakkinen K., Gonzalez-Badillo J., Ibanez J., and Gorostiaga E. Effects of long-term training specificity on maximal strength and power of the upper and lower extremities in athletes from different sports. *Eur J Appl Physiol* 2002; 87: 264–271.

14 Prevention and Management of Common Musculoskeletal Injuries in the Adult Female Athlete

Mimi Zumwalt

CONTENTS

14.1 LEARNING OBJECTIVES
14.2 INTRODUCTION
14.3 RESEARCH FINDINGS
14.4 CONCLUSIONS
14.5 SCENARIO WITH QUESTIONS AND ANSWERS

14.1. LEARNING OBJECTIVES

After completing this chapter, you should have an understanding of the following:

- The fundamental differences in the anatomy, physiology, and body composition between adolescent and adult females in terms of the musculoskeletal system.
- The relative similarities and pertinent differences between adult males and females concerning the anatomy, body composition, and biomechanics of the musculoskeletal system.
- Various more common types of orthopedic injuries sustained by adult females involved in certain athletic activities.
- Several measures for prevention of musculoskeletal injuries incurred by adult female athletes.
- Different modes of treatment for orthopedic injuries sustained by adult females participating in certain sporting activities.

14.2. INTRODUCTION

The developing female body, as growth proceeds from childhood through adolescence, then finally culminates into adulthood, carries along with it a multitude of changes. Not only does a mature woman have to face morphological challenges, she must also adapt to structural, hormonal, and metabolic alterations as well. Indeed, the musculoskeletal system is certainly no exception to this growing rule throughout the

From: *The Active Female*
Edited by: J. J. Robert-McComb, R. Norman, and M. Zumwalt © Humana Press, Totowa, NJ

different life stages. Although it is healthy to be engaged in a regular exercise program, participating in multiple athletic competitions and sporting activities can leave the adult female more vulnerable to orthopedic injuries, especially if she is not sufficiently fit to perform her best in extremely demanding levels of physical play.

This chapter will focus on several anatomical features unique to the adult female athlete, different types of musculoskeletal trauma which tend to be more common for women participating in sports, along with methods of injury prevention and how to go about treating these orthopedic issues once they do occur.

14.3. RESEARCH FINDINGS

14.3.1. The Fundamental Differences in Anatomy, Physiology, and Body Composition between Adolescent and Adult Females in Terms of the Musculoskeletal System

To recap some of the similarities and differences between young females and adult women, the following anatomical changes are found to be fairly consistent: breast tissue development and fat deposition around the thighs and buttocks, along with broadening of the pelvis all of which herald the onset of puberty under the influence of the female sex hormone, estrogen *(1,2)* (Fig. 14.1). This transition stage usually occurs between the ages of 12 and 14. Estrogen also brings about longitudinal skeletal growth, with the final bone length achieved between 2 and 4 years after pubescence. A woman's body composition continues to fluctuate during the teenage years, and somewhere in the mid-twenties, more adipose tissue is accumulated and muscle mass starts to decline, at a rate of about 3 kg per decade (greater than 0.5 lb per year). The rise in total body fat along with loss in the quantity of fat-free mass is because of lower levels of physical activity, relative lack of testosterone, and consuming the same quantity of caloric intake *(2)*. In terms of skeletal density and therefore bony integrity, calcium, among other essential minerals, continues to be deposited into the skeleton to build up bone as the body is steadily growing. This process of calcium deposition into the bone bank thus increasing the quantity of bone gained becomes accelerated during the adolescent growth spurt of puberty during the teenage years and then reaches its

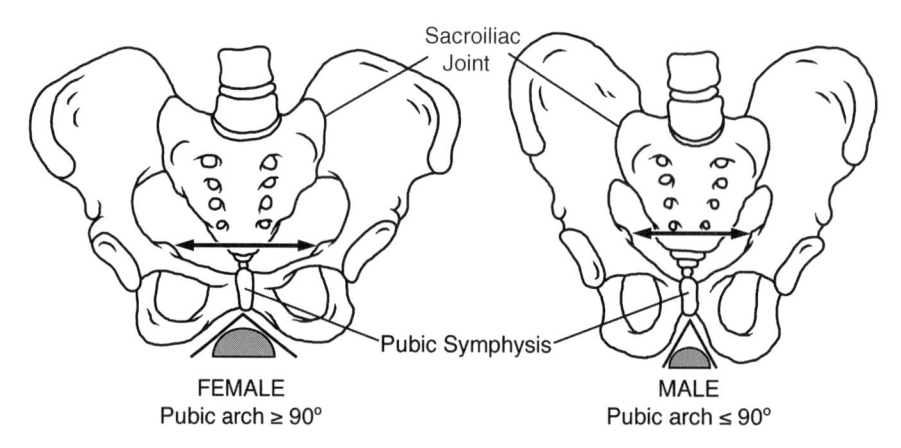

Fig. 14.1. Wider pelvic brim in females vs males.

Chapter 14 / Common Musculoskeletal Injuries in Adult Female Athletes   171

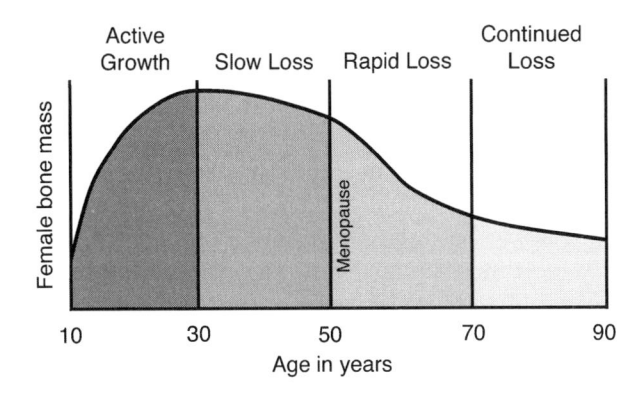

Fig. 14.2. Rate of bone loss though a woman's lifetime.

peak in terms of bone mass acquisition in the early twenties. After this period of very rapid bone deposition, bone loss begins to occur gradually during the late twenties or early thirties dependent on nutritional and hormonal status, along with the amount of applied mechanical stimuli *(3)* (*see* Chapter. 15) (Fig. 14.2).

14.3.2. *The Relative Similarities and Pertinent Differences Between Adult Males and Females Concerning the Anatomy, Body Composition, and Biomechanics of the Musculoskeletal System*

Up until the onset of puberty around 13 years of age or so, anatomical structure and body composition between males and females are fairly similar in terms of height, weight, girth, bone width, and subcutaneous fat. The body build/shape and size of both sexes then begins to diverge once the endocrine system starts to undergo some changes. Namely, two sex-specific hormones, i.e., estrogen and testosterone, begin to take over to influence the development of adolescent features, separating teenage girls and boys in terms of anatomy of the body in general and the musculoskeletal system in particular *(1,2)*. Specifically speaking, females during the adolescent growth spurt tend to deposit more fat around the breasts, hips, and thighs. This gender-related adipose tissue deposition, in addition to that surrounding internal organs, is termed essential fat, which in women composes about 9–12% of total body weight, as compared to only 3% in males. The other type of adipose tissue, i.e., storage fat, is comparable in both sexes at about 15% of body weight *(2,4–6)*. At physiological maturation, the average adult female nonathlete carries about 18–26% adipose tissue overall, whereas mature men only contain between 12–16% body fat on average *(4,5)*. As for the pattern of fat distribution, males carry more adipose tissue in the abdomen/flank and upper body versus females, who store fat mostly around their hips and lower body *(2)*. What does having more adipose tissue mean for a woman competing in certain types of sports? Well, other than performing athletic activities in water such as swimming, where increased buoyancy is deemed advantageous by decreasing drag and lowering energy expenditure by 20%, the higher fat percentage is thought to hinder physical performance in many endurance sports from having to carry the excess body weight (though hard evidence is somewhat lacking) *(4–6)*. However, for those female athletes engaged in excessive exercise, especially long-distance/marathon runners who train and/or compete over 100 miles per week, their percentage of body fat can be reduced to well below 10% and even lower than that of males, which is not entirely healthy *(4,6)*. In fact, some elite endurance athletes can eventually become too lean, with body fat around 6–8%, which in and of itself could bring on other problems such as endocrine/hormonal/skeletal dysfunction leading to the "Female Athlete Triad"—more on this later *(4–7)*. How about absolute versus relative strength between men and women? Generally speaking, between 11 and 12 years of age, girls are about 90% as strong as boys. By the time they are 15 to 16 years old, males are about 25–33% stronger than females overall *(4,6)*. When adjustments are made for body mass, men are at least one-third stronger than women *(5)*. Specifically, women are only 50% as strong in the upper body and 75% in the lower body as men *(6)*. When muscle mass is taken into account, relative leg strength is similar in both sexes, but upper extremity strength in females still lags behind that of males *(1,4,6)*. This is because of men having more muscle mass relative to body weight than women, 40% as compared to 23%. The lack of muscle strength stems from women having smaller fiber size (15–40%), making the cross-sectional area about 60–85% smaller as well *(4,6)*. In fact, this relative difference in fat-free mass tends to be lower in females from age 7 until 25 years of age. After the mid-twenties, lean muscle mass begins to decline in both sexes, 0.25–0.50 lb yearly *(2)*. Corresponding to men having more lean muscle mass, the skeletal framework in males is larger as well; they also mature later and attain more final height (taller) and weight (heavier) than females from undergoing a longer period of growth *(6)*. On the other hand, under estrogen's effect, growth rate of the female skeleton is greatly accelerated, culminating in the final bone length to be achieved only a few years after the onset of puberty. In other words, females cease to grow earlier—their body and bones reach a plateau after 2–4 years of fairly rapid growth *(2)*. Therefore, men reach skeletal maturity about 21 or 22 years old, rather than 17–19 years of age in women. In terms of bony and articular differences, females' bones are smaller as well and therefore their joints have less surface area. Additionally, men have broader shoulders, larger chests, and narrower hips; women, on the other hand, have wider pelves, more varus hip, and higher knee valgus angles (Fig. 14.3). This increased angular inclination in the lower limbs of females causes an asymmetrical force distribution/transmission through the extremity's overall malalignment from the hips to the ankle and, subsequently, can contribute to a myriad of overuse type of musculoskeletal injuries *(1,2,4–6)*. In addition, lower extremity neuromuscular recruitment in women differs from that of men after puberty as far as their landing position/attitude from a jump, putting their knees and especially the anterior cruciate ligament (ACL) more at risk for trauma when playing certain sports. In other words, female athletes as a whole tend to be quadriceps rather than hamstrings dominant, meaning they land with their knees more extended thus having a less protective mechanism against ACL tears *(4,8–12)* (*see* Chap. 13 Fig. 13.6).

14.3.3. Various more Common Types of Orthopedic Injuries Sustained by Adult Females Involved in Certain Athletic Activities

Historically speaking, after the beginning of civilization but definitely dating as far back as ancient times, women were strictly forbidden to even be spectators in the audience, much less allowed to participate in the Greek Olympic games. It was viewed

Chapter 14 / Common Musculoskeletal Injuries in Adult Female Athletes   173

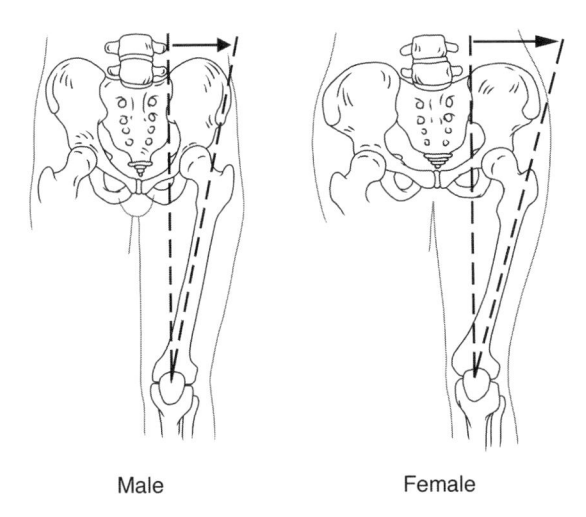

Fig. 14.3. Hip/knee angular difference.

as "against the laws of nature"! Indeed, back then females were treated as winning awards for men/competitors, i.e., the ultimate prizes for male Olympic champions. As time passes, society's view on women and exercise has come a long way since the prehistoric era *(3,6)*. As a matter of fact, not until in 1912, that women were finally able to compete in diving and swimming as part of the Olympic sporting events *(3,4)*. Later on in this century about six decades later, competition in athletics really began to explode for women. This landmark occurrence started in the early seventies as result of the Education Amendments Act by US Congress—Title IX passage in 1972, mandating absolute equality for females participating in sports supported by any federally funded schools. In other words, this law prohibited sex discrimination in athletic competitions. Subsequently, in the period from 1970s to 1990s, the role of females in athletic competition has been revolutionized, from sitting in the stands to standing in the sidelines, and of course playing in sporting events and even setting records. Alongside the tremendous rise in female sports participation comes with it the visible gender differences between men and women, especially in the different types of musculoskeletal injuries incurred among other unique orthopedic issues occurring *(3–6,13–15)*.

The following depicts musculoskeletal conditions more commonly seen in female athletes that may vary in types and incidence according to different sporting activities. Even recreational events such as running can bring about 25–65% of injuries severe enough to keep women away from training. In fact, about half of these female runners must seek medical care for their musculoskeletal problem. Other modes of physical activity which portend a similar rate of injury risk include aerobic dancers and even just walkers *(13)*. However, first and foremost starting with the lower extremity, the rate of knee and especially ACL injuries affecting women competitors can range anywhere from 2 to 10 times higher than male athletes *(3–6,8,12,13,15)* (Fig. 14.4)! Sports that seem to carry an increased risk for ACL tears include gymnastics, basketball, soccer, downhill skiing, field hockey, team handball, and lacrosse. Etiology of this gender-specific phenomenon is theorized to be multifactorial, including anatomical, hormonal,

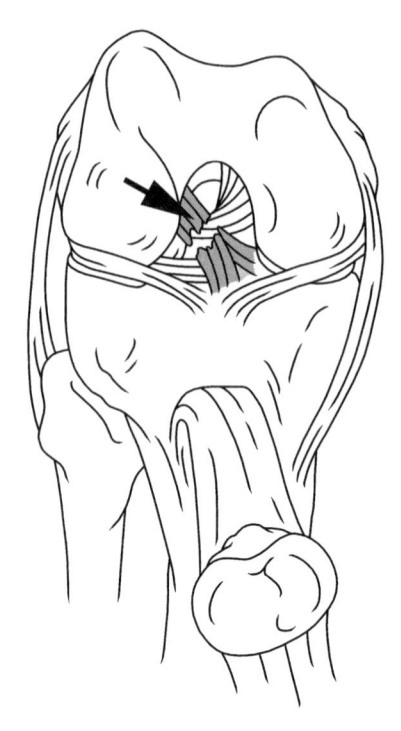

Fig. 14.4. Anterior cruciate ligament tear.

biomechanical, neuromuscular, and environmental. Risky landing positions have been previously alluded to but usually involve an off-balance body position especially on one limb—hip extended/adducted/internally rotated, knee extended/valgus, leg externally rotated, and flat foot attitude during decelerating, cutting, or planting maneuvers *(4,5,8,10,11)* (*see* Fig. 13.1). Other contributing factors may include skill level, physical condition, imbalanced thigh musculature (especially hamstring weakness as compared to the quadriceps), increased flexibility, and decreased proprioceptive capability *(4,5,9)*. Other musculoskeletal concerns more unique to female athletes around the knee include the patellofemoral joint (PFJ). Contributing factors to anterior knee pain and dysfunction stem from static structural misalignment from the pelvis to the feet, dynamic imbalance of limb muscular strength, excessive soft tissue pliability or extreme stiffness, or a combination thereof *(8)*. Specifically speaking, this entity is often named the miserable malalignment syndrome, where increased hip external rotation/femoral anteversion coupled with a higher Q angle, genu valgum, tibia vara, and a hypermobile patella all contribute to patellofemoral pain *(1,4,6,15)* (*see* Fig. 13.5). The body, as an attempt to center the knee under the hips and above the ankle, compensates by internally rotating the former proximal joint, along with externally rotating the tibia and pronating the foot, all as an effort to keep the latter anatomical structure pointing straight forward with gait. The other scenario affecting the PFJ is increased pressure from excessive tightness of the lateral patellar retinaculum (soft tissue restraint), again contributing to patellofemoral maltracking and pain (PFPS) (Fig. 14.5). Additionally, affected females may have a shallow trochlear grove of the distal femur and/or high or low riding patella, accentuating the maltracking problem *(1,6,15)*. Furthermore, as

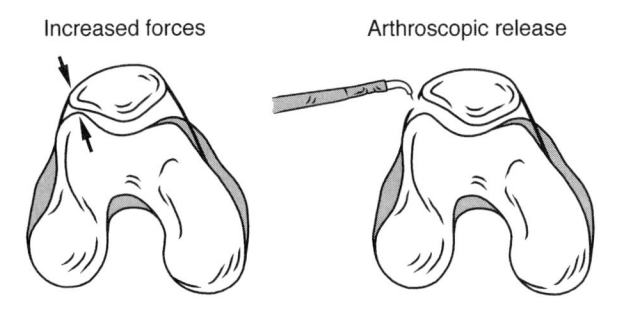

Fig. 14.5. Patellofemoral joint incongruency.

females flex their knees from full extension to about 30°, the femoral condyles do not support the patella as well as in males, contributing to the more lateral riding position of the latter. One study reported the incidence of PFPS in men is only 7.4%, as compared to 20% in women. Furthermore, sports such as cycling or running on hills or those involving repeated strikes to the anterior knee region (volleyball) tend to place more stresses across the PFJ, in turn causing an increased risk of experiencing painful symptoms (4). The third category affecting the PFJ in female athletes involves subluxation or frank dislocation, which may be a result of an acute, traumatic episode or from chronic, repetitive type of activity (1,4). Other lower extremity conditions more prevalent in athletic females involve the foot, especially if special shoe wear is deemed necessary for sports participation. Bunion formation from bursitis overlying the first metatarsal head of the great toe prominence can occur and, in fact, exceeds the incidence in men by about 9 times. This condition is because of a wide forefoot shape combined with wearing a narrow shoe toe box, causing excessive pressure and therefore more wear, resulting in inflammation and pain. This inflammatory entity is aggravated by midfoot pronation or flat arches and wearing high heels, the price women have to pay for being "fashionable" (1,4,15). The latter type of footwear can also cause Achilles tendonitis from tendon shortening and reduced flexibility, along with microtrauma at the insertion point into the heel (4). Additionally, the other factor playing a contributory role toward female foot problems lies in the type of sports consisting of frequent starts, decelerations, and sudden stops such as basketball, causing excessive forward and backward movements of the foot inside the shoe and thus increased friction around the lesser toes. Again, this overuse condition is exacerbated by wearing athletic shoes originally designed for males, not taking into account that a female's foot is shaped wider in front and narrower in the back (hindfoot) (1,4,6).

Moving on to the upper extremity, a larger proximal key articulation which seems to be more susceptible in females in terms of orthopedic injuries is the shoulder girdle joint, especially when engaging in higher risk athletic activities such as gymnastics, swimming/diving, throwing/pitching maneuvers, racquet sports such as tennis, and volleyball. Again, etiology of this upper limb trauma could also be acute or chronic, with anatomical malalignment, structural imbalance, poor posture, muscular weakness, soft tissue inflexibility, or excessive laxity contributing to the incidence and severity of injury. Biomechanically speaking, female athletes performing in sports requiring overhead activities tend to place undue stress on the soft tissue stabilizers (both static and dynamic) surrounding the shoulder girdle articulations, resulting in bursitis and

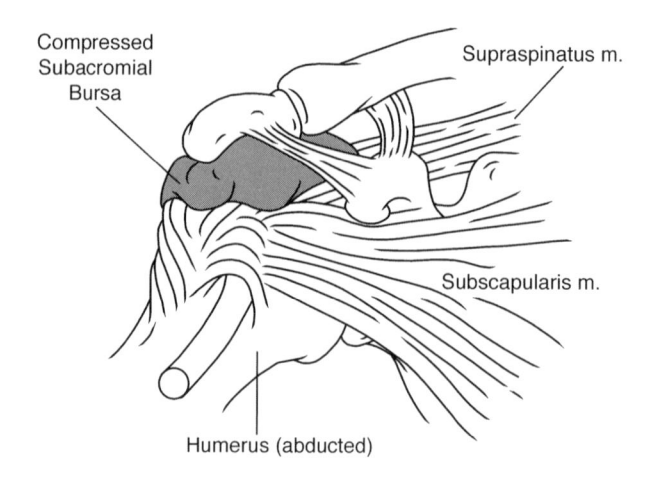

Fig. 14.6. Impingement syndrome.

rotator cuff tendinosis, leading to inflammation and pain (impingement syndrome) (Fig. 14.6). For example, female swimmers, as their bodies and arm lengths are generally shorter, must utilize more strokes across the water to cover the same distance as men, increasing the number of repetitive insults placed across the shoulder joint. Another cause of shoulder problems in overhead female athletes is because of increased laxity of the joint capsule, allowing the humeral head to "ride" out and in over the glenoid, making the dynamic stabilizers such as the rotator cuff muscles work that much harder to keep the "ball" centered in the "socket," again causing painful symptoms *(1,4,6)* (Fig. 14.7). Similarly for throwing mechanics, as humeral length is less in women but forearm length is the same as men, along with having more narrow shoulder girdles, the lever arm is shorter as well, making overhead winding motions more difficult, placing differential forces on the surrounding supportive musculature. A second upper extremity joint which is more susceptible to "wear and tear" in females is the elbow, with an incidence of 10% versus 1–5% in men between 42 and 46 years of age. This orthopedic entity is lateral epicondylitis or more appropriately termed tendinosis, a painful overuse condition because of repetitive forearm rotation affecting the extensor muscle(s) *(4)*.

Yet, an additional musculoskeletal condition which happens to appear more frequently in female athletes stems from overuse because of repeated bouts of impact activity, resulting in stress fractures. This manifestation could be one or more isolated traumatic episodes or only part of a very involved medical/psychosocial issue such as the "Female Athlete Triad," a constellation of anemia, menstrual dysfunction, and osteoporosis (*see* Chap. 5). In this case, stress fracture incidence could be as high as 12 times more than men! One study reports *(16)* that the incidence of bilateral stress fractures in females is twice that of males. Any sport involving excessive, recurring impact loading such as running or jumping and especially track and field/cross-country events could place the involved female athlete more vulnerable for stress injury. Susceptible anatomical regions for stress fractures in women include the pelvis/pubis, femur, tibia, fibula, and foot, with the tibial and femur regions being the most prevalent. In fact, tibial stress fractures account for 33–55% of all stress injuries

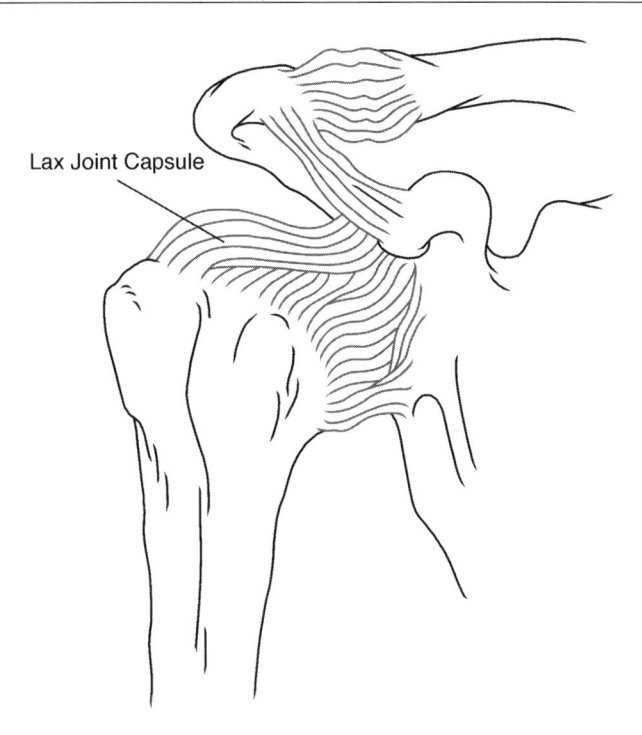

Fig. 14.7. Shoulder instability.

(Fig. 14.8). Associated risk factors can be internal to include muscle imbalance and limb malalignment or attributed to external factors such as the type of footwear, running style, and training errors, i.e., exercise mode/frequency/intensity/volume; additional contributors can involve psychological and medical issues, including inadequate/inappropriate nutrition/disordered eating, osteopenia, and hormonal/metabolic irregularity *(1,4–6,8,16)*.

14.3.4. Several Measures for Prevention of Musculoskeletal Injuries Incurred by Adult Female Athletes

As previously stated, as compared to males, both intrinsic (internal) and extrinsic (external) factors contribute to the increased incidence and varied types of orthopedic injuries incurred and musculoskeletal conditions occurring in adult female athletes. To reiterate, intrinsic factors include sex-specific changes in musculoskeletal anatomy and physiology affecting relative limb length, lower extremity alignment, body composition, muscle mass/strength, and neurological recruitment. Extrinsic factors involve physical condition/reaction/proprioception, playing level, competitive environmental conditions/terrain, sporting equipment, and athletic wear, especially shoes, along with interaction/friction between footwear and ground surfaces *(1,2,5,6,8)*. Naturally, internal issues obviously cannot be changed as this is inherent within the athlete. However, neuromuscular dominance can be modified through training/conditioning to affect the manner by which the athlete lands during certain athletic activities to help decrease lower extremity injury risk. Additionally, strengthening exercises for the hip (gluteus/flexors/abductors/adductors), thigh (quadriceps/hamstrings), and leg

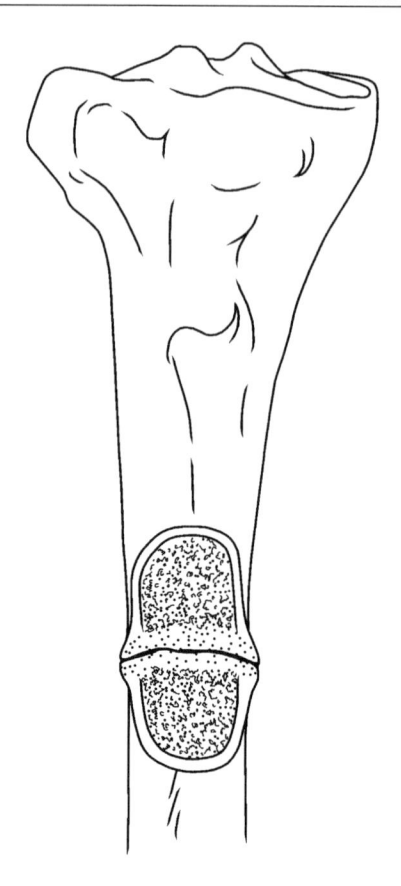

Fig. 14.8. Tibial stress fracture.

(gastrocsoleus) muscles can be instituted to help support key joints of the lower limb (hip/knee/ankle). The importance lies in balancing the anterior and posterior musculature to optimize muscular contraction with neurological activation while playing sports. Furthermore, other aspects of the workout program should also be maximized, such as stretching for flexibility, ipsilateral stances for balance/proprioception, core conditioning for trunk stabilization, plyometrics for anaerobic power, and agility drills for speed/coordination. To be extremely effective, these training programs must be at least 6 weeks long, and sessions must be more frequent than once a week *(5,6,17,18)*. Sports-specific training should also be integrated into the exercise regimen to ensure that correct/strict technique is maintained. Similar conditioning programs can also be applied to the upper body, enhancing physical fitness and minimizing injury mechanisms. Of extreme importance is addressing deficits in range of motion (ROM) and strength of the shoulder girdle region, especially the rotator cuff muscles. Other musculotendinous units surrounding this upper limb articulation, such as the pectorals, lats, deltoids, and triceps/biceps, should also be strengthened for added support *(1,4,6)*. As for external factors, these can be better modified to a certain extent by ensuring that playing equipment is adjusted to body size, and athletic gear, especially shoes, are well padded/fitted to the foot. As far as prevention of stress injury is concerned, avoidance of excessive, repeated bouts of impact activity will help to reduce this and other types of overuse orthopedic conditions. Cling to the principle of gradual moderation when increasing training intensity, duration, and frequency in order to optimize athletic performance and minimizing musculoskeletal injury *(1,6,13,15)*.

In short, the majority of extrinsic factors contributing to female musculoskeletal conditions can be modified to decrease the incidence of orthopedic injury. The mainstay of this prevention strategy lies in physical conditioning programs, training women athletes to keep their body in balance as far as strength, flexibility, and proprioception, along with optimizing other factors such as speed/agility, coordination, and power. However, the exercise regimen must be individualized to match the female athlete's fitness profile, and progression should be gradual in order to avoid overtraining, attrition, and injury rates *(5,8,17,18)*.

14.3.5. Different Modes of Treatment for Orthopedic Injuries Sustained by Adult Females Participating in Certain Sporting Activities

Of course it is always better to prevent rather than treat musculoskeletal injuries, but once they occur in adult female athletes, proper management measures should be taken promptly to ensure the best chance of recovery in order not to miss too many training sessions or games/sporting competitions. The following general principles (PRICE) hold when addressing orthopedic issues: Protection from further harm to the injured extremity; relative Rest for the affected limb(s)—keep ROM to minimize stiffness; Ice to aid in soft tissue inflammation; even/equal distribution Compression wrap (not to tight) to help with edema; and Elevation above heart level as much as possible to control swelling. In other words, utilize appropriate first aid initially to hopefully halt the progression of the injurious process and reduce painful symptoms. As far as treatment for specific injuries is concerned, tendon strains, muscle "pulls," and ligament sprains, dependent on the degree, will usually respond to avoidance of provocative maneuvers and first-aid measures initially *(19)*. An addition of anti-inflammatory (NSAID) medications is acceptable in the short term to help with painful symptoms. Similarly, electrical stimulation devices may also be used, but only as an adjunct to neuromuscular retraining. In terms of orthotics for the feet, the custom molded ones tend to be better that off the shelf to help distribute load more evenly inside the shoe *(4,15)*. As for bracing of the limbs, use these judiciously because muscles can become deconditioned if worn too frequently, and no studies have shown conclusive evidence that they can prophylactically prevent knee injuries *(1,6,15)*. Similarly, athletic taping, although helpful in terms of joint stability, tends to take away the protective proprioceptive capability. In addition, the tape loosens after about 20 minutes of play, then eventually becomes completely ineffective *(4,6,15)*. Once trauma to any part of the skeleton has happened; however, this is deemed more serious as it may represent a bone bruise, stress reaction or at the worst, frank fractures! In this case, absolute limb immobilization is of utmost importance and radiological studies are definitely warranted to rule out displaced fractures, as musculoskeletal insults of this severity should have an orthopedic consultation and could even culminate in surgical fixation *(5,8)*. As far as nonoperative management of musculoskeletal conditions in the adult female is concerned, as soon as the athlete has a pain free limb and ROM is regained, progression of activities should ensue to include therapeutic rehabilitation exercises. The rehab program should institute conditioning workouts previously outlined to include strength, flexibility, power, speed, and agility. In addition, elements of coordination, balancing/proprioception should also be integrated as part of the physical training regimen. Furthermore, faulty body biomechanics should be adjusted when engaging in movement/motion to minimize stress as part of the kinetic chain. Finally, sports-specific drills need to be added once prior physical milestones have been mastered. From this point on, functional exercises are emphasized in this phase of recovery in order to return the athlete to her preinjury status and level of play *(5,6,8,13,17)*.

14.4. CONCLUSIONS

In summary, the adult female athlete, through the growing years of development into the mature stage of life, has to face not only morphological changes of the entire musculoskeletal system, but she also has to deal with sex-specific hormonal alterations influencing body composition as well. Before the early 1970s, involvement of women in sports was sparse until Congress intervened, mandating equality for females participating in academic/athletic competitions. Paralleling the exponential rise in female sporting events, the rate of orthopedic injuries/conditions also became much more prevalent and exceeding the incidence in men, affecting especially the knee joint and namely the ACL. Different factors, both intrinsic and extrinsic, contribute to the increased occurrence of musculoskeletal injury in women. What appears to be fairly effective in preventing certain orthopedic conditions in female athletes consists of physical conditioning programs, instituting strength, flexibility, agility, speed, proprioception, coordination, plyometrics, and power. Once musculoskeletal injuries have occurred, however, it is prudent to recognize/diagnose the offending problem early in order to protect the athlete from further harm. In addition, initial first-aid type measures can be instituted to help decrease symptoms and enhance healing/recovery. Add physical rehabilitation exercises with individualized gradual progression as tolerated once the athlete responds to these modalities. However, if the injury becomes recalcitrant to conservative management, especially after a couple of weeks, orthopedic consultation should be sought after for further evaluation and treatment. The ultimate goal for any injured athlete is to provide aid promptly and hasten rapid recovery as much as possible in order to return her to play, whether it is in elite competitions or recreational sports.

14.5. SCENARIO WITH QUESTIONS AND ANSWERS

14.5.1. Scenario

You are a fairly new athletic trainer for a collegiate level female basketball team in a mid-size town. This is your first season working with these women athletes. One half is composed of rookies (junior varsity) and the other half veterans (varsity) players. Recently you have noticed that not all of these females show up every day for practice and that several have had more "off" than "on" days in terms of "scrimmage" games. You have also been providing daily "treatments" as far as local modalities and physical therapy rehab exercises in the training room for over a third of these female athletes for various musculoskeletal complaints. In fact, the majority of the problem lies in the lower extremity and most notably the tibia and knee regions. A couple of females are improving in terms of their symptoms, but the rest is still not making any appreciable gains. Alarmingly, some are actually getting worse. The regular sports season will begin in about 2 months and you are quite concerned that some players may not be able to "start" in their best physical condition.

14.5.2. Questions

1. Initially, how should you go about discovering the root of these female athletes' problem?
2. What specific questions should you pose to each woman about her complaints/symptoms?
3. What physical sign(s) of overuse injury should you look for to point you toward a potential stress fracture?
4. What further actions, if any, should be taken to manage these athletes' problems?
5. How can you ensure that these females will recover in time to start playing basketball once the regular sports season begins?

14.5.3. Plausible Answers

1. First of all, you want to gain the athletes' confidence as their trainer and confidant. Approach each female in private, chat with her as a friend, and try to find out the underlying reason, i.e., real deal behind them missing practice(s). They may be having problems with their studies at school, or issues with their family at home. Allow each athlete to air her concerns, be understanding and supportive.
2. You may want to investigate each athlete's current fitness condition, i.e., previous formal training as she may be deconditioned to begin with; also inquire about other physical activities outside of basketball practices in which she may be participating, such as running, jogging, or other impact maneuvers. Ask specific questions about timing, location, quality, intensity, duration, and associated symptoms.
3. Watch for indications of overuse injury, such as pain that does not resolve with rest, persistent pain despite activity/exercise modification, and especially pain with weight bearing.
4. Once you have determined the extent of each athlete's orthopedic injury, talk it over with your head coach before approaching/consulting the team physician about the next step in diagnosis and treatment. Institute "PRICE" measures and depending on their level of response, proceed with the physical therapy rehab program to the level of each athlete's tolerance.
5. Progress with an appropriate conditioning program once the acute symptoms subside, concentrating on body mechanics, muscle balancing, and especially running and landing techniques. If indeed the affected females are engaged in excessive training/workouts beyond their practices, educate them on the importance of not "overdoing it" so that they can maximize their chance of recovery. Gradually incorporate sports-specific drills after they master the "basics." This way, these athletes' musculoskeletal injuries can heal in time and they will be physically ready to participate/compete in with their teammates.

REFERENCES

1. Arendt E. Orthopaedic issues for active and athletic women. *Clin Sports Med* 1994 April; 13 (2): 483–503.
2. Wilmore J. and Costill D. Gender issues and the female athlete: Chapter 19. *Physiology of Sports and Exercise* Human Kinetics Champaign, IL: 1994.
3. Cassels J. and Magelssen D. General physiology the female athlete: Chapter 18. *Sports Medicine for the Primary Care Physician* Richard B Birrer, Ed. CRC Press Boca Raton, FL: 1994.

4. Beim G. Sports injuries in women: how to minimize the increased risk of certain conditions. *Women's Health* 1999. February–March; 2 (1): 27–33
5. Gill J. and Miller S. Female athletes: Chapter 4. *Sports Medicine* 2006; Philadelphia, PA Lippincott Williams & Wilkins.
6. Nattiv A., Arendt E., and Hecht S. The female athlete. *Principles and Practice of Primary Care Sports Medicine* 2001; Lippincott Williams & Wilkins.
7. Locke R. and Warren M. How to prevent bone loss in women with hypothalamic amenorrhea. *Women's Health* 2000 May–June; 3 (3): 82–88.
8. Griffin L., Hannafin J., Indelicato P., Joy E., Kibler W., Lebrun C., Pallay R., and Putukian M. Female athlete issues for the team physician: a consensus statement. *Med Sci Sports Exerc* 2003 October; 35(10):1785–1793.
9. Ahmad C., Clark M., Heilmann N., Schoeb S., Gardner T., and Levine W. Effect of gender and maturity on quadriceps-to-hamstring strength ratio and anterior cruciate ligament laxity. *Am J Sports Med* 2006; 34 (3): 370–374.
10. Chappell J., Yu B., Kirkendall D., and Garrett W. A comparison of knee kinetics between male and female recreational athletes in stop-jump tasks. *Am J Sports Med* 2002; 30 (2): 261–267.
11. Hass C., Schick E., Tillman M., Chow J., Brunt D., and Cauraugh J. Knee biomechanics during landings: comparison of pre- and postpubescent females. *Med Sci Sports Exerc* 2005; 37 (1): 100–107.
12. Hewett T., Ford K., and Myer G. Anterior cruciate ligament injuries in female athletes. Part 2, A meta-analysis of neuromuscular interventions aimed at injury. *Am J Sports Med* 2006; 34 (3): 490–498.
13. Gilchrist J. CDC recommendations for preventing exercise-related injuries in women. *Women's Health* 2000 May–June; 3 (3): 90–92.
14. Hasbrook C. Gender: its relevance to sports medicine: Chapter 15. *Principles and Practice of Primary Care Sports Medicine* 2001; Lippincott Williams & Wilkins.
15. Teitz C., Hu S., and Arendt E. The female athlete: evaluation and treatment of sports-related problems. *J Am Acad Orthop Surg* 1997 March–April;5 (2):87–96.
16. Milner C., Ferber R., Pollard C., Hamill J., and Davis I. Biomechanical factors associated with tibial stress fracture in female runners. *Med Sci Sports Exerc* 2006; 38 (2): 323–328.
17. Hewett T., Zazulak B., Myer G., and Ford K. A review of electromyographic activation levels, timing differences, and increased anterior cruciate ligament injury incidence in female athletes. *Br J Sports Med* 2005; 39: 347–350.
18. Myer G., Ford K., McLean S., and Hewett T. The effects of plyometric versus dynamic stabilization and balance training on lower extremity biomechanics. *Am J Sports Med* 2006; 34 (3): 445–455.
19. Rich B. PRICE: management of injuries: Chapter 57. *Principles and Practice of Primary Care Sports Medicine* 2001; Lippincott Williams & Wilkins.

15 Prevention and Management of Common Musculoskeletal Injuries Incurred Through Exercise During Pregnancy

Mimi Zumwalt

CONTENTS

15.1   LEARNING OBJECTIVES
15.2   INTRODUCTION
15.3   RESEARCH FINDINGS
15.4   CONCLUSIONS
15.5   SCENARIO WITH QUESTIONS AND ANSWERS

15.1. LEARNING OBJECTIVES

After completing this chapter, you should have an understanding of the following:

- The pertinent female anatomy involved in pregnancy and how the surrounding musculoskeletal structures are affected by the gravid uterus and growth of other essential organs.
- The accompanying physiological alterations along with other associated systemic changes affecting the pregnant woman's ability to exercise.
- The anatomical and structural issues that contribute to or detract from performing different types of exercises while pregnant.
- Different prevention measures for common musculoskeletal injuries incurred with exercising while pregnant, to include current recommended standard medical guidelines.
- Various modes of treatment for common musculoskeletal injuries that can be exacerbated or sustained by a woman while exercising in the pregnant state.

15.2. INTRODUCTION

Pregnancy is a time of tremendous change in a woman's body, not only involving the anatomical structures around the trunk/abdominal/pelvic area, but physiologically/systemically as well. As a matter of fact, these bodily changes persist for the entire

From: *The Active Female*
Edited by: J. J. Robert-McComb, R. Norman, and M. Zumwalt © Humana Press, Totowa, NJ

duration of pregnancy (about 9 months or so) and may continue even beyond childbirth for weeks or months. These morphological and hormonal/biochemical alterations may incite new symptoms and/or predispose the pregnant woman to a myriad of injuries involving the musculoskeletal system, or even aggravate previously preexisting orthopedic conditions. Therefore, the expecting female must try to protect her own body from physical harm, and, at the same time, she also needs to make sure that the fetus she is carrying remains safe from any injury. This does not mean that she must feel completely confined and give up exercising altogether. However, she has to be cognizant of the fact that she cannot and should not train as "hard," or intensely, as prior to conception; she must also modify the manner in which she exercises and avoid certain athletic activities because of the potential increased risk of trauma to herself and/or the fetus. Beyond these specific considerations, a pregnant woman can engage in various forms of physical activity as long as she remains strict in following current standard medical guidelines, takes measures to prevent musculoskeletal insult, and seeks appropriate orthopedic care if she ends up sustaining injuries from physical training *(1–5)*.

This chapter will focus on different potential musculoskeletal injuries and conditions that can occur and/or become exacerbated with exercising while pregnant, along with various methods of prevention and management of these more common orthopedic injuries.

15.3. RESEARCH FINDINGS

15.3.1. The Pertinent Female Anatomy Involved in Pregnancy and How the Surrounding Musculoskeletal Structures are Affected by the Gravid Uterus and Organ Growth

The important anatomical female structures involved with pregnancy include the main organ, the uterus, which is connected to a pair of ovaries by two Fallopian tubes (*see* Chap. 15 Fig. 15.1). This female sexual complex is completely contained within the bony pelvis and totally surrounded/supported by soft tissues. As compared with males, sex differences exist in terms of the female pelvis morphology in general. Although anatomically and structurally similar to the male pelvis, the female pelvis is configured just a little bit differently. In other words, the pelvis of a woman is built larger in all dimensions to accommodate a fetus and eventually allow natural childbirth. The basic shape of the female pelvis is wider, rounder, and deeper. Specifically, the pelvis is composed of three main bones: the ilium/pubis, the ischium, and the sacrum/coccyx. Two of these three large bones join anteriorly at the pubic symphysis in the midline and posteriorly via the two sacroiliac (SI) joints (*see* Fig. 14.1). Formation of the pelvic basin thus consists of bony structures joined together by ligaments, partially covered by muscles, and entirely lined by fascia. In effect, the pelvis is made similar to a room without a ceiling, containing four walls: one anterior, two laterally, one posterior, and one inferior or floor. The anterior wall of the pelvis forms from the two pubic bones connected in the midline by the symphysis. Each of the lateral walls is formed by the ilium, obturator foramen, sacrospinous/sacrotuberous ligaments, and obturator internus muscle. The posterior wall is composed of the sacrum, coccyx, and piriformis muscle. Finally, the pelvic floor is made up of the levator ani and coccygeus muscles, along

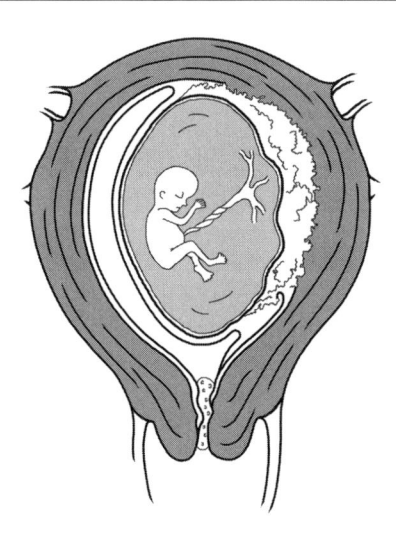

Fig. 15.1. Gravid uterus.

with fascial covering. The latter structure can only be palpated internally through a rectal examination *(2,6)*.

In addition, various ligaments attaching to the pelvis from the spine, along with different musculotendinous units originating from the pelvis and inserting onto the lower extremities, also become affected by the gravid uterus, especially over time as the fetus grows in size through the different trimesters. Specifically speaking, the iliolumbar ligaments, the iliopsoas muscle, the rectus abdominus, internal/external obliques, the quadratus lumborum, the erector spinae, and the pelvic walls/floor musculature are the main contributors to the "core" group of supporting structures surrounding the fetus, which over time will be affected gradually by fetal and other organ growth within the expanding uterus *(2,6)* (Figs. 15.1–15.3). Furthermore, another structure that also plays a role in changing a woman's anatomy after conception involves the mammary glands; as this tissue enlarges in preparation for hormone secretion and milk production, the surrounding skeletal framework, that is, rib cage, thoracic spine, chest, and upper back, along with their attached muscles, must accommodate accordingly (consequently receiving more load) to support these enlarging breasts *(7,8)*.

15.3.2. The Physiological and Other Associated Systemic Changes Affecting the Pregnant Woman's Ability to Exercise

After conception, the woman's pregnant body is not only affected by structural modifications as previously mentioned but also influenced by hormonal/biochemical changes. This combination of anatomical and systemic alterations makes it much more challenging for the mother-to-be if, and when, she decides to engage in any sort of physical activity.

Shortly after the egg is fertilized by the sperm and implantation occurs in the uterus, numerous hormones are being produced by the placenta, a unique organ of pregnancy formed to help sustain life for the fetus and also act as an endocrine gland. Human chorionic gonadotropin (HCG) is the initial key hormone secreted by

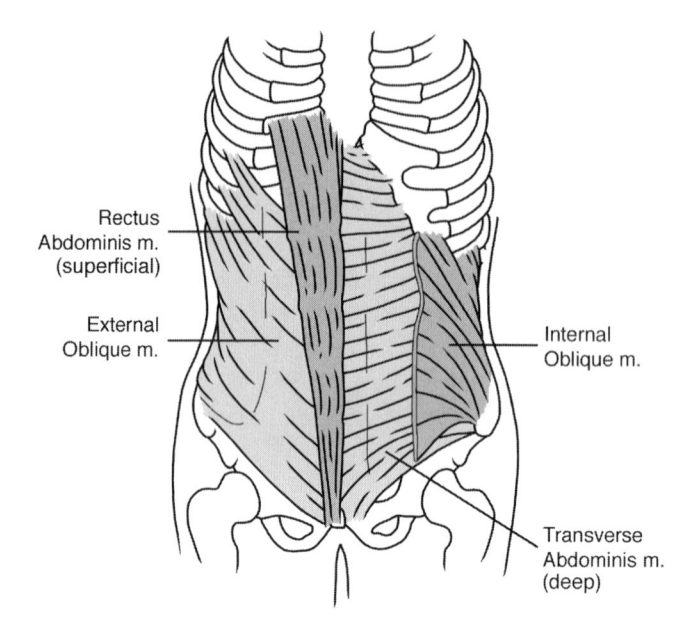

Fig. 15.2. Truncal musculature: ventral view.

the placenta, and thus, it is widely utilized as an indicator of pregnancy when tested positive in the urine or serum of any expecting mother. HCG then, in turn, stimulates the formation of relaxin. This "relaxing" hormone, originating from the corpus luteum (another special organ to support the fetus) along with estrogen from the placenta, causes softening and stretching of maternal pelvic ligaments and the pubic symphyseal joint, to allow better accommodation of the ever expansile uterus occurring as the

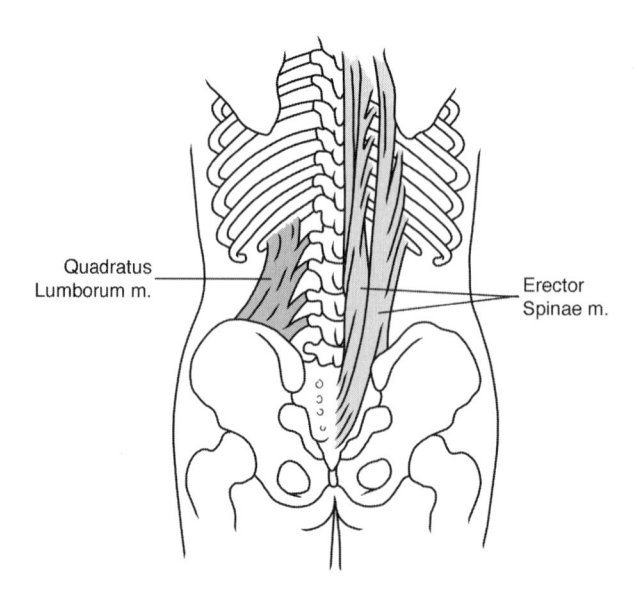

Fig. 15.3. Truncal musculature: dorsal view.

fetus grows in size. Other hormones of pregnancy originating from the corpus luteum include, but are not limited to, progesterone (establishes and sustains the fetus within the uterine cavity), estrogen (responsible for endometrial/uterine lining growth among other functions), and prolactin from the pituitary and uterus (stimulates the breast milk apparatus along with other roles). Additionally, other maternal hormonal alterations affecting a pregnant woman's ability to exercise involve insulin (affects blood glucose levels), cortisol (influences gain in adipose tissue and thinning of the skin leading to easy bruising), thyroid hormone (responsible for increased basal metabolic rate and pulse at rest), and parathyroid hormone (regulates calcium absorption from dietary sources) *(8,9)*.

Aside from numerous hormonal changes noted earlier originating from the corpus luteum/placenta and affecting the mother systemically, the interaction between fetal and maternal tissues causes metabolic changes that will also influence a pregnant woman's body and ultimately her capacity to exercise. In other words, the mother not only has to struggle with being that much heavier but also will have to deal with the altered biochemistry of the uterine/fetal unit. Specifically, the normal course of pregnancy induces weight gain on the average of about 10–15 kg (20–40 lb), with 50% from evolving maternal tissues and 50% from the placental–fetal complex. The process of fetal–maternal metabolism encompasses two phases. The first half of pregnancy is anabolic for the mother and the second half catabolic for her ("accelerated starvation"), in order to be anabolic for the fetus, thus enhancing normal fetal growth; as a result, it will be extremely draining in terms of energy for the mother as the time of delivery draws near *(8,10)*.

Another significant physiological alteration of pregnancy includes generalized/dependent soft tissue edema from retained extracellular fluid in the periphery of the body; this could cause inflammation of musculotendinous structures or entrapment of neural tissues, possibly interfering with certain movements/motion involved with different activities or exercises *(2)*.

15.3.3. The Anatomical and Structural Issues that Contribute to or Detract from Performing Different Types of Exercise While Pregnant

The main anatomical issue to consider regarding a woman's pregnant body, of course, is the physical change resulting in forward shift of the center of gravity caused by hypertrophy of breast tissue, along with the growth of the fetus inside the mother's uterus, moving her body weight more anterior and inferior *(1,3)*. These frontal tension forces are counteracted by accentuation of the thoracic kyphotic and lumbar lordotic curvature posteriorly, causing an increased upper roundback and lower swayback, respectively; the latter must be resisted by the SI joints. However, this attempt at relocating the center of mass and rebalancing the head over the pelvis in the upright position becomes less effective over time because of relaxation and stretching of the local supporting ligaments/connective tissues. This, in turn, contributes to mechanical low back strain (50% prevalence) along with pain in the pelvic area and SI joints *(3,7,11)*. Widening of the pubic symphysis (up to 1 cm) starting near the end of the first trimester is associated with tenderness and aggravated by exercising. These painful symptoms appear to correlate with serum relaxin levels, especially in the first 12 weeks when this value peaks. Between the fourth and the fifth month of gestation, concentration of this hormone decreases and then reaches a steady level about half the quantity of its previous peak during the first 3 months. Interestingly, the level of relaxin does not seem to correlate with measured quantity of laxity, however *(2,12)*. As a corollary, the accompanied weight gain also places additional mechanical strain on the low back, pelvis, and SI areas, increasing forces as much as 100% for only a 20% increase in weight, which adds to the symptoms of joint discomfort *(2)*. In fact, forces across the hips and knees during pregnancy may also reach as high as 100% compared with normal when engaged in impact activities such as running, which potentially could compound more damage to these large weight-bearing joints *(11)*. Other biomechanical alterations that accompany the pregnant body and also will be affected by exercising include increased pressure on the pelvic floor musculature; tightening/shortening of the pectoral, iliopsoas, hamstring, and lumbar extensor muscles; lengthening and weakening of the rectus abdominus and gluteal muscles; and upward flaring of the rib cage. In short, the hormone relaxin definitely contributes to soft tissue elongation, predisposing the pregnant woman to increased risk of overuse injuries with exercising, for example, strains of tendons and sprains of ligaments *(7,11,13)*.

15.3.4. Common Musculoskeletal Injuries/Conditions Which can be Incurred During and/or Aggravated by Exercise While Pregnant, Along with Preventive Measures and Current Standard Medical Guidelines

Although pregnancy induces laxity of tendinous/ligamentous structures secondary to hormonal influence as previously noted, no specific exercise-related injuries have been cited as a direct cause *(14)*. However, some more common musculoskeletal conditions do exist that can be initiated or exacerbated while exercising in the pregnant state include low back pain (LBP), groin pain, pelvic girdle, pubic symphysis, and SI joint pain, hip/lower limb pain, leg cramps, diastasis recti, meralgia paresthetica, carpal tunnel syndrome, and de Quervain's stenosing tenosynovitis *(2,3,7,12)*.

The incidence of LBP during the course of pregnancy is indeed quite high, occurring in at least 50–70% of women. This occurrence increases with decreasing age and even more so in those females with a previous history of back pain and more prior births. Almost one-third of afflicted women have to refrain from performing one or more activities of daily living because of their back symptoms. Episodes of recurring LBP tend to last longer as well. Surprisingly, the amount of weight gain while pregnant does not seem to proportionately correlate with the extent of back problems. As previously noted, as a compensatory mechanism toward the growing fetus, higher shear forces occur on the dorsal/posterior aspect of the truncal soft tissue, which, combined with abnormal pelvic rotation and poor posture in the erect position, predispose certain pregnant women to lumbar discomfort *(2,3,7)*. Similarly, increased upper spinal segment stress also occurs as a result of breast tissue gaining in size and weight, consequently causing the shoulders to sag and the neck to ache as well. Another common musculoskeletal condition (up to two-third incidence) experienced by pregnant women is diastasis recti, that is, separation of the two bands of rectus muscle on either side of the linea alba or midline caused by stretching of the waistline, to increase volume for accommodating uterine/fetal growth. This frontal/ventral core division of connective tissue causes the abdominal musculature to weaken, thus providing less support dorsally, and adding to even more chronic low backache *(7,15)*.

Variations of LBP experienced by the pregnant woman include pelvic girdle pain (PGP), SI joint dysfunction, and symphysis pubis insufficiency. Their incidence ranges from approximately 4 to 90%, with 9–15% having severe symptoms. The former entity is more generalized, and the latter two conditions describe posterior weakening and anterior pelvic widening, respectively. Earlier studies have shown that separating the pubic symphysis 1–2 cm is associated with minimal discomfort, whereas separation wider than 2 cm is associated with much more groin pain. Later studies, however, could not replicate these findings in terms of correlation between the degree of widening and clinically relevant symptoms, although approximately one in three women does have sleeping problems because of the pelvic pain *(3,12,13,16)*. Regarding the absolute width of separation, once the symphyseal diastasis approaches 5 cm, however, operative management may have to be considered from a mechanical standpoint since the pelvis has now become an unstable structure, depending on the stage of pregnancy and potential risk to the fetus. On a lighter note, osteitis pubis is another condition that can also affect the symphyseal joint during pregnancy. This results from resorption then spontaneous reossification of the adjacent pubic bones, leading to severe, rapid radicular pain from the central pubic/groin area down both thighs which progresses quickly over a few days and is worsened by limb motion/movement. This entity is usually transient and needs only symptomatic treatment, resolving spontaneously within several days to a few weeks *(2)*. Most of these conditions do have a common denominator, however, resulting from relaxation of different articulations of the pelvis, allowing more mobility of the involved bones, inciting symptoms of pressure and pain from less skeletal support and therefore lack of stability. One word of caution is that one must not attribute lumbosacral/pelvis pathology occurring during pregnancy solely to hormonal changes leading to mechanical alterations *(13)*. Keep in mind that degenerative spondylolisthesis (anterior/posterior vertebral translation) or true spinal discal problems can exist concurrently with LBP as well, which may become more clinically significant, especially in those women presenting with classic radicular signs and symptoms, all of which need to be addressed accordingly and appropriately *(2,3,13)* (Fig. 15.4). Fortunately, true lumbar disc herniation is quite rare, occurring only in 1 of 10,000 pregnant women *(2,3)*. Likewise, the majority of the aforementioned conditions do tend to resolve on their own postpartum after a few weeks, so they will not require actual medical management per se *(13)*. In the meantime, it is best to be proactive in terms of preventing potential symptoms from occurring by having the pregnant woman engage in central core conditioning exercises for strengthening the truncal musculature around the pelvis and spine using European methods such as the Swiss Pilates *(7)*.

Another temporary lower limb orthopedic condition that can occur during pregnancy and may be exacerbated by exercise is transient osteoporosis of the hip *(2,3)*. This idiopathic entity presents with an antalgic gait usually during the third trimester. Other pathologies that need to be ruled out in terms of limping with weight bearing are referred pain from the back/pelvis and true intra-articular hip problems, along with osteonecrosis of the femoral head, which could portend a worse prognosis. Similarly, if the osteoporotic hip condition is not recognized, the affected woman

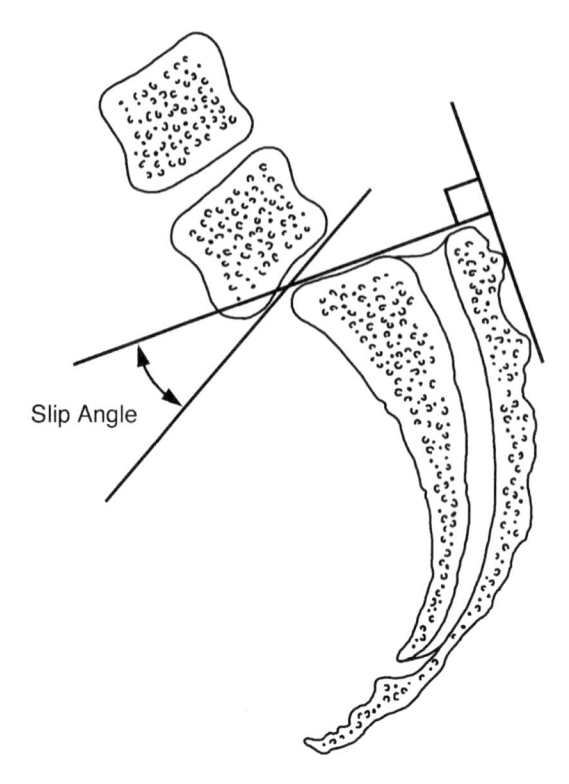

Fig. 15.4. Spondylolisthesis.

can sustain a femoral neck or intertrochanteric stress fracture from even very minor trauma, which ultimately may eventually require surgical intervention. Additionally, other skeletal sites, including the sacrum, tibia, rib cage, and spine, may be afflicted with osteopenia and subsequently increased risk of osteoporotic fractures during pregnancy, especially with more than normal loading in the third trimester. Unrelated lower extremity pathology may also occur involving the acetabular labrum in the hip or meniscus in the knee, predisposing the pregnant woman to a higher chance of injury and causing mechanical symptoms such as locking, which may end up needing orthopedic management. Again, this is dependent on the severity of symptoms and stage of pregnancy. Finally, recurring patellofemoral pain and ankle sprains are two more orthopedic considerations in terms of planning what sort of exercise program would present minimal risk to participate for the expecting mother *(2)*.

As far as musculoskeletal conditions occurring in the upper extremity and compression neuropathy from peripheral edema are concerned, these do not tend to be as serious as some of the lower limb conditions because weight bearing is not a factor. However, the quantity of retained fluid can accumulate as much as 6.5–7.5 L as the day goes on, causing dependent/pitting edema, especially in the lower limbs *(3)*. Anatomical issues predisposing the pregnant woman to peripheral nerve injury include certain areas being more prone to excessive use/pressure and/or those existing at superficial locations. Common sites of compressive neural lesions involve the median nerve at the wrist causing carpal tunnel syndrome, common peroneal nerve at the lateral leg/proximal fibula, and the lateral femoral cutaneous nerve of the anterior thigh. The latter, also known as meralgia paresthetica syndrome, occurs over 10 times more often in pregnant women; the risk is increased even further with associated obesity and diabetes *(2,3)*.

Miscellaneous pregnancy-related musculoskeletal conditions include leg cramps and de Quervain's disease. The former entity is experienced by about 15–30% of women, more often in the latter part of the second and the third trimesters, with 75% happening during nocturnal hours. These extremely strong and fast muscular contractions are severe and very painful, enough to awaken the pregnant woman from deep sleep. Causation of this leg muscle tetany may be because of electrolyte imbalances, primarily insufficient calcium and/or magnesium. de Quervain's disease occurs on the radial aspect of the wrist, is caused and aggravated by repetitive motions, and is influenced by prolactin, relaxin, and progesterone. This overuse type of injury involves inflammation of two tendons of the first dorsal wrist compartment: the abductor pollicis longus and the extensor pollicis brevis. Again, these are only temporary conditions that will resolve spontaneously once the hormonal milieu returns to normal status post delivery *(3)*.

As far as the treatment of pregnancy-related musculoskeletal conditions is concerned, measures to prevent these injuries from occurring are best because it is difficult to manage the mother with medications or invasive procedures because of the inherent potential fetal risk *(3)*. Current standard medical guidelines for prevention of orthopedic injuries and exercising during the pregnant state stem from the American College of Obstetricians and Gynecologists (ACOG). As a preface to recommending a training program, the majority of studies have found that exercising has a primarily neutral or positive effect on the pregnancy course and its outcome (labor and delivery) *(2,3,14,15)*. A balance must be sought nonetheless between possible harmful effects to the mother and fetus and potential health benefits to both when considering an exercise prescription. In terms of duration of physical activity, it is recommended that a woman having an uncomplicated pregnancy engage in moderate exercise sessions for 30–45 min up to 7 days a week. This is also supported by the American College of Sports Medicine (ACSM) and Centers for Disease Control and Prevention (CDC). Specific precautions include avoidance of visceral shunting of blood flow and therefore prolonged motionless standing (pedal blood pooling and risk of syncope) and excessive supine positioning, especially after the first trimester (to minimize hypotension and reduced heart rate from vena cava compression). Other issues to consider are exertion at extreme environmental conditions such as excessive heights (risk of altitude sickness), so pregnant women are advised to stay below 6000–8000 feet and avoid being submerged under water below 30 feet (chance of decompression sickness and venous air embolism), that is, scuba diving. Additionally, recreational activity involving other players (team sports) tends to have a high potential of collision or hard physical contact that could harm both the mother and fetus and therefore should be avoided. These athletic activities include, but are not limited to, soccer, basketball, and ice hockey. Similarly, the pregnant woman should abstain from athletic endeavors that carry an inherent risk of losing her balance and falling, thus causing abdominal trauma such as downhill skiing, horseback riding, or gymnastics. Vigorous racquet sports (tennis) and other impact activities such as jogging/running that could contribute to joint damage from repetitive high loads should be omitted as well. For the most part, those women who are excessively underweight or extremely obese should refrain from exercising altogether. Additionally, pregnancy is not an ideal time to begin a vigorous exercise program or to make substantial gains in overall fitness; rather, it is a time to maintain a healthy lifestyle *(1–5,11,14,15,17–19)*. Recommended safe and effective physical activity for the pregnant woman includes resistance training with light weights and multiple repetitions through a dynamic range of motion to enhance muscle tone, thus adding stability and minimizing injury to the already lax ligaments and joints. Lifting techniques and positioning with weights/machines are of paramount importance in terms of safety in order to adjust to morphological changes of the pregnant body. It is also prudent not to engage in repetitive heavy or excessive isometric type weight lifting to limit the Valsalva maneuver/pressor response in elevating systemic blood pressure excessively while exercising *(1,3,4,11,14,19)*. The exercise regimen should also include some sort of aerobic type conditioning to maintain cardiorespiratory fitness. Any activity involving movement of large muscle groups in a rhythmic fashion sustained for at least 15 min will suffice. The ACSM/CDC guideline for intensity of aerobic exercise is moderate, equating to a brisk walking pace of 3–4 mph (3–5 metabolic equivalents [METs]) and gaged by the Borg scale/rating of perceived exertion (RPE) between 12 and 14, with 6 being the lightest and 20 the hardest. This method is preferred because heart rate is not an accurate predictor of how hard one is working during pregnancy. The following are representative aerobic training programs that, if followed regularly, will help in accomplishing set goals: walking, stationary cycling, hiking, dancing, swimming, rowing, cross-country skiing, skating, rope skipping, and indoor/outdoor group exercise classes (on land or in water). As for flexibility exercises, because the pregnant woman's joints are already lax, she should use static stretches and be careful not to overstretch or perform ballistic movements in order to minimize incurring potential musculoskeletal injury *(10,11,14,15,18,19)*.

In short, numerous studies dealing with pregnant women participating in a regular exercise program have demonstrated substantial maternal benefits with minimal fetal risks. Positive effects include improved cardiorespiratory capacity, control of excessive weight gain and fat retention, and better mental/emotional outlook (sense of well-being). Additional beneficial effects of exercise include lessening of somatic symptoms associated with pregnancy such as insomnia, anxiety, gastrointestinal complaints, leg cramps, pelvis discomfort, and LBP. As long as standard medical guidelines are followed in terms of the "FITT" principle (frequency, intensity, type/mode, and amount of time or duration of exercise), working out during pregnancy is safe and effective in terms of maintaining health and fitness for both the mother and the baby-to-be *(3,10,14,20)*.

15.3.5. Various Modes of Treatment for Common Musculoskeletal Injuries Sustained by Exercising While in the Pregnant State

If indeed the pregnant woman, despite precautions and attempted prevention measures, still is quite symptomatic regarding associated conditions or injuries incurred through exercise, then different modalities exist to treat her pain and discomfort. One must keep in mind, however, that protection of the fetus is first and foremost so her symptoms may not be relieved completely despite medical management. She may have to modify and/or stop certain activities/exercises in order to carry her baby to term without undue damage to herself or the fetus.

The premier treatment method of pregnancy-related musculoskeletal conditions involves avoidance of positions or activities that would exacerbate symptoms. The ultimate goal, of course, is to protect the injured area from further harm while enhancing mobility, strength, and endurance. Resting the affected body part is not an absolute but a relative requirement, otherwise joint stiffness will ensue. For example, management of LBP, PGP, and SI joint pain entails the pregnant woman avoiding stairs, twisting, bending, and lifting. At the same time, she should engage in a physical therapy program concentrating on spinal realignment/stabilization and balancing of the central core muscle group *(4,12,21–24)*. In other words, the symptomatic woman during pregnancy needs to concentrate on maintaining correct posture and utilizing appropriate body mechanics. Specifically, abdominal, hip/gluteal strengthening comprises a large part of the treatment strategy. Exercising the surrounding lumbar and pelvic floor musculature, that is, "angry cat" (arch back up while on all fours) and pelvic tilts (raise up buttock while in supine position), will also help in terms of support and stability. As for ground or aquatic exercises, water has a buoyant effect in offloading joints by easing painful motions, thus is more advantageous. It also aids in controlling peripheral edema and has a lower heart rate response as well, adding to the positive effects *(2,4,22)*. One study has found that a combination of acupuncture and physiotherapy was more effective than therapy alone in treating pregnancy-related PGP *(22)*. Another study confirmed that an exercise program to strengthen the abdominal/hamstring muscles and stretching the iliopsoas and paravertebral muscles was quite successful in decreasing the intensity of LBP and increasing spinal flexibility during the third trimester *(25)*. A third study demonstrated that physical activity prior to pregnancy reduces the risk of pelvic and LBP while pregnant *(26)*. Two other studies observed that a 50% reduction in disability did indeed occur in pregnant women engaging in a regimen composing of lumbopelvic stabilizing exercises, which continued up to 2 years postpartum *(24,27)*. Yet a different study revealed that physical therapy programs with or without additional support such as a lumbosacral corset (nonelastic belt worn just proximal to the greater trochanter) are in fact beneficial in symptomatic management of LBP and PGP and improving functional status; however, other review papers found a neutral effect for both treatment modalities *(4,12,16,23)*.

As far as other local modalities used in treating musculoskeletal conditions/injuries related to pregnancy, most are contraindicated because of the potential transfer of heat and/or transcutaneous nerve stimulation to the fetus. These methods include ultrasound, superficial heat, and electrical current. Likewise, a traction apparatus for the spine may place too much pressure on the abdomen from the belt location along with too much tension on already lax ligaments, so it is not recommended as a treatment device *(2)*. Local injections of an anesthetic steroid mixture, such as those for management of de Quervain's disease, appear to be fairly safe. However, this too should be utilized judiciously as an adjunctive treatment measure, along with splinting and occupational rehabilitation exercises. As far as nonsteroidal anti-inflammatory drugs (NSAIDs) are concerned, these drugs are not recommended during pregnancy, again for the fear of potential detrimental fetal effects *(3)*.

15.4. CONCLUSIONS

In summary, a pregnant woman carries with her an additional weight, along with morphological and physiological alterations of her body. Although pregnancy may be limiting in some aspects, it need not be an entity that severely confines a woman to inactivity. In fact, numerous studies have shown that working out during pregnancy is very beneficial to both the mother and the fetus. Nonetheless, certain musculoskeletal conditions do occur with or become aggravated by exercising while pregnant, but they can be treated symptomatically. Fortunately, most of these problems do tend to resolve spontaneously fairly shortly after delivery, so there is no cause for extreme concern. The pregnant female, however, must take every precaution possible by modifying the manner in which she moves and exercises in order to minimize undue injury to herself and/or the fetus. By doing this, she can still maintain her body in tiptop shape, despite anatomical and systemic changes to optimize both physical and mental health for her, and ultimately benefit the baby to come.

15.5. SCENARIO WITH QUESTIONS AND ANSWERS

15.5.1. Scenario

You are a family practice physician working in a small town where the closest major medical center is a couple of hours away. An emergency room is available nearby, but medical resources are scarce. You have been the "family doctor" for a few decades in the community. In fact, you have made "house calls" in the middle of the night to tend to extremely ill patients or to deliver an unexpected baby! Currently, you have five different women who are in various stages of pregnancy, three of whom are fairly healthy and all desire an exercise program. One woman, unfortunately, has gestational diabetes along with being extremely overweight. In fact, she has already gained over 60 lbs nearing the end of her second trimester. She has been giving in to her "food cravings" and wants to start exercising to lose as much weight as possible. The second woman just found out she is pregnant and has been working out for several years along with competing in triathlons. She wants to keep up with her training and competitions and is very afraid of "putting on weight." The third woman is near the end of the first trimester, is gaining weight appropriately, and has begun a workout program. She needs to know whether she can continue to exercise and when she should stop working out. The fourth mother-to-be, although she has been involved with a training regimen for a few years, is so afraid that she will damage the fetus because it is her first baby, so she decides that she needs to quit exercising altogether. The fifth female patient is complaining of low back spasms and abdominal cramping when she attempts to work out, especially lower extremity exercises, but she does not want to slow down.

15.5.2. Questions

1. For the obese, diabetic patient who has never exercised before, what other concerns do you have and how will you go about helping her to control further unwanted weight gain safely?
2. As far as the competitive triathlete, what other issues must you address prior to advising her about training and nutrition?

3. In terms of the third patient, what guidelines would you recommend in order for her to keep on exercising?
4. As for the fourth patient, what advice do you have for her about potential fetal damage with exercising?
5. And finally with the last patient, what do you think is happening with her body and should she continue with the same workout program?

15.5.3. Plausible Answers

1. First of all, you must reiterate to this complicated patient that pregnancy is not a time to resort to drastic measures. Yes, she should slow down on eating and/or change her choice of foods, but she should do this under a dietician's supervision. The other issue to address is her diabetic state: limiting her simple sugar intake should help control the diabetes as well. Owing to her extreme weight, it would not be wise to have her engage in any sort of vigorous activity. However, an exception can be made to help her lose some (not all) unwanted pounds in order to improve her overall state of health. The exercise regimen should consist of very light aerobic type of conditioning, starting out at 10–15 min a day, two to three times a week, such as stationary cycling or walking on a treadmill. She can then increase gradually as tolerated to 20–30 min, four to six times weekly. It is prudent for her not to lose more that 2–3 lbs a week. She can remain on this weight control program as long as the fetus is doing well during periodic obstetrics (OB) checkups *(1,17)*.
2. The second patient represents a bit more of a challenge because most likely it would be quite difficult for her to slow down her training, but much less to stop competing altogether. If she insists on continuing with triathlons, advise her to compete in mini-type or half-type races rather than full ones so she would not have to train as intensely and not overtax herself and the fetus. If she agrees, explain to her that she should gain, not lose weight while pregnant in order to "feed" the fetus; then make sure she eats appropriately as well, taking in enough calories to meet pregnancy needs (at least 300 extra calories per day, even more so because she is still training). Additionally, she probably would have to stop competing mid-term, but she can still exercise to maintain her fitness status. She has to modify her workouts by substituting the elliptical cardio machine for running and riding the stationary bike rather than road bicycling, but she can keep on swimming because this mode of exercising is actually beneficial during pregnancy. On the contrary, if this athlete is resistant to your recommendations, you may have to consult a psychologist to help her understand the important reasons behind your medical advice *(1,17)*.
3. As for the third patient's request, it appears that she is the more reasonable out of five pregnant women. As she is already engaged in an exercise program, you can just guide her along in terms of avoiding risky sporting activities and extreme environmental conditions. Otherwise, she can continue to work out as long as she follows medical guidelines set by the ACOG/ACSM/CDC, that is, aerobic activity 30–45 min, 5–7 days a week at moderate intensity, along with resistance training using light weights and high repetitions for toning purposes only. She need not stop exercising until close to the time of delivery if both she and the fetus are tolerating the training regimen, as monitored by you and/or her obstetrician *(17)*.
4. In terms of the fourth patient's concerns, she may be overreacting a bit because she wants to stop exercising so abruptly to "save her baby from harm." Because she has been on a regular training program for the past few years, her body is already conditioned physically. By carrying a fetus, granted this will add excess weight along with imposing

further metabolic demands and physiologic stress to her body, she does not have to arrest all physical activity. No studies have demonstrated ill effects to the baby with working out, as long as she modifies her exercises to follow ACOG/ACSM/CDC guidelines *(17)*.

5. Finally with your fifth patient, you need to take a detailed history concerning the frequency, quality, duration, and associated symptoms of her back and abdominal complaints. She should avoid physical maneuvers that stress the pelvic area, including exercises that would increase internal pressure, such as repetitive squatting or heavy leg presses. She should also stay away from any sort of high impact activity because this may stress the fetus even more. She may be experiencing discomfort associated with lifting or joint relaxation because of hormones of pregnancy—that is, relaxing and/or having preterm labor. In that case, she should return to her OB for further evaluation with prior to resuming any type of athletic activity *(2)*.

REFERENCES

1. Artal R, Sherman C. Exercise during pregnancy. *Phys Sportsmed* 1999; 27 (8): 51–52, 54, 57–58, 60, 75.
2. Borg-Stein J, Dugan SA, Gruber J. Musculoskeletal aspects of pregnancy. *Am J Phys Med Rehabil* 2005; 84 (3): 180–92.
3. Ireland M, Ott S. The effects of pregnancy on the musculoskeletal system. *Clin Orthop Relat Res* 2000; (372): 169–79.
4. Teitz C. Special populations. *The Female Athlete*, American Academy of Orthopaedic Surgeons AAOS Monograph Series Rosemont, IL 1996.
5. Treuth M, Butte N, Puyau M. Pregnancy-related changes in physical activity, fitness, and strength. *Med Sci Sports Exerc* 2005; 37 (5): 832–7.
6. Agur A, Lee M. *Grant's Atlas of Anatomy*, 10 ed. Philadelphia, PA: Lippincott Williams & Wilkins, 1999.
7. Balogh A. Pilates and pregnancy. *RCM Midwives* 2005; 8 (5): 220–2.
8. Berne R, Levy M, Koeppen B, Stanton B, eds. The reproductive glands. *Physiology*, 5th ed. St. Louis, MO: Mosby, November 2004; 920–78.
9. Guyton A, Hall J. Female physiology before pregnancy; and the female hormones. In: Schmitt W, Gruliow R, Faber P, Norwitz A, Shaw P, eds. *Textbook of Medical Physiology*, 10th ed. Philadelphia, PA: Saunders, 2001: 929–43.
10. Paisley TS, Joy EA, Price RJ Jr. Exercise during pregnancy: a practical approach. *Curr Sports Med Rep* 2003; 2 (6): 325–30.
11. Artal R, O'Toole M. Guidelines of the American College of Obstetricians and Gynecologists for exercise during pregnancy and the postpartum period. *Br J Sports Med* 2003; 37: 6–12.
12. Stones W, Vits K. Pelvic girdle pain in pregnancy. *BMJ* 2005; 331: 249–50.
13. Bastiaanssen JM, de Bie RA, Bastiaenen CH, Essed GG, van den Brandt PA. A historical perspective on pregnancy-related low back and/or pelvic girdle pain. *Eur J Obstet Gynecol Reprod Biol* 2005; 120: 3–14.
14. Clapp J. Exercise during pregnancy. *Athl Woman* 2000; 19 (2): 273–86.
15. Davies G, Wolfe L, Mottola M, MacKinnon C. Joint SOGC/CSEP Clinical Practice Guideline: exercise in pregnancy and the postpartum period. *Can J Appl Physiol* 2003; 28 (3): 330–41.
16. Stuge B, Hilde G, Vollestad N. Physical therapy for pregnancy-related low back and pelvic pain: a systematic review. *Acta Obstet Gynecol Scand* 2003; 82 (11): 983–90.
17. American College of Obstetricians and Gynecologists. Exercise during pregnancy and the postpartum period. *Clin Obstet Gynecol* 2003; 46 (2): 496–9.
18. Snyder S, Pendergraph B. Exercise during pregnancy: what do we really know? *Am Fam Physician* 2004; 69 (5): 1053, 1056.
19. Wolfe L, Davies G. Canadian guidelines for exercise in pregnancy. *Clin Obstet Gynecol* 2003; 46(2): 488–95.
20. Morris S, Johnson N. Exercise during pregnancy: a critical appraisal of the literature. *J Reprod Med* 2005; 50: 181–8.

21. Bastiaenen CH, de Bie RA, Wolters PM, Vlaeyen JW, Bastiaanssen JM, Klabbers AB, Heuts A, van den Brandt PA, Essed GG. Treatment of pregnancy-related low back and/or pelvic girdle pain. *Eur J Obstet Gynecol Reprod Biol* 2005; 120 (1): 3–14.
22. Elden H, Ladfors L, Olsen MF, Ostgaard HC, Hagberg H. Effects of acupuncture and stabilizing exercises as adjunct to standard treatment in pregnant women with pelvic girdle pain: randomized single blind controlled trial. *BMJ* 2005; 330: 761.
23. Nilsson-Wikmar L, Holm K, Oijerstedt R, Harms-Ringdahl K. Effect of three different physical therapy treatments on pain and activity in pregnant women with pelvic girdle pain: a randomized clinical trial with 3, 6, and 12 months follow-up postpartum. *Spine* 2005; 30 (8): 850–6.
24. Stuge B, Laerum E, Kirkesola G, Vollestad N. The efficacy of a treatment program focusing on specific stabilizing exercises for pelvic girdle pain after pregnancy: a randomized controlled trial. *Spine* 2004; 29 (4): 351–9.
25. Garshasbi A, Faghih Zadeh S. The effect of exercise on the intensity of low back pain in pregnant women. *Int J Gynaecol Obstet* 2005; 88 (3): 271–5.
26. Mogren IM. Previous physical activity decreases the risk of low back pain and pelvic pain during pregnancy. *Scand J Public Health* 2005; 33 (4): 300–6.
27. Stuge B, Veierod MB, Laerum E, Vollestad N. The efficacy of a treatment program focusing on specific stabilizing exercises for pelvic girdle pain after pregnancy: a two-year follow-up of a randomized clinical trial. *Spine* 2004; 29 (10): E197–203.

16 Prevention and Management of Common Musculoskeletal Injuries in the Aging Female Athlete

Mimi Zumwalt

Contents

16.1 Learning Objectives
16.2 Introduction
16.3 Research Findings
16.4 Conclusions
16.5 Scenario with Questions and Answers

16.1. LEARNING OBJECTIVES

After completing this chapter, you should have an understanding of the following:

- The hormonal, physiological, neurological, and anatomical/musculoskeletal changes in female athletes as they transition from young adulthood into the middle age and beyond.
- The differences and similarities between older males and females in terms of their body composition, musculoskeletal components, and athletic performance.
- Various musculoskeletal injuries and orthopedic conditions more common to aging and fairly unique to older female athletes.
- Several measures to prevent musculoskeletal injuries in older active females from occurring.
- Different methods of treatment for orthopedic conditions incurred in aging female athletes to include exercise prescription as recommended by various nationally recognized organizations.

16.2. INTRODUCTION

Throughout the lifespan of active females, they have had to undergo a multitude of changes including structural, hormonal, physiological, neurological, and musculoskeletal. Noteworthy are the two most dramatic stages in a woman's life: one occurring early, pubescence during the teen years, and the other happening several decades later, senescence after menopause. It appears that certain aspects of a woman's

From: *The Active Female*
Edited by: J. J. Robert-McComb, R. Norman, and M. Zumwalt © Humana Press, Totowa, NJ

body come around full circle to the point where she started after birth; she eventually ends up in a similar state ultimately near the end of life. In particular are the endocrine alterations affecting the quality of her bones, making the skeleton so fragile that even the slightest amount of trauma could result in osteoporotic or fragility fractures and, if not addressed promptly and appropriately, could eventually result in dangerous demise. On a lighter note, an older woman can effectively combat some of the negative effects of aging with several positive actions, one of which involves keeping her main muscular components strong in order to protect the skeletal system. What this means is that the aging female must remain relatively active physically to guard herself from harm. However, she may pay the price of incurring potential orthopedic trauma to her body, but it is definitely worth her effort, not necessarily to reverse the effects of time, but hopefully to slow it down to the point of allowing her to more effectively cope with the inevitable changes of life.

This chapter will address different issues of biological alterations within a woman's body as she ages, as well as various musculoskeletal/orthopedic injuries more common in the master female athlete, along with strategies for prevention and management of these acute and chronic traumatic conditions.

16.3. RESEARCH FINDINGS

16.3.1. The Hormonal, Physiological, and Anatomical/Musculoskeletal Changes in Female Athletes as they Transition from Young Adulthood into the Middle Age and Beyond

To place things into perspective, let us review the continuum or stages of the human life cycle. After conception, *in utero*, the baby goes through two transformations: from being an embryo at 8 weeks then turning into a fetus until term pregnancy. After the birth, the neonate grows into an infant at approximately 1 month. Next, the childhood phase spans from 2 years old until puberty or early teens. Subsequently, the adolescent reaches physical maturity or the young adult stage, from the age of 20 to 40. The next transition is to middle age, which comprises people from 40 to 65 years old, and then finally, the elderly phase is reached from age 65 until death *(1,2)*. It is the latter, middle-to-older adult or geriatric stage that we will be concentrating on in terms of unique bodily changes regarding the female athlete. As mentioned chapter 12, after peak bone mass is reached in the mid-twenties, the quantity of bone in the skeleton then gradually declines at a rate of 0.3–2% per year until the fourth decade, when bone loss is accelerated to over 3% yearly for at least 5–10 years after menopause in women, placing them at an increased risk for osteoporosis *(1,3–7)* (*see* Fig. 14.2). This skeletal manifestation is because of a precipitous drop in the female sex hormone levels, namely, estrogen, which previously, especially during the early growing years, has been protective for bone building *(7)*.

In terms of physiological changes between the young and older, middle-aged adult athletes, endurance performance, aerobic capacity, and cardiorespiratory functions all deteriorate with time, whether caused by relative activity reduction or as the natural resultant effects of aging *(1,8,9)*. For example, maximum oxygen uptake tends to decrease approximately 10% per decade after the mid-twenties *(10)*. However, if training intensity and volume are maintained as compared with their younger counterparts, female master athletes will only lose 1–2% of their previous aerobic capacity yearly until 50 years of age. As far as body composition and other elements involving the musculoskeletal system are concerned, these parameters fluctuate in the older athlete as well *(9)*. In other words, although body weight gradually increases as one ages from 20 to 70 years old, especially during 5–10 years prior to menopause, the amount of fat-free mass undergoes a reduction while the percentage of body fat rises (5–10 kg in the mid-thirties to mid-forties), although not to the same extent as compared with more sedentary women *(1,8,9)*. For example, regarding highly trained female runners in their mid-forties, their body fat percentage is about 18%, versus 26% in females who are not active. Similarly, women swimmers also remain slimmer as well although not as marked as runners, with body fat staying around 23%, which is still lower than sedentary counterparts in the similar age group. The average woman's body fat in her youth is approximately 25%, which rises to over 35% by the age of 50. The reasons behind the increase in body fat are threefold: more dietary intake, less physical activity, and decreased ability to mobilize adipose tissue. In terms of muscular strength, the absolute level required to perform activities of daily living stays the same throughout one's lifetime. On the contrary, the maximal level of muscular strength declines at a steady rate from young adulthood (maximal achievement between the ages of 25 and 35) onto later years *(1–3,8,9)*. Quantitatively speaking, after the third decade, if one does not remain active, the loss of muscular strength starts to occur gradually, up to 15% per decade between the fifth and seventh decades, then progresses faster subsequently, at a rate of 30% per decade *(3)*. About 25% of peak force is lost by the time a woman is in her mid-sixties. This age-related loss in strength stems from reduced mass of primarily skeletal muscle (sarcopenia), which in turn affects the basal metabolic rate (10–20% decline from early adulthood to beyond middle age), and/or restricted physical activity levels *(1,4,7,9)*. The latter functional disuse condition (hypokinesia) accounts for 50% of the physiological weakness of muscles and bones *(6)*. This decline in muscular strength seems to affect the lower extremity to a greater extent than the upper extremity *(1)*. After peaking in the third to fourth decade, the loss of lean muscle mass is attributed to both a reduction in the size and in the number of muscle fibers *(1,2,8,9)*. After the fifth decade of life, about 1% of the total number of muscle fibers are lost per year from atrophic (wasting) changes *(9)*. Beyond the sixth decade, approximately 15% of muscle strength is lost every 10 years *(4)*. In fact, midlife adults not involved in strength training lose about 5–7 lb of muscle every 10 years *(8)*. In addition, the ability of the nervous system to process information and activate the musculoskeletal system is altered as well, resulting in slower response times with both complex and simple movements *(9)*. Resistance training can, however, counteract and partially offset ongoing biological muscular weakness by increasing muscle size and strength *(4,9)*. Furthermore, older adults who are more sedentary also gain a substantial amount of fat in their subcutaneous tissue, as previously mentioned, contributing to the natural decline in function associated with aging *(9)*. As a corollary, the skeletal system in an older female gradually becomes quite frail as well, leading to thinning bones and increased risk of osteoporosis *(4–6)* (Fig. 16.1). However, mechanical loading of the skeleton, especially exercising while weight bearing (standing upright), will also help build bone according to Wolff's law, thus counteracting the continuing bone loss because of inactivity and aging *(6,7,11)*. So even though arresting Mother Nature's

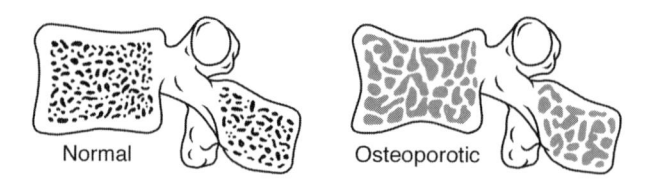

Fig. 16.1. Normal versus osteoporotic bone.

tendencies is unrealistic, by participating in regular workouts/exercise, older females can fight off some of the inevitable decrements in musculoskeletal strength/power and endurance, enabling them to lead a more meaningful lifestyle, that is, to age with grace, especially when engaging in competitive athletic activity/performance *(4,9)*.

16.3.2. The Differences and Similarities Between Older Males and Females in Terms of their Body Composition, Musculoskeletal Components, and Athletic Performance

It is estimated that the population segment that grows the fastest is those 85 years and older, most of whom are women. Although women are living longer, they must also fight the physiological bodily adaptations such as an increase in body fat along with a decline in proprioception (balance)/coordination, strength (both muscular and skeletal)/power, flexibility (pliability of soft tissues such as joint capsules, articular cartilage, tendons, and ligaments), speed of reaction, and aerobic capacity *(3,9,12,13)*. In fact, over the adult lifespan, stiffness of the hip and lower back ensues, resulting in an 8–10 cm loss of stretch ability *(1)*. In terms of body composition, those adults who remain active into middle age tend to maintain a lower percentage of adipose tissue, because fat-free mass starts to decline after the third decade. For example, as previously noted, competitive runners between the ages of 40 and 50 can keep their body fat as low as 18% in women, and even lower than that in men, averaging 11%, versus 6% and 19%, respectively, in inactive adults. A similar trend is seen for highly trained swimmers, although the amount of body fat appears to be a bit higher, at 23% in females and 15% in males. Despite being able to keep the percentage of body fat down as aging occurs, athletic performance of both endurance and strength events in males and females inevitably experiences a decline at about 1–2% per year after the third decade. A study of over 500 subjects (men and women) between the ages of 30 and 70 shows that maximal running velocity gradually decreases approximately 8.5% per decade independent of distance. Similarly, for cycling performance, speed for both male and female cyclists between the ages of 20 and 65 undergoes a gradual drop of about 0.7% per year. There are a few exceptions, such as golf, equestrian, and swimming performance, because success in these sports is dependent on athletic skill that continually improves after years of experience, along with strength and endurance. Thus, the former two types of athletes max out in the third decade, and master swimmers may peak and perform their best as late as 45–50 years of age *(1,9)*. In contrast, gymnasts usually peak in their teens as far as athletic performance, because flexibility is of paramount importance for this sport *(1)*.

In regard to osteoporotic changes, older men are affected as well, but perhaps not as early as women, fewer in numbers (20% of males versus 80% of females), and not quite to the same magnitude. After the fourth decade, men lose bone mass at a rate of 0.5–0.75% per year, which is only half that of women, whose bone density declines at a rate of 1.5–2% per year, increasing to at least 3% per year after menopause. It is not until after the sixth decade that the rate of bone loss in females begins to parallel that of males *(1,2,4,5,7,13,14)*. According to the World Health Organization (WHO), in the United States, approximately 15% of postmenopausal Caucasian women and 30–35% of those older than 65 are inflicted with osteoporosis *(5,15)*. Epidemiology also shows that about 40–60% of Caucasian females and 13% of Caucasian males after the fifth decade of life will suffer one or more clinically significant osteoporotic related fractures (from falling) in their lifetime, 7% of whom will experience permanent disability to a certain extent *(5,6,14,15)* (Fig. 16.2). About 9 out of 10 elderly individuals who sustain a hip fracture are over 70 years of age *(4)*. Alarmingly, if an osteoporotic fracture occurs between the ages of 20 and 50, then the risk for another fragility fracture goes up to almost 75% *(16)*. All is not lost, however, because it has been shown that men and women older than 65 involved in 10–12 weeks of physical activity such as T'ai Chi (Chinese exercise program consisting of repetitive, rhythmic body movements) or other low-intensity workouts can enhance athletic type performance, improve physical balance, and decrease the chances of falling *(4,12)* (Fig. 16.3). Another study has demonstrated that this martial arts form, practiced for a few months, can be very effective by reducing the risk of falls by almost 50% *(5,6)*.

As far as strength is concerned, more than 100% gains can be made in men (60–72 years of age) after lifting weights for about 3 months. Similarly, other studies in both males and females demonstrate that those older than 90 or even 100 years of age participating in strength training could also become physically stronger (approximately 125%) with a concomitant increase in lean muscle mass (10%) *(12)*. Specifically speaking, one study has shown that an 8-week resistance program in those ranging in age between 89 and 92 years old effected an increase in quadriceps strength of 175% *(1,13)*. Another similar study over the same amount of time with women 87–96 years

Fig. 16.2. Osteoporotic fracture rate versus age.

Fig. 16.3. T'ai Chi.

old effected an increase of strength three times over baseline values along with 10% gain in muscle size *(8)*. This rise in muscular strength has been shown to translate into enhancing physical function, decreasing limitations, and increasing mobility, such as speed of gait and stair climbing *(11,12)*. Although the relative strength gains are comparable for men and women engaged in the same sort of conditioning regimen, the absolute rise in females is only 50% that of males *(1,12)*. Muscular weakness also occurs earlier in females; thus, it is very important to continue a sound resistance-training regimen *(1)*. To recap, older adults can make great gains in their capacity to generate force by participating in a regular weightlifting program. In fact, even as late as the eighth decade of life, intensive muscular training can minimize or even reverse the age-related effects of ongoing weakness *(3,4,12)*. However, improvements made in muscular strength far exceed the enhancement in the quantity of lean muscle mass, meaning that neural activation is responsible for the strength increase, similar to changes occurring in the adolescent years, but never catches up to the same extent *(12)*. The exception lies in the quality of muscular integrity in men and the ability of their skeletal muscle to retain the capacity for endurance training. Despite the loss of muscle mass over time in males, the remaining muscle still retains 85–90% of the enzymatic activities required for aerobic type sports *(9)*. Unfortunately, the amount of strength gains needed to effect a meaningful change in athletic performance is still not exactly known. What has been shown lies in the fact that being stronger appears to improve physical function qualitatively and partially offset age-related bone loss *(6,11,12)*. In other words, having more muscle mass tends to positively affect bone mineral density *(6)*. As for other physiological aspects such as range of motion (ROM), coordination, reaction time, and cardiorespiratory fitness, their natural decline/alterations can also be improved with exercises specifically aimed at enhancing these parameters, as long as the age range of the subjects studied is between 60 and 80. For example, men and women in their sixties and early seventies can improve their aerobic capacity by over 20% after engaging in 9 months of endurance training *(12)*. However, another study found that older subjects 65–90 years of age can also effectively increase the flexibility of several large joints after participation in a 12-week program of dance and stretching exercises *(13)*.

16.3.3. Various Musculoskeletal Injuries and Orthopedic Conditions more Common to Aging and Fairly Unique in Older Female Athletes

Because the average life expectancy has increased by more than 30 years within the past century, from surviving the late forties to almost eighties in 2002, more and more people, especially older women, are involved in all sorts of athletic endeavors. Along with this increased participation in physical activity have come different types of musculoskeletal problems, some of which are because of the hormonal changes associated with the female sex, whereas others occur as a result of pathologic aging *(3,8)*. As previously alluded to, the collagen in ligaments and tendons in the body gradually become less elastic with age by losing water content among other ultrastructural alterations, contributing to increased stiffness. As a result, this type of connective tissue is less apt to adapt to mechanical "wear and tear," making it more susceptible to injury along with less capability for healing. A similar phenomenon happens in cartilaginous tissue, such as the menisci within the knee joint or articular cartilage surrounding bone ends. These connective tissues are unable to distribute force as well over time, especially when subjected to repetitive high-impact loading. The resultant cumulative microtrauma has an additive effect of joint destruction and cartilaginous degeneration, leading to frank osteoarthritis and disability. This type of mechanical breakdown is more marked in large weight-bearing joints such as the hip and knee. In fact, those middle-aged athletes participating in intense physical loads are more than eight times likely to develop degenerative arthrosis of their hips. The effects of arthritis are accelerated and exacerbated by previous episodes of injury and/or surgical removal of certain intra-articular structures, such as anterior cruciate ligament tear or partial meniscectomy in the knee, resulting in post-traumatic arthropathy. Fortunately, unlike the aforementioned soft tissues where one must succumb to natural progressive deterioration, the physiological decline in muscle function because of sarcopenia can be countered somewhat by engaging in an intense resistance-training program. This is possible because muscular contractility is not influenced to the same extent as muscle mass with aging, enabling maintenance of strength through exercising. However, the musculotendinous junction is affected in a similar fashion as compared with other types of connective tissues in terms of relative inflexibility, thus the importance of also incorporating a stretching routine into the workout regimen in order to decrease injury risk (3,13).

So what are a few of the more common musculoskeletal conditions besides osteoarthritis that can be experienced by master female athletes? One of the more noted orthopedic injuries occurring in older sports participants is the overuse-type injury, such as tendinosis/bursitis/tendonitis affecting the shoulder (rotator cuff disorder/impingement syndrome) and elbow in tennis players, elbow and wrist (golfers), lateral and medial epicondylitis/extensor tendonitis, respectively, and Achilles tendonitis in joggers. This type of chronic, repetitive trauma accounts for 70% of veteran athletes beyond the sixth decade versus only about 40% of younger athletes in their twenties. In addition, 20% of these overuse injuries can last much longer in older athletes (up to 2 years), affecting their ability to train and/or compete in sports (3). Another frequent injury that tends to occur in older athletes is muscle strain with only moderate exercise, because the musculotendinous unit is not as flexible as when they were younger, when much more force is needed to inflict trauma, such as eccentric muscle contraction (13).

The final, yet most prevalent kind of musculoskeletal trauma occurring in the geriatric female is fragility fracture because of osteoporosis from excessive bone thinning. The amount of bone loss can be quantified by obtaining a bone density study, with DEXA being the current gold standard. Alarmingly, once osteoporosis sets in, even minimal amounts of force inherent in activities of daily living (ADLs) such as bending over and lifting objects can subject osteopenic bone to injurious loads. Regions of the elderly body most affected by osteoporotic fractures include the hip, spine, and wrist (5,6,14). Compared with the femur of a younger person, the femur of an older person is only half as strong and has merely one-third of the energy-absorbing capability before failure (5). Over 90% of hip fractures occur in older adults beyond the seventh decade (4). As for the spinal column, height loss of more than 2 inches should raise the suspicion for one or more vertebral compression fractures resulting from osteoporotic bone (5) (Fig. 16.4). As far as wrist fractures are concerned, they herald the early onset of osteoporosis and tend to occur especially in women between the ages of 50 and 65. If they occur after this age, then the risk of other types of osteoporotic fracture definitely increases in the future (14). Fortunately, the incidence of acute orthopedic trauma in the master athlete is much lower. Only about 1% of those hospitalized for sports-related injuries are over 55 years of age (3).

16.3.4. Several Measures of Prevention to Keep Musculoskeletal Injuries in Older Active Females from Occurring

Now that we have gone through the physiological changes in the master female athlete, including decline in musculoskeletal mass/strength, power, proprioception, coordination, speed, and flexibility, how can an active older female keep from deteriorating to the point of being unable to take care of herself in terms of ADLs? Furthermore, what measures can

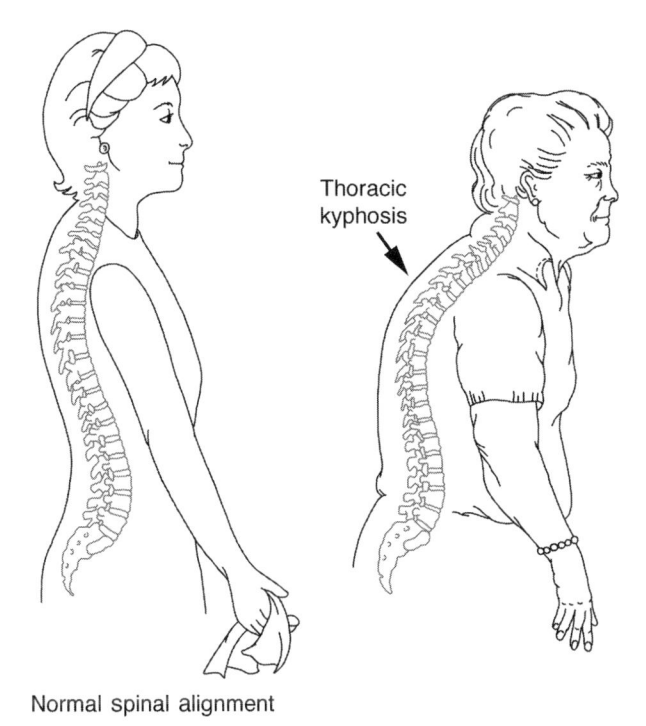

Fig. 16.4. Dowager's hump.

she take to keep from being injured when participating in different types of sports? First of all, by the time the female reaches adulthood, the physical condition of her body is the byproduct of prior years of living. The fact that she has been involved in numerous athletic activities has positive and negative consequences. So now that old age is facing her, she must do what she can to fight off some of the inevitable effects of aging. To begin with, the active elderly female must try to preserve and maintain whatever remains of her fitness shape. In other words, she has to keep on exercising and modify the way she plays and/or the type of sports in which she competes. Not only does she have to strength train to maintain some muscle mass, increase basal metabolic rate, and slow down bone loss, she also has to participate in endurance training to maximize cardiorespiratory function to support her athletic endeavors *(3–5,12)*. As far as physical activities that are recommended, any type of weight-bearing workouts such as dancing, volleyball, walking, basketball, jogging, and stair climbing would be beneficial for both muscles and bone (osteogenic), along with cardiorespiratory fitness *(4–7)*. Other moderate but lower impact exercises including calisthenics, weight-supported swimming or water aerobics, bicycling, and T'ai Chi will also help in physical fitness while protecting problematic joints *(1,4,6,7)* (Fig. 16.3). In terms of sporting activities to avoid, the older athletic female should refrain from skiing and rock climbing, among others, because these sports place the body at a higher risk of falling and causing damage to the already thinning bones and stiffer connective tissues *(13)*. Avoidance of contact team sports involving potential collision with stationary objects or opponents is also prudent and well advised *(1)*.

16.3.5. Different Methods of Treatment for Orthopedic Conditions Incurred in Aging Female Athletes, to Include Exercise Prescription as Recommended by Various Nationally Recognized Organizations

Once the master female athlete has suffered any of the aforementioned musculoskeletal injuries, that is, strains of tendons, muscle pulls, ligamentous sprains, tendinosis/tendonitis, posttraumatic arthropathy, or osteoporotic fractures, she must arrest all provocative maneuvers and proceed through the following process: PRICE—Protect/prevent the damaged area from further harm, relatively Rest the involved limb while maintaining ROM to counteract stiffness, apply Ice (not directly on skin) intermittently to help with inflammation, use Compression (snug but not tight) to minimize edema, and Elevate above the heart to aid in swelling. In other words, institute first-aid-type measures to hopefully alleviate painful symptoms *(13,17)*. Oral agents such as mild analgesics and nonsteroidal anti-inflammatory drugs (NSAIDs) should only be used as an adjunct to physical modalities *(3,13)*. As an aside, sports supplements have not been shown conclusively to improve joint condition or enhance athletic performance. In fact, the associated side effects to internal organs far outweigh the touted benefits *(3)*. In terms of medications for osteoporosis, other than treatment with calcium and vitamin D for bone building, it is best to refer to a primary care physician or endocrinologist for further workup and management *(5)*. Once pain has subsided, an intensive rehabilitation program should be incorporated, along with an exercise regimen designed to maximize recovery and return to preinjury status *(6,12,13)*. A structured, balanced exercise program with the following elements would be ideal: resistive training, cardiovascular endurance activity, flexibility, and balance. Various workout programs are endorsed by different national organizations, including the American Heart Association (AHA), American College of Sports Medicine (ACSM), and the American Academy of Orthopaedic Surgeons (AAOS) *(12)*. If indeed all conservative management strategies fail to heal the injured extremity, radiographs should be ordered and orthopedic consultation should be sought because invasive procedures may be warranted at this point *(3,13)*. Specifically speaking, the exercise prescription should involve the following three elements for each workout session: FIT(T)—frequency (how often), intensity (how hard), time or duration (how long), and type of physical activity. The ACSM and Centers for Disease Control and Prevention's (CDC) recommendations are as follows: 20–60 min of moderately intense activity 3–7 days per week *(4,6,7,10)*. The mode of exercising is dependent on a woman's most current physical fitness level and/or associated musculoskeletal problems. In older women with healthy bones, jumping in place and other plyometrics type of mechanical loading can be performed as long as attention is paid toward having correct form/strict technique *(7)*. However, for those who are fairly frail or females already inflicted with arthritic joints, low or no impact activities should be chosen, that is, exercises done in a pool or sitting/recumbent position such as bicycling *(1,4,6,7)*. In contrast, to benefit osteoporotic bone, workouts should be done in the standing or erect position (climbing stairs or walking but no running) to best load the skeleton, especially the spine *(4,6,7)*. To maximize aerobic fitness, the target heart rate range should be 40 to 80% of maximal heart rate of 220 while exercising *(4,7)* (Table 16.1).

Furthermore, aside from aerobic conditioning for cardiorespiratory fitness, resistance training (isometric exercise is safe and efficacious) needs to be incorporated into

Table 16.1
Target heart rate (THR) ranges

Age (years)	THR (bpm)
50	102 – 136
60	96 – 128
70	90 – 120
80	84 – 112

bpm, beats per minute.

the workout routine as well to minimize loss of muscle strength, stabilize joints, and maintain bone mass *(1,4,6,13)*. The effect of exercise on the latter is quite specific in terms of bone building so make sure to target all at-risk sites *(6,7)*. Start out with mild resistance/light weights for 10–15 repetitions and one to three sets, then gradually increase the amount lifted as tolerated. Work up to 6–10 repetitions for three to four sets, two to three sessions per week for all major muscle groups, concentrating on trunk extensors and the lower extremity muscles to help with body stability and dynamic balance. Again, ensure appropriate body alignment, especially on the structural type exercises using multiple muscle groups, directing force through the hip and spine such as freestanding squats, deadlifts, and lunges *(7)*. Additionally, neighboring joints and surrounding musculature close to the injured one(s) should also be included in the rehabilitation phase *(13)*. Furthermore, a regimen of stretching (several times per day is better than one long episode) to maintain flexibility and balancing activities for proprioception (body position awareness in space) will complete the workout program *(4,7,13)*. Static, passive, and active assisted stretching for major musculotendinous units should be done on most days of the week. As for balance and stability training, the following drills have been found to be effective: getting up from a chair without arm use, one-legged stance, backward walking, and negotiating various obstacles set up on a course. In addition, physio-balls can be used for strengthening of core muscles, neutral spine promotion, and posture improvement *(7)*. Therefore, a regular exercise program will promote mobility while decreasing falls, which have been shown to be the sixth leading cause of death *(6)*. In effect, exercising will maximize residual function and even reduce biological age by as much as two decades *(1)*. Just remember, for safety reasons, it is extremely crucial for an older female to obtain medical clearance from her primary care physician prior to engaging in any sort of workout regimen *(7,10)*.

16.4. CONCLUSIONS

In summary, as the active female proceeds from childhood through adolescence into adulthood, she must constantly try to slow down her biological clock as far as the body is concerned in general and the musculoskeletal system in particular. The old adage of "use it or lose it" applies when speaking about maintaining physical fitness in order to continue participating in different athletic activities. However, as time passes, both the muscular and skeletal systems start to gradually decline in terms of strength and integrity. Furthermore, higher risks of potential injuries exist to challenge the master female athlete as well as slower course of recovery after sustaining traumatic episodes. Ideally, the best method of treatment for orthopedic trauma is of course prevention, but certain situations are beyond a woman's control. In that case, measures to protect the injured extremity are of prime importance, so she can rehabilitate and physically recover in a timely fashion. It might last several days or a couple of weeks but may take as long as a few months. In addition, a regular exercise program should be instituted, not only targeting the affected limb but also incorporating resistance training and aerobic endurance workout, along with balance and flexibility elements. This way, the aging female can regain strength, function, and mobility to hopefully return to competitive sports or any other athletic endeavor of her choosing.

16.5. SCENARIO WITH QUESTIONS AND ANSWERS

16.5.1. Scenario

You are a nurse practitioner for an internist in a midsize town. Your adult patient population ranges from young adults to senior citizens. You see an array of patients with different clinical entities ranging from the simple common cold to complicated cardiopulmonary disease. One of your hobbies is playing recreational tennis in a "female over fifty" league at the YWCA. You meet with your teammates twice a week and you serve as the team captain for these female athletes. Another pastime is working out at a local health club three to five times a week. Your doubles tennis partner is 55 years of age and is an avid runner and climber, as well. She pretty much participates in some sort of sporting activity every single day of the week. Lately, she has been frequently absent, sending in a substitute player and complaining of vague, achy shoulder, and deep knee pain. Unfortunately, before you could help her sort things out, she suffered a fall, sustaining a wrist fracture on her dominant side (opposite of her involved shoulder) and finally made it to your clinic for evaluation and treatment. You have known this woman for several years, and she refuses to let anything slow her down in terms of exercise. She is extremely frustrated with her injuries and wants to heal really fast and get on with her active life again as soon as possible.

16.5.2. Questions

1. First of all, what pertinent questions should you ask about your patient's medical history and what modalities should you utilize for diagnosis?
2. What should you advise her about the shoulder/knee complaints and wrist injury?
3. What other medical specialist(s) should you consult and why?
4. What other factors should you consider prior to sending her back to athletic activity after healing?
5. How soon after she heals from her injury should you allow her to return to tennis? Will you let her go back to running; how about rock climbing?

16.5.3. Plausible Answers

1. First of all, you should take a detailed history about your patient's physiological status and physical fitness profile, including the frequency, intensity, and duration of each athletic endeavor in which she has been involved, to see if indeed she is overtrained and also how she fell to begin with: that is, was it from one of her climbs or stumbling

elsewhere? Then, order radiographs of the shoulder, knee, and wrist, and a DEXA scan to evaluate her bone density because she is most likely postmenopausal.
2. It seems as though your partner has not allowed her body to recover, jumping from one sport to another, lending herself to be involved with risky athletic activities. Her knee pain may be coming from early degenerative changes aggravated by the excessive impact of running. On the contrary, it also appears that rock climbing and tennis may not be compatible, because both sports require repetitive overhead maneuvers that may overload the shoulder joint, causing an overuse-type impingement syndrome. As for the wrist injury, you must advise her to stop all sporting activities requiring upper extremity use. This will force her to rest her shoulder and knee, as well, but have her move the surrounding joints gently so she would not get too stiff. Also make sure she institute first-aid measures, such as the PRICE principle, which was discussed in this chapter.
3. You should discuss her case with your internist, then seek a consult from an orthopedic or sports medicine surgeon, especially for the fractured wrist, because she may need casting and/or surgery. She/he can address the shoulder and knee conditions as well, which most likely will need some sort of physical rehabilitation exercises. In addition, an endocrinologist should be sought out to see about the best course of medicinal or other therapy beyond calcium and vitamin D for her bone density.
4. After clearance from her orthopedist and more information from the endocrinologist, you should sit down with your patient and make sure she understands that most likely the wrist fracture stems from osteoporosis and that the risk of other similar skeletal injury will increase several fold. Therefore, she should be extremely careful from now on and avoid risky-type sports.
5. In terms of allowing your patient to return to competitive athletic activities, she must have a pain-free ROM and good strength of both her upper and lower extremities, as documented by a physical therapist before she can even practice tennis. This may take several weeks to a few months or more. She should also modify her workouts at the gym to maximize muscular endurance, skeletal strength, and proprioception, along with cardiorespiratory capacity. In addition, she needs to perform low-impact-type aerobic activity so as not to stress her arthritic joint or osteoporotic bone. As far as running goes, if she must, she can jog on alternate days on the treadmill so as not to stress weight-bearing joints. As for rock climbing, again stress to her the importance of keeping away from activities that would increase the potential for repetitive trauma, so obviously give her an emphatic NO!

REFERENCES

1. Shephard, RJ, Aging and exercise, in *Encyclopedia of Sports Medicine and Science*, March 7, 1988 pp. 1–7, http://www.sportsci.org/encyc/agingex/agingex.html.
2. Solomon, EP, Davis, PW, *Human Anatomy and Physiology*, 1983, Chap. 26.
3. Chen, AL, Mears, SC, Hawkins, RJ, Orthopaedic care of the aging athlete, *J Am Acad Orthop Surg* 2005, pp. 407–415.
4. Curl, WW, Aging and exercise: are they compatible in women? *Clin Orthop Relat Res* 2000, pp. 151–158.
5. Ireland, ML, Nattiv, A, Osteoporosis, The Female *Athlete Osteoporosis*, 2002, pp. 249–258.
6. Lindsey, C, Fighting frailty in the elderly, *Biomechanism* July 2002, pp. 20–29, http://www.biomech.com.
7. Sorace, P, Osteoporosis and exercise, *ACSMs Certif News* 2002, pp. 6–7.
8. Kilibwa, M, Nutrition and health for women at midlife: recommended approaches for clinicians, *J Womens Health Prim Care* 2006, pp. 14–26.
9. Wilmore, JH, Costill, DL, Aging and the older athlete, in *Physiology of Sport and Exercise*, 1994, Chap. 18 Human Kinetics Champaign, IL pp. 424–441.

10. Zasadil, M, Douglas, PS, The master female athlete, *Med Sci Sports Exerc* 2005, pp. 1444–1447.
11. Rejeski, JW, Brawley, LR, Functional health: innovations in research on physical activity with older adults, *Med Sci Sports Exerc* 2006, pp. 93–99.
12. Brown, M, Rosemont, IL American Academy of Orthopaedic Surgeons In: William E Garrett Jr, Gayle E Lester, Joan McGowan, & Donald T Kirkendall eds. *Fitness and Aging: Women's Health in Sports and Exercise*, 2001, pp. 369–377.
13. Leach, RE, Orthopaedic Surgery Essentials-Sports Medicine. In: Anthony A Schepsis & Brian D Busconi eds. *Aging athletes*, pp. 81–88.
14. Ioannidis, G, Papaioannou, A, Adachi, JD, Quality of life impacts osteoporotic fracture outcomes, June 2002, pp. 27–34. http://www.biomech.com.
15. Lochmuller, E-M, Muller, R, Eckstein, F, Does bone densitometry predict mechanical competence throughout the skeleton? *Menopause Manage* 2003 March-April, pp. 21–24.
16. Wu, F, Mason, B, Horne, A, *Postmenopausal Fractures Linked to Premenopausal Fracture History: Women's Health Orthopaedic Edition*, 2002, July-August, Vol. 5(4); p. 143.
17. Rich, BS, In: Lippincott, Williams & Wilkins Philadelphia, PA 2001 ed. William E Garrett Jr, Donald T Kirkendall & Deborah L Squire *Price: Management of Injuries: Principles and Practice of Primary Care Sports Medicine*, 2001, pp. 665–667.

17 Osteoporosis and Current Therapeutic Management

Kellie F. Flood-Shaffer

CONTENTS

17.1 LEARNING OBJECTIVES
17.2 INTRODUCTION
17.3 RESEARCH FINDINGS
17.4 CONCLUSIONS

17.1. LEARNING OBJECTIVES

After completing this chapter, you should have an understanding of the following:

- The definition of osteoporosis.
- The pathophysiology of osteoporosis.
- Classifications of osteoporosis.
- Risk factors in the development of osteoporosis.
- The diagnosis of osteoporosis.
- Recommendations for bone mineral density (BMD) measurement.
- The management of osteoporosis.

17.2. INTRODUCTION

Osteoporosis, which means "porous bones," is one of the most common skeletal disorders faced by women and their health care providers today. It occurs when there is a change in the balance of bone renewal. The process of bone resorption and renewal requires three elements: (1) the regular supply of protein, calcium, and vitamins; (2) the regular use of the bones (exercise); and (3) essential hormones that harmonize and direct the breakdown and removal of old bone and the creation of new bone.

Osteoporosis is characterized by low bone mass, compromised bone strength, and structural deterioration of the bone tissue that leads to bone fragility and an increased susceptibility to fractures of the wrist, hip, and spine. These fragility fractures lead to a decreased quality of life and a staggering economic burden. There are an estimated

From: *The Active Female*
Edited by: J. J. Robert-McComb, R. Norman, and M. Zumwalt © Humana Press, Totowa, NJ

2.5 million physician visits annually, 1.5 fractures annually, and over 400,000 hospital admissions all at a cost of nearly $20 billion in the United States alone. The primary goal of treatment is to reduce the risk of fractures.

17.3. RESEARCH FINDINGS

17.3.1. The Definition of Osteoporosis

Osteoporosis is a progressive skeletal disease characterized by low bone mineral density (BMD) and deterioration of the microarchitecture of the bone tissue, the consequences of which are bone fragility and increased susceptibility to fracture *(1)*. The World Health Organization (WHO) criteria for diagnosing osteoporosis and osteopenia (decreased bone mineral content) are based on a comparison of an individual's BMD measured by dual-energy X-ray absorptiometry (DEXA) with that of a young adult reference group. BMD is expressed using a *T*-score that represents the difference in a number of standard deviations relative to the average peak bone mass of young, healthy adults of same gender. The WHO classifies osteoporosis as a *T*-score lower than -2.5 *(1,2)* and osteopenia as a *T*-score between -1 and -2.5. The National Institutes of Health has expanded the definition of osteoporosis to include compromised bone strength, independent of bone density.

17.3.2. The Pathophysiology of Osteoporosis

The process of bone remodeling occurs throughout the life cycle. The activity of osteoclasts (bone breakdown or resorption) in normal adults is balanced by the activity of osteoblasts (bone formation). Estrogen deficiency leads to overexpression of osteoclasts. As estrogen levels diminish with the onset of menopause, there is excessive bone resorption that is not fully compensated for by an increase in bone formation. Thus, osteoporosis results from an imbalance in the two processes of bone remodeling. The result is a net loss of bone tissue with associated changes in the microarchitecture of the bone. The most rapid bone loss in a woman's life cycle appears to occur during the first 5 years after menopause. Osteoclastic activity increases with aging in all humans. Bone loss can be measured in the hip and spine years prior to the last menstrual period (menopause) *(3,4)*. BMD is reduced by approximately 50% between the quantity of peak bone mass at 30 and at 90 years of age, when bone loss seems to end.

17.3.3. Classifications of Osteoporosis

Riggs and Melton *(5)* have classified osteoporosis as either primary, which has two subtypes, or secondary. Primary osteoporosis is the more common form of the disease and refers to an inadequate bone remodeling process. *Type 1* primary osteoporosis is characterized primarily by trabecular bone loss, although some measurable cortical bone loss may also occur. It is more common in women and is often termed postmenopausal osteoporosis. It results primarily from estrogen deficiency. The skeleton is more sensitive to parathyroid hormone (PTH), and this causes increased calcium resorption from the bone. *Type 2* primary osteoporosis is characterized by both trabecular and cortical bone loss. It is often called osteoporosis of aging. It occurs in adults over the age of 70 and is associated with decreased availability of vitamin D and decreased bone formation. Hallmarks of Type 2 primary osteoporosis are a decline in calcium absorption, increased PTH levels, and bone resorption.

Secondary osteoporosis is caused by calcium imbalance, genetic, endocrine, gastrointestinal, neoplastic, renal, or rheumatologic disorders or as a result of medication therapies or use of other drugs. Glucocorticoid-induced osteoporosis is the most common secondary form of the disease. Glucocorticoids suppress the osteoblast function, inhibit intestinal calcium absorption, and increase osteoclastic activity. These drugs may also increase renal excretion of calcium. BMD may be increased by as much as 60% in patients who are on glucocorticoid therapy.

Other important drugs associated with secondary bone loss include but are not limited to: alcohol, tobacco, heparin and other anticoagulants, lithium, anticonvulsants, cytotoxic drugs, gonadotropin-releasing hormone agonists, vitamin A, and tamoxifen *(6)*. Other important disorders that may cause secondary bone loss include but are not limited to: hemophilia, thalassemia, Cushing's syndrome, hypothyroidism, type 1 diabetes mellitus, chronic liver disease, malabsorption syndromes, lymphoma, leukemia, anorexia nervosa, rheumatoid arthritis, and primary hyperparathyroidism *(6)*.

17.3.4. Risk Factors in the Development of Osteoporosis

Osteoporosis is associated with a number of risk factors, some modifiable and some nonmodifiable. Nonmodifiable factors are inherited such as Asian or white race, female gender, family history of fracture, ethnicity, and thin body habitus. These also include advanced age and a personal history of fracture *(7,8)*. Modifiable factors include malnutrition, estrogen deficiency, sedentary lifestyle, smoking, and excessive alcohol intake among others.

Nonmodifiable factors require that the patient receive education on these risk factors and avoid exacerbating their risk of fracture by adopting and/or not modifying any of the modifiable risk factors. Perhaps the most difficult task that a health care provider may have to deal with is encouraging and being successful with changing behaviors that predispose patients to osteoporosis. In the United States, sedentary lifestyle, smoking, excess use of alcohol, inadequate intake of vitamin D and calcium, and the epidemic of heavy consumption of carbonated and caffeinated drinks continue to place younger and younger women at risk of not reaching their genetically determined peak bone mass *(9)*.

17.3.5. The Diagnosis of Osteoporosis

The gold standard for the diagnosis of osteopenia and osteoporosis of the hip and spine remains the DEXA. DEXA machine standard error is small, and it has the best normative database available. For trabecular bone measurements, lateral spine DEXA may be superior to anterior–posterior views. There are clinical assessments, radiographic studies, and biochemical measurements that can be used to assist in the diagnosis as well. All women should have their risk factors for osteoporosis assessed at the time of their annual physical examination. A thorough medical history should include review of hormone status, nutritional status (specifically to include vitamin D, calcium, and protein intake), personal history of low trauma fracture, lifestyle factors (to include smoking, alcohol intake, and physical activity level). Family history should also be reviewed with respect to height loss with age, spinal compression, or hip fractures. Height should be measured and compared with self-reported maximum height. Repeat measurements of height should be performed annually. Radiographic studies of the lateral spine should be included in the evaluation if a compression fracture is suspected or spinal deformity is present. Biochemical tests may be incorporated into the diagnostic evaluation particularly if a secondary cause of osteoporosis is suspected. Laboratories should include serum chemistries, complete blood count, thyroid function tests, and a 24 hour urine collection for calcium excretion. Further testing may include intact PTH, urinary-free cortisol, erythrocyte sedimentation rate, serum protein electrophoresis, serum 25-hydroxyvitamin D concentration, and other biochemical markers of bone turnover.

17.3.6. Recommendations for BMD Measurements

The National Osteoporosis Foundation (NOF) recommends BMD testing be performed on (1) all women age 65 years of more, (2) all postmenopausal women less than 65 years of age if they have any additional risk factors other than being white, postmenopausal and female, and (3) postmenopausal women who have suffered a fragility fracture (to confirm diagnosis and determine disease severity) *(10)*.

As previously mentioned, the *T*-score is used to determine the presence or absence of bone loss. However, the Z-score may be of more use in younger women and in identifying women who may need to undergo a workup for secondary causes of osteoporosis. The Z-score represents the difference in the number of standard deviations between an individual's BMD and the average BMD in age-matched women. One should suspect secondary causes of osteoporosis if the Z-score is more than 1.5 SD below the mean *(11)*.

The American College of Rheumatology has recommended that patients undergo a baseline lumbar spine or hip BMD before initiating long-term glucocorticoid therapy or other drug therapy associated with bone loss.

Conditions that may falsely elevate BMD measurements especially in the elderly patient include osteophytes, arthritis, and spinal compression fractures. In these cases, a hip or femoral neck BMD is recommended. Another alternative site that has shown to be a good predictor of hip fracture is the calcaneus, which is rich in trabecular bone. A peripheral measurement, such as the calcaneus, is performed using single absorptiometry.

17.3.7. The Management of Osteoporosis

17.3.7.1. Prevention

Maximizing peak bone mass begins early in life with adequate nutrition and exercise *(12–14)*. Other lifestyle factors, such as limiting caffeine, alcohol, and carbonated drinks and avoiding tobacco use, play an important role in the prevention of osteoporosis. Caffeine, tobacco, and excess alcohol consumption all play a role in altering calcium absorption and estrogen metabolism and have been directly linked to an increase in hip, wrist, and spine fractures.

Regular physical activity with weight-bearing exercises should be encouraged. Weight-bearing exercise such as walking, jogging, aerobics, along with helps stimulate osteoblastic activity, thus helping to maintain the balance in bone remodeling after the age of 30. Physical exercise also increases muscle strength, coordination, agility, and balance that serve to reduce the risk of falls and thereby reduce the risk of fracture.

17.3.7.2. VITAMINS AND MINERALS

17.3.7.2.1. Calcium The daily recommended dosage of calcium for premenopausal women is 1200 mg/day to maintain bone mass. As calcium absorption decreases with age, the requirement for postmenopausal women increases to 1500 mg/day *(15)*. A diet rich in calcium would include any or all of the following calcium-rich foods: dairy products such as skim milk, low-fat or non-fat yogurt, cheese, cottage cheese, and ice cream; dark green vegetables such as broccoli, kale, and turnip greens; calcium-fortified foods such as orange juice, breads, and cereals; and certain types of fish and shellfish. Sardines and salmon are a good source of calcium only if the bones are consumed with the fish. Homemade chicken or turkey soup can provide another excellent source of calcium. If the poultry is boiled with the bones in place, the calcium is then leached out of the bones and into the broth.

Some dairy sources are also rich in fat, and low-calorie sources of calcium must be considered. Diet alone commonly fails to provide the daily recommended dose of calcium for most women, so supplements are required. The two most common commercially available calcium supplements are in the form of calcium carbonate and calcium citrate. Calcium citrate is generally better tolerated than calcium carbonate, which often causes gas and constipation. Vitamin D is required for adequate intestinal calcium absorption.

17.3.7.2.2. Vitamin D The daily recommended dosage of vitamin D in premenopausal women is 200–400 IU/day, and for the postmenopausal woman, the range is 600 to 800 IU/day. The elderly, chronically ill, homebound, or institutionalized patient is at greatest risk for vitamin D deficiency. Most calcium supplement products on the market, as well as multivitamin formulations, are fortified with vitamin D. Women should be made aware that "more does not necessarily mean better" when taking multivitamins. Owing to the association of excess vitamin A intake and increased risk of hip fracture, one should not take two different multivitamins in order to get additional vitamin D. In a large trial of elderly women (mean age 84), long-term calcium and vitamin D supplementation was associated with a 44% reduction in hip fracture incidence *(16)*.

17.3.7.3. PHARMACOLOGIC THERAPY

Fracture prevention is the primary goal of pharmacologic therapy for osteoporosis. Secondary goals include stabilizing and increasing BMD, relieving of symptoms of fracture and skeletal deformity(ies), and improving the patient's mobility and functional status. Organizations such as the North American Menopause Society (NAMS), the American Association of Clinical Endocrinologists (AACE), and NOF all offer guidelines for intervening with pharmacologic therapy. Each organization bases its therapeutic recommendations on *T*-scores, risk factors, and prior fracture history *(3,10)*.

Many factors influence the use of pharmacologic therapy including but are not limited to the length of time to realize benefit, impact on quality of life, impact on breast tissue, cardiovascular status, and potential side effects. Cost of therapy is always an issue. Prevention and treatment therapeutic options include bisphosphonates and selective estrogen receptor modulators (SERMs). Prevention-only therapy consists of hormone therapy (HT). Treatment-only therapy includes calcitonin and recombinant human PTH. There are several major clinical trials that have demonstrated the efficacy of each of these therapies in the reduction of fracture risk. Hormone therapy, raloxifene (a SERM), and the bisphosphonates have been shown to reduce the risk of vertebral fracture over 3–7 years. The bisphosphonates and HT have been shown to reduce hip fracture risk.

17.3.7.3.1. Bisphosphonates The bisphosphonates, of which there are three currently available on the US market, are antiresorptive agents that bind preferentially to hydroxyapatite crystals in the mineralized bone matrix to inhibit osteoclastic activity, thus reducing bone turnover. In addition to arresting bone loss and decreasing fracture risk, the bisphosphonates have been shown to increase BMD in postmenopausal women with osteoporosis by up to 8% in the spine and up to 4% in the hip. In micro CT-scans of human bone, the bisphosphonates have preserved horizontal trabecular bone, thus additionally maintaining bone strength. Owing to the poor bioavailability of bisphosphonates of less than 1%, there are specific and important dosing instructions and precautions. The medication must be taken on an empty stomach with at least 8 ounces of water to properly dissolve and absorb the drug from the stomach. The patient must remain upright (sitting or standing) for a minimum of 30 minutes after administration of the medication. No other food or drink is to be ingested during those 30 minutes. Careful evaluation of candidates for bisphosphonates is required and should include questions concerning any history of difficulty in swallowing, or any history of gastrointestinal disorders that cause esophageal irritation (i.e., gastroesophageal reflux disorder). Not all women are good candidates for bisphosphonates.

17.3.7.3.2. Alendronate In the Fracture Intervention Trial (FIT), postmenopausal women with low bone mass were randomized to two groups on the basis of whether or not they had vertebral fractures at baseline. Alendronate showed reduction of vertebral fractures of 47% at year 3 and reduced incidence of hip fractures up to 56% *(17)*. The recommended dosage of alendronate for prevention of postmenopausal osteoporosis is 5 mg/day or 35 mg/week; for treatment, the dosage is increased to 10 mg/day or 70 mg/week. Both dosage regimens are equally effective. Alendronate is also indicated for the treatment of glucocorticoid-induced osteoporosis. Importantly, newer data demonstrate that alendronate 10 mg/day can increase or maintain BMD after discontinuation of HT *(18)*.

17.3.7.3.3. Risedronate Risedronate has the same treatment and prevention indications as alendronate as well as the glucocorticoid-induced osteoporosis prevention and treatment indication. The approved dosage of risedronate is 5 mg/day or 35 mg/week. Risedronate has been shown to decrease both vertebral and nonvertebral fractures in postmenopausal women. After of use, risedronate was associated with a 70–75% reduction in the incidence of first vertebral fracture *(19)*. This reduction appears to be sustained over at least 5 years *(20)*.

17.3.7.3.4. Ibandronate Ibandronate is currently marketed in the United States, having received FDA approval in 2003, under the trade name Boniva®. A dosage of 2.5mg daily or 150mg once monthly is indicated for the treatment and prevention of osteoporosis in postmenopausal women *(21)*.

17.3.7.3.5. Selective estrogen receptor modulators SERMs are nonhormonal agents that bind to the estrogen receptor but induce different estrogenic responses in various tissues. Raloxifene is currently the only SERM that is FDA approved for the prevention and treatment of osteoporosis.

17.3.7.3.6. Raloxifene Raloxifene has the unique property of activating estrogen receptors in the bone while not activating estrogen receptors in the breast or uterus. The recommended dosage of raloxifene is 60 mg/day for both prevention and treatment of osteoporosis. Unlike the bisphosphonates, raloxifene can be taken at any time of day, with or without food. Also, unlike the bisphosphonates, raloxifene has shown an increased risk of hot flashes, leg cramps, and thromboembolic events among users *(22,23)*. In the Multiple Outcomes of Raloxifene Evaluation (MORE), raloxifene was associated with a 30% reduction in vertebral fractures in postmenopausal women with at least one vertebral fracture and a 55% reduction in postmenopausal women with no vertebral fracture over 3 years. BMD improved 2.6% and 2.7% at the spine and hip, respectively *(24)*. Raloxifene was shown to have extra skeletal effects that are quite important to the overall health of women. In the MORE study, raloxifene was associated with a 65% relative risk reduction in all breast cancers with a 90% reduction in estrogen receptor-positive breast cancers, a 12% reduction in estrogen receptor-negative breast cancer, and a 76% relative risk reduction in invasive breast cancer. There was also evidence that raloxifene use resulted in a reduction in cardiovascular events in a subgroup of at-risk women in the study group *(25)*.

17.3.7.4. HORMONE THERAPY

17.3.7.4.1. Estrogen Bone tissue is exquisitely sensitive to estrogen. The rationale for estrogen therapy was based on the observation that the bone remodeling rate increased as estrogen levels declined resulting in greater bone resorption than bone formation. The beneficial effect of estrogen on bone is due to several factors including its ability to increase circulating calcitonin levels, which leads to decreased bone resorption. Estrogen receptors are also present on osteoblastic cells, which respond to the hormone by increasing collagen production. There are more than 50 randomized, placebo-controlled studies which have demonstrated that HT reduces the rate of bone resorption and loss and promotes an increase in BMD *(3)*. The benefits of HT on BMD are independent of estrogen preparation but are dose dependent. However, estrogen therapy even in low doses has been shown to have a beneficial effect on hip and spine BMD *(26)*. Low-dose birth control pills may be used for perimenopausal women at risk for bone loss, if no contraindications to that therapy exist.

Prior to the publication of the Heart and Estrogen/Progestin Replacement Study (HERS) *(27)* and the Women's Health Initiative (WHI) *(28)*, most of the data available on the effects of HT in fracture prevention were based on observational and epidemiologic studies *(29)*. In the HERS trial, the primary evaluation was on postmenopausal women and cardiovascular disease; fracture was a secondary endpoint.

Study participants were not selected on the basis of osteoporosis risk. HERS concluded that fracture risk following 4 years of HT (0.625 mg/day of conjugated equine estrogen plus 2.5 mg/day of medroxyprogesterone acetate) was not reduced.

The WHI estrogen plus progestin trial was conducted in postmenopausal women aged 50–79 (average age 63 years) with an intact uterus. The women were randomized to HT using the same preparation as the HERS trial versus placebo. The HT group showed a 34% lower relative risk of hip and a 34% lower relative risk of spine fracture as well as a 24% lower risk of fracture at all sites. In a study done in 2002, Greenspan et al. showed that once HT is withdrawn, accelerated bone loss occurs. Reduction of BMD was as much as 4.5% in the spine and as much as 2.4% in the hip after just 1 year after HT *(30)*.

The WHI created a paradigm shift that has almost entirely changed the way estrogens are prescribed in the United States today not because of the bone data but rather because of the other risks that were identified. Women randomized to the HT arm also showed a 29% increased risk of coronary heart disease and a 41% higher risk of cerebral vascular accident (stroke). The WHI also found another risk identified in HERS, which was an increased risk of venous thromboembolism. Unlike HERS, which found no significant increased risk of breast cancer, the WHI reported a very significant increased risk in invasive breast cancer in women treated with HT. One other positive finding of the WHI in addition to reduced fracture risk was the data that showed a significantly decreased risk of colon cancer development in women taking HT. Following the release of the WHI both, the NAMS and the US Preventive Services Task Force have recommended against the uses of estrogen for the treatment of osteoporosis, recommending instead alternative therapies.

17.3.7.4.2. Androgen Androgens in women have been studied, with regard to BMD, to a limited degree. Results of two studies that employed testosterone implants and oral methyltestosterone in addition to HT indicated significant beneficial effects on BMD *(31,32)*. However, the degree to which these effects were because of aromatization, thus representing estrogen rather than androgen effects, is unclear. The small number of participants in these studies and the lack of long-term, well-controlled trials prevent proper evaluation of the effects of testosterone treatment on BMD.

17.3.7.4.3. Progesterone To date, there is no reliable evidence that topical progesterone creams offer protection against bone loss. In a 1999 study, women were randomized to a progesterone cream 20 mg or placebo cream, to be applied once daily. BMD was assessed every 4 months for 1 year, and no significant difference was found between the treatment and control groups with respect to BMD of the hip and spine *(33)*. There is some evidence that norethindrone acetate may maintain BMD in women receiving gonadotropin-releasing hormone agonists for conditions such as endometriosis or uterine leiomyomata. Norethindrone acetate or combination estrogen progesterone regimens are used for "add-back" therapy in these women primarily to control hot flashes.

17.3.7.4.4. Calcitonin Calcitonin is a polypeptide hormone that decreases the rate of bone absorption by inhibiting resorptive activity in osteoclasts. It is available as a subcutaneous injection (100 IU/day) or as a nasal spray (200 IU/day). It is indicated for the treatment of postmenopausal osteoporosis in females more than 5 years postmenopause with low bone mass relative to healthy premenopausal females. It stabilizes or increases vertebral and nonvertebral BMD, predominantly at the forearm and lumbar spine. The Prevent Recurrence of Osteoporosis Fractures (PROOF) Study showed that the intranasal dose of calcitonin was associated with a 36% risk reduction in vertebral fracture and only a 20% risk reduction in nonvertebral fracture in women with low bone mass and history of vertebral compression fractures. The study also failed to show any effect on hip fractures *(34)*.

17.3.7.4.5. Teriparatide Teriparatide is a synthetic form of human PTH. PTH is the primary regulator of calcium and phosphate metabolism in bones. Teriparatide 20 mcg/day injected subcutaneously in the thigh or abdomen is indicated for the treatment of osteoporosis in postmenopausal women at high risk of fracture. It is the first approved treatment for osteoporosis that stimulates both cortical and trabecular bone formation. In a randomized, double-blind trial, involving nearly 1700 postmenopausal women with prior vertebral fracture, teriparatide 20 mcg/day reduced new vertebral fractures by 65% and nonvertebral fractures by 35% *(35)*. Prior to teriparatide administration, laboratory tests should be done including BMD by DEXA (spine and hip), total serum calcium, total serum alkaline phosphatase, 25-hydroxyvitamin D, PTH, and creatinine clearance. The use of teriparatide is contraindicated in patients with hypercalcemia, Paget's disease, elevated bone-specific alkaline phosphatase, osteogenic sarcoma, history of skeletal radiation, pregnancy and lactation, unfused epiphyses, bone cancer or metastatic bone cancer. Therapy with teriparatide is not recommended for more than 2 years in humans because of the development of osteosarcomas in animal studies.

17.3.7.5. COMBINATION THERAPY

Several studies have examined the effect of combining two antiresorptive agents on bone mass and have found that at the spine and total hip, the combination resulted in a greater increase than with one agent alone. A recent randomized trial by Greenspan et al. *(30)* showed that in women aged 65–90 combination therapy with HT and alendronate was more effective in increasing bone mass than either therapy alone. There is as yet, however, no evidence that combination therapy reduces fracture risk to a greater degree than monotherapy. Currently, combination therapy is not recommended.

17.3.7.6. COMPLEMENTARY AND ALTERNATIVE THERAPY

Phytoestrogens are found in various plants and foods; these substances are similar in action to estrogen. There are conflicting data on phytoestrogens and their effects on BMD. Some studies report that phytoestrogens have beneficial effect on bone resorption and BMD in postmenopausal women, whereas other studies have reported no effect *(36,37)*. Much conflicting information exists with respect to the use of soy (bean) isoflavones in the treatment of patients with osteoporosis. Although it seems likely that soy isoflavones have many health benefits, the evidence for using soy as an alternative treatment in osteoporosis remains inconclusive.

One 12-week trial of postmenopausal women revealed that black cohosh extract increased levels of bone-specific alkaline phosphatase, a metabolic marker for bone formation, compared with women on HT or placebo. However, no BMD measurements were done. This study may suggest that black cohosh is beneficial, but the data are insufficient to be conclusive *(38)*. Red clover has undergone similar evaluations, and the results may be promising but are as yet insufficient *(38)*. Wild yam has not been studied in a controlled trial for the treatment of osteoporosis; however, it has been marketed for that indication in the United States. All herbal remedies and alternative or complementary forms of therapy must be used with caution because little or no data in controlled trial are available. Currently, herbals are sold as nutritional supplements and are therefore not regulated by the FDA. Many herbal remedies have interactions with prescription medications, ranging from annoying to life-threatening. No herbal remedy should be initiated without a thorough review and discussion of all other medications that a patient is taking.

17.4. CONCLUSIONS

Early prevention of bone loss is the key to reducing a woman's lifetime risk of developing osteoporosis and fracture. Although numerous effective therapies for the treatment are now available, maintaining a healthy lifestyle with proper, balanced nutrition and sufficient, regular physical activity should be paramount for all women. Therapeutic regimens today are safe and beneficial and should be employed as soon as the need is identified.

REFERENCES

1. World Health Organization (WHO). Assessment of fracture risk and its application to screening for post menopausal osteoporosis: report of WHO Study Group. (Technical report series 843). Geneva, Switzerland: WHO; 1994.
2. Kanis JA, Gluer C-C. An update on the diagnosis and assessment of osteoporosis with densitometry. Committee of Scientific Advisors, International Osteoporosis Foundation. *Osteoporos Int* 2000; 11:192–202.
3. The North American Menopause Society. Management of postmenopausal osteoporosis: position statement of the North American Menopause Society. *Menopause* 2002; 9:84–101.
4. Recker RR, Lappe J, Davies K, et al. Characterization of perimenopausal bone loss: a prospective study. *J Bone Miner Res* 2000; 15:1963–1973.
5. Riggs BL, Melton LJ III: Involutional osteoporosis. *N Engl J Med* 1992; 326:357–362.
6. Fitzpatrick LA. Secondary causes of osteoporosis. *Mayo Clin Proc* 2002; 77:453–468.
7. Slemenda CW, Christian JC, Williams CJ, et al. Genetic determinants of bone mass in adult women: a reevaluation of the twin model and the potential importance of gene interaction of heritability estimates. *J Bone Miner Res* 1991; 6:561–567.
8. Pocock NA, Eisman JA, Hopper JL, et al. Genetic determinants of bone mass in adults: a twin study. *J Clin Invest* 1987; 80:706–710.
9. Nordin BE, Need AG, Steurer T, et al. Nutrition, osteoporosis, and aging. *Ann N Y Acad Sci* 1998; 854:336–351.
10. *Osteoporosis*. Belle Mead, NJ: Excerpta Medica, Inc; 2003.
11. Msissinger-Papport BJ, Thacker HL. Prevention for the older woman: a practical guide to prevention and treatment of osteoporosis. *Geriatrics* 2002; 57:16–27.
12. Lau EM, Woo J. Nutrition and osteoporosis (review). *Curr Opin Rheumatol* 1998; 10:368–372.
13. Nieves JW, Golden AL, Siris E, et al. Teenage and current calcium intake are related to bone mineral density of the hip and forearm in women aged 30–39 years. *Am J Epidemiol* 1995; 141:342–351.

14. Gutin B, Kasper MJ. Can vigorous exercise play a role in osteoporosis prevention? A review. *Osteoporos Int* 1992; 2:55–69.
15. Standing Committee on the Scientific Evaluation of Dietary Reference Intakes. Institute of Medicine. *Dietary Reference Intakes: Calcium, Phosphorus, Magnesium, Vitamin D, and Fluoride*. Washington, DC: National Academy Press; 1997.
16. Chapuy MC, Arlot ME, Duboeuf F, et al. Vitamin D3 and calcium to prevent hip fractures in elderly women. *N Engl J Med* 1992; 327:1637–1642.
17. Black DM, Cummings SR, Karpf DB, et al. for the Fracture Intervention Trial Research Group. Randomized trial of effect of alendronate on risk of fracture in women with existing vertebral fractures. *Lancet* 1996; 348:1535–1541.
18. Ascott-Evans BH, Guanabens N, Kivinen S, et al. Alendronate prevents loss of bone density associated with discontinuation of hormone replacement therapy: a randomized controlled trial. *Arch Intern Med* 2003; 163:789–794.
19. Reginster J-Y, Minne HW, Sorenson OH, et al. Randomized trial of the effects of risedronate on vertebral fractures in women with established postmenopausal osteoporosis. *Osteoporos Int* 2000; 11:83–91.
20. Sorensen OH, Crawford GM, Mulder AH, et al. Long-term efficacy of risedronate: a 5-year placebo-controlled clinical experience. *Bone* 2003; 32:120–126.
21. Chestnut CH III, Skag A, Christiansen C, et al. Effects of oral ibandronate administered daily or intermittently on fracture risk in postmenopausal osteoporosis. *J Bone Miner Res* 2004; 19:1241–1249.
22. Cranney A, Tugwell P, Zytaruk N, et al. Meta-analysis of therapies for postmenopausal osteoporosis. IV. Meta-analysis of raloxifene for the prevention and treatment of postmenopausal osteoporosis. *Endocr Rev* 2002; 23:524–528.
23. Ettinger B, Black DM, Mitlak BH. Reduction of vertebral fracture risk in postmenopausal women with osteoporosis treated with raloxifene: results from a 3-year randomized clinical trial. Multiple Outcomes of Raloxifene Evaluation (MORE) Investigators. *JAMA* 1999; 282:637–645.
24. Cummings SR, Eckert S, Krueger KA, et al. The effect of raloxifene on risk of breast cancer in postmenopausal women: results from the MORE randomized trial. Multiple Outcomes of Raloxifene Evaluation. *JAMA* 1999; 281:2189–2197.
25. Barrett-Connor E, Grady D, Sashegyi A, et al. Raloxifene and cardiovascular events in osteoporotic postmenopausal women: a four-year results from the MORE randomized trial. *JAMA* 2002; 287:847–857.
26. Recker FF, Davies KM, Dowd RM, et al. The effect of low-dose continuous estrogen and progesterone therapy with calcium and vitamin D on bone in elderly women: a randomized, controlled trial. *Ann Intern Med* 1999; 130:897–904.
27. Simon JA, Hsia J, Cauley JA, et al. Postmenopausal hormone therapy and risk of stroke: the Heart and Estrogen/Progestin Replacement Study (HERS). *Circulation* 2001; 133:638–642.
28. Writing Group for the Women's Heal Initiative Investigators. Risks and benefits of estrogen plus progestin in healthy postmenopausal women: principal results from the Women's Health Initiative randomized controlled trial. *JAMA* 2002; 288:321–333.
29. Wells G, Tugwell P, Shea B, et al. Meta-analysis of the efficacy of hormone replacement therapy in treatment and preventing osteoporosis in postmenopausal women. Part V. *Endocr Rev* 2002; 23:529–539.
30. Greenspan S, Resnick N, Parker R. Combination therapy with hormone replacement and alendronate for prevention of bone loss in elderly women: a randomized controlled trial. *JAMA* 2003; 289:2525–2532.
31. Davis SR, McCloud P, Strauss BJ, et al. Testosterone enhances estradiol's effects on postmenopausal bone density and sexuality. *Maturitas* 1995; 21:227–236.
32. Barrett-Connor E, Young R, Notelovitz M, et al. J. A two-year, double-blind comparison of estrogen-androgen and conjugated estrogens in surgically menopausal women. Effects on bone mineral density, symptoms and lipid profiles. *J Reprod Med* 1999; 44:1012–1020.
33. Leonetti HB, Longa S, Anasti J. Transdermal progesterone cream for vasomotor symptoms and postmenopausal bone loss. *Obstet Gynecol* 1999; 94:225–228.
34. Chestnut CH III, Silverman S, Andriano K, et al. for the PROOF Study Group. A randomized trial of nasal spray salmon calcitonin in postmenopausal women with established osteoporosis: the Prevent Recurrence of Osteoporotic Fractures Study. *Am J Med* 2000; 109:267–276.

35. Neer RM, Arnaud CD, Zanchetta JR, et al. Effect of parathyroid hormone (1–34) on fractures and bone mineral density in postmenopausal women with osteoporosis. *N Engl J Med* 2001; 344: 1434–1441.
36. Zhang X, Shu X-O, Li H, et al. Prospective cohort study of soy food consumption and risk of bone fracture among postmenopausal women. *Arch Intern Med* 2005; 165:1890–1895.
37. Supplement: Fourth International Symposium on the role of soy in preventing and treating chronic disease. The American Society for Nutritional Sciences. *J Nutr* 2002; 132:588S–619S.
38. *PDR for Herbal Medicines*, 3rd ed. Montvale, NJ: Thompson PDR; 2004:93–97, 679–681, 747–757.

V Safe Exercise Guidelines Throughout the Lifespan

18 Physical Activity Recommendations and Exercise Guidelines Established by Leading Health Organizations

Jacalyn J. Robert-McComb

CONTENTS

18.1 LEARNING OBJECTIVES
18.2 INTRODUCTION
18.3 RESEARCH FINDINGS
18.4 SUMMARY
18.5 SCENARIO WITH QUESTIONS AND ANSWERS

18.1. LEARNING OBJECTIVES

After completing this chapter, you should have an understanding of the following:

- The difference between moderate physical activity, vigorous physical activity, exercise, health-related physical fitness, and sports-related fitness.
- Physical activity recommendations for health-related fitness from leading health organizations.
- The American College and Sports Medicine's (ACSM's) exercise recommendations for cardiorespiratory fitness.
- The ACSM's exercise recommendations for muscular fitness.
- The ACSM's exercise recommendations for healthy body composition.
- The ACSM's exercise recommendations for flexibility.

18.2. INTRODUCTION

Women are generally less active than men at all ages and by the age of 75—one in three men and one in two women engage in *no* regular physical activity *(1)*. This has important public health implications. In 1995, the National Institutes of Health published a consensus statement entitled Physical Activity and Cardiovascular Health. The conclusion of this consensus statement was as follows:

All Americans should engage in regular physical activity at a level appropriate to their capacity, needs, and interest. Children and adults alike should set a goal of accumulating at least 30 min of moderate-intensity physical activity on most, and preferably, all days of the week. Most Americans have little or no physical activity in their daily lives, and accumulating evidence indicates that physical inactivity is a major risk factor for cardiovascular disease. However, moderate levels of physical activity confer significant health benefits. Even those who currently meet these daily standards may derive additional health and fitness benefits by becoming more physically active or including more vigorous activity. For those with known cardiovascular disease, cardiac rehabilitation programs that combine physical activity with reduction in other risk factors should be more widely used (2).

Likewise in 1995, the Centers for Disease Control (CDC) and the ACSM recommended that all adults in the United States should accumulate 30 min or more of moderate physical activity on most, and preferably all, days of the week *(3)*. Another landmark report on the benefits of physical activity was released on July 11, 1996, by the United States Department of Health and Human Services. It was the first Surgeon General's Report (SGR) on Physical Activity and Health *(4)*. The SGR indicated that although health benefits improve with moderate amounts of physical activity (15 min of running, 30 min of brisk walking, or 45 min of playing volleyball), greater amounts are obtained with greater amounts of physical activity *(4)*. As understanding of the benefits of physical activity grew, recommendations followed suit, because, on average, physically active people outlive those who are inactive *(5)*. The United States Department of Health and Human Services in conjunction with leading federal agencies published Healthy People 2010 in the year 2000 *(1)*. This document presents a comprehensive set of disease prevention and health promotion objectives developed to improve the

Table 18.1
Healthy People 2010 physical activity and fitness objectives: a highlight

Reduce the proportion of adults who engage in no leisure-time physical activity.
Increase the proportion of adults who engage regularly, preferably daily, in moderate physical activity for at least 30 min/day.
Increase the proportion of adults who engage in vigorous physical activity that promotes the development and maintenance of cardiorespiratory fitness 3 or more days per week for 20 min or more per occasion.
Increase the proportion of adults who perform physical activities that enhance and maintain muscular strength and endurance.
Increase the proportion of adults who perform physical activities that enhance and maintain flexibility.
Increase the proportion of adolescents who engage in moderate physical activity for at least 30 min on 5 or more of the previous 7 days.
Increase the proportion of adolescents who engage in vigorous physical activity that promotes cardiorespiratory fitness 3 or more days per week for 20 min or more per occasion.

Source: For a complete list of objectives, see http://www.healthypeople.gov/document/HTML/Volume2/22Physical.htm.

health of all people in the United States during the first decade of the twenty-first century. Co-leading agencies for the physical activity and fitness objectives were the CDC and the President's Council on Physical Fitness. Table 18.1 highlights some of the objectives designed to improve health, fitness, and quality of life through daily physical activity. For more information, visit http://www.healthypeople.gov/.

18.3. RESEARCH FINDINGS

18.3.1. The Difference Between Moderate Physical Activity, Vigorous Physical Activity, Exercise, Health-Related Physical Fitness, and Sports-Related Fitness

According to Ainsworth and colleagues *(6)*, moderate-intensity physical activity refers to any activity that burns 3.5–7 kcal/min. The ACSM defines moderate exercise intensity as 40–59% of oxygen uptake reserve (VO_2R) or heart rate reserve (HRR) *(7)*. In the 1996 SGR *(4)*, moderate exercise was defined as some increase in breathing or heart rate or a "perceived exertion" of 11–14 on the Borg rate of perceived exertion (RPE) scale. These levels are equal to the effort a healthy individual might burn while walking at a 3–4.5 mph pace on a level surface, playing golf, gardening and yard work, swimming for recreation, or bicycling. In Appendix 10, steps are outlined to determine a moderate training heart rate using the HRR method. The Borg RPE scale can be found in Appendix 11.

Ainsworth et al. *(6)* defined vigorous-intensity physical activity as any activity that burns more than 7 kcal/min. The ACSM *(7)* refers to vigorous exercise as 60% or more of VO_2R or HHR (Appendix 10). In the 1996 SGR *(4)*, vigorous exercise was defined as a large increase in breathing or heart rate (conversation is difficult or "broken") or a "perceived exertion" of 15 or greater on the Borg RPE scale (Appendix 11). These levels are equal to the effort a healthy individual might burn race walking or walking at a 5 mph pace, engaging in heavy yard work, participating in high-impact aerobic dancing, swimming continuous laps, scuba diving, or bicycling uphill.

Examples of general physical activities that meet the guidelines established by the CDC and the ACSM for moderate activity (3.0–6.0 METs or 3.5–7 kcal/min) and vigorous activity (greater than 6.0 METs or more than 7 kcal/min) can be found at http://www.cdc.gov/nccdphp/dnpa/physical/recommendations/index.htm.

What is the difference between *physical activity* and *exercise*? The United States Department of Health and Human Services refers to *physical activity* as bodily movement that is produced by the contraction of skeletal muscle that substantially increases energy expenditure *(4)*. But *exercise* is a type of physical activity that represents structured, planned activities, and repetitive bodily movement, designed to promote or enhance overall physical fitness *(8)*.

Both physical activity and exercise contribute to increases in physical fitness. *Physical fitness* is a multidimensional concept that has been defined as a set of attributes that people possess or achieve that relates to the ability to perform physical activity *(7,8)*. There are skill-related components of physical fitness (also known as sports-related forms of physical fitness) and health-related components of physical fitness. *Skill-related components of physical fitness* include balance, agility, coordination, speed, power, and reaction time. These components are associated mostly with sport performance. *Health-related components of physical fitness* include cardiovascular endurance, muscular strength and endurance, flexibility, and body composition. The focus of this chapter is on the health-related components of physical fitness.

18.3.2. Physical Activity Recommendations for Health-Related Fitness from Leading Health Organizations

Before beginning any physical activity or exercise program, it is recommended that individuals complete a self-administered questionnaire to help identify risks that may warrant further medical clearance before embarking on an activity program (Appendix 12). Individuals should consult their physician before beginning a new physical activity program if they have chronic diseases, such as cardiovascular disease and diabetes mellitus, or are at high risk for these diseases. Additionally, men over the age of 45, and women over the age of 55, should consult a physician before beginning a vigorous activity program *(7)*.

Table 18.2 lists physical activity recommendations for cardiorespiratory (CR) health-related fitness from selected leading health organizations. Public health recommendations have evolved from emphasizing vigorous activity for CR health-related fitness to including the option of moderate levels of activity for numerous health benefits.

18.3.3. The ACSM's Exercise Recommendations for CR Fitness

CR fitness is defined as the ability of the body to transport and use oxygen and is dependent on the effective integration of both the cardiac and the pulmonary systems *(9)*. The terms CR fitness, $VO_{2\,max}$, aerobic capacity, and cardiovascular fitness are used synonymously. These terms refer to the maximal capacity to produce energy aerobically and are usually expressed in METs or mL O_2/kg/min. One MET (metabolic equivalent unit) is equal to approximately 3.5 mL O_2/kg/min. Improvements in the ability of the heart to deliver oxygen to the working muscles and in the muscle's ability to generate energy aerobically result in increased CR fitness. CR endurance, aerobic endurance, or cardiovascular endurance refers to the ability to persist or continue in strenuous activity requiring large-muscle groups for prolonged time *(7)*.

Exercise recommendations must take into account the fitness level of the individual. Individuals with low level of fitness generally demonstrate the greatest improvements in CR fitness, whereas modest increases occur in healthy individuals and in those with high initial fitness levels *(10)*. Exercise recommendations to improve CR fitness include three components: frequency, intensity, and duration, also known as the FIT principles. The format for an exercise session should include a warm-up period (approximately 5–10 min), a stimulus or conditioning phase (20–60 min), and a cool-down period (5–10 min). Table 18.3 lists the recommendations proposed by the ACSM to improve and maintain CR fitness. The recommendation range is broad because of the heterogeneity in a response to an exercise stimulus.

18.3.4. The ACSM's Exercise Recommendations for Muscular Fitness

Improving muscular functioning through a resistance-training program may provide physiological benefits for girls and women of all ages. Research has clearly indicated that strength can be effectively increased with training in girls before the age of

Table 18.2
Physical activity recommendations for health-related fitness from leading health organizations

Organization: Centers for Disease Control (CDC) and the American College of Sports Medicine (http://www.cdc.gov/)
Recommendation: All adults in the United States should accumulate 30 min or more of moderate physical activity on most, and preferably all, days of the week.
Purpose: Health promotion and prevention of chronic diseases
Organization: American Heart Association (http://www.americanheart.org)
Recommendation: Older adults and people with disabilities can gain significant health benefits with a moderate amount of physical activity, preferably daily. Physical activity does not need to be strenuous to bring health benefits. What is important is to include activity as part of a regular routine.
Children and adolescents should participate in at least 60 min of moderate to vigorous physical activity every day.
All children age 2 and older should participate in at least 30 min of enjoyable, moderate-intensity activities every day. They should also perform at least 30 min of vigorous physical activities at least 3–4 days each week to achieve and maintain a good level of cardiorespiratory (heart and lung) fitness.
Purpose: Health promotion and prevention of chronic diseases
Organization: A Report of the Surgeon General (http://www.cdc.gov/nccdphp/sgr/chapcon.htm)
Recommendation: People of all ages should include a minimum of 30 min of physical activity of moderate intensity (such as brisk walking) on most, if not all, days of the week. Experts advise previously sedentary people embarking on a physical activity program to start with short durations of moderate-intensity activity and gradually increase the duration or intensity until the goal is reached. It is also acknowledged that for most people, greater health benefits can be obtained by engaging in physical activity of more vigorous intensity or of longer duration. Recent recommendations from experts also suggest that cardiorespiratory endurance activity should be supplemented with strength-developing exercises at least twice per week for adults, in order to improve musculoskeletal health, maintain independence in performing the activities of daily life, and reduce the risk of falling.
Purpose: Health promotion and prevention of chronic diseases
Organization: American College of Obstetrics and Gynecology (http://www.acog.org/)
Recommendation: For the promotion of a healthy pregnancy and postpartum recovery, women should accumulate 30 min of exercise a day on most days of the week *(28)*.
Purpose: Promotion of a healthy pregnancy and postpartum recovery.

---

puberty *(11)*. However, the specific health benefits of resistance training for girls remain to be clarified *(7)*. For middle-aged, older adult, and postmenopausal women, a reduction in the risk of osteoporosis, low back pain, hypertension, and diabetes have been associated with resistance-training programs *(12,13)*. Even though guidelines for children and adolescents are similar to those for adults, there are specific guidelines for children and adolescents. Guidelines for children and adolescents are discussed in Chap. 19.

Unlike cardiovascular activity, intensity for resistance exercise is not easily determined. Miriam E. Nelson *(14)*, the author of *Strong Women Stay Young*, suggests that women beginning a resistance-training program should use a scale similar to the RPE

Table 18.3
American College of Sports Medicine's Training Guidelines for cardiorespiratory fitness

Frequency
3–5 days/week
Intensity
Determining intensity using the HR_{max} Method
64%/70%–94% of maximum heart rate (HR_{max})
$HR_{max} = 220 -$ age
Target heart rate $= HR_{max}$ (64%/70–94%)
i.e., $HR_{max} = 220$–20 (age)
    $HR_{max} = 200$
    Target heart rate $= 200$ (64%/70–94%)
    Target heart rate may range from 128/140 to 188 bpm
Determining intensity using the HRR method
40%/50–85% of heart rate reserve (HRR)
$HRR = HR_{max}$ (i.e., 220 – age) – resting heart rate
Target heart rate $=$ [(HRR) (exercise intensity)]+resting heart rate
i.e., Target heart rate
    $= HR_{max} = 220 -$ age (e.g., 220–40 = 180)
    $= HRR = HR_{max}$ – resting heart rate; (e.g., 180–60 = 120)
    $=$ [(HRR) (exercise intensity)] + resting heart rate
    $=$ [(120) (40%/50–85%)] + 60 (example)
    $= 108/120$–162 bpm is your training heart rate range (example)
Determining intensity using the VO_2R Method
40%/50–85% of oxygen uptake reserve (VO_2R)
$VO_2R = VO_{2\,max} - VO_{2rest}$ ($VO_{2\,max}$ mL/kg/min–3.5 mL/kg/min)
Target VO_2 mL/kg/min $=$ [(VO_2R) (exercise intensity)] + VO_{2rest}
i.e., Target VO_2 mL/kg/min for a person with $VO_{2\,max}$ of 40 mL/kg/min
    $=$ [(40 mL/kg/min–3.5 mL/kg/min)(60%)] + 3.5 mL/kg/min (example)
    $=$ [(36.5 mL/kg/min)(60%)] + 3.5 mL/kg/min (example)
    $= 21.9$ mL/kg/min + 3.5 mL/kg/min (example)
    $= 25.4$ mL/kg/min (example)
Determining exercise intensity using the rate of perceived exertion (RPE) *Scale*
12–16 (in the range of "somewhat hard" to "hard") on the original 6–20
Borg RPE scale is the average range associated with physiologic adaptation.
However, there is significant interindividual variability in the psychophysiologic
relationship to $\%HR_{max}$, %HRR or $\%VO_2R$ and RPE. Therefore, RPE should
only be used as a guideline in setting the exercise intensity. The 6–20 original
RPE scale can be found in Appendix 11.
Duration
20–60 min of continuous aerobic activity. Duration is dependent on the intensity
of the activity:
thus, lower intensity should be conducted over a longer period of time.

For individuals with $VO_{2\,max}$ below 40 mL O_2 mL/kg/min, a minimal intensity of 30% VO_2R or HRR can elicit improvements in $VO_{2\,max}$. In contrast, individuals with greater cardiorespiratory (CR) fitness (> 40 mL O_2 kg/min) require a minimal threshold of 45% VO_2R or HRR. For most individuals, intensities within the range of 60–80% of VO_2R or HRR or 70–90% HR_{max} are sufficient to achieve improvements in CR fitness with the appropriate duration and frequency of training.

scale (Appendix 11) to determine the amount of weight that they should incorporate into their resistance-training program. In order for strength gains to continue to accrue, there must be a gradual increase in the stress or load placed on the body throughout the resistance-training program. The *overload principle* refers to placing greater demands on the body than what it is accustomed to, and the *principle of progression* refers to the constant application of the overload principle throughout the resistance-training program. Hence, the term *progressive overload* has been coined. The RPE scale has become a popular method to assess *progressive overload*. Nelson suggests that during the first week, individuals should focus on form and that the effort involved in lifting should be easy or moderate. When learning the exercises, an intensity of 9 (very light) to 11 (light) on the RPE scale would be appropriate. According to Nelson *(14)*, the goal should be for the exercise set (eight repetitions) to become difficult after six or seven repetitions. The ACSM recommends an initial goal of 12–13 (somewhat hard) and a final goal of 15 to 16 (hard [heavy]) on the RPE scale for submaximal training *(7,15)*. A target of 19 to 20 (extremely hard and maximal exertion, respectively) on the RPE scale is synonymous with high-intensity strength stimuli for healthy populations *(15)*. However, for people with high cardiovascular risk or those with chronic disease, the

Table 18.4
American College of Sports Medicine's resistance-training guidelines for healthy adults

Choose a mode of exercise (free weights, bands, or machines) that is comfortable throughout the full pain-free range of motion.
Perform a minimum of 8–10 separate exercises that train the major muscles of the hips, thigh, legs, back, chest, shoulders, arms, and abdomen. A primary goal of the program should be to develop total body strength and endurance in a relatively time-efficient manner. Total exercise training programs lasting longer than 1 h per session to the point of volitional fatigue for healthy individuals while maintaining good form.
Perform one set of each exercise to the point of volitional fatigue for healthy individuals while maintaining good form.
Although the traditional recommendation of 8–12 repetitions is still appropriate, choose a range of repetitions between 3 and 20 (e.g., 3–5, 8–10, or 12–15) that can be performed at a moderate repetition duration (approximately 3 s concentric, approximately 3 s eccentric).
Exercise each muscle group two to three nonconsecutive days per week and, if possible, perform a different exercise for the muscle group every two to three sessions.
Adhere as closely as possible to the specific techniques for performing a given exercise. Allow enough time between exercises to perform the next exercise in proper form.
For people with high cardiovascular risk or those with chronic disease (hypertension, diabetes), terminate each exercise as the concentric portion of the exercise becomes difficult (RPE 15–16) while maintaining good form.
Perform both the lifting (concentric phase) and lowering (eccentric phase) portion of the resistance exercises in a controlled manner.
Maintain a normal breathing pattern; breath holding can induce excessive increases in blood pressure.
If possible, exercise with a training partner who can provide feedback, assistance, and motivation.

Source: American College of Sports Medicine *(7)*.

exercise should be terminated if the lifting portion of the exercise becomes difficult corresponding to an RPE of 15–16 *(7)*. Therefore, a more modest level of exertion should be chosen. Table 18.4 provides resistance guidelines as outlined by the ACSM for healthy adults *(7)*.

Table 18.5 outlines the ACSM Position Stand on Progression Models in Resistance Training for Healthy Adults *(16)*. Although it is common to estimate intensity on repetition maximum (RM), this should only be used as a general guideline because RM differs between muscle groups *(17)*. These guidelines are appropriate for healthy

Table 18.5
Overview of American College of Sports Medicine Position Stand on progression models in resistance training for healthy adults

Recommendations should be viewed in context of individual's target goals, physical capacity, and training status. Exercise selection should include concentric and eccentric muscle actions as well as both single-joint and multiple-joint exercises
Exercise sequence
Exercise large-muscle groups before small-muscle groups
Perform multiple-joint exercises before single-joint exercises
Perform higher intensity exercises before lower intensity exercises
When training at a specific repetition maximum (RM) load
A 2–10% increase in load should be applied when one to two repetitions can be performed over the current workload
Training frequency
For novice and intermediate training: 2–3 days/week
For advanced training: 4–5 days/week
Novice training
About 8–12 RM
Intermediate to advanced training
About 1–12 RM in periodized fashion with emphasis on the 6–12 RM zone using 1-min to 2-min rest periods between sets at a moderate velocity (1–2 s concentric, 1–2 s eccentric)
Eventual emphasis on heavy loading (1–6 RM) with at least 3-min rest periods between sets at a moderate contraction velocity (1–2 s concentric, 1–2 s eccentric)
Hypertrophy training
About 1–12 RM in periodized fashion with emphasis on the 6–12 RM zone, 2-min rest periods between sets, moderate contraction velocity (1–2 s concentric, 1–2 s eccentric)
Moderate loading for novice (70–85% of 1 RM), advanced training (70–100% for 1–12 repetitions) higher volume, multiple-set programs result in greater hypertrophy than low-volume, single-set programs
Power training (novice and intermediate-trained individuals)
One to three sets per exercise using light to moderate loads (30–60% of 1 RM) at a fast contraction velocity for 3–6 repetitions with 2–3 min of rest between sets for multiple sets, with emphasis on multiple-joint exercises, especially those involving the total body, and progression of power enhancement uses various loading strategies in a periodized manner
Local muscular endurance training
Light to moderate loads, 40–60% of 1 RM, high repetitions (> -15), short rest periods (< -90 s)

Source: American College of Sports Medicine Position Stand *(16)*.

adult women who desire goal-oriented guidelines for athletic performance enhancement rather than simply health benefits. In order to more fully understand the guidelines as outlined in Table 18.5, the following terms are defined.

Concentric (shortening): Concentric muscle actions occur when the total tension developed in all the cross-bridges of a muscle is sufficient to overcome any resistance to shortening.

Eccentric (lengthening): Eccentric muscle actions occur when the tension developed in the cross-bridge is less than the external resistance, and the muscle lengthens despite contact between the myosin cross-bridge heads and the actin filament (18).

Hypertrophy: The muscular enlargement that results from resistance is called hypertrophy and is primarily a result of an increase in the cross-sectional area of the existing fibers.

Multiple-joint exercise: Multiple-joint exercises involve two or more primary joints (i.e., front or back squat, bench press, shoulder press).

Periodization: Effective program design involves the use of periodization, which is the varying or cycling of training specificity, intensity, and volume to achieve peak levels of conditioning (19).

Power: Power is precisely defined as the "time rate of doing work" (20) where work is the product of the force exerted on an object and the distance the object moves in the direction in which the force is exerted (power = work/time).

Repetitions: Repeating an identical movement for a specific number of times. To improve strength, you must do enough repetitions of each exercise to fatigue your muscles. The number of repetitions needed to cause fatigue depends on the amount of resistance. In general, a heavy weight and a low number of repetitions (1–5) build strength, a lightweight and high number of repetitions (15–20) build endurance, and, for general fitness purposes, 8–12 repetitions are usually recommended.

Repetition maximum: The maximum amount of resistance a person can move a specific number of times is referred to as a repetition maximum (RM). The RM indicates that the muscle has reached a point of fatigue in which the force-generating capacity falls below the required force to shorten the muscle against the imposed resistance (7). One RM is the maximum amount of resistance that can be lifted one time, and 5 RM is the maximum amount of weight that can be lifted five times.

Set: A set refers to a group of repetitions of an exercise followed by a rest period.

Single-joint exercise: Single-joint exercise involves only one primary joint (i.e., bicep curl).

For more specific information on developing effective resistance-training programs for specific goals, I recommend that you access online resources at http://www.nsca-lift.org/(National Strength and Conditioning Association), http://www.acsm.org/ (ACSM), or http://www.exrx.net/index.html (Exercise Prescription on the Net). These resources will also provide recommendations for texts and videos. In addition, workout templates and live video clips of the proper form for performing resistance exercise can be found at http://www.exrx.net/index.html (Exercise Prescription on the Net).

18.3.5. The ACSM's Exercise Recommendations for Healthy Body Composition

Although national standards have been developed and accepted for body mass index and waist circumference, there are no national standards for body composition (21).

Table 18.6
Recommendations for body fat percentages for women

Category	Recommended percentage
Essential	8–12
Minimal	10–12
Athletic	12–22
Recommendations for 34 years or less	20–35
Recommendations for 35–55 years	23–38
Recommendations for over 56 years	25–38

Source: Lohman et al. (22).

Lohman et al. (22) proposed a set of standards for women using data from the National Health and Nutrition Examination Survey. Table 18.6 lists Lohman et al.'s (22) body fat percentage recommendations for women.

Exercise programs to optimize lean body mass should include both cardiovascular and muscular fitness exercise components. In accordance with the United States Department of Health and Human Services, the ACSM recommends a target range of 150–400 kcal of physical activity and/or exercise a day for energy expenditure (4,7). However, recent reports indicate that 60 min or more per day may be necessary for weight loss (23). Physical activity and/or exercise expenditure in excess of 2000 kcal/week have been successful for both short-term and long-term weight control (24). Energy expenditure equivalents for activity can be found on the Fitness Partner Connection Jumpsite (http://www.primusweb.com/fitnesspartner/jumpsite/calculat.htm).

18.3.6. The ACSM's Exercise Recommendations for Flexibility

Because flexibility is believed to be transient, it is recommended that flexibility exercises should be performed 5–7 days a week (7). The greatest change in flexibility has been shown to be in the first 15 s of the stretch with no significant improvements after 30 s (25). Therefore, ACSM recommends that each stretch be held for 15–30 s. The optimal number of stretches per muscle group is two to four because no significant improvement in muscle elongation is seen in repeated stretching of 5–10 repetitions (26). Table 18.7 lists general exercise guidelines for achieving and maintaining flexibility (7,27).

Table 18.7
General exercise prescription for achieving and maintaining flexibility

Precede stretching with a warm-up to elevate muscle temperature
Stretch major muscle groups: neck, shoulders and upper arms, chest and upper arms, back and posterior thighs, thighs and hips, and legs and ankles
Perform a minimum of 2–3 days/week, ideally 5–7 days/week
Stretch to the end of the range of motion to a point of tightness, without inducing discomfort
Hold each stretch for 15–30 s
Repeat two to four repetitions for each stretch

Source: American College of Sports Medicine (7) and Knudson (27).

Table 18.8
High-risk flexibility exercises with alternative stretches

High-risk stretch	Alternative stretch
Standing toe touch	Seated toe touch or modified hurdler's stretch
Barre stretch	Seated toe touch or modified hurdler's stretch
Hurdler's stretch	Modified hurdler's stretch (bend the knee so that the tibia moves toward the torso rather than away from the torso as in the traditional hurdler's stretch)
Neck circles	Nontwisting directional stretch
Knee hyperflexion	Kneeling hip and thigh stretch
Yoga plow	Seated toe touch

Source: American College of Sports Medicine *(7)* and Knudson *(27)*.

Table 18.8 lists some common high-risk stretches and safe alternative exercises. The high-risk exercises could be appropriate for certain groups of athletes, for example, ballet dancers would perform stretches using the barre, whereas the alternative stretch would be more suitable for the average female. There are many web sources to find appropriate flexibility exercises. The American Academy of Orthopaedic Surgeons provides examples of flexibility exercises for the young athlete as well as for older individuals (*see* http://orthoinfo.aaos.org/fact/thr_report.cfm?thread_id=4&topcategory=wellness).

18.4. SUMMARY

Despite the documented benefits of regular physical activity, the majority of children, adolescents, and adult women are not engaged in physical activity consistent with public health recommendations. At one end of the continuum, we have young girls and female athletes who are not taking in enough calories to meet their energy needs associated with exercise, and at the other end of the continuum, we have the vast majority of children, adolescents, and adult women who do not get enough physical activity.

There are many resources on the web to assist with activity and exercise recommendations to achieve optimum health and fitness. The Department of Health and Human Resources Centers for Disease Control and Prevention offers a website that provides information to assist in healthy physical activity program planning and evaluation as well as ideas for healthy physical activity promotion (http://www.cdc.gov/nccdphp/dnpa/physical/health_professionals/index.htmoffers information).

18.5. SCENARIO WITH QUESTIONS AND ANSWERS

18.5.1. Scenario

You are an athletic trainer for a high school women's cross-country track team. At your school, you do not have a strength and conditioning coach for your athletes. Ashley, one of your senior track athletes, came in to see because of a sports-related injury. You questioned her regarding the origin of her injury. She stated that she has been following the coaches training guidelines for school practice; however, she has begun a resistance-training program with a friend at a local gym. She stated that she has been training for power in order to increase her speed. Her resistance-training program consists of six to eight repetitions at 70–85% of 1 RM with 1 min rest between multiple sets of 3.

18.5.2. Questions

1. Is Ashley following the recommendations of the ACSM for power?
2. What is the recommended load to develop power for novice and intermediate-trained individuals?
3. How many repetitions are suggested per set?
4. What is the recommended rest period between sets when training for power?

18.5.3. Plausible Answers

1. No, Ashley is not following the recommendations of the ACSM for power.
2. The recommended load to develop power is 30–60% of 1 RM.
3. Three and six repetitions are suggested.
4. The recommended rest period is 2–3 min of rest between sets.

REFERENCES

1. United States Department of Health and Human Services. *Healthy People 2010: Understanding and Improving Health*, 2nd ed. Washington, DC: U.S. Government Printing Office, November 2000.
2. Physical Activity and Cardiovascular Health. NIH Consensus Statement Online 1995 December 18–20 [cited January 17, 2006]; 13(3):1–33.
3. Pate RR, Pratt M, Blair SN, et al. Physical activity and public health. A recommendation from the Centers for Disease Control and Prevention and the American College of Sports Medicine. *JAMA* 1995;273:402–407.
4. U.S. Department of Health and Human Services. Physical activity and health: a report of the surgeon general. Atlanta, GA: Centers for Disease Control and Prevention (CDC), National Center for Chronic Disease Prevention and Health Promotion, 1996.
5. Franco O, De Laet C, Peeters A, et al. Effects of physical activity on life expectancy with cardiovascular disease. *Arch Intern Med* 2005;165:2355–2360.
6. Ainsworth B, Haskell WL, Whitt MC, et al. Compendium of physical activities: an update of activity codes and MET intensities. *Med Sci Sports Exerc* 2000;32(9):S498–S516.
7. American College of Sports Medicine. *ACSM's Guidelines for Exercise Testing and Prescription*, 7th ed. Philadelphia, PA: Lippincott Williams & Wilkins, 2006.
8. Dwyer G, Davis S. *ACSM's Health-Related Physical Fitness Assessment Manual*. Philadelphia, PA: Lippincott Williams & Wilkins, 2004.
9. Keteyian S, Brawner C. Cardiopulmonary adaptations to exercise. In: Kaminsky LK, ed. *ACSM's Resource Manual for Guidelines for Exercise Testing and Prescription*, 5th ed. Philadelphia, PA: Lippincott Williams & Wilkins, 2006:313–324.
10. Pollock ML, Gaesser GA, Butcher JD. The recommended quantity and quality of exercise for developing and maintaining cardiorespiratory and muscular fitness, and flexibility in healthy adults. *Med Sci Sports Exerc* 1998;30:975–991.
11. Falk B, Tenenbaum G. The effectiveness of resistance training in children. A meta-analysis. *Sports Med* 196;22:176–186.
12. Hurley BF, Hagberg JM, Goldberg AP, et al. Resistance training can reduce coronary risk factors without altering VO2max or percent body fat. *Med Sci Sports Exerc* 1988;20:150–154.
13. Kerr D, Morton A, Dick I, et al. Exercise effects on bone mass in postmenopausal women are site-specific and load-dependent. *J Bone Miner Res* 1996;11:218–225.
14. Nelson M, Wernick S. *Strong Women Stay Young*. New York: Bantam Books, 2000.
15. Faigenbaum A, Pollock ML, Ishida Y. Prescription of resistance training for health and disease. *Med Sci Sports Exerc* 1996;28:1311–1320.

16. American College of Sports Medicine Position Stand. Progression models in resistance training in healthy adults. *Med Sci Sports Exerc* 2002;34:364–380.
17. Hoeger WW, Barette SL, Hale DR. Relationship between repetitions and selected percentages of one repetition maximum. *J Appl Sport Sci Res* 1987;1(1):11–13.
18. Dragenich LF, Jarger RT, Kradji AR. Coactivation of hamstring and quadriceps during extension of the knee. *J Bone Joint Surg* 1989;71(7):1075–1081.
19. Baechle T, Earle R. *Essentials of Strength Training and Conditioning*, 2nd ed. Champaign, IL: Human Kinetics, 2000.
20. Meriam J. *Engineering Mechanics*. New York: Wiley, 1978.
21. Kaminsky L, Dwyer G. Body composition. In: Kaminsky LK, ed. *ACSM's Resource Manual for Guidelines for Exercise Testing and Prescription*, 5th ed. Philadelphia, PA: Lippincott Williams & Wilkins, 2006:313–324.
22. Lohman TG, Houtkooper LB, Going SB. Body composition assessment: body fat standards and methods in the field of exercise and sports medicine. *ACSM Health Fitness J* 1997;1:30–35.
23. American College of Sports Medicine Position Stand: Appropriate intervention strategies for weight loss and prevention of weight regain for adults. *Med Sci Sports Exerc* 2001;33:2145–2156.
24. Ross R, Janssen I. Physical activity, total and regional obesity: dose-response considerations. *Med Sci Sports Exerc* 2001;33:S521–S527;discussion S528–S529.
25. McHugh MP, Magnusson SP, Gleim GW, et al. Viscoelastic stress relaxation in human skeletal muscle. *Med Sci Sports Exerc* 1992;24:1375–1382.
26. Taylor DC, Dalton JD Jr, Seaber AV, et al. Viscoelastic properties of muscle-tendon units. The biomechanical effects of stretching. *Am J Sports Med* 1990;18:300–309.
27. Knudson D. A review of stretching research. *TAHPERD J* 1995;64:16–18.
28. American College of Obstetrics and Gynecology Committee on Obstetric Practice. Exercise during pregnancy and the postpartum period. Committee Opinion No. 267. *Int J Gynaecol Obstet* 2002;77:79–81.

19 Exercise Guidelines for Children and Adolescence

Jacalyn J. Robert-McComb and Chelsea Barker

CONTENTS

    19.1   LEARNING OBJECTIVES
    19.2   INTRODUCTION
    19.3   RESEARCH FINDINGS
    19.4   CONCLUSIONS
    19.5   SCENARIO WITH QUESTIONS AND ANSWERS

19.1. LEARNING OBJECTIVES

After completing of this chapter, you should have an understanding of the following:

- How to approach exercise with children and adolescents.
- Children's unique exercise capacity characteristics.
- Cardiorespiratory considerations in exercise with children.
- Guidelines for resistance training with children and the progression of resistance training with age.

19.2. INTRODUCTION

The idea of engaging in physical activity is not limited to adults only. Exercise is something that should be done throughout the lifetime, and establishing those habits begins early in childhood. The US government's document "Healthy People 2010" outlines current problems faced by American citizens and goals to improve health (1). One goal directed at children and adolescents was to increase the percent of children in the nation who participate in moderate activity for at least 30 min on five or more days a week from 27 to 35%. It also proposed increasing the number of children who participate in vigorous activity for 20 min or more on three days each week to 85%. In addition, the document suggests that the health benefits of exercise are not limited to adults but that children can benefit in cardiorespiratory function and weight control (1). By establishing good health habits early, children will be more likely to

From: *The Active Female*
Edited by: J. J. Robert-McComb, R. Norman, and M. Zumwalt © Humana Press, Totowa, NJ

continue an active lifestyle into their adult life. Even as early as adolescence, experts see a decline in physical activity that continues into adulthood. The Surgeon General reported that with each year of age, the level of activity decreases *(2)*. The promotion of a healthy lifestyle should begin with children and adolescents, as many habits are formed during this crucial time and the cultural trend can be combated if the youth maintain an active lifestyle.

Although the terms "physical activity" and "exercise" tend to be interchanged synonymously, the terms in fact imply different approaches to activity. The term "physical activity" describes simply moving the body in an activity. Activities of daily living can often be classified as physical activity such as working in the yard or walking while shopping and running errands. The term "exercise" takes that one step further in that the movements are purposeful and planned for a specific training goal. Activities such as going for a run, lifting weights, or playing a sport fall into the category of exercise. Both physical activity and exercise can be done by children of all ages, but the emphasis should remain on physical activity with young children, progressing to purposeful exercise with age.

19.3. RESEARCH FINDINGS

19.3.1. How to Approach Exercise with Children and Adolescents

Much of the activity that young children engage in should have an emphasis on play rather than on work or exercise. According to the National Association for Sport and Physical Education, physical activity for children between the ages of 5 and 12 should primarily be from activities incorporated into their daily lifestyle *(3)*. Being active should be enjoyable and something the children look forward to. This establishes a positive approach to an active lifestyle early that can be carried with them into their adult life. This is especially important for girls because girls are more inclined to significantly decrease their level of activity after they reach puberty.

19.3.2. Children's Unique Exercise Capacity Characteristics

Children's exercise capacity is unique when compared with adults. Early in life, the relative $VO_{2\,max}$ is higher than at any other point in life *(4)*. Therefore, a child has a high level of aerobic capacity relative to its body size. Although adults can train to increase their $VO_{2\,max}$, there is no evidence that children's $VO_{2\,max}$ can be increased to the same degree through training *(4)*. Any increases in $VO_{2\,max}$ for children will be accomplished by working at a level that maintains a heart rate of about 170–180 beats per minute *(4)*.

Absolute $VO_{2\,max}$ increases in both genders at the same rate during childhood until the age of 12. Boys continue to increase at the same rate until the age of 17 or 18, but girls begin to plateau with very small increases after the age of 12. Between the ages of 17 and 22, women show a decrease in $VO_{2\,max}$ that is attributed to a decrease in physical activity and an increase in adipose tissue during their adolescent years *(5)*.

Metabolic cost is also relatively greater in children and adolescents than in adults. At the age of 5, the excessive metabolic cost is approximately 37%, and at the age of 17, the excessive metabolic cost is 3% *(5)*. *See* Table 19.1 for factors that contribute to metabolic costs in children.

Table 19.1
Possible causes of increased metabolic costs in children

Increased resting BMR
Increase in ventilatory costs
Higher stride frequency
Lower storage of elastic recoil forces
Locomotion style is biomechanically inefficient
Higher co-contraction of antagonist muscles

The anaerobic system of children and adults also differs significantly. When engaged in an activity that utilizes the anaerobic energy system, children show a decreased performance level compared with adolescents, who in turn show a decreased level of performance when compared with adults. It is thought that these decreases are the result of a smaller muscle mass, a lower glycolytic capacity, as well as a lack of neuromuscular coordination *(5)*.

19.3.3. Cardiorespiratory Exercise Guidelines for Children

Guidelines from leading organizations for cardiovascular fitness for children are summarized in Appendix 13. Exercise should come in the form of intermittent bouts approximately 15 min that are both moderate and vigorous activities. The emphasis for children should be on activities that they enjoy, and outside play should be encouraged.

19.3.4. Guidelines for Resistance Training with Children and the Progression of Resistance Training with Age

The focus that should be maintained when working with children in resistance training is technique regardless of age or stage of development. In the past, resistance training was avoided for children because of the fear that resistance training would interfere with their growth. This fear was in reference to children's growth plates closing prematurely with the stresses of resistance exercise placed on the plates. The theory originated from research on animal models and from monitoring the effects of the intense exercise regimen maintained by gymnasts who seemed to have a corresponding decrease in growth *(6)*. Resistance training was also dismissed under the belief that children could not receive the benefit of increased strength because of their lack of androgenic hormones *(6)*. These schools of thought have faded though, with the understanding that resistance training can be done by children when the load applied is appropriate to the age and a qualified individual supervises the activity. In fact, resistance training can be beneficial to children of all ages. For those participating in organized sports, a resistance program can work to prepare the body for the demands of the sport to reduce the risk of injury *(4)*. Performing resistance exercise can also increase strength, but hypertrophy is not seen in children *(7)*. The increases in strength come from neural adaptations, just as initial strength gains in adults come from neural adaptations *(4)*.

It is important to remember that before beginning any session of resistance exercise, or really any type of exercise, children need to be given about a 10-min warm-up time, just as recommended for adults, as well as a cool-down time following the exercise session *(8)*. Overall, the resistance exercises need to be age and ability appropriate for the children. Maximal or near-maximal type lifts should not be performed with children just as explosive-type lifts should also be avoided *(8)*. The American College of Sports Medicine provides general guidelines when involving children in resistance exercises that are outlined in Appendix 14.

As children age, the resistance exercises they engage in should also progress with them. Dr. William J. Kraemer suggests specific guidelines as outlined in Appendix 15. Kraemer emphasizes that through each age division the focus should lie in teaching exercise technique because that component of a resistance-training program is of pivotal importance *(9)*.

19.4. CONCLUSIONS

Children must be physically active in order to cultivate a healthy lifestyle. Running, swimming, and playing are important activities for children. Children can also safely participate in weight-bearing activities as they grow because these activities stimulate the growth and formation of bones. Muscles attach to growth plates in the bone, and the contraction of muscles in exercise and physical activity provide force application that stimulates proper bone growth *(6)*. Research has shown that children who participate in weight-bearing activities during childhood and adolescence have a higher bone mineral content and density, which could help prevent osteoporosis later in life *(7)*. More specifically, research has shown that those positive benefits occur only at the skeletal sites where the activity the children engaged in caused skeletal stress *(6)*. Activity in children also seems to stimulate growth hormone production *(7)*.

19.5. SCENARIO WITH QUESTIONS AND ANSWERS

19.5.1. Scenario

You are an exercise physiologist at a local fitness center that has a youth sport-conditioning program. A mother and father enroll their daughter, age 10, in your youth sports program. The girl has played soccer in a city league and now wants to play on her junior high soccer team in the upcoming school year. Her parents want her to have coaching and training because she shows such an interest in playing soccer. She currently practices with her team three times per week for 1 h and plays an hour-long game each Saturday.

19.5.2. Questions

1. What is your recommendation to the parents concerning the frequency of her training sessions with you?
2. What types of exercises and at what intensity level would be appropriate for their daughter?
3. What should be the focus of the program during her sessions with you?

19.5.3. Plausible Answers

1. One to two times per week would be an acceptable number of sessions because she already has team practice three times a week as well as a game each Saturday. All of these practice/game times would accumulate to provide 5–6 days of soccer training each week.
2. The exercises performed should be weight bearing and at a low intensity and low volume. More vigorous exercises can be incorporated in but should be done in short bouts with adequate rest following these bouts.
3. The focus on the program for a 10-year-old child should be on correct form. This could possibly include increasing efficiency of movement such as running efficiency. The focus should also be on making soccer even more enjoyable. This is important because of her young age and the fact that girls have a tendency to decrease activity level as they get older. If she can find her exercise and participation in athletics enjoyable, it will help promote a healthy active lifestyle that may even carry on to adulthood.

REFERENCES

1. Physical activity in children and adolescents. (Accessed May 21, 2006, at http://healthypeople.gov/Document/HTML/volume2/22Physical.htm#_Toc490380803).
2. Physical activity and health: adolescents and young adults. A Report of the Surgeon General, 1999. (Accessed March 3, 2006, at http://www.cdc.gov/nccdphp/sgr/pdf/adoles.pdf).
3. Corbin CB, Pangrazi RP. *Physical Activity for Children: A Statement of Guidelines.* Reston, VA: NASPE Publications, 1998.
4. American College of Sports Medicine. *ACSM's Guidelines for Exercise Testing and Prescription*, 7th ed. Philadelphia, PA: Lippincott Williams & Wilkins, 2006.
5. Bar-Or O, Rowland TW. *Pediatric Exercise Medicine: From Physiologic Principles to Health Care Application.* Champaign, IL: Human Kinetics, 2004.
6. Malina RM, Bouchard C, Bar-Or O. *Growth, Maturation, and Physical Activity*, 2nd ed. Champaign, IL: Human Kinetics, 2004.
7. Rowland TW. *Children's Exercise Physiology*, 2nd ed. Champaign, IL: Human Kinetics, 2005.
8. Jenkins D. Training for strength. In Jenkins D, Reaburn P, eds. *Guiding the Young Athlete.* Australia: Allen & Unwin, 2000: 117–129.
9. Kraemer WJ, Fleck SJ. *Strength Training for Young Athletes*, 2nd ed. Champaign, IL: Human Kinetics, 2005.

20 Exercise Precautions for the Female Athlete
Signs of Overtraining

Jacalyn J. Robert-McComb and Abigail Schubert

Contents

20.1 Learning Objectives
20.2 Introduction
20.3 Research Findings
20.4 Conclusions
20.5 Scenario with Question and Answer

20.1. LEARNING OBJECTIVES

After completing this chapter, you should have an understanding of the following:

- The terms overreaching, overtraining, and staleness (the overtraining syndrome [OTS]).
- Sympathetic and parasympathetic staleness.
- Mood variations in response to overtraining and staleness.
- Hormonal responses to overtraining and staleness.
- Signs of overtraining.

20.2. INTRODUCTION

In the realm of athletics, it is a common practice to expose the body to progressive amounts of overload in order to improve performance. This type of training is most prevalent in endurance sports such as cycling, swimming, long-distance running, and cross-country skiing. Often, elite and/or competitive athletes are willing to push their bodies to extreme limits in order to reach peak performance levels. In addition, outside sources of pressure from coaches, parents, and the media can place undue stress on the athlete. Excessive amounts of physical, emotional, and mental stressors can lead to overtraining and staleness in the athlete.

Although some may say that overtraining is necessary for an athlete to reach full potential, the ramifications of overtraining are profound. Chronic fatigue, decreased

From: *The Active Female*
Edited by: J. J. Robert-McComb, R. Norman, and M. Zumwalt © Humana Press, Totowa, NJ

performance, inability to train at normal levels, and depressed mood states are extremely common in athletes diagnosed with staleness. When an athlete's performance is hindered because of this syndrome, they may view the season as a waste and give up or begin to push themselves even harder to try to compensate. Either reaction can be damaging to the athlete's emotional and physical well-being. Thus, many efforts have been made in recent years to examine the causes and markers of staleness (also referred to as the OTS) as well as interventions to treat this malady.

The term overtraining by itself is not synonymous with the term staleness. There are varying degrees of overtraining. Staleness represents an outcome or product, whereas overtraining reflects a process *(1)*. Once the athlete is diagnosed with staleness, the term OTS and staleness will be used synonymously in this chapter, because both terms are used in the literature *(1–4)*. Staleness (the OTS) is a stress-related condition that consists of alterations in physiological functions and adaptations to performance, impairment of psychological processing, immunological dysfunction, and biochemical abnormalities *(5)*.

20.3. RESEARCH FINDINGS

20.3.1. The Terms Overreaching, Overtraining, and Staleness (the OTS)

Physical exercise results in a homeostatic imbalance within cells and organs, which, if intense enough, can lead to mechanical failure and/or fatigue. There must be adequate recovery time after this occurs in order for homeostasis to be restored. Overtraining is an imbalance between training and recovery. However, there are varying degrees of overtraining. Short-term overtraining, such as what one might experience during peak training times lasting several weeks during the season, is generally termed overreaching. Overreaching refers to a brief period of heavy overload without adequate recovery *(6)*. Overreaching can actually contribute to enhanced performance because it usually lasts no more than 2 weeks. Overreaching can be remedied with decreased training intensity and volume. Often, the body will continue this regenerative process beyond the previous level, thus resulting in supercompensation. In situations where overreaching has occurred, supercompensation can be achieved as quickly as 1 or 2 weeks with decreased training.

When overreaching continues for longer bouts of time, the consequences are far greater. Overtraining refers to the point where the athlete starts to experience physiological maladaptions and chronic performance decrements *(6)*. Overtraining that occurs for extended periods of time and leads to significant decreases in performance is termed staleness, also known as the OTS. The OTS is defined as persistent underperforming, with or without other accompanying psychological and physical symptoms, despite weeks of lighter training or complete rest *(1–5,7)*. Staleness or the OTS is a much more serious condition that goes deeper than the obvious fatigue and hindered performance seen with overreaching or short-term overtraining. Many studies have been conducted to investigate biomarkers to diagnose the OTS. Some of the variables that have been analyzed to diagnose staleness are mood states, heart rate variability, hormonal changes, blood pressure, heart rate, and motor coordination. Because of the marked individual differences in the physiological and/or psychological responses to staleness, it is very hard to diagnose.

20.3.2. Sympathetic and Parasympathetic Staleness

There are generally two types of staleness that are clinically recognized: sympathetic and parasympathetic. In the sympathetic type, the athlete typically experiences an increased resting heart rate, decreased appetite, and weight loss because of increased sympathetic neural control. Sympathetic staleness is associated with changes in the adrenal medullar response system in the body. This type of overtraining is mostly identified with nonendurance-type sports. The parasympathetic type of staleness is related to an altered pituitary adrenal cortical response and is harder to identify than sympathetic staleness but is often associated with low pulse and fatigue very early in exercise as well as excessive amounts of sleep. It is found more commonly in endurance athletes (2).

20.3.3. Mood Variations in Response to Overtraining and Staleness

It is well known that exercise can be used as a way to enhance positive moods; however, it is less well known that too much exercise can bring about negative moods. This is one of the dangers of staleness. The Profile of Mood States (POMS) was administered to approximately 200 men and women on the University of Wisconsin—Madison swimming team during the period 1875–1986 (1). The POMS has been used to assess athlete's moods throughout the course of a season utilizing micro-cycles and macro-cycles or seasonal variations of training. There seems to be a dose–response to score below average on tension, depression, anger, fatigue, and confusion in relation to training stimulus, and staleness. Swimmers presented chronic fatigue and inability to complete workouts as training stimulus increased, both indicators of staleness. Scores for depression increased, and some scores indicated clinical depression. In a similar study, the global mood disturbance of the swimmers increased greatly during peak training time mid-season because of increased fatigue and decreased vigor. The swimmers were also tested during micro-cycles within a 6-month macro-cycle, and they reported an increased mood disturbance during the first 5 months in which the training was progressively more intense. It is important to note that the athlete's moods did improve with rest and/or tapering. Some athletes actually reported mood scores back at their baseline levels after rest, whereas others' moods improved but did not return to baseline scores (1).

Being in a positive mental state prior to competition is essential to an athlete's success in competition. The OTS tends to manifest itself in one's moods, and this can have potentially harmful effects on an athlete's mental state. Prolonged periods of negative moods and fatigue can lead to depression and stress, which can compound the problem. In addition, depressed moods are often associated with decreased appetite and weight loss, which can also hinder performance. The OTS is often part of a cycle with decreased performance. Typically, when an athlete has a poor performance, they will come back the next week and work even harder. However, if the athlete is suffering from staleness, this only serves to exacerbate the problem. The added work output depresses mood states even more and leads to decreased performance and eventually the inability to complete workouts and competitions. The OTS has negative effects on mood, which manifests itself in poor performance and workouts. It is important to be aware of an athlete's mental state and of the signs of staleness in order to alleviate the problem before performance and health outcomes are affected too drastically.

20.3.4. Hormonal Responses to Overtraining and Staleness

Changes in the recovery phase of the plasma level of catecholamines, epinephrine and norepinephrine, and cortisol may provide an objective measure of diagnosing the OTS *(4,5)*. However, results are inconsistent *(2)*. Increments in plasma cortisol levels immediately after exercise have been found by some authors *(8)*, whereas others have found decrements in adrenocorticotropic hormone and cortisol responses in trained runners following a bolus of ovine corticotropin-releasing hormone *(9)*.

20.3.5. Signs of Overtraining

The OTS is a frequent condition among elite athletes. Common volumes of training load, which exceed 30 h a week, expose athletes to the possibility of staleness *(10)*. A definition of the prevalence of OTS is lacking, but it has been suggested that almost all professional athletes experience this condition at least once in their career *(11)*. Signs and symptoms of overtraining have been categorized into four classes by Fry et al. *(11)* as indicated in Table 20.1.

20.4. CONCLUSIONS

Despite the research that has been conducted on the OTS known as staleness, there is a vast body of knowledge still unknown to researchers concerning the extent of this malady. Universally agreed diagnostic criteria for OTS are lacking *(5)*. However, it is easy to miss the warning signs of staleness. It can be confused with fatigue because of a hard workout or stress-related symptoms, leading the athletes to become frustrated with their performance and push themselves even harder. This can set off a very

Table 20.1
Major signs and symptoms of overtraining

Alterations in physiological functions and adaptations to performance
Decreased performance
Decreased muscular strength
Muscle soreness and tenderness
Reduced tolerance of loading
Prolonged recovery
Chronic fatigue
Headache
Sleep–wake cycle abnormalities
Changes in blood pressure and heart rate
Alterations in sexual functions
Gastrointestinal disturbances
Feelings of depression
General apathy
Difficulty in concentration
Emotional instability
Fear of competition

Loss of appetite
Psychological symptoms
Excitation and restlessness
Immunological dysfunction
Increased susceptibility to bacterial infections
Reactivation of herpes viral infections
Decreased functional activity of neutrophils
Decreased total lymphocyte count
Decreased production and secretion of immunoglobulins
Biochemical alterations
Decreased hemoglobin, serum iron, and ferritin
Negative nitrogen balance
Increased urea levels
Increased uric acid productions
Decreased glutamine concentration
Mineral depletion (Zn, Co, Al, Mn, Se, Cu, etc.)
Low free testosterone
Decreased free testosterone to cortisol ratio of more than 30%

Source: Fry *(12)*.

damaging cycle of events that can be harmful to not just the athletes' performance but more importantly their health and well-being. Therefore, it is important for coaches to monitor their athletes closely in order to avoid OTS. Table 20.2 presents indicators that coaches or health care professionals could use to alert them to the possibility of staleness in their athletes *(12)*. However, there are significant individual differences in the manifestations of OTS; therefore, variables must be examined collectively. Finally,

Table 20.2
Indicators that could alert coaches and health care professionals to the possibility of staleness or the overtraining syndrome in their athletes

Possible indicators related to parasympathetic overtraining
Low resting heart rate
Rapid heart rate recovery postexercise
Hypoglycemia during exercise
Decreased maximal plasma lactate during exercise
Decreased catecholamine levels
Depression
Possible indicators related to sympathetic overtraining
Restlessness
Excitation
Increased catecholamine levels
Increased heart rate
Increased blood pressure
Increased metabolic rate
Increased respiratory rate

Source: Hogan *(12)*.

consideration of these variables in alerting the health care professional to overtraining in the athlete should be made in conjunction with the athlete's own assessment of well-being *(4)*.

20.5. SCENARIO WITH QUESTION AND ANSWER

20.5.1. Scenario

Abby is a cross-country runner. Her resting heart rate and blood pressure is 50 bpm and 100/70 mmHg, respectively. Her performance has been stable, and she has not exhibited signs of depression.

20.5.2. Question

1. Should her coach be alerted to the signs of overtraining given her low resting heart rate?

20.5.3. Plausible Answer

1. A low resting heart rate can be a normal physiological adaptation to training because of an increase in stroke volume. Without other manifestations of the OTS, this variable is not significant.

REFERENCES

1. Morgan WP, Raglin JS, O'Connor PJ, et al. Psychological monitoring of overtraining and staleness. *Br J Sports Med* 1987;21:107–114.
2. Kuipers H, Keizer HA. Overtraining in elite athletes. *Sports Med* 1988;6:79–92.
3. Uusitalo ALT, Huttunen P, Hanin Y, et al. Hormonal responses to endurance training and overtraining in female athletes. *Clin J Sport Med* 1998;8:178–186.
4. Hooper SL, Mackinnon LT, Howard A, et al. Markers for monitoring overtraining and recovery. *Med Sci Sports Exerc* 1994;27:106–112.
5. Angeli A, Minetto M, Dovio A, et al. The overtraining syndrome in athletes: a stress related disorder. *J Endocrinol Invest* 2004;27:603–612.
6. Wilmore JH, Costill DL. *Physiology of Sport and Exercise*, 3rd ed. Champaign, IL: Human Kinetics, 2005:378.
7. Fry AC, Kraemer WJ. Resistance training and overreaching: neuroendocrine responses. *Sports Med* 1997;23(2):106–129.
8. Kuoppasaimi K, Naveri H, Harkonen M, et al. Plasma cortisol, androstenedione, testosterone, and luteinizing hormone in running exercise of different intensities. *Scand J Clin Lab Invest* 1980;40:403–409.
9. Luger A, Duester PA, Kyle SB, et al. Acute hypothalamic-pituitary-adrenal responses to the stress of treadmill exercise. Physiological adaptations to physical training. *N Engl J Med* 1987;316:1309–1315.
10. Armstrong LE, VanHeest JL. The unknown mechanism of the overtraining syndrome: clues from depression and psychoneuroimmunology. *Sports Med* 2002;32:185–209.
11. Fry FW, Morton AR, Keast D. Overtraining in athletes. An update. *Sports Med* 1991;12:32–65.
12. Hogan B. Higher, faster, stronger…are your clients overtraining. American College of Sports Medicine. *Certified News* 2006;16(1):2–5.

21 Exercise Guidelines and Recommendations During Pregnancy

Jacalyn J. Robert-McComb and Jessica Stovall

CONTENTS

21.1 LEARNING OBJECTIVES
21.2 INTRODUCTION
21.3 RESEARCH FINDINGS
21.4 CONCLUSIONS
21.5 SCENARIO WITH QUESTIONS AND ANSWERS

21.1. LEARNING OBJECTIVES

After completing this chapter, you should have an understanding of the following:

- Physiological changes of pregnancy that may affect exercise.
- Effects of exercise on the baby.
- Activities to encourage and discourage during pregnancy.
- Guidelines for exercising while pregnant.
- Postpartum exercise.

21.2. INTRODUCTION

In 2002, the American College of Obstetricians and Gynecologists (ACOG) published recommendations and guidelines for exercise during pregnancy and the postpartum period *(1)*. The basis of their recommendations was that clinical and epidemiological studies have not provided evidence that there were adverse maternal or fetal effects for women engaged in mild and moderate exercise activities during pregnancy *(2)*. On the contrary, mild and moderate exercises during pregnancy can have very positive effects on the mother and infant.

Some of the benefits include (1) a reduction in typical pregnancy-related symptoms such as back pain, heartburn, leg cramps, nausea, fatigue, varicosities, and swelling of the extremities; (2) a reduction in the "active stage" of labor (the time from 4 to 10 cm dilation); (3) an increased sense of maternal well-being; (4) decreased delivery

From: *The Active Female*
Edited by: J. J. Robert-McComb, R. Norman, and M. Zumwalt © Humana Press, Totowa, NJ

complications because of the endurance gained that is needed to get through the long hours of labor; (5) a reduction in the risk of developing pregnancy-induced hypertension (PIH); (6) a reduction in bone density loss during the lactation state; (7) an enhancement of birth weight; and (8) a reduction in the risk of developing gestational diabetes (especially in women with a BMI > 33) *(2–6)*.

Moderate exercise during pregnancy also improves glycemic control in women with gestational diabetes mellitus (GDM). Researchers found a 76% decrease in GDM risk in active women or those who spent 4.3 h or more per week engaged in recreational physical activity, compared with inactive women *(7)*. GDM is associated with long-term and short-term morbidity in the mother and offspring and affects 4–7% of pregnancies in the United States *(2)*. Perhaps one of the greatest incentives to exercise moderately during pregnancy is that children of GDM mothers are more likely to have diabetes in adolescence or become obese *(2)*.

21.3. RESEARCH FINDINGS

21.3.1. Physiological Changes of Pregnancy

Muscoloskeletal alterations occur during pregnancy, which potentially increase the risk of injury during exercise. One alteration is increased joint laxity and hypermobility caused by hormonal changes. This increase in joint laxity can lead to a higher risk of strains or sprains. Another alteration is the weight of growing breasts, and the uterus and fetus, which can cause a change in the woman's center of gravity that may cause problems with balance *(8)*.

The body temperature of nonpregnant women rises about 1.5 °F after exercising strenuously for about 20 min, but that of a fit pregnant woman decreases slightly after exercise *(9)*. This adaptation may help protect the baby, suggesting that a pregnant woman's body dissipates heat more efficiently. However, pregnant women must still take steps to avoid overheating. Hyperthermia is a concern given that both exercise and pregnancy raise the metabolic rate. It can be considered a teratogenic effect for the mother's core temperature to rise above 102.6 °F, especially in the first trimester *(10)*. Shorter exercise sessions may obviate concerns related to thermoregulation *(11)*. Additionally, exercise should be performed in a thermoneutral environment or in a controlled environmental setting (air conditioning).

Cardiac output (i.e., stroke volume × heart rate) is affected by the maternal body position during pregnancy. Cardiac output decreases 9% after the first trimester when the mother is in the supine position; thus, exercise should be avoided in the supine position after 16 weeks of gestation because of vena cava compression *(12)*. The ACOG advice is to avoid exercise in the supine position as much as possible *(1)*. Motionless standing is also associated with a significant decrease in cardiac output—a decrease of 18% is because of the mother standing for a prolonged period of time. Therefore, motionless standing should also be avoided because of the reduction in cardiac output *(13)*. Lying on the side is the best position for cardiac output to be optimal.

The change in hemodynamics during intense exercise is also a concern during pregnancy. During exercise, blood is diverted away from the abdominal viscera to the exercising muscle. Blood is also diverted away from the uterus, this can be a concern when there is a decrease in splanchnic blood flow by 50%, which can cause fetal hypoxemia *(14)*.

Another obvious physiological change during pregnancy is the increased oxygen demand. The increased oxygen requirement of the fetus causes a mild increase in tidal volume and oxygen consumption in the pregnant woman. When exercise is increased to a more strenuous level, however, pregnant women have a decreased respiratory frequency and a lower maximal oxygen consumption and tidal volume. The adaptive changes that usually occur at rest are overridden by this new oxygen demand. This can be a result of the inhibition of the diaphragm because of the woman's enlarged uterus *(9)*.

Another physiological change is the increase in energy demand. Pregnant women need to increase their caloric intake by approximately 300 kcal per fetus. If the pregnant woman exercises, she needs to increase her caloric intake to meet the energy cost of the activity *(15)*.

21.3.2. Effects of Exercise on the Baby

Recent studies have found that exercise might actually decrease the risk of premature labor. Growth of the placenta is increased if the woman moderately exercises during early pregnancy. Researchers have also discovered that by the age of 5, children of exercising mothers scored higher on tests of intelligence and language skills than children of nonexercisers. Also, the children of exercising mothers were much leaner than the children of nonexercisers. During labor, the babies of women who had exercised seemed to deal with the stress of contractions better than the controls or babies of women who had stopped exercising. These babies also had a decrease in the likeliness of getting tangled in the umbilical cord *(3)*.

21.3.3. Activities to Encourage and Discourage During Pregnancy

Sport activities that are encouraged during pregnancy are walking, stationary cycling, low-impact aerobics, and swimming. The sport activities that need to be avoided are contact sports such as hockey, wrestling, football, and soccer. High-risk sports that have an increased potential for falls or trauma should also be avoided. These include gymnastics, horseback riding, skating, skiing (snow and water), hang gliding, vigorous racquet sports, weightlifting, and scuba diving *(15)*.

21.3.4. Guidelines for Exercising While Pregnant

All prenatal populations should obtain a clearance from their physician prior to beginning an exercise program. One of the most often used screening tool, the PAREMED-X, for pregnancy is available to download from the Canadian Society for Exercise Physiologists at http://www.csep.ca/pdfs/parmed-xpreg.pdf *(5)*. Absolute and relative contraindications to exercise during pregnancy and the postpartum period as recommended by the ACOG are presented in Table 21.1 *(1)*.

ACOG currently recommends 30 min or more of moderate-intensity exercise per day for most days of the week during pregnancy in the absence of medical or obstetric complications. Two new components of the ACOG guidelines merit special attention *(2)*. First, the recommendations promote exercise for previously sedentary pregnant women and those with medical or obstetric complications, but only after having gone through an extensive medical evaluation and clearance. Second, guidelines suggest that exercise may play an important role in the prevention and management of GDM. Complications in pregnant women who have GDM include preeclampsia, infection, and postpartum hemorrhage *(2)*. Summary of recommendations for exercise during pregnancy can be found in Table 21.2 *(1,11,16,17)*. Precautionary measures

Table 21.1
American College of Obstetricians and Gynecologists relative and absolute contraindications to aerobic exercise during pregnancy

Relative
Severe anemia
Unevaluated maternal cardiac arrhythmia
Chronic bronchitis
Poorly controlled type 1 diabetes
Extreme morbid obesity
Extreme underweight (BMI < 12)
History of extremely sedentary lifestyle
Intrauterine growth restriction in current pregnancy
Poorly controlled hypertension
Orthopedic limitations
Poorly controlled seizure disorder
Poorly controlled hyperthyroidism
Heavy smoking
Absolute
Hemodynamically significant heart disease
Restrictive lung disease
Incompetent cervix/cerclage
Multiple gestation at risk for premature labor
Persistent second-trimester or third-trimester bleeding
Placenta previa after 26 weeks of gestation
Premature labor during the current pregnancy
Ruptured membranes
Preeclampsia/pregnancy-induced hypertension

Source: ACOG Committee *(1)* (with permission from American College of Obstetricians and Gynecologists).

Table 21.2
Summary of recommendations for aerobic exercise during pregnancy

The Centers for Disease Control and the American College of Sports Medicine recommendation for exercise, aimed at improving the health and well-being of nonpregnant individuals, suggest that an accumulation of 30 min or more of moderate exercise a day should occur on most, if not all, days of the week. In the absence of either medical or obstetric complications, pregnant women can also adopt this recommendation.

Moderate exercise is defined as activity with an energy requirement of 3–5 metabolic equivalents (METS). This is equivalent to brisk walking.

Given the variability in heart rate during pregnancy, ratings of perceived exertion have been found to be useful during pregnancy. Moderate exercise is defined as a "perceived exertion" of 11–14 on the 6–20 Borg scale. This feeling to total bodily exertion ranges from light to somewhat hard. *See* Appendix 11 for the Borg scale.

Because of the issues of thermoregulation and energy balance, exercise sessions longer than 45 min may not be well tolerated, and caution should be used. Women should monitor hydration status and body temperature and be sensitive to subjective feelings of heat stress.

Source: ACOG Committee *(1)*, Pate et al. *(18)*, American College of Sports Medicine *(16)*, and Artal et al. *(11)*.

for exercise as well as the warning signs to terminate exercise while pregnant can be found in Table 21.3 *(1,9,18)*.

In support of the recommendations from the ACOG, the Royal College of Obstetrics and Gynaecologists (RCOG) issued a position statement on exercise and pregnancy statement in January 2006. A summary of their key points can be found in Table 21.4 *(18)*.

Although there is less evidence on strength conditioning, weight training, and stretching exercise such as yoga and pilates in pregnancy, it seems that appropriate resistance training provides a pregnant woman with an enhanced level of muscular fitness, which may help compensate for the postural adjustments that typically occur during pregnancy *(19)*. However, heavy weightlifting should be avoided during pregnancy. An intensity of 9 (very light) to 11 (light) on the RPE scale would be appropriate for resistance training during pregnancy (Appendix 11). More research is needed in this area, because data are limited. Table 21.5 lists recommendations for strength training during pregnancy *(19)*.

Table 21.3
Precautionary guidelines for exercise during pregnancy and indications to terminate exercise during pregnancy

Precautionary guidelines for exercise during pregnancy
Avoid scuba diving while pregnant
Avoid activities with a high risk of falling
Avoid exercises in the supine position after the first trimester
Exercising at high altitudes (more than 6000 ft) should be avoided because the baby's oxygen supply will be reduced
Overheating should be avoided; this includes hot tubs, saunas, and Jacuzzis
Drink plenty of fluids before, during, and after exercise
The pregnant exercising mother needs to consume approximately 300 kcal per fetus plus the energy costs of the exercise; increased carbohydrate intake is recommended for exercise
Any kind of bouncing or jerky movements should be avoided
Prolonged periods of motionless standing should be avoided because it can reduce the amount of blood flow to the uterus
Indications to terminate exercise during pregnancy
Excessive shortness of breath
Pelvic girdle pain
Vaginal bleeding
Painful uterine contractions or preterm labor
Dyspnea prior to exertion
Presyncope or dizziness
Abdominal pain, particularly in back or pubic area
Headache
Chest pain or palpitations
Muscle weakness
Calf pain or swelling
Excessive fatigue
Decreased fetal movement
Amniotic fluid leakage

Sources: ACOG Committee *(1)*, March of Dimes *(9)*, and Royal College of Obstetrics and Gynecologists *(18)*.

Table 21.4
Royal College of Obstetricians and Gynecologists key points for exercise during pregnancy

All women should be encouraged to participate in aerobic and strength-conditioning exercise as a part of a healthy lifestyle during pregnancy.

Reasonable goals of aerobic conditioning in pregnancy should be to maintain a good fitness level throughout pregnancy without trying to reach peak fitness level or train for athletic competition.

Women should choose activities that will minimize the risk of loss of balance and fetal trauma.

Women should be advised that adverse pregnancy or neonatal outcomes are not increased for exercising women.

Initiation of pelvic floor exercises in the immediate postpartum period may reduce the risk of future urinary incontinence.

Women should be advised that moderate exercise during lactation does not affect the quantity or composition of breast milk or impact fetal growth.

These recommendations were produced on behalf of the Guidelines and Audit Committee of the Royal College of Obstetricians and Gynaecologists by Dr. BB Bell, MBChB, DipObsSA, Dorset, and Mr. MM P Dooley, FRCOG, Dorset. The final version is the responsibility of the Guidelines and Audit Committee of the RCOG. Valid until January 2009 unless otherwise indicated.
Source: Royal College of Obstetrics and Gynecologists *(18)*.

21.3.5. Postpartum Exercise

Exercising after pregnancy can be very beneficial. Benefits of postpartum exercise include increased vitality, a better overall mood, and a reduction in anxiety. However, women need to exercise with caution and gradually work up to participating in the activities that they did before pregnancy. Recommendations suggest that if pregnancy and delivery were complicated, the woman should consult a medical caregiver before getting back to exercising regularly. However, if the delivery was not complicated, mild exercising is encouraged (i.e., stretching, walking, or pelvic floor exercises) *(18)*.

21.4. CONCLUSIONS

Exercising during pregnancy is encouraged for the healthy woman and has also been shown to be very beneficial to the baby as well. Caution is advised for certain high-risk activities that might increase the risk of falling. During the later weeks of pregnancy, sports that require sudden changes in body position, extreme range of motion, or exceptional balance should be avoided. These kinds of activities can cause a fall and a risk of injury to the fetus *(15)*. She should also know the contraindications to exercise and the signs and symptoms to stop exercising.

Contraindications to exercising while pregnant are PIH, preterm rupture of membranes, preterm labor during the prior or current pregnancy, incompetent cervix or cerclage placement, persistent second-trimester or third-trimester bleeding, placenta previa, and intrauterine growth retardation. Relative contraindications are chronic hypertension, thyroid function abnormality, cardiac disease, vascular disease, and pulmonary disease. Indications to stop exercising include abdominal pain, dizziness, and vaginal bleeding.

Table 21.5
Recommendations regarding resistance training during pregnancy

Medical advice and physician recommendations should be obtained prior to resistance training during pregnancy

Resistance training for all pregnant women may not be appropriate. If women have any of the contraindications to aerobic exercise as proposed by American College of Obstetrics and Gynecology, they should not participate in resistance training.

Women who have never participated in resistance training should not initiate one during pregnancy.

Women should be encouraged to breathe normally during resistance training; breath holding reduces oxygen delivery to the placenta.

Heavy resistance should be avoided because it may expose the joints, connective tissue, and skeletal structures of an expectant woman to excessive forces. An exercise set consisting of at least 12–15 repetitions without undue fatigue is recommended.

As training occurs, overload initially by increasing number of repetitions and, subsequently, by increasing resistance.

Resistance training on machines is preferred to free weights because machines can be more easily controlled and require less skill.

Source: Byrant et al. *(19)*.

In closing, the pregnant mother must be aware of her limitations and exercise within those limitations. However, if properly educated on the appropriate exercise to perform during pregnancy, the mother and her baby can enjoy the benefits of exercise throughout her entire pregnancy and postpartum period.

21.5. SCENARIO WITH QUESTIONS AND ANSWERS

21.5.1. Scenario

Flora is 22 years old and has recently become pregnant. She wants to have a healthy pregnancy and baby. She has always been extremely physically active and physically fit. She runs 5 miles a day and performs heavy resistance training every other day. Since her pregnancy, she has maintained her normal exercise and eating habits; despite this, she has recently felt fatigued.

21.5.2. Questions

1. Are running and heavy weightlifting the best activities for Flora to continue throughout the pregnancy?
2. Given the scenario, what factor could be contributing to the feeling of fatigue that Flora is experiencing?

21.5.3. Plausible Answers

1. Low-impact activities should be encouraged during her pregnancy, provided that she does not have contraindications to exercise. However, heavy resistance training should be avoided during pregnancy. It is prudent for her to replace running with cross-training

activities such as the elliptical or the stationary bike. Water aerobics would also be a good alternative activity. Because she has been performing resistance exercise prior to her pregnancy, she should be encouraged to continue resistance training, provided that she does not have contraindications to exercise. However, she should modify her resistance-training program by decreasing the amount of resistance and increasing the number of repetitions to 12–15 that she can perform comfortably.

2. During pregnancy, women should increase their caloric intake by approximately 300 cal a day per fetus to compensate for the additional energy cost related to childbearing.

REFERENCES

1. ACOG Committee. Opinion no. 267: exercise during pregnancy and the postpartum period. *Obstet Gynecol* 2002;99:171–173.
2. Dempsey J, Butler C, Williams M. No need for a pregnant pause: physical activity may reduce the occurrence of gestational diabetes mellitus and preeclampsia. *Exerc Sport Sci Rev* 2005;33(3):141–149.
3. Clapp J. *Exercising Through Your Pregnancy*. Champaign, IL: Human Kinetics, 1998.
4. Ezmerli NM. Exercise in pregnancy. *Prim Care Update Obstet Gynecol* 2000;7(6):260–265.
5. Nordahl K, Petersen C, Jeffreys RM. *Fit to Deliver*, 2nd ed. Canada: Fit to Deliver Inc., 2005.
6. Pivarnik JM. Potential effects of maternal physical activity on birth weight: brief review. *Med Sci Sports Exerc* 1998;30(3):400–406.
7. Dempsey J, Sorensen TK, Williams M, et al. Prospective study of gestational diabetes mellitus in relation to maternal recreational physical activity before and during pregnancy. *Am J Epidemiol* 2004;159:663–670.
8. Calgueri M, Bird HA, Wright V. Changes in joint laxity occurring during pregnancy. *Ann Rheum Dis* 1982;41:126–128.
9. March of Dimes. Fitness for Two. 2006 March of Dimes Birth Defects Foundation, February 2004 (accessed July 18, 2006, at http://marchofdimes.com/printableArticles/14332_1150.asp).
10. Milunsky A, Ulcickas M, Rothman KJ, et al. Maternal heat exposure and neural tube defects. *JAMA* 1992;268:882–885.
11. Artal R, O-Toole M, White S. Guidelines of the American College of Obstetricians and gynecologists for exercise during pregnancy and the postpartum period. *Br J Sports Med* 2003;37:6–12.
12. Artal R. Exercise during pregnancy: safe and beneficial foremost. *Phys Sportsmed* 1999;27(8):51–62.
13. Clark SL, Cotton DB, Pivarnik JM, et al. Position change and central hemodynamic profile during normal third-trimester pregnancy and post partum. *Am J Obstet Gynecol* 1991;164:883–887.
14. Sternfeld B. Physical activity and pregnancy outcome: review and recommendations. *Sports Med* 1997;23:33–47.
15. Wang T, Apgar B. Exercise during pregnancy. *Am Fam Physician* 1998;57(8):1–10.
16. American College of Sports Medicine. *ACSM's Guidelines for Exercise Testing and Prescription*, 7th ed. Philadelphia, PA: Lippincott Williams & Wilkins, 2006.
17. Pate RR, Pratt M, Blair SN, et al. Physical activity and public health. A recommendation from the Centers for Disease Control and Prevention and the American College of Sports Medicine. *JAMA* 1995;273:402–407.
18. Royal College of Obstetrics and Gynecologists. Exercise in pregnancy. RCOG Statement No. 4, January 2006 (accessed July 19, 2006, from http://www.rcog.org.uk/printindex.asp?PageID=1366&Print=Yes).
19. Byrant C, Peterson J, Graves J. Muscular strength and endurance. In: Roitman J, ed. *ACSM's Resource Manual for Guidelines for Exercise Testing and Prescription*, 3rd ed. Philadelphia, PA: Williams and Wilkins, 1998:448–455.

22 Mindful Exercise, Quality of Life, and Cancer

A Mindfulness-Based Exercise Rehabilitation Program for Women with Breast Cancer

Anna M. Tacón

CONTENTS

22.1   LEARNING OBJECTIVES
22.2   INTRODUCTION
22.3   MIND-BODY MEDICINE, INTEGRATED CARE, AND MINDFULNESS
22.4   THE CONCEPT OF MINDFULNESS AND THE PROGRAM
22.5   EXERCISE THERAPY FOR CANCER
22.6   MINDFULNESS-BASED EXERCISE REHABILITATION PROGRAM
22.7   SCENARIO WITH QUESTIONS AND ANSWERS

22.1. LEARNING OBJECTIVES

After completing this chapter, you should have an understanding of the following:

- Increased use of mind-body therapies in cancer.
- Breast cancer patients' use of complementary therapies.
- The concept of mindfulness as well as the Mindfulness-Based Stress Reduction (MBSR) program.
- Quality of life (QOL) factors in cancer, especially cancer-related fatigue.
- The benefits of exercise as a complementary therapy in cancer.

22.2. INTRODUCTION

Breast cancer is the most common cancer among women, accounting for nearly one in three cancers diagnosed in the United States. In 2005, it was estimated that 269,730 new cases of breast cancer would be diagnosed and that approximately 40,410 women

From: *The Active Female*
Edited by: J. J. Robert-McComb, R. Norman, and M. Zumwalt © Humana Press, Totowa, NJ

would die from this disease *(1)*. The diagnosis of cancer is considered to elicit greater distress than any other diagnosis, regardless of the prognosis *(2)*, and the most prevalent, negative psychological consequences of this disease are depression and anxiety *(3)*. A patient's QOL is brutally attacked, so much so, that it has been suggested as being a reason for the steady increases in cancer patients seeking complementary or mind-body therapies to help them cope; indeed, thousands of patients seek unconventional or complementary therapies independent of standard health care *(4)*. One mind-body program that is increasingly being used with cancer patients is Kabat-Zinn's *(5)* well-known MBSR which includes meditation, yoga, and walking.

In addition to dealing with the disease process, there is also the formidable task of enduring harsh treatment effects. Side effects such as debilitating fatigue, decline in physical activity, and loss of muscle strength and flexibility are well-known. In an effort to ameliorate such effects, research has grown in the area of using exercise, that is, multiple studies indicate moderate physical activity to be a relatively new therapy for rehabilitating those with cancer *(6,7)*. The purpose here is to describe a mind-body or complementary intervention for breast cancer patients which integrates the MBSR with exercise.

The guiding rationale is that patients stressed biologically and psychologically with significantly reduced QOL will be compromised in all domains and systems and thus be unable to participate fully in their recovery. Lastly, women with breast cancer were selected as the initial population because women with breast cancer are more likely to use complementary therapies than both the general public and individuals with other forms of cancer *(8)*. The complementary therapies of mindfulness and exercise will be reviewed in this chapter.

22.3. MIND-BODY MEDICINE, INTEGRATED CARE, AND MINDFULNESS

22.3.1. Evidence-Based, Cost-Effective, and Quality Health Care

In 2000, Sobel noted that "mind-body medicine ... should be an integral part of evidence-based, cost-effective, quality health care" *(9)*. In the primary care setting, integrated care means that behavioral health providers are available to provide psycho-physiological, psychosocial, educational interventions as part of treatment for patient distress. Integrated care is thus aligned with the mind-body medicine approach that includes a wide range of behavioral and lifestyle interventions on even par with traditional medical interventions. Mindfulness meditation (MM), such as Kabat-Zinn's program at the Stress Reduction Clinic at the University of Massachusetts Medical Center *(5)*, is a form of inclusive meditation that involves including rather than excluding stimuli from the field of consciousness. This is more reality-based with daily life as we are constantly challenged and bombarded with multiple stimuli. Mindfulness encourages detached nonjudging observation or witnessing of thoughts, perceptions, sensations, and emotions, which provides a means of self-monitoring and self-regulating one's arousal with detached awareness.

22.4. THE CONCEPT OF MINDFULNESS AND THE PROGRAM

22.4.1. The Concept and the MBSR Program

The cognitive-behavioral MBSR program meets once per week for 2 h over 8 weeks. Mindfulness is described in different ways, that is, mindful awareness, being in the present moment, or just paying attention to *now*. Many individuals start out thinking that this is easy, that "paying attention is a piece of cake," as one woman remarked. Yet, mindfulness is deceptive, for being acutely focused and aware for a considerable amount of time is difficult. To quiet the mind to where it is empty or silent of thoughts or images or concerns is a not an easy task. Participants learn quickly how the mind gets distracted and disrupted with constant chatter, or suddenly they are jolted to the realization that they were on "automatic pilot" and just became mindful of that. The participants receive training in several basic practices of MM: the body scan, hatha yoga, and sitting and walking meditation.

The body scan is a journey through the geography of the physical form. It involves a gradual, thorough sweeping of attention through the entire body from feet to head, focusing noncritically on any sensations or feelings in body regions with suggestions of breath awareness and relaxation. Hatha yoga involves simple stretches and postures (asanas) designed to strengthen the body and increase flexibility via relaxation of the musculoskeletal system and the development of mindful movement. Yoga is considered to be "meditation in motion" where focused body awareness of each movement is gently coordinated with breathing. Sitting meditation involves attention to the breath and other psychophysical perceptions with an observational yet nonjudging awareness of cognitions and distractions that constantly flow through the mind. Walking meditation is the cognitive-behavioral practice of paying attention to all physical sensations and fluctuations in concentrative awareness during the simple activity of walking. Participants receive audiotapes to facilitate homework practice.

22.4.2. Research and the MBSR

Empirically, MBSR programs have demonstrated effectiveness in reducing levels of stress and anxiety, chronic pain, and fibromyalgia *(10–14)*. MBSR interventions with cancer populations have shown significant improvements pre-post in QOL, significant positive shifts in mood, coping, health locus of control, and mental adjustment to cancer *(11,12,15)*. Last, significant positive shifts in overall immune profile were found in breast and prostate patients after completing the program *(16)*.

22.5. EXERCISE THERAPY FOR CANCER

22.5.1. The American Cancer Society's Physical Activity Guidelines

In March of 2002, the American Cancer Society (ACS) issued new guidelines recommending that adults get at least 30 min of moderate activity 5 or more days per week to help prevent chronic diseases like cancer. Specifically, it is suggested that physical activity throughout life can help protect against breast cancer and prostate cancer because exercise may aid in regulating hormone levels *(17)*. Exercise is gaining acceptance as a major rehab intervention for cancer survivors.

22.5.2. Exercise, Quality of Life, and Cancer

The cancer process and the treatments to destroy it are notorious for eroding the QOL of cancer patients. In the past two decades, research has grown regarding the positive effects of exercise on the QOL of those with cancer *(18)*. Specific QOL-related factors include, for example, fatigue, physical functioning and emotional distress, self-esteem, and depressive and anxiety symptoms *(19–21)*. Improved designs using randomized controlled trials have shown QOL benefits from exercise during high-dose cancer treatments in the hospital (e.g., supine ergometer biking) as well as after treatments are finished (e.g., home-based walking program) *(22–24)*.

A typical exercise prescription for those with cancer is moderate-intensity exercise, 3–5 times/week, 20–30 min/session; although high-intensity exercise should be avoided during cancer treatment because of potential immunosuppressive effects, it is not usually contraindicated for cancer survivors *(22,23)*. Walking is the preferred and most common exercise for cancer patients and is generally the prescribed moderate exercise for home-based programs such as Courneya's (3–5 times/week, 20–30 min, at 65–75% of their estimated maximum heart rate as soon as safely possible) or Mock's (walk 5–6 times/week in the target heart range 50–70% of maximum heart rate) in line with the American College of Sports Medicine *(19,22,25)*.

22.5.3. Cancer-Related Fatigue and Exercise: Definition and Rationale

The most prevalent and debilitating symptom of cancer patients, in general, and of women with breast cancer receiving adjuvant chemotherapy or radiation therapy is fatigue *(19)*. In fact, 70–100% of patients being treated for cancer (radiation, cytotoxic chemotherapy, or biological response modifiers) are affected by cancer-related fatigue, which can be even more distressing and disruptive to a patient's daily activities than the pain-associated cancer *(26)*. Cancer-related fatigue is defined as "a persistent, subjective sense of tiredness related to cancer or cancer treatment that interferes with usual functioning" and can be described in terms of perceived energy, mental capacity, and psychological status *(19)*. Moreover, fatigue can be a problem for months to even years after treatment ends; effects tend to peak during treatment, yet this does not guarantee that symptoms will disappear automatically once treatment is over *(26)*.

Cancer-related fatigue is important not only for the physical and functional impairments and psychological distress it causes to patients but also it is important because this iatrogenic fatigue may affect a patient's ability to tolerate treatments. Thus, treatments may be delayed or reduced which impacts logically disease progression and long-term survival *(27)*. In the recent past, the prescription for fatigue experienced by cancer patients was to rest and conserve energy. An accumulating body of data indicates the opposite may be true and that exercise reduces fatigue *(7,19,23,28)*. Inactivity may generate its own impairment problems as well as compound morbid effects of treatments (e.g., decreased energy, and strength). The question remains as to why exercise or moderate physical activity may decrease cancer-related fatigue.

The rationale for exercise as a treatment for cancer-related fatigue, as described by Mock et al. *(19)*, is that the combined effects of toxic treatments plus decreased levels of activity during treatment reduce the capacity for physical performance. A reduced functional capacity means that the cancer patient expends greater effort relative to maximal ability to perform usual activities, leading to higher levels of fatigue (p. 465).

Exercise training attenuates the loss and even increases functional capacity which results in reduced effort and decreased fatigue at any level *(22)*. Additionally, Courneya suggests that exercise may be an effective therapy in this population because exercise enhances biopsychosocial mechanisms that underlie coping or self-efficacy, which, in turn, may reduce the occurrence of some symptoms and their impact on activities of daily living. Additionally, QOL issues such as weight gain, muscle atrophy, and fatigue have physical concomitants which likely make exercise a therapy better able to address such problems more effectively *(28)*.

22.5.4. Summary

In sum, evidence exists to support the position that both the MBSR program and exercise rehabilitation are two complementary therapies that can produce significant and positive effects for individuals diagnosed with cancer. Next, a program that integrates both of these modalities for women with breast cancer is proposed.

22.6. MINDFULNESS-BASED EXERCISE REHABILITATION PROGRAM

22.6.1. Format and Structure of the MBER

Prior to beginning the program, the women will have been cleared by their physician to participate. A baseline physical fitness assessment will have been conducted by an ACSM-certified exercise specialist knowledgeable about cancer, who also will co-facilitate the Mindfulness-Based Exercise Rehabilitation (MBER) program. The same basic format and mindfulness practices of the original program will be used during the 8-week program as well as a modified program content for cancer patients *(29)*. Participants will still learn the three core MM practices of the body scan, sitting meditation, and hatha yoga.

Participants meet for the traditional 2 h, 1 night/week for 8 weeks, where they are exposed to didactic, inductive, and experiential modes of learning regarding mindfulness strategies. Although exercise physiologists may recommend 10 weeks or more to see demonstrable fitness changes, a 6-week exercise program showed significant changes in symptoms and exercise tolerance, while breast cancer patients were undergoing radiation therapy *(20)*. In addition to the weekly meetings, the participants will be expected to perform daily "homework" that emphasizes strategies learned in the sessions as to mindful practice of exercise or physical activity.

22.6.2. Session Contents

22.6.2.1. WEEKS 1 AND 2

At the first meeting, introductions and reasons for participating are shared and an overview of the intervention is presented. To allay fear about unrealistic expectations of the word "exercise" and competitive performance anxiety among group members, the philosophy of mindfulness is discussed and reiterated in different ways. It is emphasized that being mindful or fully present in each moment is the essence of paying attention of simply being aware of one's thoughts, feelings, bodily sensations, needs, and so on. Thus, every participant will be at a different place with different physical and functional abilities, limitations, muscle strength, and flexibility as all of the women can only start from where they and their bodies are at the present time, that is the place from where each of them will start. Group rules (confidentiality, group attendance, homework, and weekly exercise logs) are discussed.

The body scan is introduced in the first session as the women lay on mats. It is a guided journey through the physical geography of body in succession and paying attention to whatever sensations—or lack of sensations—that occur. Mindfulness includes an observational, nonjudging attitude with acceptance of ourselves as we are without harsh, negative criticism. This meditation is to get the women reacquainted with their bodies, especially if treatment has changed the body landscape. It is not uncommon for a breast cancer patient to have a difficult time accepting her changed body image and struggle also with identity issues that may include the concept of "femininity." Thus, the body scan experience may be a particularly arduous task for some because mindfulness or present-centered awareness is the unequivocal opposite of denial; yet, it is one path of facilitating the reality of a new physical geography, accepting it and adjusting one's perceived (cognitive) schemas or representational forms of self.

The assigned homework for the first week is to do the guided body scan on the first audiocassette tape (45 min) at least 5 days/week as well as 15 min of walking at a moderate (brisk) pace at 50–70% of the maximum heart rate as tolerated. Pedometers and heart rate monitors are distributed to the women along with instructions on use. Also, they are instructed how to access the website to record their daily homework information (dates, times, pedometer, and heart rate readings before and immediately after finishing the walking component).

The second session begins with the participants discussing the previous week's homework (body scan and walking) and sharing their experiences. The exercise specialist responds to any questions/concerns and addresses individual issues or adjustments to exercise as needed based on having reviewed the daily homework logs from the website. The women participate in an experiential exercise of mindfulness eating known as the raisin exercise. This simple exercise has two purposes: the first of which is to reinforce the idea that meditation is the natural act of paying attention and that there is nothing mysterious about it and the second reason is to initiate the topic of food and nutrition in the context of cancer and exercise. Next, sitting meditation (the breath as an anchor) is presented where the women learn about breath-work, positions, and so on during seated meditation. Finally, they experience mindful walking where they pay close attention to all sensations involved during this daily, routine activity. Their homework is to continue the body scan as per the previous week, increase daily walking to 25 min, and now add 15 min of sitting meditation daily at least 5 days/week.

22.6.2.2. WEEKS 3 AND 4

During the third week, walking meditation is followed by an introduction to hatha yoga. Participants are presented the principles and practice of this meditation in motion and are gently guided through different postures (asanas). Hatha yoga is then followed by an increased period of sitting meditation. Participants are given Tape 1 of a 45-min guided yoga routine as daily homework. In the 4th week, the topic of stress and the physiological stress response is presented with a discussion of mindfully tuning in to the body's language. Participants are encouraged to identify early bodily/behavioral signals or areas that indicate stress or discomfort. Once one begins paying attention to and listening to the body, early detection of stress can aid in proactive behavior to reduce tension, for example, physical and mental strategies may be combined (i.e., brisk walking, yoga, body scan, and meditation), depending on what works for the individual. Participants continue with Tape 1 of yoga as daily homework, and brisk walking is now at a minimum of 40 min/day, at least 5 days/week. They also alternate 30 min of meditation daily with 30 min of mindful eating.

22.6.2.3. WEEKS 5 AND 6

Participants are introduced to different yoga asanas and provided with Tape 2 of a new 45-min guided yoga routine for daily homework. The home exercise of brisk walking is now at 45 min/day at least 5 days/week. The women alternate 30 min of sitting meditation daily with 30 min of mindfulness eating.

22.6.2.4. WEEKS 7 AND 8

The last two sessions involve discussing the primary mental attitudes of coping with cancer and exploring issues of internal versus external locus of control and irrational beliefs. All of the mindfulness-based strategies are reviewed and discussed in terms of being applied to stressful, real life situations and challenges (a doctor's appointment, undergoing procedures, etc.). The purpose is to drive home the notion that conscious (mindful) awareness, intention, and responsible action are needed for an improved QOL. That is, a "quality" of life is not passive, nor is it a lucky accident—it takes work. Homework consists of the women alternating 45 min of daily brisk walking with 45 min of yoga as well as alternating 45 min of daily sitting meditation with the body scan or with mindful eating as desired. Thus, both physical and mental skills are used each day to achieve a balanced strength and harmony of functioning. Finally, the role of conscious awareness in exercise motivation and behavioral maintenance are discussed along with the needed commitment to healthy lifestyle choices.

22.7. SCENARIO WITH QUESTIONS AND ANSWERS

22.7.1. Scenario

Your mother was diagnosed with breast cancer 8 months ago and underwent surgery as well as chemotherapy. Physically, she has poor muscle tone and strength and has been plagued by fatigue since she left the hospital. You visited home last weekend and tried talking with her about your observations of her "being stuck" in cancer and not moving on with life. She vehemently disagrees and insists her that breast was removed simply as a precaution and that she did not have cancer so she is not "stuck" at all. Yet, she refuses to look at or feel the front of her upper torso and refuses to look at herself in the mirror. Your observations lead you to think that she is seriously in denial and afraid to face the reality of her recent health condition and experience.

22.7.2. Questions

1. From the MBSR program, which core technique do you think would help to transform the mother's cognitive or mental denial into acceptance of reality?

2. From the MBSR program, which core technique do you think would assist in the mother's acceptance of to the physical/surgical changes to her body?
3. What would you recommend to improve the mother's continued fatigue and poor muscle tone and strength?

22.7.3. Plausible Answers

1. Although there are multiple options of mindfulness techniques, probably the most direct would be sitting meditation. This would address the cognitive aspect solely as sitting meditation is the formal, cognitive practice. Also, mindfulness is the opposite of denial and calmly but, directly, confronts the nonacceptance of reality.
2. The body scan would be the appropriate technique because it is a journey through all of the regions of the body; it forces one to bring attention to and focus on each region and to accept whatever sensations or lack of sensations that are experienced. This technique encourages one to accept the body and physical form as it truly is.
3. There are several options here; however, getting the mother started on an exercise regimen of brisk, moderate walking on a daily basis would be one approach. In addition, this could be combined with the practice of yoga several times per week to increase muscle tone, strength, and flexibility as well.

REFERENCES

1. American Cancer Society. *Cancer Facts & Figures*. Atlanta, American Cancer Society, 2005.
2. Shapiro SL, Lopez AM, Schwartz GE, et al. Quality of life and breast cancer: relationship to psychosocial variables. *J Clin Psychol* 2001; 57(4): 501–519.
3. National Cancer Institute. *Cancer Facts*. Washington, DC, 1997.
4. Cassileth B. Complementary therapies. *Cancer Nurs* 1999; 22(1): 85–90.
5. Kabat-Zinn J. *Full Catastrophe Living: Using the Wisdom of your Body and Mind to Face Stress, Pain, and Illness*. New York, Dell Publishing, 1990.
6. Courneya KS, Keats MR, Turner AR. Physical exercise and quality of life in cancer patients following high dose chemotherapy and autologous bone marrow transplantation. *Psychooncology* 2000; 9: 127–136.
7. Dimeo FC, Stieglitz RD, Novelli-Fischer U, et al. Effects of physical activity on the fatigue and psychologic status of cancer patients during chemotherapy. *Cancer* 1999; 85(10): 2273–2277.
8. Morris KT, Johnson N, Homer L, et al. A comparison of complementary therapy use between breast cancer patients and patients with other primary tumor sites. *Am J Surg* 2000; 179(5): 407–411.
9. Sobel D. Mind matters, money matters: Cost-effectiveness of mind/body medicine. *JAMA* 2000; 284: 1705.
10. Tacón A, McComb J, Caldera Y, et al. Mindfulness meditation, anxiety reduction and heart disease: a pilot study. *Fam Community Health* 2003; 26: 25–33.
11. Tacón A, Caldera Y, Ronaghan C. Mindfulness-based stress reduction in women with breast cancer. *Fam Syst Health* 2004; 22: 193–203.
12. Tacón A, Caldera Y, Ronaghan C. Mindfulness, psychosocial factors and breast cancer. *J Cancer Pain Symptom Pall* 2005; 1: 45–54.
13. Kaplan KH, Goldenberg DL, Galvin-Nadeau M. The impact of a meditation-based stress reduction program on fibromyalgia. *Gen Hosp Psychiatry* 1993; 15: 284–289.
14. Kabat-Zinn J, Lipworth L, Burney R, et al. Four-year follow-up of a meditation-based program for the self-regulation of chronic pain. *Clin J Pain* 1987; 2: 159–173.
15. Speca M, Carlson L, Goodey E, et al. A randomized, wait-list controlled clinical trial: the effect of a mindfulness meditation-based stress reduction program on mood and symptoms of stress in cancer outpatients. *Psychosom Med* 2000; 62: 613–622.
16. Carlson LE, Speca M, Patel, K. Mindfulness-based stress reduction in relation to quality of life, mood, symptoms of stress and levels of cortisol, DHEAS and melatonin in breast and prostate cancer outpatients. *Psychoneuroendocrinology* 2004; 29(4): 448–474.

17. American Cancer Society. *Cancer Facts & Figures*. Atlanta, American Cancer Society, 2002.
18. Cheema, BS, Gaul CA. Full-body exercise training improves fitness and quality of life in survivors of breast cancer. *J Strength Cond Res* 2006; 20(1): 14–21.
19. Mock V, et al. Exercise manages fatigue during breast cancer treatment: a randomized controlled trial. *Psychooncology* 2005; 14(6): 464–477.
20. Mock V, Dow KH, Meares CJ, et al. Effects of exercise on fatigue, physical functioning, and emotional distress during radiation therapy for breast cancer. *Oncol Nurs Forum* 1997; 24(6): 991–1000.
21. Segar ML, Katch VL, Roth RS, et al. The effect of aerobic exercise of self-esteem and depressive and anxiety symptoms among breast cancer survivors. *Oncol Nurs Forum* 1998; 25(1): 107–113.
22. Courneya KS, Friedenrich CM, Sela RA, et al. The group psychotherapy and home-based physical exercise (Group-Hope) Trial in cancer survivors: physical fitness and quality of life outcomes. *Psychooncology* 2003; 12: 357–374.
23. Dimeo FC, Stieglitz RD, Novelli-Fischer U, et al. Effects of physical activity on the fatigue and psychologic status of cancer patients during chemotherapy. *Cancer* 1999; 85(10): 2273–2277.
24. Segal R, Evans W, Johnson D, et al. Structured exercise improves physical functioning in women with stages I & II breast cancer: results of a randomized controlled clinical trial. *J Clin Oncol* 2001; 19(3): 657–665.
25. American College of Sports Medicine. *ACSM's Guidelines for Exercise Testing and Prescription*, 6th ed. Baltimore, Williams & Wilkins, 2000.
26. Curt G, Breitbart W, Cella D et al. Impact of cancer-related fatigue on the lives of patients. *Oncologist* 2000; 5: 353–360.
27. Vogelzang N, Breitbart W, Cella D, et al. Patient, caregiver, and oncologist perceptions of cancer-related fatigue: results of a tripart assessment survey: The Fatigue Coalition. *Semin Hematol* 1997; 34(3 suppl 2): 4–12.
28. Courneya, KS. Exercise in cancer survivors: an overview of research. *Med Sci Sports Exer* 2003; 35(11): 1846–1852.
29. Tacón, A. Meditation as a complementary therapy in cancer. *Fam Community Health* 2003; 26: 25–33.

23 Exercise Guidelines for the Postmenopausal Woman

Shawn Anger and Chelsea Barker

CONTENTS

23.1 LEARNING OBJECTIVES
23.2 RESEARCH FINDINGS
23.3 SUMMARY AND CONCLUSIONS
23.4 SCENARIO WITH QUESTIONS AND ANSWERS

23.1. LEARNING OBJECTIVES

After completing this chapter, you should have an understanding of the following:

- The cardiopulmonary physiology for older women when engaging in activity.
- The changes in bone mass density for postmenopausal women.
- The conformational changes of muscles and strength capabilities as women age.
- Guidelines for appropriate exercise for postmenopausal women.

23.2. RESEARCH FINDINGS

23.2.1. The Cardiopulmonary Physiology for Older Women When Engaging in Activity

General cardiovascular changes occur as an individual ages that cause a gradual decline in cardiac function. These changes are seen starting as young as age 25. For every decade, after the age of 25, there is an approximately 10% decline in $VO_{2\,max}$ for sedentary individuals. For those who continually engage in regular vigorous and endurance exercise, that rate of decline is only about 5% per decade (1). Another decline seen with age is HR_{max}. The older an individual is, there is a lower physiological maximum heart rate. Although there are a number of formulas predicting age-associated maximum heart rate, no formula is an exact prediction because each person will vary slightly from the norm. For example, one such formula gives the equation $220 - age = HR_{max}$. The stipulation given to this formula is that the maximum heart rate can be plus or minus approximately 15 beats per minute. Although an accurate prediction is hard to reach for an age-predicted maximum heart rate, general trends can be

From: *The Active Female*
Edited by: J. J. Robert-McComb, R. Norman, and M. Zumwalt © Humana Press, Totowa, NJ

seen, as there is a gradual decline in maximum heart rate with age. However, this general decline may be delayed with regular endurance exercise *(1)*. Heart rate recovery also takes longer as people age. As a result, it is important to understand that the older an individual is, the recovery time will be extended in order to allow the heart rate to return to baseline *(2)*.

All of these changes are brought about by a change in the cardiac tissue itself. With age comes a decrease in compliancy in the myocardial tissue because of an influx of elastin, reticulin, and fat. This in turn causes a general decline in cardiac output. These changes also have an effect on the heart's electrical conduction system that results in the aforementioned declines in maximum heart rate and heart rate recovery time *(2)*.

Peripheral circulation is also affected as individuals age. Atherosclerosis begins to gradually take effect causing a narrowing of the blood vessels. This in turn causes resting blood pressure to increase, thereby increasing the workload on the heart. The narrowing of blood vessels can also cause a decrease in blood flow to the organs of the body, including working muscles. This can be an important aspect of athletic performance and ability to reach a certain workload as an individual ages *(2)*.

The lungs also change to accompany the changes seen with the heart. The alveoli themselves begin to lose their elastic recoil therefore decreasing the effectiveness of breaths. Also, there is a decrease in pulmonary blood volume as well as an increase in intercostal muscle stiffness and weakness *(2)*. With the onset of exercise, physiological reactions occur in the body to increase ventilatory rate in order to provide oxygen to working muscles and expel carbon dioxide from metabolism so that the workload can be maintained. The rate of increased ventilation declines slightly with age giving another indication for a more gradual and lengthy warm-up as age increases so that the body is allowed time to induce physiological changes to achieve a desired workload *(3)*.

23.2.2. The Changes in Bone Mass Density for Postmenopausal Women

One of the most commonly associated conditions with postmenopausal women is a loss in bone density often diagnosed as osteoporosis if the loss is significant. Loss in bone mass can be a very serious issue for women as there can be about 20% loss in bone mass by the age of 65 and a 30% loss by the age of 80 *(4)*. The underlying cause of osteoporosis is an imbalance in bone reabsorption and new bone formation causing a deficit in bone density. Some of the main contributors of osteoporosis include an inadequate calcium intake and a lack of mechanical forces applied to the bone. One stimulant for bone growth is the application of external stimuli such as mechanical load, muscular stresses, and gravitational forces on the bone *(5)*. The type of bone that is most sensitive to a lack of force application is trabecular bone. This type of bone is found in the vertebrae, the neck of the femur, and the wrist, which causes these sites to be particularly susceptible to fractures in women *(5)*.

Exercise becomes a key component to maintaining bone density and even increasing it. It is important to remember that increasing bone density takes at least 24 weeks if not closer to a year *(6)*. Performing weight-bearing activities and engaging in resistance training can possibly increase bone density or at least slow bone density loss *(6)*. Research has shown that even with exercise the amount of change seen in bone mass density is very small. Oftentimes, the goal is rather to stabilize bone mass density. The most effective method of maintaining or increasing bone mass density is to combine exercise with a calcium-rich diet *(7)*.

23.2.3. The Conformational Changes of Muscles and Strength Capabilities as Women Age

A loss in strength becomes inevitable as an individual ages, and women seem to experience an even greater loss in strength with age *(2)*. The loss in strength is because of a number of physiological and conformational changes within the muscles and motor units. Research suggests that at least some of the loss in strength comes from a decrease in motor units. One study reported a 47% decrease in motor unit count from individuals in their twenties to their sixties *(8)*. The fiber-type distribution changes as well as a number of other changes cause an overall loss in muscle mass. As individuals age, there is a gradual decrease in type II fibers and an increase in type I fibers. Type II fibers are responsible for short bursts of higher intensity exercise, whereas type I fibers are responsible for low-intensity endurance exercise that even includes postural control *(2)*.

There is a loss in overall muscle mass, more substantially seen in the lower extremities, that begins as early as age 30, and then around the age of 50, this decrease in muscle mass becomes even more pronounced *(9)*. Having more muscle mass lost in the lower extremities would logically make it more difficult to perform general locomotion and would present a balance challenge making older individuals more susceptible to falls. A gradual replacement of muscle tissue with fat and connective tissue also occurs with age, a phenomenon that is heightened in women *(10)*. Other changes that occur causing a loss in muscle mass include a decrease in the actual size of each individual fiber and a decrease in the number of fibers *(11,12)*. An average reduction was seen between the ages of 20 and 80 by 40% for muscle mass and 39% loss in muscle fibers *(12)*. There is also evidence that loss in muscle fibers can be the result of muscle cell death from a loss of contact with their motor unit and nervous system *(6)*. If a motor neuron were to become inactive, all of the muscle fibers innervated by that nerve in the motor unit would consequently die off, resulting in a loss of muscle. Research data suggest that, to a great degree, the loss in muscular strength with age is more of a result of a loss in muscle mass rather than a loss in muscular function *(11)*.

23.2.4. Guidelines for Appropriate Exercise for Postmenopausal Women

Appropriate workloads should be determined based on each individual's needs, but there are some general guidelines that can be used when beginning an exercise program. For cardiovascular exercise, ACSM recommends exercising 3–5 days at an intensity ranging from 40 to 85% using heart rate reserve, for 30–60 min. *See* Chap. 18 for more specific guidelines for cardiovascular fitness. Also the Borg scale and details how to exercise at a moderate (perceived exertion 11–14) or vigorous pace (15 or greater) using this scale can be found in Appendix 11.

At any age, initial increases in strength when beginning a resistance-training program come from neural adaptations, and postmenopausal women are no different. A vast majority of strength gains seen will be from neural adaptations because there is a limited ability for changes in the muscle fibers and muscular hypertrophy. It is also suggested that the positions in which exercises are performed are varied in order to incorporate balance challenges on the body. This training technique includes progressing from stable to unstable, which may include progressing from a hard surface to a softer foam surface or progressing from a machine where the plane of motion is fixed to free weights where the plane of motion must be controlled by the exerciser. Appendix 16 provides a sample exercise program appropriate for postmenopausal women while Appendix 17 provides descriptions as well as photographs of those sample exercises.

23.3. SUMMARY AND CONCLUSIONS

The importance of exercise is apparent throughout the lifespan, and women who have reached their postmenopausal years are no different. Exercise will provide women with a number of benefits that include but are no limited to increasing resting metabolic rate, improving balance, and a reduced risk of falling, which is important with bones becoming more fragile following menopause. Throughout the exercise program, it is important to remember that there needs to be more time allowance for a sufficiently long warm-up/cool-down as well as an extended recovery time between exercises during the workout.

23.4. SCENARIO WITH QUESTIONS AND ANSWERS

23.4.1. Scenario

A 75-year-old woman visits her physician for her yearly physical, and her doctor recommends that she start an exercise program to maintain strength and bone mass density and to help prevent a fall. The woman has not exercised in 30 years and is very intimidated by the thought of "working out." Her physician recommends that she see you, an exercise physiologist at a local fitness center, to visit with her about starting an exercise program.

23.4.2. Questions

1. As part of your visit with her, you want to make it clear that starting an exercise program, no matter at what age, has benefits. What benefits would you discuss with her to make the exercise program a priority for her?
2. What physiological changes would you need to consider when designing an exercise program for her?
3. What areas would you target with resistance exercise to promote stronger bone health?

23.4.3. Plausible Answers

1. The discussion should include cardiovascular benefits such as lower blood pressure and a stronger heart, bone mass density maintenance to reduce the risk of fractures, and strength in the skeletal muscles that will improve balance and work to prevent falls.

2. Cardiovascular considerations include changes such as decreased max HR and increased time needed for the warm-up and cool-down portions of the exercise session. Skeletal muscle changes include a decrease in type II fibers and an overall loss in muscle mass. Bone mass density decreases should also be considered when designing the program.
3. Women are particularly prone to a loss in bone mass density, and subsequently fractures at the sites of the hip, the spine, and the wrist.

REFERENCES

1. Rogers MA, Hagberg JM, Martin WH, et al. Decline in VO$_2$ max with aging in master athletes and sedentary men. *J Appl Physiol* 1990;68:2195–2199.
2. Harris SA, Coven E. The older female athlete. In: Swedan N, ed. *Women's Sports Medicine and Rehabilitation*. Gaithersburg, MD: Aspen, 2001:181–188.
3. Babcock MA. Exercise on transient gas kinetics are slowed as a function of age. *Med Sci Sports Exerc* 1994;26:440.
4. Roberts RA, Roberts SO. Exercise and aging. In: Roberts RA, Roberts SO, eds. *Exercise Physiology: Exercise, Performance, and Clinical Applications*. St. Louis, MO: Mosby, 1997:578–599.
5. Notelovitz M, Shangold MM. Menopause. In: Shangold MM, Mirkin G, eds. *Women and Exercise: Physiology and Sports Medicine*, 2nd ed. Philadelphia, PA: F.A. Davis Company, 1994:187–213.
6. Fleck SJ, Kraemer WJ. *Designing Resistance Training Programs*, 3rd ed. Champaign: Human Kinetics, 2004.
7. Borer KT. Physical activity in the prevention and amelioration of osteoporosis in women: interaction of mechanical, hormonal and dietary factors. *Sports Med* 2005;35:779–830.
8. Doherty TJ, Vandervoort AA, Taylor AW, et al. Effects of motor unit losses on strength on older men and women. *J Appl Physiol* 1993;74:868–874.
9. Janssen I, Heymsfield SB, Wang Z, et al. Skeletal muscle mass and distribution in 468 men and women aged 18–80 yr. *J Appl Physiol* 2000;89:81–88.
10. Imamura K, Ashida H, Ishikawa T, et al. Human major psoas muscle and sacrospinalis muscle in relation to age: a study by computed tomography. *J Gerontol* 1983;38:678–681.
11. Frontera WR, Hughes VA, Lutz KJ, et al. A cross-sectional study of muscle strength and mass in 45 to 78-year-old men and women. *J Appl Physiol* 1991;71:644–650.
12. Lexell J, Taylor CC, Sjostrom M. What is the cause the aging atrophy? Total number, size and proportion of different fiber types studied in whole vastus lateralis muscle from 15 to 83-year-old men. *J Neurol Sci* 1988;84:275–294.

VI Nutrition, Energy Balance, and Weight Control

24 Estimating Energy Requirements

Jacalyn J. Robert-McComb

CONTENTS

24.1 LEARNING OBJECTIVES
24.2 INTRODUCTION
24.3 RESEARCH FINDINGS
24.4 CONCLUSIONS
24.5 SCENARIO WITH QUESTIONS AND ANSWERS

24.1. LEARNING OBJECTIVES

After completing this chapter, you should have an understanding of the following:

- The components of energy expenditure.
- Methods of measuring energy expenditure.
- Average dietary reference intake values of energy for active individuals.
- Estimated energy expenditure prediction equations.

24.2. INTRODUCTION

Energy is defined as "the capacity to do work." Energy is expended by the human body in the form of resting energy expenditure (REE), the thermic effect of food (TEF), and energy expended in physical activity (EEPA) *(1,2)*. These three components make up a person's daily total energy expenditure (TEE). The REE constitutes the largest portion (60–75%) of the TEE *(3)*. The only exception to this would be in extremely active individuals. The TEF represents approximately 10% of the daily TEE. The most variable component of TEE is the contribution of physical activity (EEPA). Criterion methods such as doubly labeled water (DLW), indirect calorimetry, and direct observation are the most reliable and valid measurements against which all EEPA should be validated, but they are sophisticated and expensive techniques that are not available to all.

Traditionally, recommendations for energy requirements have been based on self-recorded estimates (e.g., diet records) or self-reported estimates (e.g., 24-h recalls) of food intake. However, it is now well accepted that these methods are biased and

From: *The Active Female*
Edited by: J. J. Robert-McComb, R. Norman, and M. Zumwalt © Humana Press, Totowa, NJ

do not provide an accurate account of food intake *(4)*. The World Health Organization (WHO) *(5)* states that "as a matter of principle, we believe the estimates of energy requirements should, as far as possible, be based on estimates of energy expenditure." The National Academy of Sciences, Institute of Medicine (IOM), and Food and Nutrition Board in partnership with Health Canada developed the estimated energy requirements (EER) for men, women, children, and infants and for pregnant and lactating women *(2)*. Knowledge of energy requirements for women is essential for reproductive health as well as optimum health status throughout the life cycle. The Food and Agriculture Organization/World Health Organization/United Nations University (FAO/WHO/UNU) Expert Consultation on Energy and Protein requirements defines energy requirements as follows *(5)*:

The energy requirement of an individual is the level of energy intake from food that will balance energy expenditure when the individual has a body size and composition, and level of physical activity, consistent with long-term good health; and that will allow for the maintenance of economically necessary and socially desirable physical activity. In children and pregnant or lactating women the energy requirement includes the energy needs associated with the deposition of tissues or the secretion of milk at rates consistent with good health.

24.3. RESEARCH FINDINGS

24.3.1. The Components of Energy Expenditure

Figure 24.1 depicts the components of TEE. The components, REE, the TEF, and the EEPA, will be discussed separately.

24.3.1.1. RESTING ENERGY EXPENDITURE

REE is the energy expended in the activities necessary to sustain normal body functions and homeostasis. This energy encompasses respiration, circulation, the synthesis of organic compounds, the pumping of ions across membranes, the energy required by the central nervous system, and the maintenance of body temperature. Basal energy expenditure (BEE) can be defined as the minimal amount of energy expended to sustain life and is usually expressed as kilocalories per 24 hours (kcal/24 h). Basal metabolic rate (BMR) remains relatively constant and typically represents approximately 60–70% of the TEE. These measurements are made before engaging in any physical activity and 10–12 h after the ingestion of any food, drink, or nicotine. If any of the conditions are not met, energy expenditure should be referred to as the resting metabolic rate (RMR). Practically speaking, RMR is used more often than BMR and may be 10–20% higher than the BMR because of the energy spent as a result of the TEF of excess postexercise oxygen consumption.

Numerous factors cause the REE to vary among individuals. Contributing variables include but are not limited to body size and weight (body surface area [BSA]), body composition, age, sex, and hormonal status *(1,2)*.

BSA is computed from height and weight—taller individuals who weigh more will have the greatest surface area. Individuals with greater surface area will have the greatest metabolic rate. Various body surface formulas have been developed over the

Chapter 24 / Estimating Energy Requirements

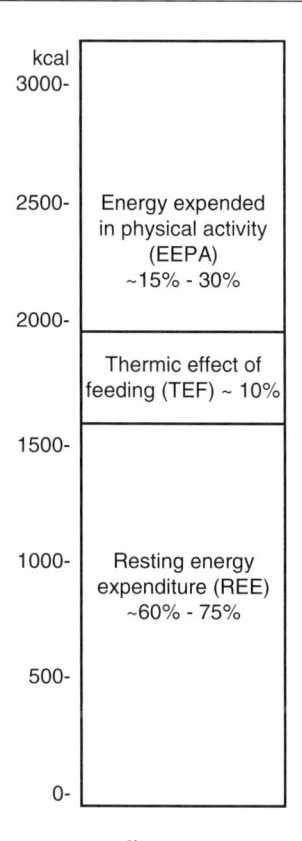

Fig. 24.1. Components of total energy expenditure.

years, and these formulas all give slightly different results; however, the importance is that the formulas are standardized (http://www.halls.md/body-surface-area/bsa.htm) to determine BSA.

The main determinant of REE is fat-free mass (FFM) or lean body mass (LBM). Because of their greater FFM, athletes or individuals who are extremely fit have approximately a 5% higher BMR than nonathletic individuals. One of most practical and inexpensive ways to assess FFM is to use skinfold anthropometry. However, to gain valid and reliable data, the technician must be well trained. Standardized techniques to measure skinfold thickness can be found in ACSM Guidelines for Exercise Testing and Prescription *(6)*.

The effect age has on REE is highly correlated with FFM. REE is highest during periods of rapid growth, chiefly during the first and second years of life, and peaks throughout adolescence and puberty *(7)*. As a child becomes older, the caloric requirement for growth is reduced to about 1% of the TEE. REE continues to decline with increasing age in adulthood. The loss of FFM with aging can be attenuated with exercise; however, exercise cannot completely negate the effects of age. There is an approximately 2–3% decline in REE after early adulthood largely because of the loss of FFM *(3)*.

Sex differences in metabolic rates are primarily contributed to differences in body size and composition. Women have approximately 5–10% lower REE than men primarily because of differences in LBM *(1)*.

Hormonal status also has an effect on the metabolic rate. The hormones associated with the sympathetic nervous system or those involved in the fight or flight response such as epinephrine and norepinephrine increase metabolic rate. Probably the hormones most closely aligned with REE are the thyroid hormones because these hormones are considered to be the permissive hormones and allow other hormones to exert their full effect. Also the metabolic rate of women fluctuates with the menstrual cycle. An average of 359 kcal/day difference in the BMR has been measured from 1 week before ovulation and just before the onset of menstruation. The average increase in energy expenditure is about 150 kcal/day during the second half of the menstrual cycle (8).

24.3.1.2. THERMIC EFFECT OF FOOD

The TEF accounts for approximately 10% of the TEE. The TEF varies with the composition of the diet and is greater after the consumption of carbohydrate and proteins than after fat. Spicy foods enhance and prolong the effect of TEF. Caffeine and nicotine also stimulate the TEF (1).

24.3.1.3. ENERGY EXPENDED IN PHYSICAL ACTIVITY

The EEPA is the most variable component of TEE. EEPA includes energy expended in physical activity and exercise and during activities such as shivering and maintaining postural control (1,2). The energy cost of the physical activity is related to intensity, duration, skill level, and FFM. As the intensity of the physical activity and/or duration of the activity increases, so does the energy expenditure. All else being equal, individuals with less skill in performing an activity will expend more energy in performing the motion. Also, individuals with greater FFM will expend more energy at the same intensity and/or duration of the exercise.

24.3.2. Methods of Measuring Energy Expenditure

There are many available methods to measure energy expenditure. Each method has its advantages and disadvantages. Some of the more common methods include the DLW technique, direct and indirect calorimetry, uniaxial and triaxial accelerometers, heart rate monitors, and physical activity questionnaires.

The DLW technique for measuring TEE in free-living people uses two stable isotopes of water (deuterium [^2H_2O] and oxygen-18 [$H_2^{18}O$]); the difference in the turnover rates of the two isotopes measures the carbon dioxide production rate, from which TEE can be calculated. The DLW technique has many advantages; however, the expense of the technique makes this technique impractical for daily use by clinicians. The primary advantage of this technique is its accuracy (2–8% precision) and that it provides a measure of energy expenditure that incorporates all the components of TEE (i.e., REE, TEF, and EEPA) (1,9). It is for these reasons that the DLW technique has been used for validation studies and to generate data to develop prediction equations (2).

Direct calorimetry is a method for measuring the amount of energy expended by monitoring the rate at which a person loses heat from the body using a structure called a whole-room calorimeter. Direct calorimetry provides a measure of energy expended in the form of heat but does not provide information on the type of fuel being oxidized.

Indirect calorimetry is a method for estimating energy production by measuring oxygen consumption and carbon dioxide utilizing a respirator gas-exchange canopy or ventilation hood. Data are obtained from indirect calorimetry that permit calculation of the respiratory quotient (RQ = moles CO_2 expired/moles O_2 consumed). Depending on the RQ value, energy per liter of oxygen will be converted to kilocalories of heat produced and extrapolated to energy expenditure in 24 h.

Estimates of EEPA can be made using the DLW technique in conjunction with indirect calorimetry, uniaxial and triaxial accelerometers, heart rate monitors, and physical activity questionnaires. Normally, estimates of EEPA are validated using the DLW technique. The triaxial monitor uses three different planes to measure movement rather than a single vertical plane, as in the uniaxial monitor, and has been found to be more accurate than the uniaxial monitor *(10)*. Minute-by-minute heart rate monitors have been found to be valid in estimating habitual TEE in certain populations but not in individuals *(11)*. The Seven-Day Recall Physical Activity Questionnaires and the Yale Activity Survey are two questionnaires that have been shown to be valid in older men using the DLW technique *(12)*. However, validation studies have not been done for these surveys using women and the DLW technique.

24.3.3. Average Dietary Reference Intake Values for Energy for Active Individuals

The National Academy of Sciences, Institute of Medicine (IOM), and Food and Nutrition Board in partnership with Health Canada developed the EER for lifespan ranging from infants to adult men and women (Table 24.1). These are estimated values for energy in healthy, active people of reference height, weight, and age for each life-stage group *(2)*.

24.3.4. Estimated Energy Expenditure Prediction Equations

The Harris–Benedict formula published in 1919 *(13)* is one of the most widely used formulas to determine REE; however, it has been estimated that it overestimates REE by 7–24% *(14)*. The Mifflin–St. Jeor equation is more likely to estimate RMR within 10% of that measured and is presented in Chap. 29; however, noteworthy limitations exist when it is generalized to certain age and ethnic groups *(15)*. The National Academy of Sciences, Institute of Medicine (IOM), and Food and Nutrition Board in partnership with Health Canada developed new prediction equations to EER for people according to their life-stage group *(2)*. The EER incorporates age, weight, height, gender, and level of physical activity for individuals in various life stages. These equations should only be used as a guideline to promote optimal nutritional status; the individual should be monitored closely to adjust caloric intake based on target goals and changes in body mass. As in all prediction equations, standard errors are inherent *(14–16)*. Appendices 18 and 19 list the appropriate physical activity level (PAL) coefficients and EER prediction equations for people of normal weight and for various overweight and obese groups. Although all equations have been developed to maintain current body weight at the designated PAL, these equations do not to promote weight loss.

Table 24.1
Dietary reference intake values for energy for active individuals:[a] active PAL EER (kcal/day)

Life-stage group	Criterion	Male	Female
Infants (months)			
0–6	Energy expenditure + energy deposition	570	520 (3 months)
7–12	Energy expenditure + energy deposition	743	676 (9 months)
Children (years)			
1–2	Energy expenditure + energy deposition	1046	992 (24 months)
3–8	Energy expenditure + energy deposition	1742	1642 (6 years)
9–13	Energy expenditure + energy deposition	2279	2071 (11 years)
14–18	Energy expenditure + energy deposition	3152	2368 (16 years)
Adults (years)			
> 18	Energy expenditure	3067[b]	2403[b] (19 years)
Pregnant women			
14–18 years	Adolescent female EER + change in TEE + pregnancy energy		
First trimester	Deposition		2368 (16 years)
Second trimester	2708 (16 years)		
Third trimester	2820 (16 years)		
19–30 years	Adult female EER + change in TEE + pregnancy energy		
First trimester	Deposition		2403[b] (19 years)
Second trimester	2743[b] (19 years)		
Third trimester	2855[b] (19 years)		
Lactating women			
14–18 years	Adolescent female EER + milk energy output – weight loss		
First 6 months	2698 (16 years)		
Second 6 months	2768 (16 years)		
19–50 years	Adult female EER + milk energy output – weight loss		
First 6 months	2733[b] (19 years)		
Second 6 months	2803[b] (19 years)		

EER, estimated energy requirement; PAL, physical activity level; TEE, total energy expenditure.
[a] For healthy active Americans and Canadians at the reference height and weight.
[b] Subtract 10 kcal/day for men and 7 kcal/day for women for each year of age above 19 years.
Source: From Institute of Medicine of the National Academics (2).

24.4. CONCLUSIONS

The equations discussed in this chapter should be used only as a guideline in maintaining energy balance. Even though numerous energy prediction equations to promote energy balance are widely cited in the literature, there are limitations in the use of these equations. Prediction errors are inherent when using any estimated equations. Further validation studies of predictive equations are needed to minimize prediction error in certain age and ethnic groups. Older adults and US-residing ethnic minorities have been underrepresented both in the development of predictive equations and in validation studies *(15)*.

Another methodological problem in measuring energy balance, in particular energy intake, is the phenomenon of subjects reporting lower energy intake than physiologically required, noted as underreporting. Underreporting of energy intake is expressed as a ratio of reported energy intake to estimated BMR *(16)*. Underreporting is especially problematic in the obese but also occurs in the relatively lean population *(17)*.

A clinical decision of whether an accurate metabolic rate by measurement is required to provide nutritional care and counseling should be made on a case-by-case basis. If the target goals are not being met, the client should be monitored closely utilizing any dietary intake records, energy expenditure logs, and physiological measurements of body mass and/or weight change. Indirect calorimetry may be an important tool when, in the judgment of the clinician, the predictive methods fail an individual in a clinically relevant way *(15)*.

24.5. SCENARIO WITH QUESTIONS AND ANSWERS

24.5.1. Scenario

Claudia is an active female, age 25 with a BMI of $22 \, kg/m^2$. She weighs 65 kg and is 1.7 m tall. She does not want to lose weight; however, she would like to know approximately how many calories she should eat from fats, carbohydrates, and proteins using 25, 60, and 15% of total calories, respectively, from the specific food groups. To calculate this, she needs to know her estimated energy expenditure.

24.5.2. Questions

1. Using the prediction equations from the Institute of Medicine in Appendices 18 and 19, what is her estimated energy expenditure if she exercises equivalently to walking 10 miles a day?
2. What would you tell Claudia about the accuracy of these equations?

24.5.3. Plausible Answers

1. Her estimated energy expenditure can be calculated using the following formula (her unique individual data are italicized): $354 - (6.91 \times 25) + 1.9(9.36 \times 65 + 726 \times 1.7) = 3682 \, kcal$. This is a very large energy expenditure for a female; however, Claudia is 5′6″, weighs 143 lb, and is very active.
2. Prediction errors are inherent when using any estimated equations. These equations should only be used as a guideline in maintaining energy balance. If weight is lost, the equations are underestimating energy needs, and if weight is gained, the equations

are overestimating energy needs. Adjustments to energy needs must be made based on target goals and physiological data.

REFERENCES

1. Fray CD, Johnson RK. Energy. In: Mahan LK, Escott-Stump S, eds. *Krause's Food, Nutrition, & Diet Therapy*, 11th ed. Philadelphia, PA: Saunders, 2004:21–36.
2. Institute of Medicine (U.S.), Standing Committee on the Scientific Evaluation of Dietary Reference Intakes. *Dietary Reference Intakes for Energy Carbohydrate, Fiber, Fat, Fatty Acids, Cholesterol, Protein, and Amino Acids*. Washington, DC: National Academy Press, 2002.
3. Poleman ET. Regulation of energy expenditure in aging humans. *J Am Geriatr Soc* 1993;41:552–559.
4. Johnson RK. What are people really eating, and why does it matter? *Nutr Today* 2000;35(2):40–46.
5. World Health Organization: Energy and protein requirements. Report of a Joint Food and Agriculture Organization/World Health Organization/United Nations University (FAO/WHO/UNU) Expert Consultation, Technical Report Series 724, Geneva, WHO, 1985.
6. American College of Sports Medicine. *ACSM's Guidelines for Exercise Testing and Prescription*, 7th ed. Philadelphia, PA: Lippincott Williams & Wilkins, 2006.
7. Torun B, Davies PS, Livingstone MB, et al. Energy requirements and dietary energy recommendations for children and adolescents 1 to 18 years old. *Eur J Clin Nutr* 1996;50 Suppl 1:S37–S80.
8. Webb P. 24-Hour energy expenditure and the menstrual cycle. *Am J Clin Nutr* 1986;44(5):614–619.
9. Schoeller DA. Measurement of energy expenditure in free-living humans by using double labeled water. *J Nutr* 1988;118:1278–1289.
10. Bouten CV, Verboeket-Van Venne WP, Westerterp KR, et al. Daily physical activity assessment: comparison between movement registration and doubly labeled water. *J Appl Physiol* 1996;81(2):1019–1026.
11. Livingstone MB, Coward WA, Prentice AM, et al. Daily energy expenditure in free-living children: comparison of heart rate monitoring with the doubly labeled water method. *Am J Clin Nutr* 1992;56:343–352.
12. Bonnefoy M, Normand S, Pachiaudi C, et al. Simultaneous validation of ten physical activity questionnaires in older men: a doubly labeled water study. *J Am Geriatr Soc* 2001;49(1):28–35.
13. Harris JA, Benedict FG. *A Biometric Study of Basal Metabolism in Man*, Pub. No. 279. Washington, DC: Carnegie Institute of Medicine, 1919.
14. Daly JM, Heymsfield SB, Head CA, et al. Human energy requirements: overestimation by widely used prediction equations. *Am J Clin Nutr* 1985;42:1170–1174.
15. Frankenfield D, Roth-Yousey L, Compher C, et al. Comparison of predictive equations for resting metabolic rate in healthy nonobese and obese adults: a systematic review. *J Am Diet Assoc* 2005;105(5):775–789.
16. Goldberg GR, Black A, Jebb SA, et al. Critical evaluation of energy intake data using fundamental principles of energy physiology: 1. derivation of cut-off limits to identify under-recording. *Eur J Clin Nutr* 1991;45:569–581.
17. Okubo H, Sassaki S. Underreporting of energy intake among Japanese women aged 18–20 years and its association with reported nutrient and food group intakes. *Public Health Nutr* 2004;7:911–917.

25 Nutritional Guidelines and Energy Needs for Active Children

Karen S. Meaney, Kelcie Kopf, and Megan Simons

Contents
- 25.1 Learning Objectives
- 25.2 Introduction
- 25.3 Research Findings
- 25.4 Conclusions
- 25.5 Scenario with Questions and Answers

25.1. LEARNING OBJECTIVES

After completing this chapter, you should have an understanding of the following:

- The purpose and goals of the Dietary Guidelines for Americans.
- MyPyramid Food Guidance System, its calorie criteria, and the six main food groups and the foods that are found in each of those groups.
- The positive impact family meals have on children's overall nutrition.
- The daily recommended intake of calories for children.
- The role of carbohydrates, proteins, and fats as sources of energy for children.
- The importance of vitamins, minerals, fibers, and water in children's diets.

25.2. INTRODUCTION

Participation in regular physical activity during childhood will positively contribute to the development of a healthy lifestyle. Children who participate in physical activity are more likely to engage in physical activity throughout adulthood *(1)*. Being physically active has numerous physiological and psychological benefits including but not limited to lowering the risk of heart disease, type 2 diabetes, and some cancers; maintaining optimal weight and body fat; assisting in building bones and muscles; increasing overall fitness level; enhancing self-esteem; and reducing stress and feelings of depression *(2)*. Children should participate daily in 60 min of moderate to vigorous physical activity *(3)*. Additionally, these physical activity experiences should include a wide array of

From: *The Active Female*
Edited by: J. J. Robert-McComb, R. Norman, and M. Zumwalt © Humana Press, Totowa, NJ

skills and behaviors, be designed and implemented in a developmentally appropriate environment, and contribute to the children's enjoyment of participation in physical activity. Currently, it is estimated that more than 35 million children in our country participate in physical activity through engagement in organized youth sports *(4)*.

Active children require more energy to maintain normal growth and development. Energy balance is maintained when the energy one takes in is equal to the energy one expends. Individuals take in or consume energy through food and beverages they eat. Likewise, individuals expend or use energy through engagement in activity. Imbalance in the energy equation occurs when energy input does not equal energy output. Weight loss occurs when energy output is greater than energy input. Weight gain takes place when energy input is greater than energy output *(5)*.

Energy imbalance has contributed to the epidemic proportion of children in our country who are at risk of overweight and obesity. Recent estimates suggest that 16% or more than 9 million of our children and teenagers are overweight *(6)*. Sedentary lifestyles and poor dietary habits have negatively impacted many children's energy balance. Our high-tech society has in part contributed to children's lack of physical activity. Screen time (i.e., computers, television, and video games) often detracts from children's time that could be spent participating in physical activities *(7)*. Children's dietary habits are also significantly influencing our country's childhood obesity epidemic. Where and what children are eating have changed over the years. For example, our children are consuming more foods prepared outside of the home (preschool, school, and fast food). We have also witnessed a rise in the intake of calories children take in from snacks, a sharp increase in calories consumed from sweetened beverages, and an overall decrease in children's consumption of fruits and vegetables *(8)*. Portion sizes continue to increase in restaurants, grocery stores, and even vending machines. Expanded portion sizes contribute to our increase in daily total caloric intake *(9–11)*.

Understanding nutrition guidelines to provide energy needs for children is critical in today's world. Adhering to evidence-based guidelines will promote children's optimal growth and development as well as combat and prevent our current childhood obesity epidemic.

25.3. RESEARCH FINDINGS

25.3.1. Dietary Guidelines

Every 5 years, a report is issued by the United States Department of Agriculture (USDA) and the Department of Health and Human Services that serves as a guideline for healthy eating and living for the citizens of the United States. The Dietary Guidelines for Americans 2005 *(12)* differs from previous reports in that it is aimed toward individuals who are responsible for educating the public about nutrition, diet, and exercise; therefore, it is more scientific in its presentation than previous editions and may be accessed at http://www.health.gov/dietaryguidelines/dga2005/document/.

Many Americans are unaware of how much physical activity is required to manage weight, how many calories should individuals consume, or how to lose/gain weight in a healthy manner. Therefore, the Dietary Guidelines for Americans 2005 *(12)* outlines these very topics in its initial chapters. This report gives examples of caloric needs for men, women, and children separately, in addition to addressing the caloric needs of active versus sedentary individuals. With these thoughts in mind, the report discusses the importance of balancing caloric intake with the consumption of appropriate nutrients. Finally, physical activity is stressed as the mediator between the consumption of calories and the maintenance of a healthy weight.

25.3.2. MyPyramid

The Dietary Guidelines for Americans 2005 *(12)* was used along with the Dietary Reference Intakes from the National Academy of Sciences and current consumption patterns of food in the United States to create the MyPyramid Food Guidance System *(13)*. MyPyramid was created in 2005 by the USDA to replace the Food Guide Pyramid. The USDA replaced the existing Food Guide Pyramid because its dietary guidelines failed to create an individualized dietary program for Americans. MyPyramid differs from the Food Guide Pyramid in that it takes into account the age, sex, and activity level of an individual. This is significant because it provides individuals with the daily recommended servings from each of the six food groups as well as a recommended daily caloric intake based on specific characteristics of each individual. MyPyramid may be accessed at http://www.mypyramid.gov/index.html.

The MyPyramid Food Guidance System is made up of six food groups: *(1)* grains, *(2)* vegetables, *(3)* fruits, *(4)* milk, *(5)* meats and beans, and *(6)* fats, oils, and sweets. From each of these groups, MyPyramid provides a recommended food intake to an individual based on their allotted daily calorie intake. The educational framework for MyPyramid recommends that consumers have variety, proportionality, and moderation in the foods that they consume as well as in the activity in their life to achieve or maintain a healthy diet. MyPyramid also gives consumers basic nutritional guidelines to follow in order to educate them about the importance and benefits of having a healthy diet.

The MyPyramid Food Guidance System is used to inform children about nutritional guidelines and provide valuable health terms that they will need in order to develop a healthy lifestyle. In fact, the MyPyramid website includes an entire section entitled For Kids that may be accessed at http://www.mypyramid.gov/kids/index.html.

MyPyramid is also beneficial to children because it teaches them about having a well-rounded diet as well as developing specific dietary guidelines in their daily lifestyles. This is important because it causes children to be more cautious about the foods they consume on their personal calorie program. The daily caloric intake for females between the ages of 5 and 12 who are physically active ranges from 1400 to 2000 cal/day, whereas the daily caloric intake for males between the ages of 5 and 12 who are physically active ranges from 1400 to 2200 cal/day. Although these estimates might not be perfect for every child, they provide children and their parents with a strong base for nutrition which can then be used to create a healthy diet that encourages physical activity and a healthy lifestyle.

25.3.3. Discretionary Calories

An individual's total daily caloric intake is composed of "essential" and "discretionary" calories *(14)*. Essential calories are those we need to meet daily recommended nutrient guidelines. Discretionary, additional, or extra calories are those we need and use to meet energy needs. Consequently, active children have a larger allotment of discretionary calories as compared with sedentary children. Appendix 20 lists the estimated caloric requirements for children based on activity levels. As can be seen from Appendix 24, as caloric requirements increase, so does the allowance for discretionary calories.

Providing children guidance in regard to healthy food choices enables children to consume more discretionary calories. For example, modeling consumption of lower fat foods and beverages, selecting minimal to no added sugar groceries, instilling a wide array of fruits and vegetables in children's daily diet, and participating in regular physical activity will result in an increase in children's daily ingestion of discretionary calories *(8,14)*.

25.3.4. Family Meals

Throughout the last several decades, the structure and the makeup of families in our country have drastically changed. Currently, we have more families with both parents working, and a significantly great number of families are headed by a single parent *(15)*. In addition, we presently have 27 million children growing up in low-income families *(16)*. Limited financial resources often impact a low-income family's ability to provide healthy food choices; additionally, children from low-income families regularly participate in the US government's free and reduced breakfast and lunch programs *(17)*. Unfortunately, meals served via this program are at times of high fat and limited in nutritional value *(18)*. Moreover, the same families may reside in high-crime and at-risk neighborhoods that limit children's opportunities to participate in outdoor physical activities after school *(19)*. These changes in family life often influence the occurrence of families sitting down together and sharing meals. This trend is disturbing, as family meals have shown to be an integral component of nutritionally sound dietary intake *(15,20,21)*.

As children progress toward adolescence, family meals decline *(20,21)*. Movement away from family meals during adolescence is troubling because adolescence is a time of rapid growth when energy and nutritional needs are peaking. Studies examining the relationship between family meals and dietary intake during adolescence have collectively shown the positive correlation between frequency of family meals and adolescent's consumption of fruits, vegetables, grains, and calcium rich foods, and breakfast *(20,21)*. The same studies *(20,21)* also revealed the negative association between the frequency of family meals and increase in soft drink and snacking frequency. Family meals also may assist in minimizing unhealthy eating behavior in teens (i.e., binge eating and crash dieting). Research also suggests that the presence of at least one parent during mealtimes decreased the odds of children having poor consumption of fruits, vegetables, and dairy products *(21)*.

Given the positive impact family meals have on children and adolescent's diet and nutrition, it is critical that families attempt to dine more frequently together. This goal is challenging given current factors impacting families' time together (i.e., two parents working and single-parent household). Several viable approaches to increasing family meals have been offered by experts in the field *(15,20,21)*. These strategies are presented in Table 25.1.

Table 25.1
Strategies to enhance the frequency of family meals[a]

Set realistic goals for the number of meals shared per week
Identify meals that work best for your family
Share meal preparation and cooking responsibilities
Create a pleasant atmosphere around mealtime
Avoid controversial topics at mealtime
Keep meals easy and uncomplicated
Focus on nutrition
Have at least one adult present at mealtime

[a] Strategies were collectively presented in Story *(15)*, Newmark-Sztainer *(20)*, and Videon and Manning *(21)*.

25.3.5. Nutrients that Provide Energy

It is essential that children eat a diet that is composed of nutrients. Nutrients are vital to children's diets because they supply building blocks for the body, regulators, and energy that are needed to maintain health in the body. Three nutrients, carbohydrates, proteins, and fats, are the primary sources of energy to the body. The next section will describe the importance of these energy providers.

25.3.5.1. CARBOHYDRATES

Carbohydrates are found in the body in the form of simple sugars, complex carbohydrates, starches, and other substances such as glycogen and cellulose *(22)*. They consist of carbon, hydrogen, and oxygen molecules *(23)*. Carbohydrates are important nutrients in the body because they supply the fuel that is needed to provide energy to the cells. Simple sugars, or simple carbohydrates, can be found in the body in the form of monosaccharides and disaccharides *(24)*. Monosaccharides are composed of one sugar, whereas disaccharides are developed when two monosaccharides join together and make two sugars. These sugars are important to the body because they include the monosaccharides glucose, fructose, and galactose, as well as the disaccharides lactose, maltose, and sucrose. Of these simple sugars, glucose is the most important carbohydrate in the blood because it is used to provide fuel to many cells in the body and is the only fuel supply to the nervous system, brain, and red blood cells *(25)*. In comparison, complex carbohydrates are made of polysaccharides, which are made when many chains of glucose join together. From these polysaccharides, storage forms of glucose in plants and animals known as starches and glycogen can be found as well as in most types of fibers, which are indigestible carbohydrates that consist primarily of cellulose.

Because carbohydrates are a part of a healthy diet that supplies energy to cells and muscle tissues, the Dietary Guidelines for Americans 2005 suggests that carbohydrates are consumed as 45–65% of total calories in children's diets. Foods that are loaded with carbohydrates include grains and starches such as breads, pasta, beans, rice, potatoes, and cereal. Carbohydrates are also found in four of the basic food groups where fruits, vegetables, milk, and whole grain products not only provide energy to the body but the essential vitamins and minerals that it needs to function properly and remain healthy. By eating a diverse diet of carbohydrates that are within the context of a calorie-controlled program, children can reduce the risk of chronic disease in their future. However, overconsumption of carbohydrates, especially ones with added sugars, can cause extra calories to be consumed, which can provide children with a small quantity of essential nutrients and consequently lead to weight gain.

25.3.5.2. PROTEINS

Proteins are nutrients that supply energy to the body through the foods that we eat. Carbon, hydrogen, nitrogen, and oxygen are found in all proteins, whereas other proteins also contain iron, phosphorus, and sulfur *(26)*. They are important to a healthy lifestyle because they contain amino acids that are the building blocks for all the cells of the body. From these amino acids, we find two types: essential and nonessential. The diet that a child consumes must supply his or her essential amino acids because the body often fails to make them on its own. However, nonessential amino acids can be produced in the body by itself if the child's diet is adequate. Proteins are also vital to the body because they are used as enzymes, in membranes, as transport carriers, and they also produce some hormones *(27)*. They help build as well as repair tissue in the body, hold some responsibility in the maintenance of body fluids, and can be used as a backup energy source in the body if there are not enough carbohydrates and fats to supply the energy that is needed. But as a nutrient, proteins were not made to function as a main energy source. However, if there is not enough present in the body from the foods that we eat, serious health complications can occur.

The 2005 Dietary Reference Intakes recommended dietary allowance (RDA) suggests that children between the ages of 4 and 8 consume 19 g of protein per day, whereas students between the ages of 9 and 12 should consume 34 g of protein per day to meet adequate nutrition needs. Proteins are found in several types of foods with the best sources being found in cheese, eggs, fish, meat, and milk. Proteins can also be supplied to the body through consumption of cereal grains, legumes, nuts, and vegetables.

25.3.5.3. FATS/LIPIDS

Contrary to many beliefs in today's society, lipids need to and can be present in a person's daily diet in order to have a healthy lifestyle *(28)*. Lipids, more commonly known as fats, provide important energy sources and essential fatty acids to the body. These fatty acids are essential to an individual's diet because they provide carriers that help the body to absorb fat-soluble vitamins A, D, E, and K and carotenoids. Lipids act as insulators that control body temperature and help form protective cushioning around the eyeballs and the body's organs. They are also the building blocks for the development of membranes and play a substantial role in many biological functions of the body. Lipids can come from both plants and animals and are found in the form of triglycerides, phospholipids, and sterols. However, most of the fats that are consumed in our daily diets are triglycerides. Three types of fatty acids make up triglycerides. They include polyunsaturated fats, monounsaturated fats, and saturated fats. Each type of fatty acid contributes positively or negatively to one's cholesterol. Cholesterol is the waxy substance that is produced by the liver that comes from animal byproducts and can be defined by two main types. High-density lipoprotein (HDL) blood cholesterol provides positive benefits for one's health, whereas low-density lipoprotein (LDL) blood cholesterol provides negative effects to the overall health of an individual. With this knowledge, let us overview fatty acids that positively affect the health of the body.

Polyunsaturated fats are liquid at room temperature and are made up of omega-6 and omega-3 types. Omega-6 polyunsaturated fats are found in corn, soybean, and cottonseed oils that are used in products such as margarine, mayonnaise, and salad dressings. These fats help lower LDL cholesterol levels and reduce the risk of heart disease. Omega-3 polyunsaturated fats are found in fatty fish, with small amounts being found in walnut, canola, and soybean oils. Dark leafy greens also contain omega-3 polyunsaturated fats that help lower triglyceride levels and are important in maintaining cardiovascular health.

Monounsaturated fats are also liquid at room temperature and are found in olive, canola, and safflower oils. Many types of nuts, peanut butter, avocados, and olives are also good sources of monounsaturated fats. Monounsaturated fats lower LDL cholesterol levels and may help in the reduction of blood pressure and triglyceride levels as well as reducing the risk of heart disease. Although unsaturated fatty acids positively affect the health of the body, there are also fats that can be detrimental to one's diet.

Saturated fats are often solid at room temperature and include foods such as butter, cheese, and full-fat dairy products. They are also found in animal fats and palm and coconut oils. Saturated fats increase LDL cholesterol levels as well as the risk of unhealthy blood lipid levels. Because of these medical problems, the risk of heart disease also increases in these individuals. It is important to limit the consumption of fats in our daily diet, especially those that are saturated. Individuals should also do this with regard to trans fats.

Trans fats come from oils that have undergone chemical alterations in order to become hydrogenated. These fats are not healthy for individuals to have in their daily diets because they lower HDL ("good") cholesterol levels and raise LDL ("bad") cholesterol levels in the body. Foods that consist of trans fatty acids include stick margarine, shortening, french fries, and other deep-fried fast foods, packaged cookies and crackers, and processed snacks and sweets. Overconsumption of trans fat foods also increases the risk of heart disease and cancer and has drastically changed the amount of fat calories that are consumed in our daily diets.

The Dietary Guidelines for Americans 2005 recommends that children between the ages of 4 and 18 consume 25–35% of their calories from fat in order to maintain a healthy diet. If a child does not consume enough fat in its daily diets (less than 20% of calories), it can cause many unwanted health problems. These include changes in HDL cholesterol levels and triglycerides in the body, as well as health deficiencies that are related to the amounts of vitamin E and essential fatty acids that the body absorbs. However, consuming too much fat can be detrimental to one's health as well. High intake of saturated fats, trans fats, and cholesterol not only increases the risk of cardiovascular disease and health problems but also leads to weight gain that is caused by excessive calorie intake in one's diet. This is important because obesity is rapidly becoming an epidemic and serious health dilemma in the United States.

25.3.6. Vitamins, Minerals, Fibers, and Water

25.3.6.1. VITAMINS

Vitamins act to maintain a healthy body and are vital in many bodily processes. Vitamins are naturally occurring and can be found in vegetation as well as in animals. There are two types of vitamins: water soluble and fat soluble. The former are absorbed by the body and excreted through urine; whereas the latter are stored in fat for later use and eventually absorbed by bile acid (29). We need both types of vitamins to maintain our daily bodily processes, development, and growth. Consequently, it is critical that children's diets include foods that provide a wide array of vitamin supply.

Vitamins are found in many different types of food: milk contains vitamin A, tuna provides a source of vitamin D, and whole grain breads are a good source of vitamin E (30). According to the National Academy of Sciences (31), the recommended amounts of each vitamin differ based on sex and age, as well as for pregnant women. The effects of a vitamin-deficient diet are numerous and varied depending on the specific vitamin that one is lacking. Examples of vitamin-deficient symptoms include night blindness, rickets, leg cramps, depression, and decreased resistance to infection and irritability (30).

25.3.6.2. MINERALS

Minerals affect the body in a variety of ways and can be found in the earth, water, and food (29) and are then absorbed by the body. The body must have minerals to perform a plethora of tasks ranging from maintaining the acid–base balance and blood clotting to regulating the rhythm of the heart (30). Food sources of various minerals are milk and sardines for calcium, seafood and nuts for iron, table salt for sodium, and bananas and orange juice for potassium. Again, the recommended intake of each vitamin varies based on age, sex, and women's pregnancy. The National Academy of Sciences provides a table with the recommended doses of each vitamin based on these characteristics (31). If an individual consumes too much of a certain mineral, the adverse effects could be seen in symptoms such as kidney stones, increased thyroid-stimulating hormone, upset stomach, mental changes, diarrhea, low blood pressure, and decreased resistance to infection/illness (29). However, there are side effects to consuming an insufficient amount of vitamins as well. Deficiency symptoms could include bone pain, anemia, weakness, weight loss, irregular heartbeat, and possible hair loss (30). Research has shown that children who eat breakfast had higher consumption of most vitamins and minerals, as compared with nonbreakfast eaters (32). This finding corroborates the importance of promoting a variety of family meals.

25.3.6.3. FIBERS

Fiber refers to a specific type of carbohydrate that aids in the gastrointestinal processes including the excretion of fat (30). According to the Harvard School of Public Health, adults should consume 20–35 g of fiber per day; in addition, children over the age of 2 should have their age plus 2 g of fiber daily (33). Fiber is found in foods such as fruits, vegetables, nuts, whole grains, and legumes. A fiber-rich diet has been shown to decrease the risk of heart disease, type 2 diabetes, and diverticulitis. However, the consumption of too much fiber can lead to the loss of other equally important nutrients such as calcium, phosphorus, and iron; additionally, too much fiber will inevitably create intestinal discomfort *(30)*.

25.3.6.4. WATER

The importance of adequate consumption of water cannot be overstated. Water comprises 60% of total body weight and is necessary for most vital bodily activities *(34)*. Humans consume water through beverages, fruits, vegetables, eggs, some fish, and various other food sources. The amount of water required for proper functioning depends on the activity level of the individual; as activity increases, so does the need for additional water consumption. On average, consuming 64 oz of water daily is the minimum amount necessary to satisfy the body needs. However, consuming more than 64 oz is suggested and even required if strenuous activity is being performed.

Active children are at a higher risk for dehydration as compared with adults. This is particularly important during hot and humid summer months when many children are participating in outdoor sports and activities *(35)*. Children's sweat mechanisms are less efficient than adults. This inefficient system does not enable children to consistently relieve built-up body heat. Moreover, at times, children may be less likely to take in ample supplies of water during physical activity. It is imperative that youth sport coaches and adult supervisors of children's activity not only make water available to children but encourage, promote, and oversee children's regular consumption of water. Children should consume water prior to, during, and immediately following participation in sport and physical activity in hot environments. Dehydration may lead to cramps, heat exhaustion, and/or heat stroke. Children who report feeling dizzy or lightheaded or having a dry mouth are demonstrating early signs of dehydration and should receive immediate attention.

25.4. CONCLUSIONS

A nutritionally sound and balanced diet will assist in promoting children's optimal growth and development. The Dietary Guidelines for Americans 2005 *(12)* suggests that our daily diets consist of plenty of fresh fruits, vegetables, whole grains, and fat-free and/or low-fat milk; contain poultry, fish, lean meats, beans, eggs, and nuts; and are low in saturated fats, trans fats, cholesterol, salts, and sugars. These guiding principles include children as well as adults. To assist children in comprehending and implementing healthy well-balanced diets the USDA developed a children's worksheet that includes the six main food groups, and opportunities for children to use dietary recall to set goals and make sound food choices. Parents, caregivers, teachers, and youth sport coaches may find this worksheet a valuable tool when instructing children about healthy diets. The worksheet may be accessed at http://teamnutrition.usda.gov/resources/mpk_worksheet.pdf.

Parents and caregivers should model sound nutritional behaviors. Promoting family meals will benefit children's overall health. Although active children require additional energy needs, we as a society need to pay close attention to our country's everexpanding portion sizes. Ironically, we have millions of children participating in sport while our country is battling a childhood obesity epidemic. Daily diets that are founded on moderation, balance, and variety combined with participation in physical activity will enhance children's and adults' overall quality of life.

25.5. SCENARIO WITH QUESTIONS AND ANSWERS

25.5.1. Scenario

Aleshia is a 10-year-old African-American girl growing up in a low-income family. Aleshia lives with her grandmother, a younger brother, and elder sister and attends the neighborhood public school. The school nurse annually calculates children's body mass index (BMI). Aleshia's BMI places her at the 95th percentile for her age and sex, which classifies her as overweight. The school nurse has scheduled a home visit with Aleshia and her grandmother to discuss strategies to assist Aleshia with minimizing additional weight gain.

25.5.2. Questions

1. Identify potential barriers Aleshia and her family are facing in regard to healthy diet and participation ion physical activity.
2. Develop two to three strategies that the school nurse can suggest to Aleshia and her grandmother to prevent further weight gain for Aleshia.

25.5.3. Plausible Answers

1. Potential barriers encompass the following: (1) low income family, (2) participation in the federally sponsored school lunch program, and (3) living in a neighborhood that is not always conducive to children playing outside.
2. Possible strategies include but are not limited to the following: (1) introducing Aleshia and her grandmother to the MyPyramid Food Guide, (2) inquiring as to the regularity of family meals and trying to increase the number of family meals throughout the week, (3) suggesting simple changes in Aleshia's daily diet (i.e., substituting water for cola and sugar drinks, substituting fruit for high-caloric snacks), (4) monitoring Aleshia's choices from the school lunch menu, (5) teaching Aleshia and her family about portion sizes, and (6) attempting to increase daily physical activity; this may include inquiry into the school's physical education program as well as providing opportunities for safe play after school.

REFERENCES

1. Pate RR, Baranowski T, Dowda M, Trost SG. Tracking of physical activity in young children. *Med Sci Sports Exerc* 2001;3:S1–S10.
2. Centers for Disease Control and Prevention. Physical activity for everyone: the importance of physical activity. Division of Nutrition and Physical Activity, National Center for Chronic Disease Prevention and Health Promotion (2006). Available at: http://www.cdc.gov/nccdphp/dnpa/physical/importance/index.htm. Accessed May 28, 2006.
3. Strong WB, Malina RM, Blimkie CJR, Daniels SR, Dishman RK, Gutin B, Hergenroeder AC, Must A, Nixon PA, Pivarnik JM, Rowland T, Trost S, Trudeau F. Evidenced based physical activity for school-age youth. *J Pediatr* 2005;146:732–737.
4. Smoll FE, Smoll RE. History and current status. In: Smith FE, Smoll RE, eds. *Children and Youth in Sport: A Biopsychosocial Perspective*. Dubuque, IA: Kendall/Hunt; 2002:1–4.
5. McCardle WD, Katch FI, Katch VL. Optimal nutrition for the physically active person: making informed and healthful choices. In: McCardle WD, Katch FI, Katch VL, eds. *Sports & Exercise Nutrition*, 2nd ed. Philadelphia, PA: Lippincott, Williams & Wilkins; 2005:188–227.
6. Centers for Disease Control and Prevention. Overweight and obesity. Division of Nutrition and Physical Activity. Available at: http://www.cdc.gov/nccdphp/dnpa/obesity/. Accessed May 28, 2006.

7. Anderson RE, Crespo CJ, Barteltt SJ, Cheskin LJ, Pratt M. Relationship of physical activity and television watching with body weight and level of fatness among children; results from the Third National Health and Nutrition Examination Survey. *JAMA* 1998;279:938–942.
8. Gidding SS, Dennison BA, Birch LL, Daniels SR, Gilman MW, Lichtenstein AH, Rattya KT, Steinberger J, Stettler N, Van Horn L. Dietary recommendations for children and adolescents: a guide for practitioners. Consensus Statement from the American Heart Association. *Circulation* 2005;112:2061–2075.
9. Young LR, Nestle M. The contribution of expanding portion sizes to the U.S. obesity epidemic. *Am J Public Health* 2002;92(2):246–249.
10. Nielsen SJ, Popkin BM. Patterns and trends in food portion sizes, 1977–1998. *JAMA* 2003;289(4):450–453.
11. McConahy KL, Smiciklas-Wright H, Birth LL, Mitchell DC, Picciano MF. Food portions are positively related to energy intake and body weight in early childhood. *J Pediatr* 2002;140(3): 340–347.
12. Dietary Guidelines for Americans 2005. United States Department of Agriculture, United States Department of Health and Human Services joint website. Available at: http://www.health.gov/ dietaryguidelines/ dga2005/document/. Accessed June 6, 2006.
13. United States Department of Agriculture. MyPyramid.gov. Steps to a healthier you. Available at: http://www.mypyramid.gov/index.html. Accessed February 17, 2005.
14. United Sates Department of Agriculture. Discretionary calories. What are discretionary calories? Available at: http://www.mypyramid.gov/pyramid/discretionary_calories.html. Accessed June 4, 2006.
15. Story M. A perspective on family meals: do they matter? *Nutr Today* 2005;40(6):261–266.
16. National Center for Children Living in Poverty (2004, September). Rate of children in low-income families varies widely by state. Available at: http://www.nccp.org/ state_detail_demographic_ TX.html. Accessed October 5, 2005.
17. Center on Hunger and Poverty. (2003, July). The paradox of hunger and obesity in America. Available at: http://www.centeronhunger.org/pubs.html. Accessed October 5, 2005.
18. Physician's Committee for Responsible Medicine. (2004, August). School lunch report card. Available at: http://healthyschoollunches.org/reports/report2004_intro.html. Accessed November 16, 2005.
19. Lumeng JC, Appugliese D, Cabral HJ, Bradley RH, Zuckerman B. Neighborhood safety and overweight status in children. *Arch Pediatr Adolesc Med* 2006;160:25–31.
20. Newmark-Sztainer D. Eating among teens: do family mealtimes make a difference for adolescent's nutrition? *New Dir Child Adolesc Dev* 2006;111:91–105.
21. Videon TM, Manning CK. Influences on adolescent eating patterns: the importance of family meals. *J Adolesc Health* 2003;32:365–373.
22. Institute of Medicine of the National Academies. Dietary carbohydrates: sugars and starches. In: Spears GE, staff ed. *Dietary Reference Intakes for Energy, Carbohydrate, Fiber, Fat, Fatty Acids, Cholesterol, Protein, and Amino Acids*. Washington, DC: The National Academy Press; 2005: 265–338.
23. Karp S. Nutrition. In: Rosen E, Weinstein E, eds. *Teaching Children About Health: A Multidisciplinary Approach*, 2nd ed. Belmont, CA: Wadsworth/Thompson Learning; 2003:70–71.
24. Ahrens RA. Carbohydrates. In: *The World Book Encyclopedia*, vol. 3. Chicago, IL: World Book Inc.; 2003:203–204.
25. U.S. Dept. of Health and Human Services, U.S. Dept. of Agriculture. Carbohydrates. In: *Dietary Guidelines for Americans June 8, 2006*. Available at: http://www.health.gov/dietaryguidelines/ dga2005/document/html/chapter7.htm.
26. Ahrens RA. Protein. In: *The World Book Encyclopedia*, vol. 15. Chicago, IL: World Book Inc.; 2003:832–833.
27. Institute of Medicine of the National Academies. Proteins and amino acids. In: Spears GE, staff ed. *Dietary Reference Intakes for Energy, Carbohydrate, Fiber, Fat, Fatty Acids, Cholesterol, Protein, and Amino Acids*. Washington, DC: The National Academic Press; 2005.
28. U.S. Dept. of Health and Human Services, U.S. Dept. of Agriculture. Fats. In: *Dietary Guidelines for Americans 2005*. Available at: http://www.health.gov/dietaryguidelines/dga2005/document/ html/chapter6.htm. Accessed June 4, 2006.

29. Vitamins and Minerals: What You Should Know. American Academy of Family Physicians website. 2005. Available at: http://familydoctor.org/863.xml. Accessed June 6, 2006.
30. Hoeger W, Hoeger S. *Lifetime Physical Fitness & Wellness*. Englewood, CO: Morton Publishing Company; 2000.
31. Dietary Intakes Reference Table: Vitamins Table. Institute of Medicine Website. 2002. Available at: http://www.iom.edu/cms/3788/4574.aspx. Accessed June 6, 2006.
32. Dwyer JT, Ebzery MK, Nicklas TA, Feldman HA, Evans MA, Zive MM, Lytle LA, Montgomery DH, Clesi AL, Garceau A, Nichaman MZ. Do third graders eat healthful breakfasts? *Family Econ Nutr Rev* 1998;11(4):3–18.
33. Fiber: Start Roughing It. Harvard School of Public Health website. 2006. Available at: http://www.hsph.harvard.edu/nutritionsource/fiber.html. Accessed June 6, 2006.
34. Nutrition Fact Sheet: Water. Northwestern University: Feinberg School of Medicine website. 2005. Available at: http://www.feinberg.northwestern.edu/nutrition/factsheets/water.html. Accessed June 6, 2006.
35. Carey R. Tips for parents: food and drink for young athletes. *Sports Suppl Exch* 2000;13:77S.

26 Nutritional Guidelines and Energy Needs for the Female Athlete:
Determining Energy and Nutritional Needs to Alleviate the Consequences of Functional Amenorrhea Caused by Energy Imbalance

Jacalyn J. Robert-McComb

CONTENTS

26.1   LEARNING OBJECTIVES
26.2   INTRODUCTION
26.3   RESEARCH FINDINGS
26.4   CONCLUSIONS
26.5   NUTRITION COUNSELING SCENARIO

26.1. LEARNING OBJECTIVES

After completing this chapter, you should have an understanding of the following:

- The energy availability hypothesis and functional amenorrhea.
- Energy and nutritional intake estimations.
- Nutritional guidelines for the female athlete.
- The importance of hydration before, during, and after exercise.

26.2. INTRODUCTION

The prevalence of amenorrhea is elevated in women who restrict their diets and are intensely physically active *(1–4)*. Warren *(5)* was the first to suggest that menstrual disorders in female athletes are disrupted by an energy drain. Winterer et al. *(6)* proposed the hypothesis that failure to provide sufficient metabolic fuels to meet the energy requirements of the brain causes an alteration in brain function that disrupts the gonadotropin-releasing hormone (GnRH) pulse mechanism. Reproductive function critically depends on the pulsatile release of GnRH from GnRH neurons in the arcuate nucleus of the hypothalamus and on the consequent pulsatile release of luteinizing hormone (LH) from the pituitary *(7)*.

In the athletic female, energy drain can occur either by not taking in enough calories to meet the metabolic needs of the body or by overexercising and not compensating for the energy cost of the exercise by taking in additional calories. Hence, LH pulsatility can be suppressed by a combination of strenuous exercise and caloric restriction *(8)* if energy availability falls below a critical threshold. This has been termed "functional amenorrhea," because it is a functional problem (altered hormonal patterns) rather than an anatomical problem *(1)*. The restoration of normal menstrual cycling has been demonstrated to reoccur when the individual's threshold for energy availability is sequentially met *(9)*.

A series of well-controlled studies by Loucks at Ohio University *(7–11)* have demonstrated that normal menstrual cycling is altered if there is a restriction in energy availability (operationally defined and behaviorally controlled as dietary energy intake minus energy expenditure during exercise). As so defined, the term energy availability refers to the behavior of the subjects and not to the cellular availability of metabolic fuels inside of them. However, it would be speculated that there would be a correlation between behaviorally controlled restricted dietary energy intake in humans and cellular availability of oxidizable metabolic fuels because reproductive function in mammals does depend on the cellular availability of oxidizable metabolic fuels *(12)*.

The adult female human brain oxidizes approximately 80 g of glucose each day at a continuous rate, and this must be provided daily by dietary carbohydrate, because the brain's rate of energy expenditure can deplete liver glycogen stores in less than 1 day *(13)*. Moderate exercise oxidizes that much glucose in an hour *(7)*. On the basis of respiratory quotients measured during exercise training *(11)*, 62–88% of the energy expended during exercise was derived from carbohydrates, principally glucose. Thus, the special demand that aerobic exercise places on glucose stores suggests that the failure of women to sufficiently increase dietary glucose intake, specifically, in compensation for the energy cost of the exercise, may lower glucose availability to the brain below a critical threshold necessary for the normal neuroendocrine function of the thyroid, reproductive, and other endocrine axes *(14)*. Loucks et al. *(8)* found that exercise energy expenditure may compromise brain glucose availability less than the corresponding amount of dietary energy restriction alone. In their experiment *(8)*, they found that skeletal muscle derived much less energy from carbohydrate oxidation in the deprived energy availability treatment than in the balanced energy availability treatment (49 versus 73%). This alteration in fuel utilization during the deprived energy state conserved approximately 70% of the brain's daily glucose requirement. Their conclusion from this study was that prolonged exercise had no disruptive effect on LH pulsatility in women apart from the impact of its energy cost on energy availability or glucose availability and that LH pulsatility is disturbed less by exercise energy expenditure than by dietary energy restriction alone *(8)*.

Loucks and Thuma *(7)* found that LH pulsatility was disrupted abruptly at a threshold of energy availability not higher than 30 kcal/kg of lean body mass per day in regularly menstruating, habitually sedentary young women of normal body composition. They also found that the disruptive effects of subthreshold energy availability were bimodal, with substantially larger effects occurring in subjects with the shortest luteal phases. Not all females have the same threshold for energy availability; but if this threshold falls below a critical threshold, hormonal alterations will result. Their results suggested that women with short luteal phases (11 days) might be at a higher risk than others for the suppression of ovarian function and skeletal demineralization by energy deficiency. Lastly, they found that the incremental effects of restricted energy availability on LH pulse frequency and amplitude most closely resembled those on plasma glucose, cortisol, and growth hormone. However, this association does not imply that any of the metabolic substrates and hormones are involved in the mechanism mediating the effects of energy availability on LH pulsatility.

In conclusion, during extended periods of reduced energy availability, the body prioritizes by fueling the activities necessary for survival such as thermoregulation and locomotion; therefore, less critical processes such as reproductive function may be compromised. The restoration of normal menstrual cycling has been demonstrated to reoccur when the individual's threshold for energy availability is sequentially met *(9)*. However, observations suggest that appetite may be an inadequate indicator of energy balance during athletic training, just as thirst is an insensitive indicator of water balance during athletic competition. Athletes may need to eat by discipline rather than by appetite during training to prevent reproductive disorders *(8)*.

26.3. RESEARCH FINDINGS

26.3.1. Estimating Energy and Nutritional Intake

Studies consistently show that female athletes are not consuming enough energy to support their activity levels *(15)*. Research with elite female swimmers, using the doubly labeled water technique, noted that total daily energy increased to 5593 kcal daily during high-volume training. This is the highest reported caloric expenditure of female athletes reported *(16)*. However, their intake averaged only 3136 kcal, implying a negative energy balance. Energy intake of well-trained female athletes ranges from 1931 to 3573 kcal *(17)*. Consequently, their intake of essential vitamins and minerals is lower than the recommended daily allowance *(18)*. Female athletes' diets have been found to be low in iron, calcium, zinc, vitamin B_6, and folate *(19)*. Low energy and nutrient intake places these athletes at a greater risk for nutrition-related disorders such as amenorrhea, osteoporosis, iron deficiency anemia, and eating disorders.

Traditionally, recommendations for energy requirements have been based on self-recorded estimates (e.g., diet records) of food intake. However, it is thought that these records are misleading *(20)*. The percentage of people who underestimate their food intake ranges from 10 to 45% *(21)*. Since the advent of the doubly labeled water technique for measuring total energy expenditure, scientists have established energy requirements based on the actual measurement of total energy expenditure in free-living individuals *(20)*.

It has been found that some of the commonly used formulas, such as the Harris–Benedict equation, to estimate energy requirements are not accurate and underestimate or overestimate requirements *(22,23,24)*. The Institute of Medicine has developed caloric expenditure equations depending on activity level, gender, and age, which can be found in Appendices 18 and 19 *(22)*. Appendix 18 lists physical activity level categories and walking equivalents to use in the equations from the Institute of Medicine to estimate caloric expenditure, which can be found in Appendix 19. Appendix 20 provides general estimates of calorie requirements (in kilocalories) for females at three different levels of physical activity using Institute of Medicine equations *(22)*. These estimations should only be used as a guideline to determine energy balance in conjunction with close monitoring of weight status and a nutritional assessment.

Table 26.1 lists ways that an exercise nutritionist can help physically active individuals, particularly competitive athletes, achieve energy balance. Although a self-reported food nutritional assessment may misrepresent total caloric intake because of underreporting, it does provide valuable information to aid in nutritional counseling.

A nutritional assessment consists of collecting and evaluating a number of types of information. These include a brief patient history, results of a physical examination (performed by a physician), anthropometric data such as height, weight, body mass index, and percentage body fat, and finally, some biochemical data that are obtained through blood evaluation. This laboratory testing focuses on serum proteins such as albumin, prealbumin, and retinol-binding protein, creatinine height index, and overall

Table 26.1
How exercise nutritionists can help female athletes maintain energy balance

---

Athletes should be educated about energy requirements for their sport and the role of food in fueling the body. Female athletes should be educated about the Female Athlete Triad and the long-term health consequences of inadequate energy intake. Unrealistic weight and body composition goals should be discouraged.

The athlete's typical dietary and supplement intake during training, competition, and the off-season should be assessed. This assessment should be used to provide appropriate recommendations for energy and nutrient intakes for the maintenance of good health, appropriate body weight and composition, and optimal sport performance throughout the year.

Body size and composition of an athlete should be assessed for the determination of an appropriate weight and composition for the sports in which she participates. Minimum body composition for good health for the female athlete is 12%. Provide the athlete with nutritionally sound techniques for maintaining an appropriate body weight and composition without the use of severe diets or nutritionally unbalanced macronutrient choices.

The fluid intake and weight loss of athletes during exercise should be assessed. Appropriate recommendations regarding total fluid intake before, during, and after exercise should be made based on this assessment and the most current scientific literature.

Carefully evaluate any vitamin/mineral or herbal supplements, ergogenic aids, or performance-enhancing drugs an athlete wants to use. These products should be used only after a careful review of their legality and the current literature pertaining to the ingredients listed on the product label. Caution should be used in recommending these products, and these products should only be recommended after evaluating the athlete's health, diet, nutrition needs, current supplement and drug use, and energy requirements.

Source: Adapted from the American College of Sports Medicine *(25)*.

immune status. The initial review process serves a number of functions. It allows for the identification of nutritional and medical risk factors, existing nutritional deficiencies, and past nutritional problems.

The next step is a dietary assessment. The purpose of the dietary assessment is to identify a person's eating habits and to estimate their average daily nutrient intake. Through a variety of methods, information should be obtained on the amount and variety of foods eaten. The simplest method of assessment is to have the individual keep a daily dietary intake record. Because intake tends to vary from day to day, a 3-day food record is more accurate than the 24-h recall. Many nutritionists even recommend a 7-day food record. However, these methods only reflect the person's current diet, not eating habits established over a long period of time. It is important to recognize that self-reported estimates of food intake are biased and many times do not provide an accurate estimation of food intake; therefore, these should only be used as a guide. A food frequency questionnaire is also helpful to determine eating habits. Appendices 21 and 22 provide examples of nutrition and food frequency questionnaires, respectively, that can be used to gather information.

Another method of obtaining information on dietary intake is to use a computer program. There are numerous nutritional assessment software programs on the market—most are under $60.00 each. Most of these programs contain more than 23,000 food items and are upgradable to allow new food items to be entered by the user. One such program is Diet Analysis Plus© for Windows. This program allows for a 7-day food intake per individual. It takes into account height, weight, and activity level. It also computes daily and weekly values for recommended daily allowance (RDA) and energy expenditure. In addition, it gives a fairly specific breakdown of nutrients and has the ability to generate graphs, charts, and reports. These programs allow a nutritional novice to enter their own data concerning their individual diet and nutrition and receive easy-to-understand information. The use of nutritional assessment software is relatively widespread because of its low cost and availability. For more information and access to the FDA food database, contact the US Food and Drug Administration at http://www.FDA.gov.

Even though the determination of energy needs is relatively straightforward, behavior change is complex and not so readily understood. For a more complete understanding of the variables related to behavior change, the authors refer you to behavior change links (http://www.social-marketing.com/BCLinks.html). Additionally, Chap. 8 lists resources for helping athletes with eating disorders (*see also* Appendix 9). If the athlete is exhibiting signs and symptoms of energy imbalance, such as weight loss, amenorrhea, loss of concentration, and irritability, the athlete should be encouraged to seek professional counseling.

26.3.2. Nutritional Guidelines for the Female Athlete

Meeting energy needs is the first priority for the female athlete. Studies consistently state that female athletes do not take in enough calories. Energy balance is defined as a state when energy intake (the sum of energy from foods, fluids, and supplement products) equals energy expenditure (the sum of energy expended as basal metabolism, the thermic effect of food, and any voluntary physical activity). Research in exercise nutrition indicates that the large number of teenagers and adults, including competitive athletes, who exercise regularly to keep fit do not require additional nutrients beyond those obtained through the regular intake of a nutritionally well-balanced diet if energy needs are being met *(17)*. A joint position statement by the American College of Sports Medicine, the American Dietetic Association, and the Dietitians of Canada *(25)* stated that a diet substantially different from that recommended in the Dietary Guidelines for Americans *(26)* or the Nutrition Recommendations for Canadians *(27)* (55–58% of energy from carbohydrate, 12–15% of energy from protein, and 25–30% of energy from fat) is not needed for athletes. It is generally recommended that an athlete's diet should be composed of approximately 55–60% carbohydrate, 20–25% fat, and 12–15% protein *(25)*. Although, some experts in sports nutrition recommend broader ranges of the macronutrients depending on the needs of the athlete *(17)*.

The Dietary Guidelines for Americans is published jointly every 5 years by the Department of Health and Human Services (HHS) and the Department of Agriculture (USDA). The 2005 Dietary Guidelines for Americans *(28)* was released January 12, 2005, and can be downloaded at (http://www.health.gov/dietaryguidelines/dga2005/document/) or purchased from the US Government Printing Office by calling toll-free at (866)512-1800. The Dietary Guidelines describe a healthy diet as one that (1) emphasizes fruits, vegetables, whole grains, and fat-free or low-fat milk and milk products; (2) includes lean meats, poultry, fish, beans, eggs, and nuts; and (3) is low in saturated fats, trans fats, cholesterol, salt (sodium), and sugar. Although there are general food group categories, there are specific recommendations based on age, gender, and activity level. The USDA provides a website in which the pyramid plan can be customized (http://www.mypyramid.gov/mypyramid/index.aspx). Appendix 23 outlines the six general food group categories, recommendations, and major nutrients supplied by each category.

Appendix 24 lists the suggested amounts of food to consume from the basic food groups, subgroups, and oils to meet recommended nutrient intakes at 12 different calorie levels. Using the website provided by the USDA, recommendations for active females of age 19–30 are listed in Table 26.2 (*see* http://www.mypyramid.gov/mypyramid/index.aspx for specific guidelines for age, gender, and activity level).

A basic premise of the Dietary Guidelines is that food guidance should encourage individuals to achieve the most recent nutrient intake recommendations of the Institute of Medicine, referred to collectively as the Dietary Reference Intakes (DRIs). Tables of the DRIs are provided at http://www.iom.edu/Object.File/Master/21/372/0.pdf for all age groups and can be found in Appendices 25–29. Table 26.3 lists the nutrient intake levels recommended by the Institute of Medicine for females 19–30 years of age *(22)*. These nutrient intake levels should be achieved if there is macronutrient balance (approximately 55–60% carbohydrate, 20–25% fat, and 12–15% protein) in the female athlete's diet and the foods that are chosen are nutrient dense.

Carbohydrate intake is important for the athlete (male or female). Their diet should contain at least 55–60% of calories as carbohydrates, predominantly starches from fiber-rich, unprocessed grains, fruits, and vegetables. According to the joint position statement on nutrition and athletic performance, high-carbohydrate diets (above 60%) are not advocated for most athletes; however, specific recommendations must be made on body size, weight, and the sport being performed *(25)*. For athletes undergoing heavy endurance training, 10 g/kg of body mass is recommended which is approximately 65%

Table 26.2
Example of specific recommendations for physically active females from 19 to 30 years from the US Department of Health and Human Services and the US Department of Agriculture, Dietary Guidelines for Americans

Food group	Daily amount	Additional suggestions
Grains	8 ounces	Aim for at least 4 whole grains a day
Vegetables	3 cups	Dark green vegetables = 3 cups weekly
		Orange vegetables = 2 cups weekly
		Dry beans and peas = 3 cups weekly
		Starchy vegetables = 6 cups weekly
		Other vegetables = 7 cups weekly
Fruits	2 cups	
Milk	3 cups	
Meat and beans	6.5 ounces	
Oils and discretionary calories	Aim for 7 teaspoons of oils a day	Limit your extras (extra fats and sugars) to 360 cal

Note: These results are based on an estimated 2400-cal pattern. This caloric level is only an estimate of needs based on age, gender, and activity level. Body weight should be monitored to adjust caloric intake (http://www.mypyramid.gov/mypyramid/index.aspx).

Source: US Department of Health and Human Services, US Department of Agriculture, Dietary Guidelines for Americans (28).

of intake from carbohydrate (17). However, general recommendations for carbohydrates range between 6 and 10 g/kg of body mass per day (17,25). Even though the fuel burned during exercise depends on the intensity and duration of the exercise, the sex of the athlete, and prior nutritional status, an increase in the intensity of the exercise will increase the contribution from carbohydrate. A low-carbohydrate diet rapidly compromises energy reserves for vigorous physical activity or regular training. As the length of the exercise continues, the source of the carbohydrate may shift from the muscle glycogen pool to circulating blood glucose, but if the blood glucose cannot be maintained, the intensity of the exercise will decrease (29). Additionally, carbohydrate plays an important role as a protein sparer during exercise. Carbohydrate availability inhibits protein catabolism in exercise (30). If an athlete consumes 60–65% of their kilocalories from carbohydrate, and energy balance is being maintained, sufficient muscle glycogen stores should be maintained from day to day.

Protein requirements are slightly increased in highly active people. Protein requirements for endurance athletes are 1.2–1.4 g/kg body mass per day, whereas those for resistance-trained and strength-trained athletes may be as high as 1.6–1.7 g/kg of body mass per day (25). Acceptable levels for protein intake for physically active persons may range from 10 to 35% of calories (17).

Fat intake should not be restricted provided that the fat intake is low in saturated fats and trans fats; there is no benefit in consuming a diet with less than 15% of energy from fat as compared with 20–25% (25). For athletes, McCardle et al. (17) stated that the acceptable range of lipid intake is from 10 to 35% of caloric intake. Fat

Table 26.3
Nutrient intakes recommended per day by the Institute of Medicine (IOM)[a] for females 19–30 years of age

Nutrient	IOM recommendations for females 19–30
Protein, g	RDA: 46
Protein, %kcal	AMDR: 10–35
Carbohydrate, g	RDA: 130
Carbohydrate, %kcal	AMDR: 45–65
Total fat, g	–
Total fat, %kcal	AMDR: 20–35
Saturated fat, g	–
Saturated fat, %kcal	ALAP[b]
Monounsaturated fat, g	–
Monounsaturated fat, %kcal	–
Polyunsaturated fat, g	–
Polyunsaturated fat, %kcal	–
Linoleic acid, g	AI: 12
Alpha-linolenic acid, g	AI: 1.1
Cholesterol, mg	ALAP
Total dietary fiber, g	AI: 28[c]
Potassium, mg	AI: 4700
Sodium, mg	AI: 1500, UL: < 2300
Calcium, mg	AI: 1000
Magnesium, mg	RDA: 310
Copper, mg	RDA: 0.9
Iron, mg	RDA: 18
Phosphorus, mg	RDA: 700
Zinc, mg	RDA: 8
Thiamin, mg	RDA: 1.1
Riboflavin, mg	RDA: 1.1
Niacin equivalents, mg	RDA: 14
Vitamin B_6, mg	RDA: 1.3
Vitamin B_{12}, ug	RDA: 2.4
Vitamin C, mg	RDA: 75
Vitamin E (AT)	RDA: 15.0
Vitamin A, ug (RAE)	RDA: 700

[a] Recommended intakes for adult females of age 19–30: RDA, recommended dietary allowance; AI, adequate intake; AMDR, acceptable macronutrient distribution range; UL, upper limit.

[b] As low as possible while consuming a nutritionally adequate diet.

[c] Amount listed is based on 14 g dietary fiber/1000 kcal.

is important in the diets of athletes, as it provides energy, fat-soluble vitamins, and essential fatty acids.

For a more in-depth discussion of the specific nutritional recommendations for the physically active person and nutritional considerations for intense training and sports competition, the authors recommend McCardle et al.'s *Sports and Exercise Nutrition (17)*. Table 26.4 summarizes key points regarding nutritional recommendations for the physically active person.

Table 26.4
Summary of recommendations for macronutrients and energy intake for the physically active female

---

The Food Guide Pyramid provides broad recommendations for healthful nutrition for the physically active individual. Diets should emphasize fruits and vegetables, cereals and whole grains, nonfat and low-fat dairy products, legumes, nuts, fish, poultry, and lean meats. Female athletes' diets have been found to be low in iron, calcium, zinc, vitamin B_6, and folate. They should make sure that their diet contains foods that contain these vitamins and minerals.

Intensity of daily physical activity largely determines energy intake requirements. Studies consistently show that female athletes are not consuming enough energy to support their activity levels. Energy intakes of well-trained female athletes range from 1931 to 3573 kcal. However, during high-volume training, such as in swimming, total daily energy may increase to 5593 kcal daily. Low energy and nutrient intake places athletes at a greater risk for nutrition-related disorders such as amenorrhea, osteoporosis, iron deficiency anemia, and eating disorders.

Precise recommendations do not exist for daily lipid and carbohydrate intake. Fat intake should not be restricted provided that the fat intake is low in saturated fats and trans fats; there is no benefit in consuming a diet with less than 15% of energy from fat as compared with 20–25%. Acceptable lipid intakes for physically active individuals range from 20 to 35% of caloric intake.

Carbohydrate intake is important for the physically active person. General recommendations for carbohydrates range between 6 and 10 g/kg of body mass per day. This range represents approximately 55–65% carbohydrate intake. Carbohydrates should be predominantly starches from fiber-rich, unprocessed grains, fruits, and vegetables. A low-carbohydrate diet rapidly compromises energy reserves for vigorous physical activity or regular training. Successive days of hard training gradually deplete carbohydrate reserves, even when maintaining the recommended carbohydrate intake. This could lead to "staleness," making continued training more difficult.

Protein requirements are slightly increased in highly active people. Protein requirements for endurance athletes are 1.2–1.4 g/kg of body mass per day, whereas those for resistance and strength trained athletes may be as high as 1.6–1.7 g/kg of body mass per day. According to the Dietary Reference Intakes, acceptable macronutrient distribution ranges of protein for adults are 10–35%.

Excessive sweating during exercise causes loss of body water and related minerals. Mineral loss should be replaced following exercise through well-balanced meals. Athletes should be well hydrated before beginning to exercise. During exercise, optimal hydration can be facilitated by drinking 150–350 mL (6–12 oz) of fluid at 15–20-min intervals, beginning at the start of the exercise *(31,32)*. Consuming up to 150% of the weight lost during an exercise session may be necessary to cover losses in sweat and urine excretion.

26.3.3. The Importance of Hydration Before, During, and After Exercise

Athletes should be well hydrated before beginning to exercise. In addition to drinking generous amounts of fluid in the 24 h before an exercise session, the ACSM and NATA recommend drinking 400–600 mL (16–24 oz) of fluid 2–3 h before the exercise *(31,32)*. During exercise, optimal hydration can be facilitated by drinking 150–350 mL (6–12 oz) of fluid at 15- to 20-min intervals, beginning at the start of the exercise *(31,32)*. In most cases, athletes do not consume enough fluids during exercise to balance fluid losses and thus complete their exercise sessions dehydrated to some extent. Consuming up to 150% of the weight lost during an exercise session may be necessary to cover losses in sweat and urine excretion *(33)*.

26.4. CONCLUSIONS

For female athletes, energy balance must be maintained for optimum health and athletic performance. Historically, female athletes, competing in sports where leanness increases performance, have not taken in enough calories to meet their energy needs. A joint position statement by the American College of Sports Medicine, the American Dietetic Association, and the Dietitians of Canada *(25)* states that a diet substantially different from that recommended in the Dietary Guidelines for Americans *(26)* or the Nutrition Recommendations for Canadians *(27)* is not needed for athletes. It is recommended that an athlete's diet should consist of nutrient-dense food and beverages within and among the basic good groups while choosing foods that limit the intake of saturated and trans fats, cholesterol, added sugars, salt, and alcohol (55–60% carbohydrate, 20–25% fat, and 12–15% protein). Athletes should also be well hydrated before beginning to exercise and should drink enough fluid during and after exercise to balance fluid losses.

26.5. NUTRITION COUNSELING SCENARIO

26.5.1. Scenario

You are a sports nutritionist for a University Female Cross Country Track Team. One of your athletes has come to you asking your advice about the benefits of a high-protein, low-fat, low-carbohydrate diet.

26.5.2. Question

1. What advice would you give her?

26.5.3. Plausible Answer

1. This type of diet is not recommended for achieving optimal performance levels. It is recommended that an athlete's diet should consist of nutrient-dense food and beverages within and among the basic good groups while choosing foods that limit the intake of saturated and trans fats, cholesterol, added sugars, salt, and alcohol (55–60% carbohydrate, 20–25% fat, and 12–15% protein).

REFERENCES

1. Loucks A, Horvarth S. Athletic amenorrhea: a review. *Med Sci Sports Exerc* 1985;17(1):56–72.
2. Otis C, Drinkwater B, Johnson M, et al. American College of Sports Medicine position stand on the female athlete triad. *Med Sci Sports Exerc* 1997;29:i–ix.
3. Sundgot-Borgen J, Torstveit MK. Prevalence of eating disorders in elite athletes is higher than in the general population. *Clin J Sports Med* 2004;14:25–32.

4. Torstveit MK, Sundgot-Borgen J. The female athlete triad: are elite athletes at increased risk? *Med Sci Sports Exerc* 2005;37(2):184–193.
5. Warren MP. The effects of exercise on pubertal progression and reproductive function in girls. *J Clin Endocrinol Metab* 1980;51:1150–1157.
6. Winterer J, Cutler GB Jr, Loriaux DL. Caloric balance, brain to body ratio, and the timing of menarche. *Med Hypotheses* 1984;15:87–91.
7. Loucks A, Thuma J. Luteinizing hormone pulsatility is disrupted at a threshold of energy availability in regularly menstruating women. *J Clin Endocrinol Metab* 2003;88(1):297–311.
8. Louks AB, Verdun M, Heath EM. Low energy availability, not stress of exercise, alters LH pulsatility in exercising women. *J Appl Physiol* 1998;84(1):37–46.
9. Loucks A, Callister R. Induction and prevention of low-T3 syndrome in exercising women. *Am J Physiol* 1993;264:R924–R930.
10. Loucks A, Laughlin J, Mortola L, et al. Hypothalamic-pituitary-thyroidal function in eumenorrheic and amenorrheic athletes. *J Clin Endocrinol Metab* 1992;75(2):514–518.
11. Loucks A, Heath E. Induction of low-T3 syndrome in exercising women occurs at the threshold of energy availability. *Am J Physiol* 1994;66:R817–R823.
12. Wade GN, Schneider JE. Metabolic fuels and reproduction in female mammals. *Neurosci Biobehav Rev* 1992;16:235–272.
13. Bursztein S, Elwyn DH, Askanazi J, et al. *Energy Metabolism, Indirect Calorimetry, and Nutrition*. Baltimore, MD: Williams & Wilkins, 1989:59.
14. I'Anson H, Foster DL, Foxcroft GR, et al. Nutrition and reproduction. *Oxf Rev Reprod Biol* 1991;13:239–311.
15. Pate RR, Sargent RG, Baldwin C, et al. Dietary intake of women runners. *Int J Sports Med* 1990;11:461–466.
16. Trappe TA, Gastaldelli A, Jozsi AC, et al. Energy expenditure of swimmers during high volume training. *Med Sci Sports Exerc* 1997;29:950–954.
17. McCardle K, Katch F, Katch F. *Sports and Exercise Nutrition*, 2nd ed. Philadelphia, PA: Lippincott Williams and Wilkins, 2005.
18. Deuster PA, Kyle SB, Moser PB, et al. Nutritional survey of highly trained women runners. *Am J Clin Nutr* 1986;44:954–962.
19. Rudd J. *Nutrition and the Female Athlete*. Boca Raton, FL: CRC Press, 1996:11–19.
20. Frary C, Johnson R. Components of energy expenditure. In: Mahan LK, Escott-Stump S, eds. *Krauses's Food, Nutrition, & Diet Therapy*, 11th ed. Philadelphia, PA: Saunders, 2004:21–36.
21. Johnson RK. What are people really eating, and why does it matter? *Nutr Today* 2000;35(2):40–46.
22. Institute of Medicine (U.S.), Standing Committee on the Scientific Evaluation of Dietary Reference Intakes. *Dietary Reference Intakes for Energy Carbohydrate, Fiber, Fat, Fatty Acids, Cholesterol, Protein, and Amino Acids*. Washington, DC: National Academy Press, 2002.
23. Harris JA, Benedict FG. *A Biometric Study of Basal Metabolism in Man*, Pub. No. 279. Washington, DC: Carnegie Institute of Medicine, 1919.
24. Daly JM, Heymsfield SB, Head CA, et al. Human energy requirements: overestimation by widely used prediction equations. *Am J Clin Nutr* 1985;42:1170–1174.
25. American College of Sports Medicine, American Dietetic Association, and Dietitians of Canada. Joint position statement. Nutrition and athletic performance. *Med Sci Sports Exerc* 2000;32:2130–2145.
26. US Department of Agriculture, Human Nutrition Information Service, Dietary Guidelines Advisory Committee. Report of the Dietary Guidelines Advisory Committee on the Dietary Guidelines for Americans, 1995.
27. Authority of the Minister of Health and Welfare. *Nutrition Recommendations*. The Report on the Scientific Review Committee. Ottawa, Canada: Health and Welfare Canada, 1990.
28. US Department of Health and Human Services, US Department of Agriculture, Dietary Guidelines for Americans, Home and Garden Bulletin No. 232, 2005.
29. Coyle EF, Coggan AR, Hemmert MK, et al. Muscle glycogen utilization during prolonged strenuous exercise when fed carbohydrate. *J Appl Physiol* 1986;61:165–172.
30. Wagenmakers AJ, Beckers EJ, Brouns F, et al. Carbohydrate supplementation, glycogen depletion, ands amino acid metabolism. *Am J Physiol* 1991;260:E883.

31. American College of Sports Medicine. Position stand: exercise and fluid replacement. *Med Sci Sports Exerc* 1996;28:i–vii.
32. Casa DJ, Armstrong LE, Hillman SK, et al. National athletic trainer's association position statement. Fluid replacement for athletes. *J Athl Train* 2000;35(2):212–224.
33. Shirreffs SM, Taylor AJ, Leiper JB, et al. Post-exercise rehydration in man: effects of volume consumed and drink sodium content. *Med Sci Sports Exerc* 1996;28:1260–1271.

27 Ergogenic Aids and the Female Athlete

Jacalyn J. Robert-McComb and Shannon L. Jordan

CONTENTS
- 27.1 LEARNING OBJECTIVES
- 27.2 INTRODUCTION
- 27.3 RESEARCH FINDINGS
- 27.4 CONCLUSIONS
- 27.5 SCENARIO WITH QUESTION AND ANSWER

27.1. LEARNING OBJECTIVES

After completing this chapter, you should have an understanding of the following:

- Reasons why female athletes take supplements.
- Various supplements female athletes are likely to take.
- Ergogenic and ergolytic effects of these supplements.
- Variations within certain supplements and standard dosages.

27.2. INTRODUCTION

Many things can be considered ergogenic. Ergogenic aids, by definition, are items or substances that enhance performance. Female athletes tend to choose their supplements for different reasons than their male counterparts. Many collegiate female athletes report taking supplements "for their health, to make up for an inadequate diet, or to have more energy" *(1,2)*. Interestingly, although many athletes report using energy drinks and calorie replacement bars or drinks, most of them did not consider these products to be supplements *(2)*.

27.3. RESEARCH FINDINGS

27.3.1. What Are Female Athletes Taking?

According to Kristiansen et al. *(1)* and Froiland et al. *(2)*, female athletes use energy drinks and carbohydrate/meal replacement products the most. Protein products, amino acids, creatine, fat burners, caffeine, multivitamins, iron, and calcium are some of the most frequently used products. Neither study reported female steroid use; however, Congeni and Miller *(3)* state female adolescent steroid use to be up to 2.5%. Other researchers have found that 5.1% of middle school female students have reported steroid use by the age of 12 *(4)*.

27.3.2. Protein and Amino Acids

Female athletes tend to take their protein supplements as powders, as bars, or in meal replacements *(1,2)*. Some female athletes use amino acid supplements *(2)*. When classified by the type of sport, athletes (men and women) who play a sport involving a mixture of aerobic and anaerobic energy systems consume the most protein supplements *(2)*.

Aside from listing protein powers, protein bars, and amino acids, the most prevalent protein supplement listed by females were meal replacement drinks. Many of these meal replacements contain carbohydrates, vitamins, and minerals, as well as protein. In some cases, they will also contain other supplements, some of which may be banned in sport competitions. Typically, the female athlete should follow the protein requirement guidelines discussed in Chap. 26. Most athletes who consume more protein than necessary to maintain a positive nitrogen balance do not see additional benefits *(5,6)*. In certain situations, when there is excessive protein intake, the calcium:protein ratio may be out of balance and could lead to bone loss *(7)*. This will be covered more in Sect. 27.3.8.

When asked why they take protein supplements, females responded "for enhanced recovery, just like the taste, provide more energy, to meet nutritional needs, enhanced performance, greater muscle strength, and don't know" *(1)*. When the same athletes were asked whether they had questions about protein supplement use, there were several questions or concerns: increasing lean mass and losing weight, adverse effects on the kidneys, positive versus negative effects, how much to consume, and whey versus soy protein.

27.3.2.1. WHEY VERSUS SOY PROTEIN

When comparing the two, it is important to have an understanding of the essential amino acids (Table 27.1). Soy protein lacks methionine and lysine; it also contains less of the branched-chain amino acids (BCAAs) *(8)*. Results from Luiking et al. *(8)* demonstrated that net protein synthesis is higher with casein protein than with soy protein. Casein protein is a milk protein as is whey protein.

Casein protein has a higher biological value because the slow release property keeps the amino acids from being released rapidly and degraded in the liver into urea *(8)*. The same rationale would apply to whey versus casein as well. Whey is also considered to be a rapid release protein supplement when compared with casein. Interestingly, whey protein is often marketed as a rapid release formula, implying faster is better.

27.3.2.2. GLUTAMINE

Although amino acid use tends to be more prevalent in male athletes, female athletes are consuming amino acid supplements *(2)*. Aside from amino stack supplements, female athletes also listed taking glutamine individually. Glutamine is commonly a

Table 27.1
Essential and nonessential amino acids

Essential	Nonessential
Isoleucine[a]	Alanine
Leucine[a]	Arginine[b]
Valine[a]	Asparagine
Lysine	Aspartic acid
Methionine	Cysteine
Phenylalanine	Glutamic acid
Threonine	Glutamine
Tryptophan	Glycine
Histidine	Proline
	Serine
	Tyrosine

[a] Branched chain amino acids.
[b] Not considered essential as most humans synthesize arginine.
Source: Reeds *(31)*.

component of weight-gain products. A small percentage of female athletes report taking weight gainers as well *(2)*.

Glutamine has been reported to enhance protein synthesis and offset immunosuppression. However, there are no long-term studies to date substantiating these claims nor confirming or dismissing possible adverse effects *(5,8,9)*.

27.3.2.3. CARNITINE

Carnitine is synthesized from lysine and methionine, which are the amino acids soy protein is lacking. Mammals synthesize carnitine, and most healthy humans can synthesize it even with a diet lacking in animal protein. In times of low dietary consumption, daily losses through excretion are lowered to compensate *(10)*.

When marketed as L-carnitine, an isomer of carnitine, it has been touted as a fat burner. Therefore, athletes have taken it for weight loss, increased muscle mass, and enhanced β-oxidation. Although L-carnitine is involved in the transportation of fatty acid chains across the membrane, studies have generally shown it to be an effective weight loss agent for obese subjects and inconclusive for healthy nonobese subjects *(10)*.

Athletes may also take L-carnitine to enhance recovery from high-intensity exercise. Research findings include decreased creatine kinase, catabolism of purines, decreased free radical formation, and decreased reported muscle soreness *(10)*.

L-Carnitine is available as a supplement and also as a prescription. Recommendations for daily supplementation are 2–3.5 g/day. Amounts in excess of 4 g/day may cause gastrointestinal distress. A lethal dose (LD) of 630 g/day in humans has been determined from studies using rats *(10)*.

27.3.3. Creatine

Creatine is definitely taken by men more than women *(1,2)*. A large enough percentage of women are taking it, making creatine worthy of discussion.

Creatine is composed of two nonessential amino acids (arginine and glycine) and one essential amino acid (methionine) *(11)*. The average daily requirement is 2 g/day *(3,11)*. The body makes 1–2 g/day via the liver, kidneys, and pancreas. Meat and fish also contain creatine. Vegetarian athletes have lower muscle creatine than omnivores *(12)*.

Creatine is an important part of the ATP-phosphocreatine (ATP-PC) energy system. PC is used to phosphorylate ADP to ATP during high-intensity maximal muscle contractions, resulting in ATP and free creatine. Creatine is rephosphorylated during periods of rest via aerobic pathways (mitochondrial creatine kinase). Generally speaking, an average person has enough phosphocreatine stores to supply ATP for anaerobic activities up to 10–15 s *(3,11)*. Because the ATP-PC system is short-lived, glycolysis becomes the dominant energy system as PC stores are depleted and ADP builds up. Rest periods required to regenerate PC are typically greater than 3 min *(3)*.

Creatine supplementation increases muscle creatine stores in subjects who do not have maximal stores already. Some athletes who consume high quantities of meat or fish have been thought to have maximal or near-maximal stores already *(11)*. Increased muscle creatine stores may lead to more PC to regenerate ATP, a faster recovery time to rephosphorylate free creatine and in some cases buffer lactic acid *(3,6,11,13)*. Some researchers have reported no ergogenic effect on anaerobic exercise *(3,13,14)*. Furthermore, aerobic endurance-type activities and submaximal efforts will not show improvement with creatine supplementation *(3)*.

Creatine is usually taken in the form of creatine monohydrate with a loading phase (5 g/day, four times a day) up to 7 days. A maintenance phase of 2 g/day is recommended for 3 months. This protocol has shown increases of 10–25% in muscle creatine *(11,14)*. Consumption of creatine with a carbohydrate enhances absorption, whereas caffeine will inhibit absorption *(11)*.

Consumption of creatine above the recommended dosages may result in excretion. Because creatine has been indicated in dehydration cases, it is best to advise the athlete to consume plenty of water while taking creatine, especially when exercising in the heat *(11)*.

27.3.3.1. BENEFITS OF CREATINE USING THE RECOMMENDED LOADING–MAINTENANCE CREATINE PROTOCOL

There are reports of increased ability to perform more repetitions at a given percent of 1 repetition maximum (RM), enhanced power output in cyclists, track sprint times improvement in 1–2%, increased muscle mass, and lower percent body fat *(3,6,11,13,15)*. Other studies have not demonstrated improvement of the same variables *(3,6,11,13,15)*. Creatine supplementation was ineffective in subjects 60 years and older or in people with high creatine levels before supplementation. These subjects are referred to as nonresponders *(11,13)*. Although tennis has various components (i.e., serve or sprint to the net) that would utilize the ATP-PC system, no benefits of creatine supplementation have been reported with the stroke or sprint performance *(3)*. Some ergogenic findings in exercise laboratory environments may not translate into actual competition.

Side effects of creatine include weight gain from water retention, gastrointestinal discomfort, and muscle cramps *(3,11,13)*. It is speculated that long-term use may downregulate the creatine transporter protein and render creatine supplementation less effective *(13)*.

27.3.4. Fat Burners and Energy Supplements

27.3.4.1. GINSENG

Ginseng is a widely used herbal in many fat burners and energy drinks. It is reported to improve mood, performance, and alertness and increase fat utilization. There are two prevalent types of ginseng used: Chinese ginseng (*Panax ginseng*) or Siberian ginseng (*Eleutherococcus ginseng*). Other less commonly used ginseng plants include Japanese ginseng (*Panax japonica*), Tiengi ginseng (*Panax notoginseng* or *Panax pseudoginseng*), Dong Quai (*Angelicia sinesis*), or American ginseng (*Panax quinquefolium*) *(16)*.

Chinese ginseng has been evaluated in human performance studies and found to show promising effects to improve strength and aerobic capacity *(16,17)*.

The general dosage guidelines for Chinese ginseng are 1–2 g/day. This may vary based on preparation (powder versus root extract). Chinese ginseng usually requires duration of at least 8 weeks of supplementation along with the athletic training to see ergogenic benefits *(16)*.

There have been some adverse effects reported with Chinese ginseng use. As ginseng is a stimulant, it may cause sleeplessness or nervousness. Because many supplements contain ginseng and caffeine, this side effect may not be because of only ginseng. Hypertension, dermatological problems, morning diarrhea, and euphoria have also been reported. The only reported drug interaction with Chinese ginseng is that with Phenelzine, a monoamine oxidase inhibitor *(16,17)*.

Siberian ginseng is in the same plant family as Chinese ginseng. However, the two are distantly related. It does contain saponins, but they are different from the ones found in Chinese ginseng. Many of the early studies involving Siberian ginseng did not give experimental design details or data, nor were they peer reviewed. Subsequent studies have failed to show an ergogenic effect *(16)*. Without further research demonstrating effect, it would be difficult to recommend supplements containing Siberian ginseng.

27.3.4.2. EPHEDRA

Ephedra contains ephedrine and other alkaloids that are sympathomimetic alkaloids possessing α-agonistic and β-agonistic properties. It facilitates catecholamine release and stimulates the central nervous system (CNS) *(11,16,18)*. Ephedra is used as an energy booster, a fat burner, and an athletic performance enhancer. Ephedrine is also a component of cold remedies and some prescription drugs *(16,18)*. Ephedra may also go by the name of ma huang. It is often paired with caffeine in supplements to achieve an enhanced response *(11,18,19)*.

Unfortunately, many of the studies involving ephedra and athletic performance also included caffeine. Shekelle et al. *(19)* found few controlled studies with or without caffeine to include in their meta-analysis concerning the efficacy of ephedra. Even with the small number of studies included, they felt confident to conclude there is not enough evidence to support the claim that ephedra enhances athletic performance. Other reviews of ephedra and athletic performance confirm this as well *(11,16)*.

Ephedra has been reported to aid in weight loss and was a popular ingredient before the 2004 ban by the Federal Drug Administration (FDA). With the recent, 2006, Federal court ruling, ephedra may still be a component of some fat-burning or thermogenesis products (< 10 mg). Most studies have found claims of weight loss to be substantiated *(18,19)*. Again, many of these studies involve a synergistic relationship between ephedra and caffeine.

Reported side effects of ephedra are summarized in Table 27.2. Ephedra has also been linked to several deaths *(11,17–19)*.

When ephedra was banned in 2004, a product by the common name of bitter orange took its place as a fat-burning agent. Bitter orange (*Citrus aurantium*) contains synephrine and has similar effects as ephedra. It is also used in combination with caffeine *(16)*.

27.3.4.3. CAFFEINE

Caffeine is widely used by athletes and nonathletes for various reasons. Varsity females listed enhanced performance, more energy, alertness, and taste as reasons why they take caffeine. Athletes generally consume their caffeine via beverages such as coffee, tea, or soda. It is also consumed in caffeine tablets, energy bars, and even chocolate *(1)*. Many energy drinks and gels contain caffeine.

Caffeine's main mode of action is because of its structure. The structure resembles that of adenosine and will bind to adenosine receptors. Caffeine also stimulates the release of epinephrine *(20)*. Caffeine has also been reported to enhance fat oxidation and spare muscle glycogen. It is widely consumed although it is not necessary for metabolic functions. People who consume caffeine regularly may become habituated to it, and the effects will be blunted. Habituated subjects would be classified as nonresponders.

One important question asked by athletes was how does caffeine tablet ingestion compare with drinking coffee *(1)*. Although coffee does contain caffeine, it also contains many other compounds that form other metabolites in the body *(20,21)*. Coffee has shown some ergogenic effects also, just not to the same magnitude as caffeine alone *(20)*. Coffee would have to be consumed in large amounts to equal the caffeine typically found in tablets. When answering the athlete's question of coffee or tablets, coffee would most likely result in less of an ergogenic effect unless the subject was naïve to caffeine.

27.3.4.3.1. High-Intensity Exercise During high-intensity exercise, caffeine could lower the rate of perceived exertion (RPE) and increase glycolytic activity performance. Conversely, caffeine is also associated with an increase in blood lactate *(22)*. However,

Table 27.2
Side effects of ephedra

Headache
Tremors
Hypertension
Arrhythmias
Insomnia
Nervousness
Increased heart rate

Sources: Calfee and Fadale *(11)* and Powers *(18)*.

not all anaerobic studies are in agreement. Greer et al. *(23)* found no ergogenic effect or increase in blood lactate when performing the Wingate test.

27.3.4.3.2. Strength The effect of caffeine on strength activities is not conclusive. Previous studies have suggested caffeine may have a direct mode of action on muscle function. This proposed mode of action suggests that caffeine causes the sarcoplasmic reticulum to release more calcium, allowing the muscle to sustain force production for a longer period *(20,21,23)*.

27.3.4.3.3. Endurance Endurance exercise and caffeine have been studied extensively. The main concept is that of fatty acid mobilization and glycogen sparing *(20,21)*. Many of the studies investigating glycogen sparing were not designed to cause sufficient glycogen depletion with or without caffeine. It is unlikely the ergogenic effects of caffeine on endurance exercise are because of glycogen sparing *(20,21)*. Likewise, research has not supported the theory of enhanced fatty acid oxidation during exercise. Although many studies report no decrease in respiratory exchange ratio (RER) with caffeine and exercise, caffeine does promote lipolysis at rest *(20)*. It is clear that caffeine prolongs endurance exercise. The mechanism behind that is not clear. Graham *(20)* suggests that future research explore alternative hypotheses.

27.3.4.3.4. Adverse Effects It is often suggested that caffeine use will result in dehydration. Caffeine is a mild-to-moderate diuretic. Its effect is seen mainly at rest. Studies have found no effects of caffeine and dehydration after prolonged exercise so long as athletes are replacing fluid adequately *(20,21)*. During prolonged exercise, epinephrine-induced renal vasoconstriction causes blood flow to be diverted from the kidneys to the exercising muscle *(21)*. The diuretic effect of caffeine is minimized during exercise.

Other side effects reported include tachycardia with exercise, increased blood pressure, gastrointestinal distress, and habituation/addiction to caffeine *(20,21)*. There are also side effects associated with discontinuing caffeine use. Headache, fatigue, and possible flu-like symptoms may occur *(20,21)*. Combining caffeine with ephedra (ma huang) is potentially dangerous and should be avoided *(18,21)*. Caffeine also inhibits the ergogenic effect of creatine *(11,20)*.

27.3.4.3.5. Dosage For endurance exercise, 3–5 mg/kg is sufficient to produce an ergogenic effect. Because many beverages are variable in the amount of caffeine they contain, tablets are probably the most effective way to get the dosage required to get an ergogenic effect *(21)*.

27.3.5. Anabolic-Androgenic Steroids

Anabolic-androgenic steroids (AASs) are synthetic derivatives of testosterone *(3,13, 24)*. Adolescent female athlete steroid use is reported to be between 2 and 5% *(3,4)*. Collegiate use may be even higher *(24)*.

Women have considerably less testosterone than men, about 10% that of men *(24)*. AASs act beneficially on the athlete in three ways: anabolism, anticatabolism, and promoting aggression *(3,11,13,24)*. Testosterone and AASs bind to androgen receptors inside the cytoplasm and are transported to the nucleus. This stimulates an increased mRNA transcription and leads to increased synthesis of structural and contractile proteins *(3,11,24)*. It is widely reported that testosterone leaves few androgen receptors open for AASs to attach to, and therefore, AASs possess an indirect anticatabolic effect. This is thought to be accomplished by displacing glucocorticoids from their receptor sites *(3,11,13)*. A recent review by Evans *(24)* suggests that this may not always be the case. It is possible that the number of androgen receptors is upregulated with AAS use.

The aggression reported by AAS use is generally deemed beneficial from an athletic training viewpoint. Although this may lead to more intense training, the emotional side effects may not be as beneficial *(3,11,13)*.

Ergogenic effects appear to be limited to muscle hypertrophy and strength gains when performance is concerned along with lean body mass increases *(3,13)*. AASs have not been found to produce an ergogenic effect in endurance activities *(3)*. Steroids have been used to speed up recovery from workouts *(24)*.

27.3.5.1. ADVERSE EFFECTS

Far outweighing the benefits of AASs are the multiple adverse effects. Many steroid users report experiencing adverse effects *(11)*. These side effects are different for women than for men *(3,24)*.

The most notable effects of AASs on women are the virilizing effects: hirsutism, voice deepening, male pattern baldness, and enlargement of the clitoris. Females may also experience menstrual irregularities and reduced breast size *(3,24)*. Some of these adverse effects are irreversible.

The cardiovascular system is also adversely affected by AASs. Blood pressure increases have been associated with AAS use *(3,11,24)*. Left ventricular hypertrophy (LVH) was indicated as a side effect also *(11)*. Because increases in LV wall thickness (and/or increased blood pressure) can be a side effect of chronic athletic training (i.e., heavy resistance training, rowing, etc.), this effect may be a cumulative effect of AASs and training routine *(25,26)*. Decreased high-density lipoprotein (HDL) is also associated with steroid use, although not all studies have reported decreased HDL *(3,11,13,24)*.

Other adverse effects include hepatic abnormalities, dermatological problems, and psychological effects. Aggressive behavior is reported frequently along with mood swings. Depression and anger are reported as withdrawal symptoms *(3,11,13,24)*.

There are other serious repercussions of AAS use that are related to injectable steroids, such as boldenone, trenbolone, or stanozolol (ester) *(11,24)*. Injection-related complications include bacterial infection from nonsterile techniques (sharing multidose vials or needles). Hepatitis B and C and HIV may also be spread this way. Inflammation from repeat use of the same injections site can also occur *(24)*.

27.3.5.2. DOSAGE

Because AASs are illegal, well-designed controlled studies in athletic performance are lacking. Suffice it to say, most steroids users "stack" several anabolic steroid combinations (oral or injectable) in cycles followed by cycles of nonuse for a washout period *(3,11,24)*.

27.3.6. Multivitamins

Vitamin and mineral requirements can be found in Appendices 25–29, and specific needs for populations have been discussed in Chaps. 25, 26, 28, and 29. The focus of this section is on the reasons why female athletes take multivitamins, iron status, and calcium balance.

When surveyed, multivitamins were in the top five of dietary supplements taken by female athletes (1,2). The reasons females gave for using multivitamins were as follows: "to meet nutritional needs, boost the immune system, boost energy, alertness, and habit from childhood" (1,2).

However, not all multivitamins are just multivitamins. Several multivitamins contain herbal extracts (phyto extracts), soy proteins, BCAAs, ginseng, and ginko biloba, to name a few. These are usually marketed as a multivitamin labeled for performance and are widely available on drug store shelves or health marts.

It is important to remember that multivitamins fall under the dietary supplement category and are not subject to FDA regulation. Allied health professionals and coaches who work with athletes should be aware of the brand their athletes are taking and be able to counsel them to make informed purchases and encourage them to read labels. As an alternative, there are several prescription vitamins that are subject to FDA regulations that could be prescribed by the Team Physician.

27.3.7. Iron

Women in general tend to have more iron deficiency issues than men (27,28). Menstrual blood loss varies from woman to woman and can be underestimated in women with heavier cycles. Other factors include inadequate dietary intake, increased loss from sweat, or gastrointestinal blood loss in runners (27,28).

Iron status should be determined by plasma ferritin. Normal levels vary, but levels less than $35\mu g/L$ are below normal (27). Even slight anemia in athletes will negatively impact performance (28). If athletes are borderline, low normal, or below normal, iron supplementation may be prudent. As iron comes in different forms, which are absorbed differently, and is unregulated on the store shelf, it could be more beneficial to have the Team Physician prescribe iron. A multivitamin with iron could be a good choice.

27.3.7.1. SUPPLEMENTING IRON WITH NORMAL FERRITIN LEVELS

High dose of over-the-counter (OTC) iron supplements can cause gastrointestinal distress or constipation, both of which would prove to be ergolytic to athletic performance (28). In addition, studies of iron supplementation in nonanemic athletes have not shown improvement in performance (27).

27.3.8. Calcium

Calcium should be consumed in the diet as should other nutrients. Most female athletes choose to take a calcium supplement to strengthen bones, because of lactose intolerance, or low dietary intake (1,2). It may also be worthy of taking if the athlete is consuming high amount of protein.

Increased dietary protein may produce a lower urinary pH and a higher calcium excretion *(7)*. Increased calcium intake may offset the increased protein-induced urinary excretion of calcium *(29)*.

Typically, calcium:protein ratio of 20:1 is recommended for middle-aged women *(30)*. This ratio is most likely different for an athlete as dietary protein requirements are elevated to maintain a positive nitrogen balance.

27.4. CONCLUSIONS

It is of utmost importance that allied health professionals or coaches who work with female athletes be familiar with the ergogenic aids female athletes are likely to use and the products that may contain them. Athletes will be more inclined to discuss supplements and seek your advice if you address them with an open mind and explain the pros and cons. Many athletes list family and other athletes as major sources of information on supplements *(2)*.

When surveyed, athletes indicated that they do not use a certain supplement and usually list a product containing it under the "other" category *(2)*. This demonstrates that athletes lack knowledge of supplements, regulations, and how to read labels.

Finally, a supplement may be legal for OTC sales, but illegal in certain competitions. The Sports Care Team should be aware of the regulations regarding banned substances for the governing body of athletes such as the International Olympic Committee (IOC) or the National Collegiate Athletic Association (NCAA).

27.5. SCENARIO WITH QUESTION AND ANSWER

27.5.1. Scenario

Your athlete returns to fall workouts and has not followed the recommended summer workouts. She reports to training a little overweight. Her coach is concerned about her underperformance. Her friends suggested she take ginseng to gain more energy and burn fat.

27.5.2. Question

1. How do you advise her?

27.5.3. Plausible Answer

1. The first concept the athlete needs to grasp is FDA regulation. She needs to know how to read the label and realize that it may not contain what it says it does. You should counsel her on the different types of ginseng and that research has shown some improvements in strength and aerobic capacity. She should also be aware of side effects and that it may be combined with other supplements in a product. Instead of trying a supplement to "burn fat," she should be encouraged to follow a healthy diet, get plenty of rest, and ask her coach to help her by tailoring her workouts to get back to a competitive baseline and then begin performance specific workouts.

REFERENCES

1. Kristiansen M, Levy-Milne R, Barr S, et al. Dietary supplements use by varsity athletes at a Canadian university. *Int J Sport Nutr Exerc Metab.* 2005;15:195–210.
2. Froiland K, Koszewski W, Hingst J, et al. Nutritional supplement use among college athletes and their sources of information. *Int J Sport Nutr Exerc Metab.* 2004;14:104–120.
3. Congeni J, Miller S. Supplements and drugs used to enhance athletic performance. *Pediatr Clin N Am.* 2002;49:435–461.
4. Faigenbaum AD, Zaichkowsky LD, Gardner DE, et al. Anabolic steroid use by male and female middle school students. *Pediatrics.* 1998;101(5):e6.
5. Kreider RB. Dietary supplements and the promotion of muscle growth with resistance exercise. *Sports Med.* 1999;27(2):97–110.
6. Ahrendt DM. Ergogenic aids: counseling the athlete. *Am Fam Physician.* 2001;63:913–922.
7. Barzel US, Massey LK. Excess dietary protein can adversely affect bone. *J Nutr.* 1998;128:1051–1053.
8. Luiking YC, Deutz NEP, Jäkel M, et al. Casein and soy protein meals differentially affect whole-body and splanchnic protein metabolism in healthy humans. *J Nutr.* 2005;135:1080–1087.
9. Garlick PJ. The nature of human hazards associated with excessive intake of amino acids. *J Nutr.* 2004;134:1633S–1639S.
10. Karlic H, Lohninger A. Supplementation of L-carnitine in athletes: does it make sense? *Nutrition.* 2004;20:709–715.
11. Calfee R, Fadale P. Popular ergogenic drugs and supplements in young athletes. *Pediatrics.* 2006;117(3):e577–e589.
12. Barr SI, Rideout CA. Nutritional consideration for vegetarian athletes. *Nutrition.* 2004;20:696–703.
13. Tokish JM, Kocher MS, Hawkins RJ. Ergogenic aids: a review of basic science, performance, side effects, and status in sports. *Am J Sports Med.* 2004;32(6):1543–1553.
14. Stevenson SW, Dudley GA. Creatine loading, resistance exercise performance, and muscle mechanics. *J Strength Cond Res.* 2001;15(4):413–419.
15. Vandenberghe K, Goris M, Van Hecke P, et al. Long-term creatine intake is beneficial to muscle performance during resistance training. *J Appl Physiol.* 1997;83(6):2055–2063.
16. Bucci LR. Selected herbals and human exercise performance. *Am J Clin Nutr.* 2000;72:624S–636S.
17. Winterstein AP, Storrs CM. Herbal supplements: considerations for the athletic trainer. *J Athl Train.* 2001;36(4):425–432.
18. Powers ME. Ephedra and its application to sport performance: another concern for the athletic trainer? *J Athl Train.* 2001;36(4):420–424.
19. Shekelle PG, Hardy ML, Morton SC, et al. Efficacy and safety of ephedra and ephedrine for weight loss and athletic performance. *JAMA.* 2003;289(12):1537–1545.
20. Graham TE. Caffeine and exercise. *Sports Med.* 2001;31(11):785–807.
21. Mangus BC, Trowbridge CA. Will caffeine work as an ergogenic aid? The latest research. *Athl Ther Today.* 2005;10(3):57–62.
22. Doherty M, Smith PM, Hughes MG, et al. Caffeine lowers perceptual response and increases power output during high-intensity cycling. *J Sports Sci.* 2004;22:634–643.
23. Greer F, McLean C, Graham TE. Caffeine, performance, and metabolism during repeated Wingate exercise tests. *J Appl Physiol.* 1998;85(4):1502–1508.
24. Evans NA. Current concepts in anabolic-androgenic steroids. *Am J Sports Med.* 2004;32(2):534–542.
25. Pluim BM, Zwinderman AH, van der Laarse A, et al. The athlete's heart: a meta-analysis of cardiac structure and function. *Circulation.* 1999;100:336–344.
26. Whyte GP, Geroge K, Nevill A, et al. Left ventricular morphology and function in female athletes: a meta-analysis. *Int J Sports Med.* 2004;25:380–383.
27. Nielsen P, Nachtigall D. Iron supplementation in athletes. *Sports Med.* 1998;26(4):207–216.
28. Beard J, Tobin B. Iron status and exercise. *Am J Clin Nutr.* 2001;72:594S–597S.
29. Dawson-Hughes B. Interaction of dietary calcium and protein in bone health in humans. *J Nutr.* 2003;133:852S–854S.
30. Heaney RP. Excess dietary protein may not adversely affect bone. *J Nutr.* 1998;128:1054–1057.
31. Reeds PJ. Dispensable and indispensable amino acids for humans. *J Nutr.* 2000:130:1835S–1840S.

28 Nutritional Guidelines and Energy Needs During Pregnancy and Lactation

Jacalyn J. Robert-McComb

CONTENTS

28.1 LEARNING OBJECTIVES
28.2 INTRODUCTION
28.3 RESEARCH FINDINGS
28.4 CONCLUSIONS
28.5 SCENARIO WITH QUESTIONS AND ANSWERS

28.1. LEARNING OBJECTIVES

After completing this chapter, you should have an understanding of the following:

- Energy requirements during pregnancy and lactation.
- Nutritional guidelines during pregnancy and lactation.
- Important nutrients, vitamins, and minerals for optimal pregnancy/infancy outcomes.

28.2. INTRODUCTION

The optimum recommended levels of folic acid, iron, essential fatty acids, and other vitamins for pregnant women has been a topic of discussion in the research literature *(1–4)*. A UNICEF/UNU/WHO/MI technical work (see www.inffoundation.org/pdf/prevent-iron-def.pdf) held in 1998 noted that folic acid substantially decreased the risk of neural tube defects (NTD) and the adverse effect of iron - deficiency anemia on health and human development is now beyond question *(4–5)*. Additionally, essential fatty acids in the fetus and preterm infant are important in the neurodevelopmental processes *(1)*. It has also been found that women of dark complexion or who have limited sun exposure have vitamin D deficiency *(6)*, and this can lead to rickets, hypocalcemia, delayed ossification, and abnormal enamel formation in children.

From: *The Active Female*
Edited by: J. J. Robert-McComb, R. Norman, and M. Zumwalt © Humana Press, Totowa, NJ

However, large doses of fat-soluble vitamins such as vitamins A and D have been shown to cause birth defects in humans and animals *(7)*. Similar to inadequacy, there are issues of nutrient displacement when vitamin, mineral, and nutrient supplementation are excessive.

The potential for adverse effect of "inadequate" or "excess" intake of any nutrient on any organ system is based on the timing, dose, and duration of exposure. The vulnerability of a particular organ thus depends on the occurrence of two factors: (1) the presence of a critical period of growth and development that is dependent on the nutrient in question and (2) a deficiency of the nutrient in the population at the time of this sensitive period *(1)*.

General dietary guidelines for pregnant women are similar to guidelines recommended for nonpregnant women for optimum health (Appendix 23). However, there are higher recommended levels of essential nutrients, vitamins, and minerals that pregnant women should consume is higher than that for the non pregnant woman. Appendices 25–29 list the Dietary Reference Intakes (DRIs) for both pregnant and nonpregnant females for vitamins, minerals, and macronutrients.

The overall quality of a woman's diet affects her need for supplementation. If all of the necessary nutrients can be consumed in the daily diet, then supplements are not needed *(7)*. However, studies consistently show that many women, even in industrialized countries, have vitamin and mineral deficiencies that could have an adverse effect on infant development *(2,4,6,7)*.

The most desirable approach is a preventive food-based dietary approach within a public health context *(4)*. However, in developing countries and populations with low socioeconomic status, this approach may be impractical; therefore, food fortification and supplementation may be necessary.

28.3. RESEARCH FINDINGS

28.3.1. Energy Requirements During Pregnancy and Lactation

Estimated energy expenditure prediction equations for pregnant and lactating women 14–18 and 19–50 years developed by the National Academy of Sciences, Institute of Medicine (IOM), and Food and Nutrition Board in partnership with Health Canada *(8)* can be found in Appendices 18 and 19. In order to estimate energy needs during pregnancy and lactation, energy needs in a nonpregnant state must be computed using age, weight, height, gender, and physical activity level (PAL). Appendix 18 should be used to determine PAL.

Although all equations have been developed to maintain current body weight at the designated PAL, these equations do not promote weight loss. Doubly labeled water studies have also been conducted using well-nourished pregnant and lactating women *(9–14)*. Results of the energy needs during pregnancy and lactation can be found in Tables 28.1 and 28.2, respectively.

Because energy requirements in pregnancy are increased by approximately 17% over the nonpregnant state, a woman of normal weight should consume an additional 300 kcal/day, and those calories should be concentrated in foods of high-nutrient density, a value based on the percent protein, vitamins, and minerals per 100 kcal *(7,15)*. Multifetal pregnancies in the United States increased by 77% from 1980 to 2003

Table 28.1
Doubly labeled water pregnancy studies

Reference	n	Gestation week	Pregravid weight (kg)	Gestational weight gain (kg)	Total energy expenditure (kcal/day)	Physical activity level	Activity energy expenditure (kcal/day)
Goldberg et al. (9)	10	10	–	–	2470	1.42	731
Forsum et al. (10)	22	0	60.8	13.5	2484	1.87	1147
	22	16–18			2293	1.65	860
	22	30			2986	1.82	1338
	19	36			2914	1.66	1171
Goldberg et al. (11)	12	0	61.7	11.91	2274	1.58	835
		6			2322	1.54	818
		12			2426	1.64	939
		18			2456	1.65	964
		24			2621	1.66	1042
		30			2675	1.62	1026
		36			2688	1.50	885
Kopp-Hoolihan et al. (12)	10	0	–	11.6	2205	1.68	892
		8–10			2047	1.57	743
		24–26			2410	1.56	867
		34–36			2728	1.61	1038

Source: Institute of Medicine (U.S.), Standing Committee on the Scientific Evaluation of Dietary Reference Intakes (12).

largely because of assisted reproduction treatments (16). Although the exact caloric requirements for multiple gestations have not been well described, it is generally recommended that an additional 300 kcal and 10 g of protein per fetus beyond singleton are standard (7,17).

The optimal amount of prenatal weight gain is modified by a woman's prepregnancy weight for height (body mass index [BMI]). Total weight gain varies widely among women. For normal-weight women, the mean rate of weight gain is 1.6 kg in the first trimester and 0.44 kg/week in the second and third trimesters. For underweight women, the mean rate of weight gain is 2.3 kg in the first trimester and 0.49 kg/week in the second and third trimesters. For overweight women, the mean rate of weight gain is 0.9 kg in the first trimester and 0.30 kg/week in the second and third trimesters (18).

Recommended total weight gain ranges are 11–15 kg (25–35 lb) for normal-weight women, 12.7–18 kg (28–40 lb) for underweight women (BMI < 19.8), and 7–11 kg (15–25 lb) for an overweight woman (BMI > 26). This translates to 0.4 kg/week for normal-weight women, 0.5 kg/week for underweight women, and 0.3 kg/week for overweight women. The impact of maternal weight gain in the second trimester is most important for fetal development and is protective of fetal growth even if the overall weight gain is poor (7).

Table 28.2
Doubly labeled water lactation studies

Reference	n	Stage of lactation (months)	Total energy expenditure (kcal/day)	Total energy expenditure (kcal/kg/day)	Physical activity level	Activity energy expenditure (kcal/day)	Milk energy output (kcal/day)	Energy mobilization (kcal/day)	Energy requirement (kcal/day)
Goldberg et al. (9)	10	1	2109	35.8	1.50	703	536	Gained fat mass	2646
		2	2171	36.9	1.55	774	532		2702
		3	2138	36.5	1.59	793	530		2667
Forsum et al. (10)	23	2	2532	39.3	1.82	1123	502	72	2962
		6	2580	41.0	1.79	1123			
Lovelady et al. (14)	9	3–6	2413	37.2	1.75	1037	538	287	2663
Kopp-Hoolihan et al. (12)	10	1	2146	—	1.62	816	—	—	—
Butte et al. (13)	24	3	2391	38.1	1.79	1061	483	155	2719

Source: Institute of Medicine (U.S.), Standing Committee on the Scientific Evaluation of Dietary Reference Intakes (8).

28.3.2. Nutritional Guidelines During Pregnancy and Lactation

There are essential nutrients, minerals, and vitamins that pregnant and lactating women must obtain through either a well-balanced nutrient-dense diet or a diet supplemented with vitamins, minerals, and essential nutrients and fortified foods. Appendices 25–29 list the recommended levels of these nutrients for pregnant women. General dietary guidelines for healthy pregnant women include the following recommendations *(7)*. Protein should comprise 20% of a normal pregnancy diet. Fat should comprise approximately 30% of diet in pregnancy and carbohydrates the remaining 50%. A sample diet in pregnancy based on the food pyramid should include 6–11 servings of fruits, 3–5 servings of vegetables, 2–4 servings of meats, beans, or nuts, and 1 serving of sweets. Total caloric intake will vary according to BMI, but the average recommendation is 2500 kcal/day *(7)*.

It is suggested that the additional calories needed for pregnancy (300 kcal and 10 g of protein per fetus) come from adding one protein and one dairy serving *(19)*. However, women who are active during their pregnancy may need extra calories for exercise. Ideally, this additional energy should come from added servings of carbohydrate because carbohydrate intake meets the growth needs of the fetus and provides energy for exercise *(20)*.

The overall quality of a woman's diet affects her need for supplementation. According to the Department of Health and Human Services and the Department of Agriculture *(21)*, a healthy diet includes the following key recommendations (http://www.health.gov/dietaryguidelines/dga2005/document/).

- *Consume a variety of nutrient-dense foods and beverages within and among the basic food groups while choosing foods that limit the intake of saturated and trans fats, cholesterol, added sugars, salt, and alcohol.*
- *Meet recommended intakes for energy requirements by adopting a balanced eating pattern.*
- *Women of childbearing age who may become pregnant should eat foods high in heme-iron and/or consume iron-rich plant foods or iron-fortified foods with an enhancer of iron absorption, such as vitamin C-rich foods.* Numerous dietary surveys indicate that iron intake is suboptimal in a high percentage of the population. Women should be encouraged to eat foods with high iron levels that include oysters, lean red meats, especially liver, and enriched grain products and cereals. In vegetarian diets, tofu, legumes, and beans when consumed with an acidic beverage such as orange, grapefruit, or tomato juice should be encouraged. Iron, whether dietary or supplemented, should be consumed within 1 h of calcium intake *(7)*.
- *Women of childbearing age who may become pregnant and those in the first trimester of pregnancy should consume adequate synthetic folic acid daily (from fortified foods or supplements) in addition to food forms of folate from a varied diet.* The major dietary sources of folates are fresh and frozen green leafy vegetables, citrus fruits and juices, liver, wheat bread, and legumes *(5)*. It is well established that periconceptional (3 months prior to conception and continues 3 months in the postconception period) supplementation with folic acid can reduce the incidence of neural tube defects by 50%, and this finding is the basis for the CDC's recommendation that women of childbearing age who have a chance of becoming pregnant consume 0.4 mg of folate per day from 1 month preconception until the end of the first trimester *(22)*. However, there is not a universal consensus regarding the type of supplementation and the dose of folic acid.

It is recommended that women with diabetes be counseled preconceptionally. There is a fourfold increased risk of congenital malformations related to poor control of diabetes during embryogenesis. Also, girls under the age of 17 are at increased risk for preterm delivery, perinatal mortality, and low body weight. Specifically, girls within 2 years of menarche may require additional energy, protein, and calcium, to meet their nutritional needs for maternal and fetal growth. The erratic use of vitamin supplementation, poor nutrition, and body image issues for this specific population suggests that additional counseling may be warranted *(7)*.

28.3.3. Important Nutrients, Vitamins, and Minerals for Optimal Pregnancy/Infant Outcomes

28.3.3.1. Importance of Iron Reserves Prior to Conception

Maternal anemia is defined by a hematocrit of less than 32% and a hemoglobin level of less than 11 g/dL *(23)*. Representative data from the United States indicate that 5% of nonpregnant women are anemic, that the prevalence of anemia increases to 17% among pregnant women, and that the prevalence is as high as 33% among pregnant women of low socioeconomic groups *(24,25)*. Despite fortification of flour with iron, prevalence rates of anemia have increased from 2 to 4% from the periods 1884–1988 to 1999–2000 *(25)*.

Ideally, to meet iron needs during gestation, women should have 300 mg or more of iron reserves prior to conception *(26)*. Studies have shown that the best outcome conditions (birth weight, delivery time, and maternal health) have been reported to occur when hemoglobin level at term is between 95 and 125 g/L *(27,28)*. Correction of anemia during pregnancy is difficult and should be prevented if possible. Adequate prepregnancy iron reserves can be achieved by an adequate diet, food fortification, or preventive iron supplementation (low daily doses or weekly doses). Ideally, the last two types of intervention should include folate and, if needed, other nutrients such as vitamin A and zinc *(2)*.

A pregnant woman must consume an additional 700–800 mg of iron throughout her pregnancy: 500 mg for hematopoiesis and 250–300 mg for fetal and placental tissues *(23)*. For many years, national and international agencies have proposed high daily doses or iron (60, 120, 240, and even 300 mg or more) for populations in whom the prevalence of gestational iron deficiency anemia is above 40% *(29)*. However, many studies have confirmed that a weekly dose of iron in addition to small daily doses could be effectively used for the prevention of iron deficiency in populations at risk, including during gestation. Eckstrom et al. *(30)* found that 20 doses of 60 mg of iron administered either daily or weekly to anemic Bangladeshi women during gestational weeks 18–24 were enough to elevate hemoglobin to desirable levels. In populations where iron deficiency is prevalent, the intake of small doses plus vitamins early in pregnancy (especially if combined with prepregnancy supplement intakes) reduced low birth weight and preterm delivery *(31)*. Even among women who appear to be nonanemic and iron sufficient by midpregnancy, the intake of small doses of iron compared with placebo starting at week 20 had a positive effect on reducing low birth weight and preterm delivery *(2)*.

The concept of weekly iron–folic acid supplementation as a public health approach to prevent iron anemia was presented at the 1993 World Health Organization/United

Nations University meeting *(29)*. It appears that small daily doses as recommended by the Food and Drug Administration *(32)* and the Food and Nutrition Board and IOM *(33)* as well as weekly dosing starting early in pregnancy are safer and essentially as efficacious as daily iron in preventing iron deficiency and improving iron nutrition when adherence is satisfactory *(2)*. The IOM recommends that all pregnant females who are consuming a well-balanced diet should take 30 mg in divided doses of ferrous iron supplementation daily during the second and third trimesters. However, for females with iron deficiency anemia, therapy consists of 60–120 mg of ferrous iron in divided doses throughout the day *(32)*. The therapeutic dose depends on hemoglobin level of the pregnant woman. For more complete guidelines, please refer to the Guidelines for the Assessment of Iron Deficiency in Woman of Childbearing Age and the IOM Recommended Guidelines for the Prevention, Detection, and Management Among US Children and Women of Childbearing Age for Iron Deficiency Anemia *(32,33)*. Even for nonpregnant women, long-term weekly supplementation with iron and folic acid can bring benefits in terms of the prevention of neural tube defects and hyperhomocysteinemia early in pregnancy *(34)*.

28.3.3.2. FOLIC ACID SUPPLEMENTATION

Epidemiological evidence suggests that the of NTDs are not primarily because of the lack of sufficient folate in the diet but arises from genetically determined changes in the uptake, in metabolism, or in maternal and, particularly fetal cells *(35)*. Therefore, the gene–environmental interaction between vitamin dependency and nutrition may have a causal role in NTDs *(5)*. A sensitive and vulnerable period for fetuses is from the third to the eighth week after conception. Supplementation with folic acid-containing multivitamins or folic acid alone may cause an increase in folate metabolite concentrations of tissue fluids, and it may overcome the failure of the local metabolite supply. Data are available to advise all women who are capable of becoming pregnant to have periconception (i.e., at least 1 month before and until 3 months after conception) folic acid or multivitamin (including 0.4–0.8 mg of folic acid) supplementation to reduce the occurrence of NTDs and other major congenital abnormalities *(5)*. Additionally, the use of multivitamins containing folic acid and other B vitamin *(36–38)* showed a higher efficacy (90%) in the reduction of NTDs than using a high dose of folic acid alone (70%) *(39)* or a low dose of folic acid (41–79%) *(40)*.

Food fortification with folic acid and B vitamins has also been shown to be effective in preventing NTDs. In Hungary, three vitamins (folic acid [160 μg], vitamins B_{12} [0.8 μg], and vitamin B_6 [0.864 μg]) were added to 100 g of flour *(41)*. From a public health perspective, food fortification with folic acid and some other B vitamins (B_{12} and B_6) may be the most effective method to prevent NTDs in unplanned pregnancies *(5)*.

28.3.3.3. VITAMIN D DEFICIENCY IN WOMEN WITH DARK COMPLEXIONS AND LIMITED SUN EXPOSURE

Vitamin D deficiency among pregnant women is not restricted to poor countries but also occurs in highly industrialized countries, more specifically to disadvantaged subpopulations with limited sun exposure *(6)*. Vitamin D deficiencies can result from inadequate cutaneous synthesis, limited dietary intake of vitamin D, or vitamin D pathway impairment *(42)*. Such deficiency in pregnant women can lead to rickets, hypocalcemia, delayed ossification, and abnormal enamel formation in children, and osteoporosis, osteomalacia, and bone fractures in adults *(6,43)*. The lower limit of normal for 25(OH)D is controversial with suggested values in the literature ranging from 15 to 40 nmol/L *(6,44,45)*. Results from a review of vitamin D deficiency in pregnant women indicated that 35 of the 76 studies reviewed in the literature reported that the mean or median maternal concentrations of 25(OH)D levels were 35 nmol/L or less that represents the lower limits of normal. Low concentrations were reported among different ethnic groups in many regions including North America, Europe, the United Kingdom, Africa, the Middle East, and Asia. Recommendations have now surfaced for all expectant mothers with dark complexion or limited sun exposure to be routinely assessed for 25(OH)D levels during their pregnancies. Evidence also indicates that regular intake of multivitamins and fortified foods may not produce increases beyond the 40 nmol/L threshold in some populations; therefore, clinical trials are needed to identify effective preventive strategies during pregnancy to achieve vitamin D adequacy *(6,46)*.

28.4. CONCLUSIONS

The overall quality of a woman's diet affects her needs for supplementation. The DRIs of nutrients, vitamins, and minerals for pregnant women can be found in Appendices 25–29. Excess or deficiency of any nutrient, vitamin, or mineral is a concern during pregnancy. Optimal pregnancy outcomes occur when women maintain a balanced diet prior to conception. Therefore, all women of childbearing age should be encouraged to consume a variety of nutrient-dense foods and beverages within and among the basic food groups while choosing foods that limit the intake of saturated and trans fats, cholesterol, added sugars, salt, and alcohol. Specifically, women of childbearing age who may become pregnant should eat foods high in heme-iron and/or consume iron-rich plant foods or iron-fortified foods with an enhancer of iron absorption, such as vitamin C-rich foods. Additionally, women of childbearing age who may become pregnant and those in the first trimester of pregnancy should consume adequate synthetic folic acid daily (from fortified foods or supplements) in addition to food forms of folate from a varied diet. During pregnancy, women should consume an additional 300 kcal and 10 g of protein per fetus. It is suggested that the additional calories needed for pregnancy come from adding one protein and one dairy serving *(19)*. However, women who are active during their pregnancy may need extra calories for exercise. Ideally, this additional energy should come from added servings of carbohydrate because carbohydrate intake meets the growth needs of the fetus and provides energy for exercise *(20)*.

28.5. SCENARIO WITH QUESTIONS AND ANSWERS

28.5.1. Scenario

You are employed as a fitness and health consultant for the hospital. Your office is in the Lifestyle Center at the hospital. The Lifestyle Center is an employee wellness center that offers advice in nutrition, exercise, and lifestyle management. The Center has a complete workout facility, conference rooms for lectures, and a teaching kitchen to teach clients how to cook healthy foods. One of the clients, Maria, is trying to conceive. She has made an appointment with you to ask you some questions about nutrition during pregnancy.

28.5.2. Questions

1. What type of diet do you recommend for optimum pregnancy/infancy outcomes?
2. Are there specific vitamins and minerals that I should be concerned with for optimum pregnancy/infancy outcomes prior to conception?
3. After I conceive, how many additional calories will I need to consume for the fetus?

28.5.3. Plausible Answers

1. Optimum pregnancy/infant outcomes should begin before conception. It is important that you follow general dietary guidelines for health (http://www.health.gov/dietaryguidelines/dga2005/document/). Total caloric intake will vary according to BMI, but for an active woman in the age range 19–30, it is recommended that she consume 2400 cal a day. This diet would include 2 cups of fruits, 3 cups of vegetables, 8 oz of grains, 6.5 oz of meat and beans, and 3 cups of milk.
2. You should eat foods high in heme-iron and/or consume iron-rich plant foods or iron-fortified foods with an enhancer of iron absorption, such as vitamin C-rich foods. You should also consume adequate synthetic folic acid daily (from fortified foods or supplements) in addition to food forms of folate from a varied diet. Adequate intake of B_6 and B_{12} along with folic acid has also been shown to be effective in preventing NTDs.
3. You should consume an additional 300 kcal and 10 g of protein per fetus. It is suggested that the additional calories needed for pregnancy come from adding one protein and one dairy serving.

REFERENCES

1. Georgieff MK, Innin S. Controversial nutrients that potentially affect preterm neurodevelopment: essential fatty acids and iron. *Pediatr Res* 2005;57:99R–103R.
2. Viteri F, Berger J. Importance of pre-pregnancy and pregnancy iron status: can long-term weekly preventive iron and folic acid supplementation achieve desirable and safe status? *Nutr Rev* 2005;63:S65–S76.
3. Cornel MC, de Smit DJ, de Jong-van den Berg LTW. Folic acid—the scientific debate as a public health policy. *Reprod Toxicol* 2005;20:411–415.
4. Smitasiri S, Solon F. Implementing preventive iron-folic acid supplementation among women of reproductive age in some western pacific countries: possibilities and challenges. *Nutr Rev* 2005;63:S81–S86.
5. Czeizel AE. Optimal pregnancy/infancy outcomes. In: Bendich A, Deckelbaum R, eds. *Preventive Nutrition*, 3rd ed. Totowa, NJ: Humana Press, 2005:603–628.
6. Schroth RJ, Lavelle C, Moffatt M. A review of vitamin D deficiency during pregnancy: who is affected? *Int J Circumpolar Health* 2005;64:112–120.
7. Bader TJ. *OB/GYN Secrets, Updated*, 3rd ed. Philadelphia, PA: Elsevier Mosby, 2005.
8. Institute of Medicine (U.S.), Standing Committee on the Scientific Evaluation of Dietary Reference Intakes. *Dietary Reference Intakes for Energy Carbohydrate, Fiber, Fat, Fatty Acids, Cholesterol, Protein, and Amino Acids*. Washington, DC: National Academy Press, 2002.
9. Goldberg GR, Prentice AM, Coward WA, et al. Longitudinal assessment of the components in energy balance in well-nourished lactating women. *Am J Clin Nutr* 1991;54:778–798.

10. Forsum E, Kabir N, Sadurskis A, et al. Total energy expenditure of healthy Swedish women during pregnancy and lactation. *Am J Clin Nutr* 1992;56:334–342.
11. Goldberg GR, Prentice AM, Coward WA, et al. Longitudinal assessment of energy expenditure in pregnancy using the doubly labeled water method. *Am J Clin Nutr* 1993;57:494–505.
12. Kopp-Hoolihan LE, Van Loan MD, Wong WW, et al. Longitudinal assessment of energy balance in well-nourished, pregnant women. *Am J Clin Nutr* 1999;69:697–704.
13. Butte NF, Wong WW, Hopkinson JM. Energy requirements for lactating women derived from doubly labeled water and mild energy output. *J Nutr* 2001;131:53–58.
14. Lovelady CA, Meredith CN, McCroy MA, et al. Energy expenditure in lactating women: a comparison of doubly labeled water and heart-rate-monitoring methods. *Am J Clin Nutr* 1993;57:512–518.
15. Kullick K, Dugan L. Exercise and pregnancy. In: Rosenbloom CA, ed. *Sports Nutrition*, 3rd ed. Chicago, IL: American Dietetic Association, 2000:463–476.
16. Luke B, Martin J. The rise in multiple births in the United States: who, what, when, where, and why. *Clin Obstet Gynecol* 2004;47:118–133.
17. Rosello-Soberon ME, Fuentes-Chaparro L, Casanueva E. Twin pregnancies: eating for three: maternal nutrition update. *Nutr Rev* 2005;63:295–302.
18. Institute of Medicine (IOM). *Nutrition During Pregnancy*. Washington, DC: National Academy Press, 1990.
19. Worthington-Roberts B, Williams SR. *Nutrition in Pregnancy and Lactation*, 6th ed. Dubuque, IA: Brown and Benchmark Publishers, 1997.
20. Soultanakis HN, Artal R, Wiswell RA. Prolonged exercise in pregnancy: glucose homeostasis, ventilatory and cardiovascular responses. *Semin Perinatol* 1996;20:315–327.
21. US Department of Health and Human Services, US Department of Agriculture, Dietary Guidelines for Americans, Home and Garden Bulletin No. 232, 2005.
22. CDC. Use of folic acid for prevention of spina bifida and other neural tube defects. *JAMA* 1991;266:1190–1191.
23. Shabert JK. Nutrition during pregnancy and lactation. In: Mahan LK, Escott-Stump S, eds. *Krause's Food, Nutrition, & Diet Therapy*, 11th ed. Philadelphia, PA: Saunders, 2004:182–209.
24. Kim I, Hungerford DW, Yip R, et al. Pregnancy nutrition surveillancy system—United States, 1979–1990. *MMWR CDC Surveill Summ* 1992;41:25–41.
25. Bodnar LM, Scaznlon KS, Freedman DS, et al. High prevalence of postpartum anemia among low-income women in the United States. *Am J Obstet Gynecol* 2001;185:438–443.
26. Fernandez-Ballart J. Iron metabolism during pregnancy. *Clin Drug Investig* 2000;19(suppl 1):9–19.
27. Steer PJ. Maternal hemoglobin concentration and birth weight. *Am J Clin Nutr* 2000;71(suppl 5):1285S–1287S.
28. Xiong X, Buekens P, Alexander S, et al. Anemia during pregnancy and birth outcome: a meta-analysis. *Am J Perinatol* 2000;17:137–146.
29. WHO/UNICEF/UNU. *Iron Deficiency Anemia. Assessment, Prevention, and Control. A Guide for Programme Managers*. Geneva: World Health Organization, 2001.
30. Ekstrom EC, Hyder SM, Chowdhury AM, et al. Efficacy and trail effectiveness of weekly and daily iron supplementation among pregnant women in rural Bangladesh: disentangling the issues. *Am J Clin Nutr* 2002;76:1392–1400.
31. Berger J, Thanh HTK, Cavalli-Sforza T, et al. Community mobilization and social marketing to promote weekly iron-folic acid supplementation in women of reproductive age in Vietnam: impact on anemia and iron status. *Nutr Rev* 2005;63(12 part 2):S95–S108.
32. Anderson SA, ed. *Guidelines for the Assessment and Management of Iron Deficiency in Women of Childbearing Age*. Bethesda, MD: Federation of American Societies for Experimental Biology, 1991.
33. Institute of Medicine. *Iron-Deficiency Anemia: Recommended Guidelines for the Prevention, Detection, and Management Among U.S. Children and Women of Childbearing Age*. Washington, DC: National Academy Press, 1993.
34. Adank C, Green TJ, Skeaff CM, et al. Weekly high-dose folic acid supplementation is effective in lowering serum homocysteine concentrations in women. *Ann Nutr Metab* 2003;47:55–59.
35. Yates RW, Ferfuson-Smith MA, Shenkin A, et al. Is disordered folate metabolism the basis for genetic predisposition to neural tube defects? *Clin Genet* 1987;31:279–287.
36. Czeizel AE, Dudas I. Prevention of the first occurrence of neural-tube defects by periconceptional vitamin supplementation. *N Engl J Med* 1992;327:1832–1835.

37. Czeizel AE, Dob'o MC, Vargha P. Hungarian two-cohort controlled study of periconceptional multivitamin supplementation to prevent certain congenital abnormalities. *Birth Defects Res A Clin Mol Teratol* 2004;70:853–861.
38. Nevin NC, Seller MJ. Prevention of neural tube defect recurrences. *Lancet* 1990;1:178–179.
39. MRC Vitamin Study Research Group. Prevention of neural tube defects: results of the Medical Research Council Vitamin Study. *Lancet* 1991;338:131–137.
40. Berry RJ, Li Z, Erickson JD, et al. Prevention of neural-tube defects with folic acid in China. China US Collaborative Project for Neural-Tube Defect Prevention. *N Engl J Med* 1999;341:1485–1490.
41. Czeizel AE, Merhala Z. Bread fortification with folic acid, vitamin B12 and vitamin B6 in Hungary. *Lancet* 1998;352:1225.
42. Combs GF. Vitamin D. In: *The Vitamins: Fundamental Aspects in Nutrition and Health*. San Diego, CA: Academic Press, 1998.
43. Grover SR, Morley R. Vitamin D deficiency in veiled or dark-skinned pregnant women. *Med J Aust* 2001;175(5):251–252.
44. Utiger RD. The need for more vitamin D. *N Engl J Med* 1998;338(12):828–829.
45. Abrams SA. Nutritional rickets: an old disease returns. *Nutr Rev* 2002;60(4):111–115.
46. Zeghoud F, Garabedian M, Jardel A. Administration of a single dose of 100,000 U.I. of vitamin D3 in the pregnant woman in winter: the effect on blood calcium level of the newborn infant. *J Gynecol Obstet Biol Reprod (Paris)* 1988;17(8):1099–1105.

29 Nutritional Guidelines, Energy Balance, and Weight Control Issues for the Mature Physically Active Woman

Jacalyn J. Robert-McComb

CONTENTS
- 29.1 LEARNING OBJECTIVES
- 29.2 INTRODUCTION
- 29.3 RESEARCH FINDINGS
- 29.4 CONCLUSIONS
- 29.5 SCENARIO WITH QUESTIONS AND ANSWERS

29.1. LEARNING OBJECTIVES

After completing this chapter, you should have an understanding of the following:

- Energy balance and weight control.
- Dietary guidelines for the mature woman.
- Important nutrients, vitamins, and minerals for physically active women.

29.2. INTRODUCTION

Worldwide, micronutrient status of women is inadequate for several micronutrients. The rationale for micronutrient adequacy in the individual woman has been well defined for many micronutrients such as iron, calcium, iodine, folate, and vitamins A and D. However, for older women, especially those living beyond their eighties, more research about nutritional requirements is needed.

Female nutritional needs are best achieved safely through a balanced intake of nutrient-rich food, including fortified foods, and nutrient supplements when necessary (1). Although a food-based approach is ideal, fortification with micronutrients is emerging as a powerful tool in preventing or correcting inadequacies for a number of micronutrients. Multiple-micronutrient-containing supplements are widely available, and after initial survey for adequacy, women should be encouraged to take supplements when their needs are not met by food-based or fortification approaches.

29.3. RESEARCH FINDINGS

29.3.1. Energy Balance and Weight Control

Energy balance is affected by many variables, and the solution is not as simple as it would seem strictly from a mathematical standpoint; energy balance = energy input – energy output. Albeit, accurate assessment of energy needs is a necessary component to maintain energy balance. With a few extreme exceptions, the resting metabolic rate (RMR) is by far the largest single component of total caloric expenditure. The most accurate way to assess RMR is through the doubly labeled water technique in conjunction with direct calorimetry. However, this method is expensive and impractical in many health care settings; therefore, predictive equations are used to estimate RMR in most clinical and inpatient care practices.

A systematic review of four commonly used predictive equations to estimate RMR (Mifflin–St. Jeor, Harris–Benedict, Owen, and World Health Organization/Food and Agriculture Organization/United Nations University) revealed that the Miffin–St. Jeor equation is more likely than the other equations to estimate RMR within 10% of that measured (2).

The Mifflin–St. Jeor equation was derived from a sample of 498 normal-weight, overweight, obese, and severely obese individuals aged 19–78 (44.5 ± 14.1) (3). If the individual does not share important characteristics with the group of people from whom the equation was developed (for RMR these would include age, sex, body composition, and possible ethnicity, the chance of clinically important errors increases). Underestimates occur more commonly than overestimates. The maximum underestimate was 20% and the maximum overestimate was 15% of measured RMR in nonobese and obese subjects 20–82 years old (4). In elderly women, 60–82 years old, the maximal underestimation and underestimation was 31 and 7%, respectively (5). Multiple regression analyses were employed to derive relationships between REE and weight, height, and age for both sexes ($R^2 = 0.71$). Separation by sex did not affect its predictive value. Therefore, the best predictor equation for REE is as follows: REE = $9.99 \times$ weight (kg) $+ 6.25 \times$ height (cm) $- 4.92 \times$ age (years) $+ 166 \times$ sex (males, 1; females, 0) $- 161$. Fat-free mass (FFM) was the best single predictor of REE ($R^2 = 0.64$) : REE = $19.7 \times$ FFM $+ 413$ (3). Table 29.1 details the process to estimate total caloric needs depending on activity level using the Miffin–St. Jeor multiple regression equation to estimate RMR. This equation is very practical because the variables in the equation are easily derived.

A discussion of factors that contribute to RMR can be found in Chap. 24. The Institute of Medicine and Food and Nutrition Board equations to estimate energy expenditure can be found in Appendices 18 and 19 (6). Also, the estimated amount of calories to maintain energy levels for gender, age, and activity level based on the Institute of Medicine and Food and Nutrition Board equations can be found in Appendix 20 (6). As can be seen in Appendix 20, caloric expenditure decreases because of the aging process; this is due in part to the decrease in FFM. Furthermore, caloric expenditure is less in women than in men, again, largely because of differences

Table 29.1
An estimated energy expenditure prediction equation using the Miffin–St. Jeor equation to determine resting metabolic rate

STEP 1: Estimate resting metabolic rate (RMR) using the Miffin–St. Jeor equation
REE = 9.99 × weight (kg) + 6.25 × height (cm) − 4.92 × age (year) + 166 × sex (males, 1; females, 0) − 161.

$$\overline{REE} = 9.99 \times \overline{Weight}(kg) + 6.25 \times \overline{Height}(cm) - 4.92 \times \overline{Age}(years) + 166 \times \overline{Sex}(males, 1; females, 0) - 161.$$

STEP 2: Determine additional caloric requirements based on the level of activity

Activity level	Percentage above resting level
Bed rest	10
Quiet rest	30
Light activity	40–60
Moderate activity	60–80
Heavy activity	100

Additional caloric requirements = REE × percentage above resting level

$$\overline{\begin{array}{c}Additional\,caloric\\Requirements\end{array}} = \overline{REE} \times \overline{\begin{array}{c}Percentage\,above\\resting\,level\end{array}}$$

(Source: From Sharkey *(20)*.)

STEP 3: Determine predicted total energy expenditure
Total energy expenditure = REE + additional caloric requirements based on activity

$$\overline{\begin{array}{c}Total\,energy\\expenditure\end{array}}(TEE) = \overline{\begin{array}{c}Resting\,energy\\expenditure\end{array}}(REE) + \overline{\begin{array}{c}Additional\,caloric\\requirements\\based\,on\,activity\end{array}}$$

---

in FFM between genders. Women should be strongly encouraged to increase FFM through exercise throughout the aging process to maintain energy balance. Guidelines for exercise can be found in Part V.

29.3.2. Dietary Guidelines for the Mature Woman

The Dietary Guidelines for Americans is published jointly every 5 years by the Department of Health and Human Services (HHS) and the Department of Agriculture (USDA). The 2005 Dietary Guidelines for Americans *(7)* was released January 12, 2005, and can be downloaded at (http://www.health.gov/dietaryguidelines/dga2005/document/) or purchased from the US Government Printing Office by calling toll-free at (866)512–1800. The Dietary Guidelines for Americans describes a healthy diet as one that (1) emphasizes fruits, vegetables, whole grains, and fat-free or low-fat milk and milk products; (2) includes lean meats, poultry, fish, beans, eggs, and nuts; and (3) is low in saturated fats, trans fats, cholesterol, salt (sodium), and sugar. Appendix 23 outlines general recommendations for a healthy diet for most Americans. Although there are general food group categories, there are specific recommendations based on age, gender, and activity level. The USDA provides a website in which a food plan can be customized (http://www.mypyramid.gov/mypyramid/index.aspx).

Table 29.2
Guidelines for females ranging from ages 50 to 80 from the US Department of Health and Human Services and the US Department of Agriculture

	Active female (Age 50)	Active female (age 80)
Grains (oz)	7	6
Vegetables (cups)	3	2.5
Fruits (cups)	2	2
Milk (cups)	3	3
Meat and beans (oz)	6	5.5

Cholesterol: As low as possible while consuming a nutritionally adequate diet.
Added sugars: Limit to no more than 25% of total energy.
Trans fatty acids and saturated fatty acids: As low as possible while consuming a nutritionally adequate diet.
Sources of trans fatty acids and saturated fats: Trans fatty acids are found in many commercially packaged foods, commercially fried food such as french fries from some fast food chains, other packaged snacks such as microwave popcorn, as well as vegetable shortening and hard stick margarine. Saturated fats are mainly found in animal products such as meat, dairy, eggs, and seafood. Some plant foods are also high in saturated fats such as coconut, palm oil, and palm kernel oil.
Source: US Department of Health and Human Services, US Department of Agriculture, Dietary Guidelines for Americans *(7)*.

Recommendations for an active female, age 50, and an active female, age 80, can be found in Table 29.2.

Key recommendations for women who are in the childbearing years and women over the age of 50 according to the 2005 Dietary Guidelines for Americans also *(7)* can be found in Table 29.3.

Because hypertension is a major public health problem affecting millions of adult women, a Dietary Approach to Stop Hypertension (DASH) Eating Plan is compared with the USDA Food Guide at the 2000-cal-a-day level in Table 29.4. Both the USDA Food Guide and the DASH Eating Plan exemplify the 2005 Dietary Guidelines for Americans *(7)*.

29.3.3. Important Nutrients, Vitamins, and Minerals for Physically Active Aging Women

The micronutrient and macronutrient needs of individuals, men and women alike, who are physically active, have always been a subject of debate. The intensity, duration, and frequency of the physical activity, as well as the overall nutrient intake of the individual, all have an impact on whether micronutrients and macronutrients are required in greater amounts. Appendices 25–29 list the Dietary Reference Intakes (DRIs) for macronutrients, vitamins, and minerals for females of all ages, regardless of the level of physical activity, as established by the Food and Nutrition Board, Institute of Medicine, and National Academy of Sciences *(6)*. Generally, the vitamin and mineral needs of active individuals are not greater than those who are not active, if the DRIs are being met. However, it has been shown that frequently women, of all ages, do

Table 29.3
Key recommendations for women who are in the childbearing years and women over age 50 according to the 2005 Dietary Guidelines for Americans

- People over age 50 should consume vitamin B_{12} in its crystalline form (i.e., fortified foods or supplements). All individuals over the age of 50 should be encouraged to meet their Recommended Dietary Allowance (RDA) (2.4 µg/day) for vitamin B_{12} by eating foods fortified with vitamin B_{12} such as fortified cereals, or by taking the crystalline form of B_{12} supplements.
- Older adults, people with dark skin, and people exposed to insufficient ultraviolet band radiation (i.e., sunlight) should consume extra vitamin D from vitamin D-fortified foods and/or supplements.
- Women of childbearing age who may become pregnant should eat foods high in heme-iron and/or consume iron-rich plant foods or iron-fortified foods with an enhancer of iron absorption, such as vitamin C-rich foods.
- Women of childbearing age who may become pregnant and those in the first trimester of pregnancy should consume adequate synthetic folic acid daily (from fortified foods or supplements) in addition to food forms of folate from a varied diet.

not meet their nutrient needs through diet alone; therefore, supplementation may be necessary. Nonetheless, megadosing with one vitamin and/or mineral can impair the functions of other vitamins and minerals.

Older adults are at a greater risk for nutritional deficiencies than younger adults because of physiologic changes associated with aging. Among the significant age-associated changes in nutrient requirements, the need for energy decreases and the need for protein increases with age. Among the micronutrients, the significant ones that may be associated with deficiencies in elderly women include vitamins B_{12}, A, and C, calcium, iron, zinc, and other trace minerals *(8)*. Fortified foods and supplements may be needed in order to meet the DRIs of these micronutrients in the older woman. With regard to the effects of exercise on vitamin and mineral needs, irrespective of age, thiamin, vitamins C and E, calcium, and iron are significant because research has shown that the DRIs for these micronutrients may be greater in exercising individuals *(9)*. Vitamin B_{12} and folate are also discussed because of the importance of these micronutrients from a health perspective.

29.3.3.1. VITAMIN B_{12} AND FOLATE

The benefits of vitamin B_{12} and folic acid fortification/supplementation are not only applicable to women throughout the life cycle but also to all sectors of the population because of their ability to lower homocysteine level *(10)*. Folic acid is also emerging as important in lowering the risk of certain types of cancers *(11)*.

29.3.3.2. THIAMIN

Thiamin requirements may be greater for individuals who train several hours a day compared with individuals who are moderately active.

Table 29.4
Comparison of the USDA Food Guide and in the Dietary Approach to Stop Hypertension (DASH) Eating Plan at the 2000-cal level[a]

Food groups and subgroups	USDA Food Guide amount	[b]DASH Eating Plan amount	Equivalent amounts
Fruit group	2 cups (4 servings)	2–2.5 cups (4–5 servings)	1/2 cup equivalent is a. 1/2 cup fresh, frozen, or canned fruit b. 1 med fruit c. USDA: 1/2 cup fruit juice d. DASH: 3/4 cup fruit juice
Vegetable group a. Dark green vegetables b. Orange vegetables c. Legumes (dry beans) d. Starchy vegetables e. Other vegetables	2.5 cups (5 servings) 3 cups/week 2 cups/week 3 cups/week 3 cups/week 6.5 cups/week	2–2.5 cups (4–5 servings)	1/2 cup equivalent is a. 1/2 cup of cut-up raw or or cooked vegetable b. 1 cup raw leafy vegetable c. USDA: 1/2 cup vegetable juice d. DASH: 3/4 cup vegetable juice
Grain group a. Whole grains b. Other grains	6 oz equivalents 3 oz equivalents 3 oz equivalents	7–8 oz equivalents (7–8 servings)	1 oz equivalent is a. 1 slice bread b. 1 cup dry cereal c. 1/2 cup cooked rice, pasta, cereal DASH: 1 oz dry cereal (1/2 to 1.25 cup depending on cereal type—check label)
Meat and beans group	5.5 oz equivalents	6 oz or less meat, poultry, fish 4–5 servings per week nuts, seeds, and dry beans[c]	1 oz equivalent is a. 1 oz cooked lean meats, poultry, fish b. 1 egg c. USDA 1/4 cup cooked dry beans or tofu, 1 tbsp peanut butter, 1/2 oz nuts or seeds e. DASH 1.5 oz nuts, 1/2 oz seeds, 1/2 cup cooked dry beans
Milk group	3 cups	2–3 cups	1 cup equivalent is a. 1 cup low-fat/fat-free milk yogurt b. 1.5 oz of low-fat/fat-free natural cheese c. 2 oz of low-fat/fat-free processed cheese

Oils	27 g (6 tsp)	8–12 g (2–3 tsp)	1 tsp equivalent is a. DASH: 1 tsp soft margarine b. 1 tbsp low-fat mayo c. 2 tbsp light salad dressing d. 1 tsp vegetable oil
Discretionary calorie allowance			1 tbsp added sugar equivalent is
	267 cal	~ 2 tsp of added sugar (5 tbsp per week)	a. DASH: 1 tbsp jelly or jam
Example of distribution:			b. 1/2 oz jelly beans
Solid fat[d]	18 g		c. 8 oz lemonade
Added sugars	8 tsp		

[a] All servings are per day unless otherwise noted. USDA vegetable subgroup amounts and amounts of DASH nuts, seeds, and dry beans are per week.

[b] The 2000-cal USDA Food Guide is appropriate for many sedentary males 51–70 years of age, sedentary females 19–30 years of age, and for some other gender/age groups who are more physically active.

[c] In the DASH Eating Plan, nuts, seeds, and dry beans are a separate food group from meat, poultry, and fish.

[d] The oils listed in the table are not considered to be part of discretionary calories because they are a major source of the vitamin E and polyunsaturated fatty acids, including the essential fatty acids, in the food pattern. In contrast, solid fats (i.e., saturated and trans fats) are listed separately as a source of discretional calories.
Source: http://www.health.gov/dietaryguidelines/dga2005/document/html/chapter2.htm.

29.3.3.3. VITAMIN C

If an individual has adequate vitamin C status, supplementation with vitamin C does not enhance performance. However, individuals who exercise may require at least 100 mg/day of vitamin C to maintain normal vitamin C status. Additionally, individuals who are competing in ultra marathons may require up to 500 mg/day or more of vitamin C *(12)*. Both of these values are greater than the established DRIs (Appendices 26–29).

29.3.3.4. VITAMIN E

Because endurance exercise results in increased oxygen consumption and thus increased oxidant stress, it seems logical that vitamin E supplementation might be beneficial to people who exercise *(9)*. Although vitamin E has been shown to sequester free radicals in exercising individuals by decreasing membrane disruption *(13)*, there have not been reports that vitamin E actually improves exercise performance. Nonetheless, the role of vitamin E in the prevention of oxidative damage because of exercise may be significant, and more long-term research is needed *(9)*.

29.3.3.5. CALCIUM

Appendices 25–29 list adequate intake levels for calcium for various life stages for females. Presently, the DRI for calcium for adult women is 1200 mg of calcium per day, but there have been suggestions that daily intakes of 1500 mg/day may by appropriate for postmenopausal women or women over the age of 65 *(14)*. Typical calcium intakes for women are listed.

- One-half of teen-age girls consume less than two-thirds of the recommended intake *(15)*.
- One-half of adult women consume less than 70% of the recommended intake *(15)*.
- Average intake for calcium for women 20–29 years of age is 778 mg/day *(16)*.
- For women more than 65 years of age, 600 mg/day is the common daily intake *(15)*.

If an individual is not consuming enough calcium in her diet, supplementation with calcium citrate or calcium carbonate is best *(9)*. Individuals should avoid calcium supplements containing bone meal, oyster shell, and shark cartilage because of the increased lead content in these supplements, which can result in toxic effects in the body. Calcium supplements are best absorbed if taken in 500 mg or less between meals *(17)*. In older individuals who may suffer from achlorhydria, calcium carbonate is better absorbed with meals *(17)*. For older women, calcium citrate is the best supplement *(18)*.

29.3.3.6. IRON

Because female athletes often do not consume proper amounts of dietary iron as a result of lower kilocalorie consumption and/or a reduction in meat content in the diet, iron supplementation may be necessary. Ferrous sulfate is the least expensive and most widely used form of iron supplementation *(19)*. For adults diagnosed with iron deficiency anemia, a daily dose of at least 60 mg of elemental iron taken between meals is recommended *(19)*.

29.4. CONCLUSIONS

A food-based approach is ideal for meeting macronutrient and micronutrient needs of women. However, food-based approaches are still not attaining adequate intakes in most women, both in the United States and worldwide *(1)*. Consideration must be given to fortified foods and/or supplements to meet the recommended daily allowances for optimal health and performance. Among the micronutrients, the significant ones that may be associated with deficiencies in the older women include vitamins B_{12}, A, and C, calcium, iron, zinc, and other trace minerals *(8)*.

29.5. SCENARIO WITH QUESTIONS AND ANSWERS

29.5.1. Scenario

Agnes is 70 years old and is in good health. She exercises at least 60 min a day. She has contacted you for advice concerning her diet.

29.5.2. Questions

1. How many calories should I consume per day?
2. What types of food should my diet consist of and in what amounts?
3. What online resources would you recommend for reliable dietary advice?

29.5.3. Plausible Answers

1. For your age and activity level, the Institute of Medicine recommends 2000–2200 cal a day. Alternately, you may use an equation to determine your energy needs based on your gender, level of activity, height, and weight.
2. Your daily diet should consist of 6 oz of grains, 2.5 cups of vegetables, 2 cups of fruits, 3 cups of milk, and 5.5 oz meat and beans.
3. The USDA provides a website that provides valuable information about food groups and dietary needs (http://www.mypyramid.gov) based on reliable research.

REFERENCES

1. Bartley KA, Underwood BA, Deckelbaum RJ. A life cycle micronutrient perspective for women's health. *Am J Clin Nutr* 2005;81(suppl):1188S–93S.
2. Frankenfield D, Roth-House L, Compher C, et al. Comparison of predictive equations for resting metabolic rate in healthy nonobese and obese adults: a systematic review. *J Am Diet Assoc* 2005;105:775–89.
3. Miffin MD, St Jeor ST, Hill LA, et al. A new predictive equation for resting energy expenditure in healthy individuals. *Am J Clin Nutr* 1990;51:241–7.
4. Frankenfield DC, Rowe WA, Smith JS, et al. Validation of several established equations for resting metabolic rate in obese and nonobese people. *J Am Diet Assoc* 2003;103:1152–9.
5. Arciero PJ, Goran MI, Gardner AM, et al. A practical equation to predict resting metabolic rate in older females. *J Am Geriatr Soc* 1993;41:389–95.
6. Institute of Medicine (U.S.), Standing Committee on the Scientific Evaluation of Dietary Reference Intakes. *Dietary Reference Intakes for Energy Carbohydrate, Fiber, Fat, Fatty Acids, Cholesterol, Protein, and Amino Acids*. Washington, DC: National Academy Press, 2002.
7. US Department of Health and Human Services, US Department of Agriculture, Dietary Guidelines for Americans, Home and Garden Bulletin No. 232, 2005.
8. Chernoff R. Micronutrient requirements in older women. *Am J Clin Nutr* 2005;81(suppl):1240S–5S.
9. Volpe S. Vitamins and minerals for active people In: Rosenbloom C, ed. *Sports Nutrition: A Guide for the Professional Working with Active People*, 3rd ed. Chicago, IL: The American Dietetic Association, 2000:63–93.
10. Stover PJ. Physiology of folate and vitamin B_{12} in health and disease. *Nutr Rev* 2004;62(suppl):S3–12.
11. Kim YI. Role of folate in colon cancer development and progression. *J Nutr* 2003;133(11 suppl 1):3731S–9S.
12. Keith RE. Ascorbic acid. In: Wolinsky I, Driskell JA, eds. *Sports Nutrition*. Boca Raton, FL: CRC Press, 1997:29–45.
13. McBride JM, Kraemer WJ, Triplett-McBride T, et al. Effect of resistance exercise on free radical production. *Med Sci Sports Exerc* 1998;30(1):67–72.
14. Finch S, Doyle W, Lowe C, et al. National diet and nutrition survey: people aged 65 or over. Vol. 1. In: *Report of the Diet and Nutrition Survey*. London: Her Majesty's Stationary Office, 1998.
15. Looker AC, Bilezikian JP, Baily L, et al. Calcium intake in the United States. In: *Optimal Calcium Intake*. Program and abstracts of the NIH Consensus Panel Development Conference. June 6–8, 1994. Bethesda, MD.
16. Hendricks KM, Herbold NH. Diet, activity, and other health-related behaviors in college-age women. *Nutr Rev* 1998;56:65–75.
17. Wardlaw GM. *Perspectives in Nutrition*, 4th ed. Boston, MA: WCB McGraw-Hill, 1999.

18. Dawson-Hughes B, Dallal GE, Krall EA, et al. A controlled trial of the effect of calcium supplementation on bone density in postmenopausal women. *N Engl J Med* 1990;323:878–83.
19. Yip R, Dallman PR. Iron. In: Ziegler EE, Filer LJ Jr, eds. *Present Knowledge in Nutrition*, 7th ed. Washington, DC: ILSI Press, 1996:277–92.
20. Sharkey BJ. *Physiology of Fitness*, 3rd ed. Champaign, IL: Human Kinetics, 1990:359.

APPENDICES

APPENDIX 1 BODY IMAGE QUALITY OF LIFE INVENTORY (BIQLI)

Different people have different feelings about their physical appearances. These feelings are called "body image." Some people are generally satisfied with their looks, whereas others are dissatisfied. At the same time, people differ in terms of how their body image experiences affect other aspects of their lives. Body image may have positive effects, negative effects, or no effects at all. Listed below are various ways by which your own body image may or may not influence your life. For each item, circle how much our feelings about your affect that aspect of your life. Before answering each item, think carefully about the answer that is most accurate about how your body image usually affects you.

BIQLI items	−3 Very negative effect	−2 Moderate negative effect	−1 Slight negative effect	0 No effect	+1 Slight positive effect	+2 Moderate positive effect	+3 Very positive effect
1. My basic feelings about myself—feelings of personal adequacy and self-worth	−3	−2	−1	0	1	2	3
2. My feelings about my adequacy as a man or woman—feelings of masculinity or femininity	−3	−2	−1	0	1	2	3
3. My interactions with people of my own sex	−3	−2	−1	0	1	2	3
4. My interactions with people of the other sex	−3	−2	−1	0	1	2	3
5. My experiences when I meet new people	−3	−2	−1	0	1	2	3
6. My experiences at work or at school	−3	−2	−1	0	1	2	3
7. My relationships with friends	−3	−2	−1	0	1	2	3
8. My relationships with family members	−3	−2	−1	0	1	2	3
9. My day-to-day emotions	−3	−2	−1	0	1	2	3

(Continued)

Appendix 1
(Continued)

BIQLI items	−3 Very negative effect	−2 Moderate negative effect	−1 Slight negative effect	0 No effect	+1 Slight positive effect	+2 Moderate positive effect	+3 Very positive effect
10. My satisfaction with my life in general	−3	−2	−1	0	1	2	3
11. My feelings of acceptability as a sexual partner	−3	−2	−1	0	1	2	3
12. My enjoyment of my sex life	−3	−2	−1	0	1	2	3
13. My ability to control what and how much I eat	−3	−2	−1	0	1	2	3
14. My ability to control my weight	−3	−2	−1	0	1	2	3
15. My activities for physical exercise	−3	−2	−1	0	1	2	3
16. My willingness to do things that might call attention to my appearance	−3	−2	−1	0	1	2	3
17. My daily "grooming" activities (i.e., getting dressed and physically ready for the day)	−3	−2	−1	0	1	2	3
18. How confident I feel in my everyday life	−3	−2	−1	0	1	2	3
19. How happy I feel in my everyday life	−3	−2	−1	0	1	2	3

Source: From Cash T.F., Fleming E.C. Impact of body image experiences: development of the body image quality of Life Inventory. *Int J Eat Disord* 31: 445–460, 2002. Reprinted with permission of John Wiley & Sons, Inc.

APPENDIX 2 BODY IMAGE CONCERN INVENTORY

Please rate how often you have had the described feeling or performed the described behavior on a Likert scale anchored by 1= "never" and 5= "always"

	1 Never	2 Seldom	3 Sometimes	4 Often	5 Always
1. I am dissatisfied with some aspect of my appearance	1	2	3	4	5
2. I spend a significant amount of time checking my appearance in the mirror	1	2	3	4	5
3. I feel others are speaking negatively of my appearance	1	2	3	4	5
4. I am reluctant to engage in social activities when my appearance does not meet my satisfaction	1	2	3	4	5
5. I feel there are certain aspects of my appearance that are extremely unattractive	1	2	3	4	5
6. I buy cosmetic products to try to improve my appearance	1	2	3	4	5
7. I seek reassurance from others about my appearance	1	2	3	4	5
8. I feel there are certain aspects of my appearance that I would like to change	1	2	3	4	5
9. I am ashamed of some part of my body	1	2	3	4	5
10. I compare my appearance to that of fashion models or others	1	2	3	4	5

(Continued)

Appendix 2
(Continued)

		1 Never	2 Seldom	3 Sometimes	4 Often	5 Always
11.	I try to camouflage certain flaws in my appearance	1	2	3	4	5
12.	I examine flaws in my appearance	1	2	3	4	5
13.	I have bought clothing to hide a certain aspect of my appearance	1	2	3	4	5
14.	I feel others are move physically attractive than me	1	2	3	4	5
15.	I have considered consulting/consulted some sort of medical expert regarding flaws in my appearance	1	2	3	4	5
16.	I have missed social activities because of my appearance	1	2	3	4	5
17.	I have been embarrassed to leave the house because of my appearance	1	2	3	4	5
18.	I fear that others will discover my flaws in appearance	1	2	3	4	5
19.	I have avoided looking at my appearance in the mirror	1	2	3	4	5

Source: From Littleton H., Axsom D., Pury C.L.S. Development of the Body Image Concern Inventory. *Behav Res Ther* 43: 229–241, 2005. With permission from Elsevier.

APPENDIX 3 PHYSICAL APPEARANCE STATE AND TRAIT ANXIETY SCALE: TRAIT

The statements listed below are to be used to describe how anxious, tense, or nervous you feel in general (i.e., usually) about your body or specific parts of your body.

Please read each statement and circle the number that best indicates the extent to which each statement holds true in general. Remember, there are no right or wrong answers.

	1 Never	2 Seldom	3 Sometimes	4 Often	5 Always
In general: I feel *anxious*: *tense*: or *nervous* about					
1. The extent to which I look overweight	1	2	3	4	5
2. My thighs	1	2	3	4	5
3. My buttocks	1	2	3	4	5
4. My hips	1	2	3	4	5
5. My stomach	1	2	3	4	5
6. My legs	1	2	3	4	5
7. My waist	1	2	3	4	5
8. My muscle tone	1	2	3	4	5
9. My ears	1	2	3	4	5
10. My lips	1	2	3	4	5
11. My wrists	1	2	3	4	5
12. My hands	1	2	3	4	5
13. My forehead	1	2	3	4	5
14. My neck	1	2	3	4	5
15. My chin	1	2	3	4	5
16. My feet	1	2	3	4	5

Source: From Reed D.L., Thompson J.K., Brannick M.T. Development and validation of the Physical Appearance State and Trait Anxiety Scale (PASTAS). *J Anxiety Disord* 5(4): 323–332, 1991. With permission from Elsevier.

APPENDIX 4 THE SCOFF QUESTIONNAIRE

SCOFF questions

Do you make yourself **S**ick (induce vomiting) because you feel uncomfortably full?
Do you worry that you have lost **C**ontrol over how much you eat?
Have you recently lost more than **O**ne stone (14 lb [6.4 kg]) in a 3-month period?
Do you think you are too **F**at, even though others say you are too thin?
Would you say that **F**ood dominates your life?

One point for every yes answer; a score ≥ 2 indicates a likely case of anorexia nervosa or bulimia nervosa (sensitivity: 100%; specificity: 87.5%).

Source: From Morgan J.F., Reid F., Lacey J.H. The SCOFF questionnaire: assessment of a new screening tool for eating disorders. *BMJ* 319: 1467–1468, 1999. With permission.

APPENDIX 5 EATING ATTITUDES TEST (EAT-26)

	✓ Please choose one response by marking a check to the right for each of the following statements:	Always	Usually	Often	Sometimes	Rarely	Never	Score
1.	Am terrified about being overweight.	☐	☐	☐	☐	☐	☐	☐
2.	Avoid eating when I am hungry.	☐	☐	☐	☐	☐	☐	☐
3.	Find myself preoccupied with food.	☐	☐	☐	☐	☐	☐	☐
4.	Have gone on eating binges where I feel that I may not be able to stop.	☐	☐	☐	☐	☐	☐	☐
5.	Cut my food into small pieces.	☐	☐	☐	☐	☐	☐	☐
6.	Aware of the calorie content of foods that I eat.	☐	☐	☐	☐	☐	☐	☐
7.	Particularly avoid food with a high carbohydrate content (bread, rice, potatoes, etc.).	☐	☐	☐	☐	☐	☐	☐
8.	Feel that others would prefer if I ate more.	☐	☐	☐	☐	☐	☐	☐
9.	Vomit after I have eaten.	☐	☐	☐	☐	☐	☐	☐
10.	Feel extremely guilty after eating.	☐	☐	☐	☐	☐	☐	☐
11.	Am preoccupied with a desire to be thinner.	☐	☐	☐	☐	☐	☐	☐
12.	Think about burning up calories when I exercise.	☐	☐	☐	☐	☐	☐	☐
13.	Other people think that I am too thin.	☐	☐	☐	☐	☐	☐	☐
14.	Am preoccupied with the thought of having fat on my body.	☐	☐	☐	☐	☐	☐	☐
15.	Take longer than others to eat my meals.	☐	☐	☐	☐	☐	☐	☐
16.	Avoid foods with sugar in them.	☐	☐	☐	☐	☐	☐	☐
17.	Eat diet foods.	☐	☐	☐	☐	☐	☐	☐

(Continued)

Appendix 5
(Continued)

√ Please choose one response by marking an × in the box under the correct response	Always	Usually	Often	Sometimes	Rarely	Never	Score
18. Feel that food controls my life.	☐	☐	☐	☐	☐	☐	
19. Display self-control around food.	☐	☐	☐	☐	☐	☐	
20. Feel that others pressure me to eat.	☐	☐	☐	☐	☐	☐	
21. Give too much time and thought to food.	☐	☐	☐	☐	☐	☐	
22. Feel uncomfortable after eating sweets.	☐	☐	☐	☐	☐	☐	
23. Engage in dieting behavior.	☐	☐	☐	☐	☐	☐	
24. Like my stomach to be empty.	☐	☐	☐	☐	☐	☐	
25. Have the impulse to vomit after meals.	☐	☐	☐	☐	☐	☐	
26. Enjoy trying new rich foods.	☐	☐	☐	☐	☐	☐	

Total Score=

Behavioral Questions:

In the past 6 months have you: Yes   No

A. Gone on eating binges where you feel that you may not be able to stop? (Eating much more than most people would eat under the same circumstances) If you answered yes, how often during the worst week: _____

B. Ever made yourself sick (vomited) to control your weight or shape? If you answered yes, how often during the worst week: _____

C. Ever used laxatives, diet pills or diuretics (water pills) to control your weight or shape? If you answered yes, how often during the worst week? _____

D. Ever been treated for an eating disorder? When: _____

Note: For more information on the EAT-26, *see* http://www.river-centre.org.

Source: From Garner D.M., Olmsted M.P., Bohr Y., et al. The Eating Attitudes Test: psychometric features and clinical correlates. *Psycholl Med*, 12: 871–878, 1982; and adapted by D. Garner with permission.

Appendix 5 / Eating Attitudes Test (EAT-26)

Step 1: EAT-26 Item Scoring:

Score each item as indicated below and put score in box to the right of each item

Items 1–25:			Item 26 only:		
Always	=	3	=		0
Usually	=	2	=		0
Often	=	1	=		0
Sometimes	=	0	=		1
Rarely	=	0	=		2
Never	=	0	=		2

Step 2: Total EAT-26 Score

Add item scores together for a Total EAT-26 score: Total = ☐

Step 3: Behavioral Questions

Did you score Yes on Questions A, B, C, or D? (please circle)   No   Yes

Step 4: Underweight

Determine if you are "significantly underweight" according to the table on page 356

Step 5: Referral	No	Yes
If your EAT-26 score is **20 or more**	☐	☐
or if you answered **YES** to any questions A–D	☐	☐
or if your **weight** is below the number on the weight chart to the right,	☐	☐
Please discuss your results with your physician or therapist	☐	☐

Significantly Underweight According to Height (Body Mass Index of 18)*				
Height (inches)	Weight (pounds)	Height (inches)	Weight (pounds)
58	86	68	118
$58\frac{1}{2}$	88	$68\frac{1}{2}$	120
59	89	69	121
$59\frac{1}{2}$	90	$69\frac{1}{2}$	124
60	91	70	125
$60\frac{1}{2}$	93	$70\frac{1}{2}$	127
61	95	71	128
$61\frac{1}{2}$	96	$71\frac{1}{2}$	131
62	99	72	132
$62\frac{1}{2}$	100	$72\frac{1}{2}$	134
63	101	73	135
$63\frac{1}{2}$	103	$73\frac{1}{2}$	138
64	105	74	140
$64\frac{1}{2}$	106	$74\frac{1}{2}$	141
65	108	75	144
$65\frac{1}{2}$	109	$75\frac{1}{2}$	146
66	112	76	147
$66\frac{1}{2}$	113	$76\frac{1}{2}$	149
67	114	77	152
$67\frac{1}{2}$	117	$77\frac{1}{2}$	154

Note: This table indicates the body weights for heights considered to be "significantly underweight" according to a body mass index (BMI) of 18. BMI is a simple method of evaluating body weight taking height into consideration. It applies to both men and women. There is some controversy regarding whether or not BMI is the best method of determining relative body weight, and it is important to recognize that it is possible for someone to be quite malnourished even though they are above the weight listed in the table. In order to determine if you are "significantly underweight," locate your height (without shoes) on the table and see if the corresponding body weight (in light indoor clothing) is below that listed. If so, you are considered "significantly underweight" and should speak to your physician or therapist about your weight. To calculate BMI exactly: Weight (pounds) divided by Height (inches) Divide this again by Height (inches) and multiply by 703

$$BMI = (lbs) \div (inches) \div (inches) \times 703$$

SCORING AND INTERPRETATION

The Eating Attitudes Test (EAT) *(1,2)* is a widely used, standardized, self-report measure of eating disorder symptoms. The 26-item version *(2)* reproduced above is highly reliable and valid *(2,3,4)*. The EAT has been used as the first-step tool in a two-stage screening process for eating disorders *(5,6)*. According to this methodology, individuals who score 20 or more on the test should be interviewed by a qualified professional to determine if they meet the diagnostic criteria for an eating disorder. The two-stage screening process using the EAT-26 has been particularly useful in samples of athletes *(7)*. The EAT-26 items form three subscales (i.e., Dieting, Bulimia and Food Preoccupation, and Oral Control) and subscale scores are computed by summing all items assigned to that particular scale (*Dieting scale items*: 1, 6, 7, 10, 11, 12, 14,

16, 17, 22, 23, 24, 25; *Bulimia and Food Preoccupation scale items*: 3, 4, 9, 18, 21, 26; *Oral Control subscale items*: 2, 5, 8, 13, 19, 20).

Behavioral Questions: Following the methodology described for the Eating Disorder Inventory Referral Form (EDI-RF) *(8)*, four behavioral questions are included on above version of the EAT-26 aimed at determining the presence of extreme weight-control behaviors as well as providing an estimate of their frequency. These questions assess self-reported binge eating, self-induced vomiting, use of laxatives, and exercise to lose or control weight over the preceding 6 months. Although these content areas could be assessed in the same format as other items, this would not provide the type of frequency data required to evaluate the extent of the problem.

Because denial can be a problem on self-report screening instruments, low scores should not be taken to mean that an eating disorder is not present. Collateral information from parents, teammates, and coaches is useful information that can correct for denial, limited self-disclosure, and social desirability. High scores on self-report measures do not necessarily mean the respondent has an eating disorder; however, it does denote concerns regarding body weight, body shape, and eating.

REFERENCES

1. Garner, D.M., Garfinkel, P.E. (1979). The Eating Attitudes Test: an index of the symptoms of anorexia nervosa. *Psychol Med, 9*, 273–279.
2. Garner, D.M., Olmsted, M.P., Bohr, Y., Garfinkel, P.E. (1982). The eating attitudes test: Psychometric features and clinical correlates. *Psychol Med, 12*, 871–878.
3. Lee, S., Kwok, K., Liau, C., Leung, T. (2002). Screening Chinese patients with eating disorders using the Eating Attitudes Test in Hong Kong. *Int J Eat Disord, 32*, 91–97.
4. Mintz, L.B., O'Halloran, M.S. (2000). The Eating Attitudes Test: validation with DSM-IV eating disorder criteria. *J Pers Assess, 74*, 489–503.
5. Dotti, A., Lazzari, R. (1998). Validation and reliability of the Italian EAT-26. *Eat Weight Disord, 3*, 188–194.
6. Patton, G.C., Johnson-Sabine, E., Wood, K., Mann, A.H., Wakeling, A. (1990). Abnormal eating attitudes in London schoolgirls: A prospective epidemiological study-outcome at twelve month follow-up. *Psychol Med, 20*, 383–394.
7. Garner, D.M., Rosen, L., Barry, D. (1998). Eating disorders in athletes (839–857). In: *Child and Adolescent Psychiatric Clinics of North America*. 7 New York: W.B. Saunders.
8. Garner, D.M. (2004). *The Eating Disorder Inventory-3 Professional Manual*. Odessa, FL: Psychological Assessment Resources Inc.

APPENDIX 6 BULIMIA TEST—REVISED (BULIT-R)

Answer each question by circling the appropriate response. Please respond to each item as honestly as possible; remember all of the information you provide will be kept strictly confidential.

1. I am satisfied with my eating patterns.

    1. agree
    2. neutral
    3. disagree a little
    4. disagree
    +5. disagree strongly

2. Would you presently call yourself a "binge eater"?

    +1. yes, absolutely
    2. yes
    3. yes, probably
    4. yes, possibly
    5. no, probably not

3. Do you feel you have control over the amount of food you consume?

    1. most or all of the time
    2. a lot of the time
    3. occasionally
    4. rarely
    +5. never

4. I am satisfied with the shape and size of my body.

    1. frequently or always
    2. sometimes
    3. occasionally
    4. rarely
    +5. seldom or never

5. When I feel that my eating behavior is out of control, I try to take rather extreme measures to get back on course (strict dieting, fasting, laxatives, diuretics, self-induced vomiting, or vigorous exercise).

    +1. always
    2. almost always
    3. frequently
    4. sometimes
    5. never or my eating behavior is never out of control

Appendix 6 / Bulimia Test—Revised (BULIT-R) 359

6. X I use laxatives or suppositories to help control my weight.

   1. once a day or more
   2. 3–6 times a week
   3. once or twice a week
   4. 2–3 times a month
   5. once a month or less (or never)

7. I am obsessed about the size and shape of my body.

   +1. always
   2. almost always
   3. frequently
   4. sometimes
   5. seldom or never

8. There are times when I rapidly eat a very large amount of food.

   +1. more than twice a week
   2. twice a week
   3. once a week
   4. 2–3 times a month
   5. once a month or less (or never)

9. How long have you been binge eating (eating uncontrollably to the point of stuffing yourself)?

   1. not applicable; I do not binge eat
   2. less than 3 months
   3. 3 months to 1 year
   4. 1–3 years
   +5. 3 or more years

10. Most people I know would be amazed if they knew how much food I can consume at one sitting.

    +1. without a doubt
    2. very probably
    3. probably
    4. possibly
    5. no

11. X I exercise in order to burn calories

    1. more than 2 h per day
    2. about 2 h per day
    3. more than 1 but less than 2 h per day
    4. one hour or less per day
    5. I exercise but not to burn calories or I do not exercise

12. Compared with women your age, how preoccupied are you about your weight and body shape?

    +1. a great deal more than average
    2. much more than average

3. more than average
4. a little more than average
5. average or less than average

13. I am afraid to eat anything for fear that I would not be able to stop.

    +1. always
    2. almost always
    3. frequently
    4. sometimes
    5. seldom or never

14. I feel tormented by the idea that I am fat or might gain weight.

    +1. always
    2. almost always
    3. frequently
    4. sometimes
    5. seldom or never

15. How often do you intentionally vomit after eating?

    +1. 2 or more times a week
    2. once a week
    3. 2–3 times a month
    4. once a month
    5. less than once a month or never

16. I eat a lot of food when I am not even hungry.

    +1. very frequently
    2. frequently
    3. occasionally
    4. sometimes
    5. seldom or never

17. My eating patterns are different from the eating patterns of most people.

    +1. always
    2. almost always
    3. frequently
    4. sometimes
    5. seldom or never

18. After I binge eat, I turn to one of several strict methods to try to keep from gaining weight (vigorous exercise, strict dieting, fasting, self-induced vomiting, laxatives, or diuretics).

    1. never or I do not binge eat
    2. rarely
    3. occasionally
    4. a lot of the time
    +5. most or all of the time

Appendix 6 / Bulimia Test—Revised (BULIT-R) 361

19. X I have tried to lose weight by fasting or going on strict diets.

    1. not in the past year
    2. once in the past year
    3. 2–3 times in the past year
    4. 4–5 times in the past year
    5. more than 5 times in the past year

20. X I exercise vigorously and for long periods of time in order to burn calories.

    1. average or less than average
    2. a little more than average
    3. more than average
    4. much more than average
    5. a great deal more than average

21. When engaged in an eating binge, I tend to eat foods that are high in carbohydrates (sweets and starches).

    +1. always
    2. almost always
    3. frequently
    4. sometimes
    5. seldom, I do not binge

22. Compared with most people, my ability to control my eating behavior seems to be

    1. greater than others' ability
    2. about the same
    3. less
    4. much less
    +5. I have absolutely no control

23. I would presently label myself a "compulsive eater," (one who engages in episodes of uncontrolled eating).

    +1. absolutely
    2. yes
    3. yes, probably
    4. yes, possibly
    5. no, probably not

24. I hate the way my body looks after I eat too much.

    1. seldom or never
    2. sometimes
    3. frequently
    4. almost always
    +5. always

25. When I am trying to keep from gaining weight, I feel that I have to resort to vigorous exercise, strict dieting, fasting, self-induced vomiting, laxatives, or diuretics.

    1. never
    2. rarely

3. occasionally
4. a lot of the time
+5. most or all of the time

26. Do you believe that it is easier for you to vomit than it is for most people?

    +1. yes, it is no problem at all for me
    2. yes, it is easier
    3. yes, it is a little easier
    4. about the same
    5. no, it is less easy

27. X I use diuretics (water pills) to help control my weight.

    1. never
    2. seldom
    3. sometimes
    4. frequently
    5. very frequently

28. I feel that food controls my life.

    +1. always
    2. almost always
    3. frequently
    4. sometimes
    5. seldom or never

29. X I try to control my weight by eating little or no food for a day or longer.

    1. never
    2. seldom
    3. sometimes
    4. frequently
    5. very frequently

30. When consuming a large quantity of food, at what rate of speed do you usually eat?

    +1. more rapidly than most people have ever eaten in their lives
    2. a lot more rapidly than most people
    3. a little more rapidly than most people
    4. about the same rate as most people
    5. more slowly than most people (or not applicable)

31. X I use laxatives or suppositories to help control my weight.

    1. never
    2. seldom
    3. sometimes
    4. frequently
    5. very frequently

32. Right after I binge eat I feel:

    +1. so fat and bloated I cannot stand it
    2. extremely fat

Appendix 6 / Bulimia Test—Revised (BULIT-R) 363

    3. fat
    4. a little fat
    5. Okay about how my body looks or I never binge eat

33. Compared with other people of my sex, my ability to always feel in control of how much I eat is:

    1. about the same or greater
    2. a little less
    3. less
    4. much less
  +5. a great deal less

34. In the last 3 months, on the average how often did you binge eat (eat uncontrollably to the point of stuffing yourself)?

    1. once a month or less (or never)
    2. 2–3 times a month
    3. once a week
    4. twice a week
  +5. more than twice a week

35. Most people I know would be surprised at how fat I look after I eat a lot of food.

  +1. yes, definitely
    2. yes
    3. yes, probably
    4. yes, possibly
    5. no, probably not or I never eat a lot of food

36. X I use diuretics (water pills) to help control my weight.

    1. 3 times a week or more
    2. once or twice a week
    3. 2–3 times a month
    4. once a month
    5. never

Scoring: X denotes questions whose answers are not added to determine the total BULIT-R score; + denotes the most strongly symptomatic response, which receives a score of five points.

The BULIT-R may not be copied. All copies of this test must be obtained from Mark H. Thelen, Ph.D., ThelenM@missouri.edu, Department of Psychology, 210 McAlester Hall, University of Missouri, Columbia, Missouri, 65211. With permission.

APPENDIX 7 STUDENT-ATHLETE NUTRITIONAL HEALTH QUESTIONNAIRE

Stress Fracture History

1. Have you ever had a stress fracture or a stress reaction? — Yes   No
2. Have you ever had X-rays to rule out a stress fracture or a stress reaction? — Yes   No
3. Have you ever had a bone scan or a bone density test? — Yes   No
4. Do you talk calcium? — Yes   No
5. Are you a vegetarian?

Eating/Weight History

1. What is your highest and lowest weight in the past year?

    Highest weight _____ lbs.
    Lowest weight _____ lbs.

2. Have you had any recent changes in weight? — Yes   No
3. What is your desired weight? — _____ lbs.
4. Do you weigh yourself often? — Yes   No
5. Do you consciously watch your weight? — Yes   No
6. Would your weight be different if you were not exercising vigorously? — Yes   No
7. How many times a year do you lose weight intentionally?
8. When your season is over and you stop or reduce training, do you gain or lose weight?

    If "gain," how much weight? — _____ lbs.
    If "lose," how much weight? — _____ lbs.
    What is your weight in season at the peak of training? — _____ lbs.

9. Do you have to restrict your food intake more or less than in the past to be at your competitive weight? *(please circle the appropriate answer)*

    Much less than before   Somewhat less than before   No change   Somewhat more than before   Much more than before

10. Are you preoccupied with weight? — Yes   No
11. Does worrying about weight take up a significant amount of your time? — Yes   No

Appendix 7 / Student-Athlete Nutritional Health Questionnaire

Menstrual History (Females Only)

1. At what age did you have your first period?   Month _____ Year _____
2. When was your last period?   Month _____ Year _____
3. How many periods have you had in the last 12 months. _____
4. Are you on any form of estrogen/birth control?   Yes   No
   - If yes, what form? _____
   - How long? _____
   - Why? (Control of period, medical prescription, other) _____
   - If it has been recommended and you are not taking it, why? _____
5. Have you ever been diagnosed with anemia?   Yes   No
6. Do you eat red meat?   Yes   No
7. Do you have heavy menses?   Yes   No
8. Were you aware of any effect of regular training for sport on the occurrence of your first menstrual period?   Yes   No
   - If yes, briefly explain _____
9. Are menstrual problems such as cramps and irregularity common in your family?   Yes   No

Reprinted with permission from the University of Colorado Sports Medicine Department at Colorado Springs (http://web.uccs.edu/sports_medicine/).

APPENDIX 8 FEMALE ATHLETE SCREENING TOOL

Please answer as completely as possible
The key is used to quantify and define activity level for further clarification of the questions.

Key[a]: Exercise = Physical activity ≥ 20 min
Practice = Scheduled time allotted by coach to work as a team or individually in order to improve performance.
Training = Intense physical activity. The goal is to improve fitness level in order to perform optimally.

1. I participate in additional physical activity ≥ 20 min in length on days that I have practice or competition.
   1) Frequently 2) Sometimes 3) Rarely 4) Never
2. If I cannot exercise, I find myself worrying that I will gain weight.
   1) Frequently 2) Sometimes 3) Rarely 4) Never
3. I believe that most female athletes have some form of disordered eating habits.
   1) Strongly Agree 2) Agree 3) Disagree 4) Strongly disagree
4. During training, I control my fat and calorie intake carefully.
   1) Frequently 2) Sometimes 3) Rarely 4) Never
5. I do not eat foods that have more than 3 g of fat.
   1) Strongly agree 2) Agree 3) Disagree 4) Strongly disagree
6. My performance would improve if I lost weight.
   1) Strongly Agree 2) Agree 3) Disagree 4) Strongly disagree
7. If I got on the scale tomorrow and gained 2 lb, I would practice or exercise harder or longer than usual.
   1) Frequently 2) Sometimes 3) Rarely 4) Never
8. I weigh myself_____.
   1) Daily 2) 2 or more times a week 3) Weekly 4) Monthly or less
9. If I chose to exercise on the day of competition (game/meet), I exercise for
   1) 2 or more hours 2) 45 min to 1 h 3) 30–45 min 4) Less than 30 min
10. If I know that I will be consuming alcoholic beverages, I will skip meals on that day or the following day.
    1) Frequently 2) Sometimes 3) Rarely 4) Never
11. I feel guilty if I choose fried foods for a meal.
    1) Frequently 2) Sometimes 3) Rarely 4) Never
12. If I were to be injured, I would still exercise even if I was instructed not to do so by my athletic trainer or physician.
    1) Strongly agree 2) Agree 3) Disagree 4) Strongly disagree
13. I take dietary or herbal supplements in order to increase my metabolism and/or to assist in burning fat.
    1) Frequently 2) Sometimes 3) Rarely 4) Never

Appendix 8 / Female Athlete Screening Tool

14. I am concerned about my percent body fat.
    1) Frequently 2) Sometimes 3) Rarely 4) Never
15. Being an athlete, I am very conscious about consuming adequate calories and nutrients on a daily basis.
    1) Frequently 2) Sometimes 3) Rarely 4) Never
16. I am worried that if I were to gain weight, my performance would decrease.
    1) Strongly agree 2) Agree 3) Disagree 4) Strongly disagree
17. I think that being thin is associated with winning.
    1) Strongly agree 2) Agree 3) Disagree 4) Strongly disagree
18. I train intensely for my sport so I will not gain weight.
    1) Frequently 2) Sometimes 3) Rarely 4) Never
19. During season, I choose to exercise on my one day off from practice or competition.
    1) Frequently 2) Sometimes 3) Rarely 4) Never
20. My friends tell me that I am thin but I feel fat.
    1) Frequently 2) Sometimes 3) Rarely 4) Never
21. I feel uncomfortable eating around others.
    1) Frequently 2) Sometimes 3) Rarely 4) Never
22. I limit the amount of carbohydrates that I eat.
    1) Frequently 2) Sometimes 3) Rarely 4) Never
23. I try to lose weight to please others.
    1) Frequently 2) Sometimes 3) Rarely 4) Never
24. If I were unable to compete in my sport, I would not feel good about myself.
    1) Strongly agree 2) Agree 3) Disagree 4) Strongly disagree
25. If I were injured and unable to exercise, I would restrict my calorie intake.
    1) Strongly agree 2) Agree 3) Disagree 4) Strongly disagree
26. In the past 2 years, I have been unable to compete due to an injury
    1) 7 or more times 2) 4–6 times 3) 1–3 times 4) No significant injuries
27. During practice, I have trouble concentrating due to feelings of guilt about what I have eaten that day.
    1) Frequently 2) Sometimes 3) Rarely 4) Never
28. I feel that I have a lot of good qualities.
    1) Strongly agree 2) Agree 3) Disagree 4) Strongly disagree
29. At times I feel that I am no good at all.
    1) Strongly agree 2) Agree 3) Disagree 4) Strongly disagree
30. I strive for perfection in all aspects of my life.
    1) Strongly agree 2) Agree 3) Disagree 4) Strongly disagree
31. I avoid eating meat in order to stay thin.
    1) Strongly agree 2) Agree 3) Disagree 4) Strongly disagree
32. I am happy with my present weight.
    1) Yes 2) No
33. I have done things to keep my weight down that I believe are unhealthy.
    1) Frequently 2) Sometimes 3) Rarely 4) Never

Source: From McNulty K.Y., Adams C.H., Anderson J.M., Affenito S.G. Development and validation of a screening tool to identify eating disorders in female athletes. *J Am Diet Assoc* 101(8): 886–892, 2001. With permission.

Scoring Instructions: The FAST is scored on a 4-point Likert scale. The higher the number the higher the probability (i.e., 4 points = frequently; 3 points = sometimes;

2 points = rarely; and 1 point = never). Questions 15, 28, and 32 are reversed scored. For question 32, a response of yes receives 1 point and a response of no receives 2 points. In a small group of female athletes ($N = 41$), subclinical scores ranged from 77 to 94, and clinical scores were > 94. *Note*: This is a screening tool, not a diagnostic tool.

APPENDIX 9 EATING DISORDER ORGANIZATIONS AND RESOURCES

Anorexia Nervosa and Related Eating Disorders, Inc. (ANRED)
Internet: http://www.anred.com/
ANRED's mission is to provide easily accessible information on anorexia nervosa, bulimia nervosa, binge eating, and other food and weight disorders. ANDRED, a nonprofit organization, distributes materials on topics such as anorexia nervosa, bulimia nervosa, binge eating disorder, and other less well-known food and weight disorders.

Soy Unica! Soy Latina!
Internet: http://www.soyunica.org/mybody/default.htm
An excellent bilingual website for young Latinas with a good section on eating disorders.

Eating Disorder Information and Referral Center
Internet: http://www.EDreferral.com
This website is a resource for information and treatment options for all forms of eating disorders. It includes referrals to local treatment centers nationwide.

Harvard Eating Disorders Center (HEDC)
WACC 725
15 Parkman Street
Boston, MA 02114
Tel: (617) 236-7766
Email: info@hedc.org
Internet: http://www.hedc.org/
The HEDC is a national nonprofit organization dedicated to research and education, and gaining new knowledge of eating disorders, their detection, treatment, and prevention to share with the community-at-large. The website includes information about eating disorders, help for family and friends, resources, and a listing of events and programs.

Overeaters Anonymous (OA)
World Service Office
PO Box 44020
Rio Rancho, NM 87174-4020
Tel: (505) 891-2664
Email: info@overeatersanonymous.org
Internet: http://www.overeatersanonymous.org/
OA is a nonprofit international organization that provides volunteer support groups worldwide. Modeled after the 12-step Alcoholics Anonymous program, the OA recovery program addresses physical, emotional, and spiritual recovery aspects of compulsive

overeating. Members are encouraged to seek professional help for individual diet and nutrition plans and for any emotional or physical problems.

The Renfrew Center Foundation
475 Spring Lane
Philadelphia, PA 19128
Tel: 1-800-RENFREW
Email: foundation@renfrew.org
Internet: http://www.renfrew.org/
The Renfrew Center Foundation is a tax-exempt, nonprofit organization promoting the education, prevention, treatment, and research of eating disorders. The Renfrew Center Foundation is funded by private donations and by the Renfrew Center, the nation's first freestanding facility committed to the treatment of eating disorders.

APPENDIX 10 DETERMINING MODERATE AND VIGOROUS EXERCISE INTENSITY USING THE HEART RATE RESERVE (HRR) METHOD

STEP 1:
Take your resting heart rate. In order for this to be the most accurate, it is suggested that you count the number of beats per minute three mornings in a row before arising and average the three readings.
Reading 1____bpm Reading 2____bpm Reading 3____bpm Average____bpm
Resting heart rate =____ (*i.e., resting heart rate = 72*)

STEP 2
Determine your maximum heart rate (MHR) 220 – age =____ (*i.e., 220 – 35 = 185*)

Moderate Exercise Intensity

STEP 3
Target heart rate = [(MHR – resting heart rate) (40–59%)] + resting heart rate
i.e., [(185–72) (40–59%)] + 72 = 117–138 bpm

Vigorous Exercise Intensity

STEP 3
Target heart rate = [(MHR – resting heart rate) ($\geq 60\%$)] + resting heart rate
i.e., [(185–72) ($\geq 60\%$)] + 72 = 139 bpm or greater

Note: The intensity range to increase and maintain cardiorespiratory fitness (CR) is broad. For individuals with low levels of CR fitness, the lower range will result in improvements in CR fitness. For individuals who are already fit, exercise intensities at the high end of the continuum (60–85% HRR) may be needed to improve and maintain CR fitness.

APPENDIX 11 DETERMINING MODERATE AND VIGOROUS EXERCISE INTENSITY USING THE BORG RATING OF PERCEIVED EXERTION (RPE) SCALE

	13 Somewhat hard
6 No exertion at all	14
7	15 Hard (heavy)
8 Extremely light (7.5)	16
9 Very light	17 Very hard
10	18
11 Light	19 Extremely hard
12	20 Maximal exertion

Instructions: While doing physical activity, rate your perception of exertion. This feeling should reflect how heavy and strenuous the exercise feels to you, combining all sensations and feelings of physical stress, effort, and fatigue. Do not concern yourself with any one factor such as leg pain or shortness of breath, but try to focus on your total feeling of exertion. Choose the number that best describes your level of exertion. This will give you a good idea of the intensity level of your activity. *Moderate exercise* is defined as a "perceived exertion" **of 11–14**. *Vigorous exercise* is defined as a perceived exertion" **of 15 or greater**. The *average RPE* range associated with *physiologic adaptation* **is 12–16**. However, there is significant interindividual variability.

Note:

9 corresponds to "very light" exercise; for a healthy person, it is like walking slowly at his or her own pace

13 on the scale is "somewhat hard" exercise, but it still feels okay to continue.

17 "very hard" is very strenuous, a healthy really has to push himself or herself, and you will probably feel very tired

19 on the scale is an extremely strenuous exercise level; for most people, this is the most strenuous exercise they have ever experienced

Borg RPE scale © Gunnar Borg, 1970, 1985, 1994, 1998, Division of Nutrition and Physical Activity, National Center for Chronic Disease Prevention and Health Promotion. http://www.cdc.gov/nccdphp/dnpa/physical/measuring/perceived_exertion.htm

APPENDIX 12 THE PHYSICAL ACTIVITY READINESS QUESTIONNAIRE (PAR-Q) C. 2002.

Regular physical activity is fun and healthy, and increasingly more people are starting to become more active every day. Being more active is very safe for most people. However, some people should check with their doctor before they start becoming much more physically active.

If you are planning to become much more physically active than you are now, start by answering the seven questions in the box below. If you are between the ages of 15 and 69, the PAR-Q will tell you if you should check with your doctor before you start. If you are over 69 years of age and you are not used to being very active, check with your doctor.

Commons sense is your best guide when you answer these questions. Please read the questions carefully and answer each one honestly.

NO	YES	
☐	☐	1. Has your doctor ever said that you have a heart condition <u>and</u> that you should only do physical activity recommended by a doctor?
☐	☐	2. Do you feel pain in your chest when you do physical activity?
☐	☐	3. In the past month, have you had chest pain when you were not doing physical activity?
☐	☐	4. Do you lose your balance because of dizziness or do you ever lose consciousness?
☐	☐	5. Do you have a bone or joint problem that would be mode worse by a change in your physical activity?
☐	☐	6. Is your doctor currently prescribing drugs (for example water pills) for your blood pressure or heart condition?
☐	☐	7. Do you know of <u>any other reason</u> why you should not do physical activity?

Please note: If your health changes so that you then answer YES to any of the above questions, tell your fitness or health professional. Ask whether you should change your physical activity plan.

If you answered YES to one or more questions
Talk to your doctor by phone or in person BEFORE you start becoming much more physically active or BEFORE you have a fitness appraisal. Tell you doctor about the PAR-Q and which questions you answered YES.
- You may be able to do any activity you want as long as you start slowly and build up gradually. Or you may need to restrict you activities to those which are safe for you. Talk with your doctor about the kinds of activities you wish to participate in and follow his/her advice.
- Find out which community programs are safe and helpful for you.

If you answered NO to all questions
If you answered NO honestly to <u>all</u> PAR-Q questions, you can be reasonably sure that you can:
- Start becoming much more physically active – begin slowly and build up gradually. This is the safest and easiest way to go.
- Take part in a fitness appraisal – this is an excellent way to determine your basic fitness so that 7you can plan the best way for your to live actively.

Delay becoming much more active if:
- You are not feeling well because of a temporary illness such as a cold or a fever – wait until you feel better or
- If you are or may be pregnant – talk to your doctor before you start becoming more active.

I understand my signature signifies that I have read and understand all the information on the questionnaire, that I have truthfully answered all the questions, and that any questions/concerns I may have had have been addressed to my complete satisfaction.

Name (please print) _____ Date _____

Signature _____ Witness _____

Source: 'Reproduced with permission of the Canadian Society for Exercise Physiology'

APPENDIX 13 GENERAL ORGANIZATIONAL GUIDELINES FOR EXERCISE IN CHILDREN AND ADOLESCENTS

National Association for Sport and Physical Education (ages 5–12)	• Accumulate at least 60 min of activity but no more than 2 h a day on most, if not all days of the week • More than 60 min, and up to 2 h encouraged • Exercise should come in the form of intermittent bouts approximately 15 min in length that are both moderate and vigorous activities • 10–15 min bouts of vigorous activity is recommended for children over the age of 13 • Brief periods of rest need to be included between bouts of exercise or activity
American College of Sports Medicine	• Moderate levels of physical activity totaling 30–60 min a day on most days of the week • Rest and recovery periods need to be included between bouts of exercise • For older children, 20–30 minute bouts of vigorous activity should occur three times per week • For weight loss, activity time should total 30–60 min a day 6–7 days per week

Sources: Adapted from American College of Sports Medicine. *ACSM's Guidelines for Exercise Testing and Prescription,* 7th ed. Philadelphia, PA: Lippincott Williams & Wilkins, 2006.
Adapted from Corbin C.B., Pangrazi R.P. Physical activity for children: a statement of guidelines. Reston, VA: NASPE Publications, 1998.

APPENDIX 14 AMERICAN COLLEGE OF SPORTS MEDICINE GUIDELINES FOR RESISTANCE TRAINING WITH CHILDREN

American College of Sports Medicine Guidelines for Resistance Exercise in Children and Adolescent Populations

- Children should be supervised by a qualified instructor when performing exercises.
- High-intensity exercises such as 1 RM should be avoided, and progressive loading should be utilized instead.
- Equipment should be appropriate for the size and skill level of the child.
- The goals of the resistance program should work to increase motor skill and fitness level.
- The child should perform each exercise between 8 and 15 repetitions, and weight should only be increased when they can perform this number of repetitions with correct form. If eight repetitions cannot be performed, then the resistance weight needs to be lowered so the child can perform the eight repetitions with correct form.
- Young children should not perform below eight repetitions. A training load of 8 or below should be utilized only for older adolescents.
- The focus should be on developing correct form rather than maximizing weight.

Source: Adapted from American College of Sports Medicine. *ACSM's Guidelines for Exercise Testing and Prescription*, 7th ed. Philadelphia, PA: Lippincott Williams & Wilkins, 2006.

APPENDIX 15 KRAEMER'S AGE-SPECIFIC EXERCISE GUIDELINES FOR RESISTANCE TRAINING

Resistance Training Exercise for Guidelines Children by Age Group

7 years or younger	• Use little or no weight • Focus on technique • Volume should be low
8–10 years of age	• Can increase number of exercises as well as resistance and volume • Important to monitor progression and tolerance of increases
11–13 years of age	• Continue slow progression of resistance and volume • Begin to introduce advanced exercises using little or no weight • Add sport-specific exercises
14–15 years of age	• Continue resistance progression • Advance sport-specific components
16 years and older	• After demonstrating mastery of proper technique the child should progress to entry-level adult programs

Source: Adapted from Kraemer W.J., Fleck S.J. *Strength Training for Young Athletes*, 2nd ed. Champaign, IL: Human Kinetics, 2005.

APPENDIX 16 SAMPLE EXERCISE RESISTANCE PROGRAM FOR POSTMENOPAUSAL WOMEN: 4-WEEK, 6-WEEK, 8-WEEK, AND 12-WEEK PROGRAMS

A safe and effective training load for postmenopausal women is from 50 to 80% of a 1 RM. This level is enough to allow for the development of strength but light enough to reduce the risk of injury. Progress gradually by starting off with 50% 1 RM using two sets of 6 repetitions for the first 2 weeks. During weeks 3 and 4, increase the lower body intensity by 10% and the upper body intensity by 5% as well as increasing the reps to two sets of seven repetitions. When reaching weeks 5 and 6, make similar increases in upper and lower body intensity by increasing intensity by 5 and 10%, respectively, as well as increasing the repetitions in each set to 8. Weeks 7 and 8 have increasing intensity, but the repetitions are not increased. Finally, for those who have completed the 8th week, the following weeks have an increased intensity of 70–80% of the 1 RM for all exercises as well as an additional set of repetitions. The 4-week program is intended for as an introduction to resistance training that can then progress to the 6-week, 8-week, and 12-week programs for beginners, intermediates, and advanced exercisers, respectively. All programs are performed three times per week.

Introduction Part I: Weeks 1 and 2

Exercise	1 RM	% 1 RM	Weight	Set 1	Set 2
Wall squats	10 lbs	50%	5 lbs	6	6
Machine bench press	20 lbs	50%	10 lbs	6	6
Leg press	100 lbs	50%	50 lbs	6	6
Low rows	40 lbs	50%	20 lbs	6	6
Lat pulldown	40 lbs	50%	20 lbs	6	6
Back extension	60 lbs	50%	30 lbs	6	6

Introduction Part II: Weeks 3 and 4

Exercise	1 RM	% 1 RM	Weight	Set 1	Set 2
Wall squats	10 lbs	60%	6 lbs	7	7
Machine bench press	20 lbs	55%	12 lbs	7	7
Leg press	100 lbs	60%	60 lbs	7	7
Low rows	40 lbs	55%	22.5 lbs	7	7
Lat pulldown	40 lbs	55%	22.5 lbs	7	7
Back extension	60 lbs	55%	32.5 lbs	7	7

Beginner: Weeks 5 and 6

Exercise	1 RM	% 1 RM	Weight	Set 1	Set 2
Wall squats	10 lbs	70%	7 lbs	8	8
Machine bench press	20 lbs	60%	12.5 lbs	8	8
Leg press	100 lbs	70%	70 lbs	8	8
Low rows	40 lbs	60%	25 lbs	8	8
Lat pulldown	40 lbs	60%	25 lbs	8	8
Back extension	60 lbs	60%	35 lbs	8	8

Intermediate: Weeks 7 and 8

Exercise	1 RM	% 1 RM	Weight	Set 1	Set 2
Wall squats	10 lbs	80%	8 lbs	8	8
Machine bench press	20 lbs	70%	15 lbs	8	8
Leg press	100 lbs	80%	80 lbs	8	8
Low rows	40 lbs	65%	27.5 lbs	8	8
Lat pulldown	40 lbs	65%	27.5 lbs	8	8
Arm curl	20 lbs	70%	15 lbs	8	8
Arm extension	40 lbs	70%	27.5 lbs	8	8
Back extension	60 lbs	65%	40 lbs	8	8

Advanced: Weeks 9–12

Exercise	1 RM	% 1 RM	Weight	Set 1	Set 2	Set 3
Wall squats	10 lbs	80%	8 lbs	8	8	8
Machine bench press	20 lbs	70%	14 lbs	8	8	8
Leg press	100 lbs	80%	80 lbs	8	8	8
Low rows	40 lbs	70%	28 lbs	8	8	8
Lat pulldown	40 lbs	70%	28 lbs	8	8	8
Arm curl	20 lbs	75%	15 lbs	8	8	8
Arm extension	40 lbs	75%	30 lbs	8	8	8
Back extension	60 lbs	70%	42 lbs	8	8	8

Appendix 17 / Illustrations of Exercises from Sample Resistance Program 379

APPENDIX 17 ILLUSTRATIONS OF EXERCISES FROM SAMPLE RESISTANCE PROGRAM FOR POSTMENOPAUSAL WOMEN

Wall squats

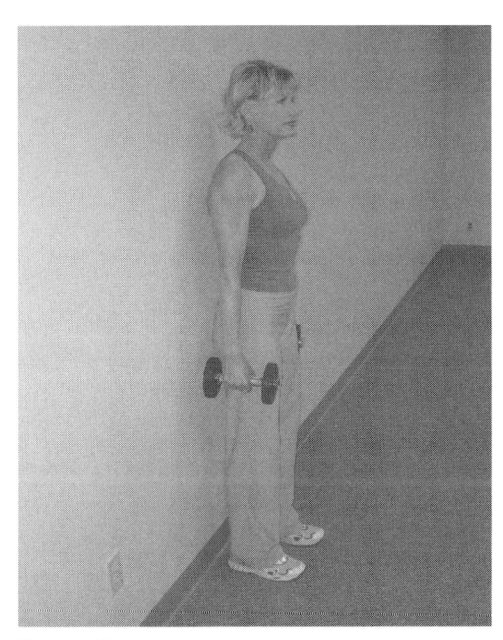

Start position

End position

Wall Squats with Ball

Wall Squats with Ball

Wall squats with ball

Wall Squats with Ball

Wall Squats with Ball

Machine Chest Press

Machine Chest Press

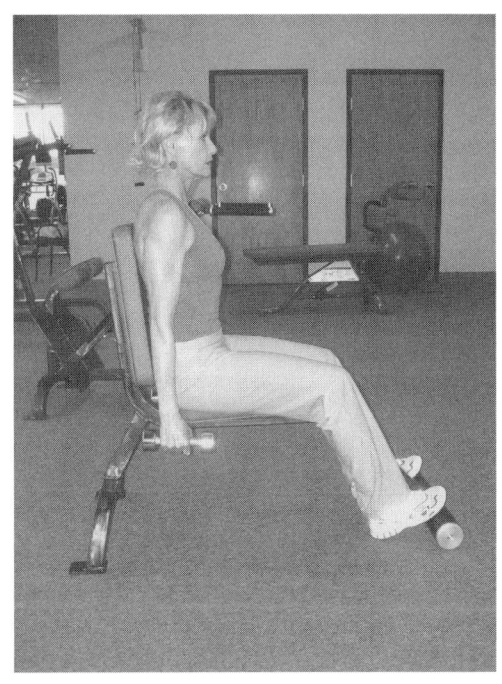

Appendix 17 / Illustrations of Exercises from Sample Resistance Program 381

Seated Dumbbell Arm Curls

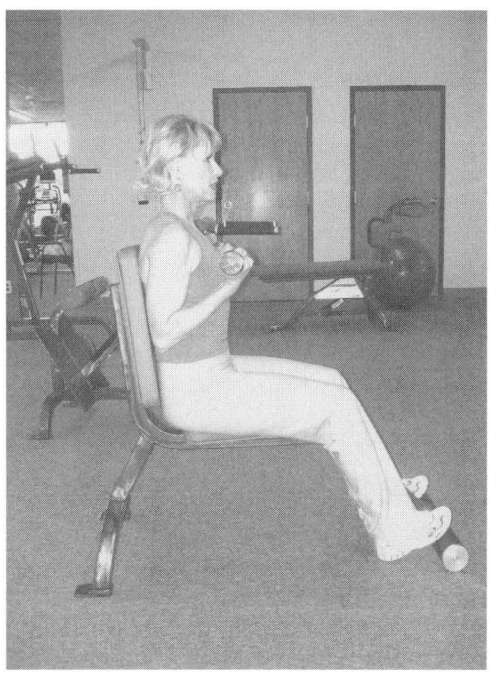

Seated Dumbbell Arm Curls

Machine Arm Curl

Machine Arm Curl

Triceps Push Down

Start position

End position

Lat Pull Down

Lat Pull Down

Appendix 17 / Illustrations of Exercises from Sample Resistance Program 383

Seated Calf Machine

Start position

End position

Seated Leg Press

Seated Leg Press

Low Seated Row

Start position

End position

Appendix 18 / Physical Activity Level (PAL) Categories and Walking Equivalence

APPENDIX 18 PHYSICAL ACTIVITY LEVEL (PAL) CATEGORIES AND WALKING EQUIVALENCE

			Walking equivalence (miles/day at 3–4 mph)*		
PAL category	PAL range	PAL	Light-weight individual (44 kg)	Middle-weight individual (70 kg)	Heavy-weight individual (120 kg)
Sedentary	1.0–1.39	1.25	∼0	∼0	∼0
Low active	1.4–1.59				
Mean		1.5	2.9	2.2	1.5
Active	1.6–1.89				
Minimum		1.6	5.8	4.4	3.0
Mean		1.75	9.9	7.3	5.3
Very active	1.9–2.49				
Minimum		1.9	14.0	10.3	17.5
Mean		2.2	22.5	16.7	12.3
Maximum		2.5	31.0	23.0	17.0

Source: From Institute of Medicine of The National Academies. *Dietary Reference Intakes for Energy, Carbohydrate, Fiber, Fat, Fatty Acids, Cholesterol, Protein, and Amino Acids.* Washington, DC: The National Academies Press, 2000.

*In addition to energy spent for the generally unscheduled activities that are part of a normal daily life. The low, middle, and high miles/day values apply to relatively heavyweight (120-kg), midweight (70-kg), and lightweight (44-kg) individuals, respectively.

APPENDIX 19 ESTIMATED ENERGY EXPENDITURE[a] PREDICTION EQUATIONS AT FOUR PHYSICAL ACTIVITY LEVELS

EER for infants and young children 0–2 years (within the 3rd–97th percentile for weight-for-height)

EER = TEE[c] + Energy deposition
0–3 months (89 × Weight of infant [kg] − 100) + 175 (kcal for energy deposition)
4–6 months (89 × Weight of infant [kg] − 100) + 56 (kcal for energy deposition)
7–12 months (89 × Weight of infant [kg] − 100) + 22 (kcal for energy deposition)
13–35 months (89 × Weight of child [kg] − 100) + 20 (kcal for energy deposition)

EER for boys 3–8 years (within the 5th–85th percentile for BMI[d])

EER = TEE + energy deposition
EER = 88.5 − (61.9 × Age [year]) + PA × (26.7 × Weight [kg] + 903 × Height [m]) + 20 (kcal for energy deposition)

EER for boys 9–18 years (within the 5th–85th Percentile for BMI[d])

EER = TEE + energy deposition
EER = 88.5 − (61.9 × Age [year]) + PA × (26.7 × Weight [kg] + 903 × Height [m]) + 25 (kcal for energy deposition)
where PA = Physical activity coefficient for boys 3–18 years:
PA = 1.0 if PAL is estimated to be ≥ 1.0 < 1.4 (sedentary)
PA = 1.13 if PAL is estimated to be ≥ 1.4 < 1.6 (low active)
PA = 1.26 if PAL is estimated to be ≥ 1.6 < 1.9 (active)
PA = 1.42 if PAL is estimated to be ≥ 1.9 < 2.5 (very active)

EER for girls 3–8 Years (within the 5th–85th percentile for BMI)

EER = TEE + energy deposition
EER = 135.3 − (30.8 × Age [year]) + PA × (10 × Weight [kg] + 934 × Height [m]) + 20 (kcal for energy deposition)

EER for girls 9–18 Years (within the 5th–85th percentile for BMI)

EER = TEE + energy deposition
EER = 135.3 − (30.8 × Age [year]) + PA × (19 × Weight [kg] + 934 × Height [m]) + 25 (kcal for energy deposition)
where PA = Physical activity coefficient for boys 3–18 years:
PA = 1.0 if PAL is estimated to be ≥ 1.0 < 1.4 (sedentary)
PA = 1.16 if PAL is estimated to be ≥ 1.4 < 1.6 (low active)
PA = 1.31 if PAL is estimated to be ≥ 1.6 < 1.9 (active)
PA = 1.56 if PAL is estimated to be ≥ 1.9 < 2.5 (very active)

EER for men 19 years and older (BMI 18.5–25 kg/m^2)

EER = TEE
EER = 662 − (9.53 × Age [year]) + PA × (15.91 × Weight [kg] + 539.6 × Height [m])
where PA = physical activity coefficient:
PA = 1.0 if PAL is estimated to be ≥ 1.0 < 1.4 (sedentary)

Appendix 19 / Estimated Energy Expenditure Prediction Equations

PA = 1.11 if PAL is estimated to be ≥ 1.4 < 1.6 (low active)
PA = 1.25 if PAL is estimated to be ≥ 1.6 < 1.9 (active)
PA = 1.48 if PAL is estimated to be ≥ 1.9 < 2.5 (very active)

EER for women 19 years and older (BMI 18.5–25 kg/m²)

EER = TEE
EER = 354 − (6.91 × Age [year]) + PA × (9.36 × Weight [kg] + 726 × Height [m])
where PA = physical activity coefficient:
PA = 1.0 if PAL is estimated to be ≥ 1.0 < 1.4 (sedentary)
PA = 1.12 if PAL is estimated to be ≥ 1.4 < 1.6 (low active)
PA = 1.27 if PAL is estimated to be ≥ 1.6 < 1.9 (active)
PA = 1.45 if PAL is estimated to be ≥ 1.9 < 2.5 (very active)

EER for pregnant women (14–18 years)

$EER_{pregnant}$ = Adolescent $EER_{nonpregnant}$ + Pregnancy Energy Deposition
First trimester = Adolescent EER + 0 (Pregnancy Energy Deposition)
Second trimester = Adolescent EER + 160 kcal (8 kcal/week × 20 week) + 180 kcal
Third trimester = Adolescent EER + 272 kcal (8 kcal/week × 24 week) + 180 kcal

EER for pregnant women (19–50 years)

$EER_{pregnant}$ = Adult $EER_{nonpregnant}$ + Pregnancy Energy Deposition
First trimester = Adult EER + 0 (Pregnancy Energy Deposition)
Second trimester = Adult EER + 160 kcal (8 kcal/week × 20 week) + 180 kcal
Third trimester = Adult EER + 272 kcal (8 kcal/week × 34 week) + 180 kcal

EER for lactating women (14–18 years)

$EER_{lactation}$ = Adolescent $EER_{prepregnancy}$ + Milk energy output − Weight loss
First 6 months = Adolescent EER + 500 − 170
Second 6 months = Adolescent EER + 400 − 0

EER for lactating women (19–50 years)

$EER_{lactation}$ = Adult $EER_{prepregnancy}$ + Milk energy output − Weight loss
First 6 months = Adult EER + 500 − 170
Second 6 months = Adult EER + 400 − 0

Weight maintenance TEE in overweight boys 3–18 years or at risk of a high BMI (BMI > 85th percentile for overweight)

TEE = 114 − (50.9 × Age [year]) + PA × (19.5 × Weight [kg] + 1161.4 × Height [m])
where PA = Physical activity coefficient:
PA = 1.0 if PAL is estimated to be ≥ 1.0 < 1.4 (sedentary)
PA = 1.12 if PAL is estimated to be ≥ 1.4 < 1.6 (low active)
PA = 1.24 if PAL is estimated to be ≥ 1.6 < 1.9 (active)
PA = 1.45 if PAL is estimated to be ≥ 1.9 < 2.5 (very active)

Weight maintenance TEE in overweight girls 3–18 years or at risk of a high BMI (BMI > 85th percentile for overweight)

TEE = 389 − (41.2 × Age [year]) + PA × (15.0 × Weight [kg] + 701.6 × Height [m])
where PA = physical activity coefficient:
PA = 1.0 if PAL is estimated to be ≥ 1.0 < 1.4 (sedentary)
PA = 1.12 if PAL is estimated to be ≥ 1.4 < 1.6 (low active)

PA = 1.24 if PAL is estimated to be ≥ 1.6 < 1.9 (active)
PA = 1.45 if PAL is estimated to be ≥ 1.9 < 2.5 (very active)

Overweight and obese men 19 years and older (BMI≥25 kg/m^2)

TEE = 1086 − (10.1 × Age [year]) + PA × (13.7 × Weight [kg] + 416 × Height [m])
where PA = 1.0 if PAL is estimated to be ≥ 1.0 < 1.4 (sedentary)
PA = 1.16 if PAL is estimated to be ≥ 1.4 < 1.6 (low active)
PA = 1.27 if PAL is estimated to be ≥ 1.6 < 1.9 (active)
PA = 1.44 if PAL is estimated to be ≥ 1.9 < 2.5 (very active)

Overweight and obese women 19 years and older (BMI≥25 kg/m^2)

TEE = 448 − (7.95 × Age [year]) + PA × (11.4 × Weight [kg] + 619 × Height [m])
where PA = 1.0 if PAL is estimated to be ≥ 1.0 < 1.4 (sedentary)
PA = 1.16 if PAL is estimated to be ≥ 1.4 < 1.6 (low active)
PA = 1.27 if PAL is estimated to be ≥ 1.6 < 1.9 (active)
PA = 1.44 if PAL is estimated to be ≥ 1.9 < 2.5 (very active)

Normal and overweight or obese men 19 years and older (BMI≥18.5 kg/m^2)

TEE = 864 − (9.72 × Age [year]) + PA × (14.2 × Weight [kg] + 503 × Height [m])
where PA = 1.0 if PAL is estimated to be ≥ 1.0 < 1.4 (sedentary)
PA = 1.12 if PAL is estimated to be ≥ 1.4 < 1.6 (low active)
PA = 1.27 if PAL is estimated to be ≥ 1.6 < 1.9 (active)
PA = 1.54 if PAL is estimated to be ≥ 1.9 < 2.5 (very active)

Normal and overweight or obese women 19 years and older (BMI≥18.5 kg/m^2)

TEE = 387 − (7.31 × Age [year]) + PA × (10.9 × Weight [kg] + 600.7 × Height [m])
where PA = 1.0 if PAL is estimated to be ≥ 1.0 < 1.4 (sedentary)
PA = 1.14 if PAL is estimated to be ≥ 1.4 < 1.6 (low active)
PA = 1.27 if PAL is estimated to be ≥ 1.6 < 1.9 (active)
PA = 1.45 if PAL is estimated to be ≥ 1.9 < 2.5 (very active)

Source: From Institute of Medicine, Food and Nutrition Board. Dietary *Reference Intakes for Energy, Carbohydrate, Fiber, Fat, Fatty Acids, Cholesterol, Protein, and Amino Acids*. Washington, DC: The National Academies Press, 2002. http:www.nap.edu.

[a] Estimated energy expenditure (EER) is the average dietary energy intake that is predicted to maintain energy balance in a healthy adult of a defined age, gender, weight, height, and level of physical activity consistent with good health. In children and pregnant and lactating women, the EER includes the needs associated with the deposition of tissues or the secretion of milk at rates consistent with good health.

[b] Physical activity level (PAL) is the physical activity level that is the ratio of the total energy expenditure to the basal energy expenditure.

[c] Total energy expenditure (TEE) is the sum of the resting energy expenditure, energy expended in physical activity, and the thermic effect of food.

[d] Body mass index (BMI) is determined by dividing the weight (in kilograms) by the square of the height (in meters).

APPENDIX 20 ESTIMATED CALORIE REQUIREMENTS (IN KILOCALORIES) FOR SPECIFIC AGE GROUPS AT THREE LEVELS OF PHYSICAL ACTIVITY[a] USING THE INSTITUTE OF MEDICINE (IOM) EQUATIONS

		Activity level		
Gender	Age (years)	Sedentary[b]	Moderately active[c]	Active[d]
Child	2–3	1000	1000–1400[e]	1000–1400[e]
Female	4–8	1200	1400–1600	1400–1800
	9–13	1600	1600–2000	1800–2000
	14–18	1800	2200	2400
	19–30	2000	2000–2200	2400
	31–50	1800	2000	2200
	51+	1600	1800	2000–2200
Male	4–8	1400	1400–1600	1600–2000
	9–13	1800	1800–2200	2000–2600
	14–18	2200	2400–2800	2800–3200
	19–30	2400	2600–2800	3000
	31–50	2200	2200–2600	2800–3000
	51+	2000	2200–2400	2400–2800

[a] These levels are based on Estimated Energy Requirements EERs from the Institute of Medicine Dietary Reference Intakes macronutrients report, 2002, calculated by gender, age, and activity level for reference-sized individuals. "Reference size," as determined by IOM, is based on median height and weight for ages up to age 18 years of age and median height and weight for that height to give a BMI of 21.5 for adult females and 22.5 for adult males. The estimates are rounded to the nearest 200 cal.

[b] Sedentary means a lifestyle that includes only the light physical activity associated with typical day-to-day life.

[c] Moderately active means a lifestyle that includes physical activity equivalent to walking about 1.5–3 miles per day at 3–4 miles per hour, in addition to the light physical activity associated with typical day-to-day life.

[d] Active means a lifestyle that includes physical activity equivalent to walking more than 3 miles per day at 3–4 miles per hour, in addition to the light physical activity associated with typical day-to-day life.

[e] The calorie ranges shown are to accommodate needs of different ages within the group. For children and adolescents, more calories are needed at older ages. For adults, fewer calories are needed at older ages.

APPENDIX 21 NUTRITION QUESTIONNAIRE WITH 3-DAY RECALL

1. Name_____
2. Age_____ Circle: Male or Female
3. Height_____ Weight_____
4. Usual weight_____Desired weight_____
5. Have you lost weight in the last month? Yes/No
   If yes how much_____
6. Have you gained weight in the last month? Yes/No
   If yes how much_____
7. Please circle activity level:
   Sedentary (sit most of the day)
   Active (move most of the day)
   Very active (aerobic exercise three times a week)
   Athletic (aerobic exercise five times a week)
8. Do you have any food allergies or intolerance? If yes, what are they?
   _____

7. Does your medical history include any of the following: (circle any that apply) Diabetes, ulcers, heart disease, hypertension, stroke, vascular disease, renal disease, or gastrointestinal disease?
   _____

8. Has your physician recommended any special diet? If so, what type of diet?
   _____

9. What type of surgeries have you had?
   _____

10. Are you currently taking any medication(s)? Please circle them all (please include all over-the-counter medication and/or all vitamin, mineral, or herbal supplements):
    _____
    _____

11. Do you restrict any foods due to religious beliefs? If yes, what type?
    _____

12. Please write a food diary of everything that you ate in the past 3 days (or the last 72 h period). Include portion size of foods eaten, the type of food, and how it was prepared. Be specific: 1 cup skim milk, baked chicken breast, $^1/_2$ cup green beans, etc.

Appendix 21 / Nutrition Questionnaire with 3-Day Recall

Day 1:

Item:_____Amount:_____
Item:_____Amount:_____
Item:_____Amount:_____
Item:_____Amount:_____
Item:_____Amount:_____
Item:_____Amount:_____
Item:_____Amount:_____
Item:_____Amount:_____
Item:_____Amount:_____
Item:_____Amount:_____
Item:_____Amount:_____
Item:_____Amount:_____
Item:_____Amount:_____
Item:_____Amount:_____
Item:_____Amount:_____
Item:_____Amount:_____
Item:_____Amount:_____

Day 2:

Item:_____Amount:_____
Item:_____Amount:_____
Item:_____Amount:_____
Item:_____Amount:_____
Item:_____Amount:_____
Item:_____Amount:_____
Item:_____Amount:_____
Item:_____Amount:_____
Item:_____Amount:_____
Item:_____Amount:_____
Item:_____Amount:_____
Item:_____Amount:_____
Item:_____Amount:_____
Item:_____Amount:_____
Item:_____Amount:_____
Item:_____Amount:_____
Item:_____Amount:_____

Day 3:

Item:_____Amount:_____
Item:_____Amount:_____
Item:_____Amount:_____
Item:_____Amount:_____
Item:_____Amount:_____
Item:_____Amount:_____
Item:_____Amount:_____

Item:_____Amount:_____
Item:_____Amount:_____
Item:_____Amount:_____
Item:_____Amount:_____
Item:_____Amount:_____
Item:_____Amount:_____
Item:_____Amount:_____
Item:_____Amount:_____
Item:_____Amount:_____
Item:_____Amount:_____

APPENDIX 22 FOOD FREQUENCY QUESTIONNAIRE

The following questionnaire is designed to help the dietician determine the frequency of food use. It should be used in conjunction with a 3-day recall of food intake. Record as accurately as possible. Amounts should be recorded in measurable amounts (e.g., cups, pounds, and teaspoons) and frequencies should be recorded in measurable amounts of time (e.g., 1 day, 3 months, and 2 weeks).

1. Do you drink milk? If so, what kind?
   _____ whole
   _____ 2%
   _____ skim
   _____ other
   How much? _____

2. Do you use fats? If so, what kinds?
   _____ butter
   _____ margarine
   _____ oil
   _____ other
   How much? _____

3. How often do you eat meat?_____ poultry?_____ fish?_____ eggs?_____ cheese?_____ cold cuts?_____ peanut butter?_____ dried beans?_____

4. Do you eat snack foods? If so which ones?_____
   _____ _____
   How often?_____
   How much?_____

5. What vegetables do you eat? How often?
   a. Broccoli _____,
      green peppers _____,
      cooked greens _____,
      carrots _____
      sweet potatoes _____,
      winter squash _____.
   b. Tomatoes _____,
      raw cabbage _____,
      white potatoes _____,
      other raw vegetables _____.
   c. Asparagus_____,
      beets_____,
      cauliflower _____,
      corn_____,
      celery _____,

peas _____,
lettuce_____,
green beans_____.

6. What fruits do you eat? How often?

   a. Apples/applesauce_____, apricots_____, bananas _____, berries ___, cherries ___, grapes ___, peaches ___, pears ___, plums ___, pineapple ___, raisins ___, other_____
   b. Oranges/orange juice _____, grapefruits/grapefruit juice_____

7. Bread and cereal products?

   a. How much bread do you eat with meals?_____
      Between meals?_____
   b. Do you eat cereal daily? _____ weekly?_____ cooked?_____ dry?_____
   c. How often do you eat foods (rice, macaroni, spaghetti, etc.)_____

8. Do you eat canned soup?_____
   Homemade soup? _____
   What kinds? _____
   How often? _____

9. Do you use salt?_____ Do you salt foods before you taste them? _____
   Do you cook with salt?_____ Do you crave salt/salty foods?_____

10. How many teaspoons of sugar or honey do you use perday? (include sugar used on cereals, toast, fruit, and beverages)_____

11. Do you drink any of the following?
    Water?_____ how much?_____
    Coffee, tea, decaf, etc.?_____
    howmuch?_____
    Carbonated beverages?_____ how much?_____
    Beer, wine, liquor?_____ how much?_____
    Other?_____ how much?_____

12. Are there any other foods not listed that you eat frequently?
    _____
    _____
    _____

13. Please rank how you determine your choice of frequently eaten foods?
    Caloric content_____
    Taste_____
    Availability_____
    Price_____
    Nutrient Content (fat, carbohydrate, protein)_____

Source: Adapted from Krey S.H., Murray R.L. *Dynamics of Nutrition Support: Assessment, Evaluation, and Implementation.* Norwalk, CT: Appleton-Century-Crofts, 1986.

APPENDIX 23 US DEPARTMENT OF HEALTH AND HUMAN SERVICES AND THE US DEPARTMENT OF AGRICULTURE, 2005 DIETARY GUIDELINES FOR AMERICANS

Food group	Major nutrients	Suggestions
Grains	Carbohydrate, thiamin, riboflavin (if enriched), niacin, folate (whole grains and enriched), magnesium (whole grains and enriched), iron (if enriched), zinc, dietary fiber	Make half your grains whole. Eat at least 3 oz of whole-grain cereals, breads, crackers, rice, or pasta every day. 1 oz is about 1 slice of bread, about 1 cup of breakfast cereal, or 1/2 cup of cooked rice, cereal, or pasta cereal
Vegetables	Carbohydrate, vitamin C, vitamin A, folate, magnesium, potassium, dietary fiber	Vary your vegetables Eat more dark green vegetables such as broccoli, spinach, and other dark leafy vegetables Eat more orange vegetables such carrots and sweet potatoes Eat more dry beans and peas such as pinto beans, kidney beans, and lentils
Fruits	Carbohydrate, vitamin C, folate, magnesium, potassium, dietary fiber	Eat a variety of fruit Choose fresh, frozen, canned, or dried fruit Go easy on fruit juice
Milk	Calcium, phosphorus, carbohydrate, protein, riboflavin, vitamin D, magnesium, zinc	Get you calcium-rich foods Go low-fat or fat-free when you choose milk, yogurt, and other milk products
Meat and Beans	Protein, thiamin, riboflavin, niacin, vitamin B_6, folate (plant protein), vitamin B_{12} (only in animals), phosphorus, magnesium (plant sources), iron, zinc	Go lean with protein Choose low-fat or lean meats and poultry Bake it, broil it, or grill it Vary your protein routine—choose more fish, beans, peas, nuts, and seeds

Note: These are general guidelines, and the guidelines should be adjusted for your age, gender, and activity level (http://www.mypyramid.gov/mypyramid/index.aspx). Appendix 24 lists the recommended serving sizes based on energy needs. For example, for a 2000-cal diet, you should eat 6 oz of grains every day, $2\frac{1}{2}$ cups of vegetables every day, 2 cups of fruit every day, 3 cups of milk every day (for kids aged 2–8, it is 2) $5\frac{1}{2}$ oz of meat and beans every day.

APPENDIX 24 MYPYRAMID FOOD INTAKE PATTERNS AT VARYING CALORIE LEVELS WITH DISCRETIONARY CALORIES

The suggested amounts of food to consume from the basic food groups, subgroups, and oils to meet recommended nutrient intakes at 12 different calorie levels. Nutrient and energy contributions from each group are calculated according to the nutrient-dense forms of foods in each group (e.g., lean meats and fat-free milk). The table also shows the discretionary calorie allowance that can be accommodated within each calorie level, in addition to the suggested amounts of nutrient-dense forms of foods in each group.

	Daily amount of food from each group											
Calorie level[1]	1000	1200	1400	1600	1800	2000	2200	2400	2600	2800	3000	3200
Fruits[2] (cups)	1	1	1.5	1.5	1.5	2	2	2	2	2.5	2.5	2.5
Vegetables[3] (cups)	1	1.5	1.5	2	2.5	2.5	3	3	3.5	3.5	4	4
Grains[4] (oz-eq)	3	4	5	5	6	6	7	8	9	10	10	10
Meat and beans[5] (oz-eq)	2	3	4	5	5	5.5	6	6.5	6.5	7	7	7
Milk[6] (cups)	2	2	2	3	3	3	3	3	3	3	3	3
Oils[7] (tsp)	3	4	4	5	5	6	6	7	8	8	10	11
Discretionary calorie allowance[8]	165	171	171	182	195	267	29	362	410	426	512	648

1. **Calorie levels** are set across a wide range to accommodate the needs of different individuals. The attached table "Estimated Daily Calorie Needs" can be used to help assign individuals to the food intake pattern at a particular calorie level.

2. **Fruit group** includes all fresh, frozen, canned, and dried fruits and fruit juices. In general, 1 cup of fruit or 100% fruit juice, or $\frac{1}{2}$ cup of dried fruit can be considered as 1 cup from the fruit group.

Appendix 24 / MyPyramid Food Intake Patterns at Varying Calorie Levels 397

3 **Vegetable group** includes all fresh, frozen, canned, and dried vegetables and vegetable juices. In general, 1 cup of raw or cooked vegetables or vegetable juice, or 2 cups of raw leafy greens can be considered as 1 cup from the vegetable group.

	Vegetable subgroup amounts are per week (cal/week)											
Calorie level	1000	1200	1400	1600	1800	2000	2200	2400	2600	2800	3000	3200
Dark green vegetables	1	1.5	1.5	2	3	3	3	3	3	3	3	3
Orange vegetables	0.5	1	1	1.5	2	2	2	2	2.5	2.5	2.5	2.5
Legumes	0.5	1	1	2.5	3	3	3	3	3.5	3.5	3.5	3.5
Starchy vegetables	1.5	2.5	2.5	2.5	3	3	6	6	7	7	9	9
Other vegetables	3.5	4.5	4.5	5.5	6.5	6.5	7	7	8.5	8.5	10	10

4 **Grains group** includes all foods made from wheat, rice, oats, and cornmeal, barley, such as bread, pasta, oatmeal, breakfast cereals, tortillas, and grits. In general, 1 slice of bread, 1 cup of ready-to-eat cereal, or $\frac{1}{2}$ cup of cooked rice, pasta, or cooked cereal can be considered as 1 ounce equivalent from the grain group. **At least half of all grains consumed should be whole grains**.

5 **Meat and beans group** includes, in general, 1 ounce of lean meat, poultry, or fish, 1 egg, 1 tbsp peanut butter, $\frac{1}{4}$ cup cooked dry beans, or $\frac{1}{2}$ ounce of nuts or seeds that can be considered as 1 ounce equivalent from the meat and beans group.

6 **Milk group** includes all fluid milk products and foods made from milk that retain their calcium content, such as yogurt and cheese. Foods made from milk that have little to no calcium, such as cream cheese, cream, and butter, are not part of the group. Most milk group choices should be fat-free or low in fat. In general, 1 cup of milk or yogurt, $1\frac{1}{2}$ ounces of natural cheese, or 2 ounces of processed cheese can be considered as 1 cup from the milk group.

7 **Oils** include fats from many different plants and from fish that are liquid at room temperature, such as canola, corn, olive, soybean, and sunflower oil. Some foods are naturally high in oils, like nuts, olives, some fish, and avocados. Foods that are mainly oil include mayonnaise, certain salad dressings, and soft margarine.

8 **Discretionary calorie allowance** is the remaining amount of calories in a food intake pattern after accounting for the calories needed for all food groups—using forms of foods that are fat-free or low in fat and with no added sugars.

Appendix 25 Dietary Reference Intakes (DRIs): Recommended Intakes for Individuals, Macronutrients (from Food and Nutrition Board, Institute of Medicine, and National Academies)

Life-stage group	Total water[a] (L/day)	Carbohydrate (g/day)	Total fiber (g/day)	Fat (g/day)	Linoleic acid (g/day)	α-Linoleic acid (g/day)	Protein[b] (g/day)
Females							
9–13 years	2.1*	**130**	26*	ND	10*	1.0*	**34**
14–18 years	2.3*	**130**	26*	ND	11*	1.1*	**46**
19–30 years	2.7*	**130**	25*	ND	12*	1.1*	**46**
31–50 years	2.7*	**130**	25*	ND	12*	1.1*	**46**
51–70 years	2.7*	**130**	21*	ND	11*	1.1*	**46**
>70 years	2.7*	**130**	21*	ND	11*	1.1*	**46**
Pregnancy							
14–18 years	3.0*	**175**	28*	ND	13*	1.4*	**71**
19–30 years	3.0*	**175**	28*	ND	13*	1.4*	**71**
31–50 years	3.0*	**175**	28*	ND	13*	1.4*	**71**
Lactation							
14–18 years	3.8*	**210**	29*	ND	13*	1.3*	**71**
19–30 years	3.8*	**210**	29*	ND	13*	1.3*	**71**
31–50 years	3.8*	**210**	29*	ND	13*	1.3*	**71**

Note: This table presents recommended dietary allowances (RDAs) in **bold** face type and adequate intakes (AIs) in ordinary type followed by an asterisk (*). RDAs and AIs may both be used as goals for individual intake. RDAs are set to meet the needs of almost all (97–98%) individuals in a group. For healthy breastfed infants, the AI is the mean intake. The AI for other life stage and gender groups is believed to cover the needs of all individuals in the group, but the lack of data or uncertainty in the data prevent one from being able to specify with confidence the percentage of individuals covered by this intake.

[a] Total water includes all water contained in food, beverages, and drinking water.
[b] Based on 0.8 g/kg body weight for the reference body weight.

APPENDIX 26 DIETARY REFERENCE INTAKES (DRIs): ACCEPTABLE MACRONUTRIENT DISTRIBUTION RANGES

	Range (% of energy)		
Macronutrient	Children (1–3 years)	Children (4–18 years)	Adults
Fat	30–40	23–35	20–35
n-6 polyunsaturated fatty acids[a] (linoleic acid)	5–10	5–10	5–10
n-3 polyunsaturated fatty acids[a] (linoleic acid)	0.6–1.2	0.6–1.2	0.6–1.2
Carbohydrate	45–65	45–65	45–65
Protein	5–20	10–30	10–35

[a] Approximately 10% of the total can come from longer-chain n-3 or n-6 fatty acids.

Source: Food and Nutrition Board, Institute of Medicine of the National Academies. *Dietary Reference Intakes for Energy, Carbohydrate, Fiber, Fat, Fatty Acids, Cholesterol, Protein, and Amino Acids.* Washington, DC: The National Academy Press, 2002.

APPENDIX 27 DIETARY REFERENCE INTAKES (DRIs): ESTIMATED AVERAGE REQUIREMENTS FOR GROUPS (FROM FOOD AND NUTRITION BOARD, INSTITUTE OF MEDICINE, AND NATIONAL ACADEMIES)

Life-stage group	CHO (g/day)	Protein (g/day)[a]	Vitamin A (mg/day)[b]	Vitamin C (mg/day)	Vitamin E (mg/day)[c]	Thiamin (mg/day)	Riboflavin (mg/day)	Niacin (mg/day)[d]	Vitamin B$_6$ (mg/day)	Folate (mg/day)[e]	Vitamin B$_{12}$ (mg/day)	Copper (mg/day)	Iodine (mg/day)	Iron (mg/day)	Magnesium (mg/day)	Molybdenum (mg/day)	Phosphorus (mg/day)	Selenium (mg/day)	Zinc (mg/day)
Females																			
9–13 years	100	28	420	39	9	0.7	0.8	9	0.8	250	1.5	540	73	5.7	200	26	1055	35	7.0
14–18 years	100	38	485	56	12	0.9	0.9	11	1.0	330	2.0	685	95	7.9	300	33	1055	45	7.3
19–30 years	100	38	500	60	12	0.9	0.9	11	1.1	320	2.0	700	95	8.1	255	34	580	45	6.8
31–50 years	100	38	500	60	12	0.9	0.9	11	1.1	320	2.0	700	95	8.1	265	34	580	45	6.8
51–70 years	100	38	500	60	12	0.9	0.9	11	1.3	320	2.0	700	95	5	265	34	580	45	6.8
>70 years	100	38	500	60	12	0.9	0.9	11	1.3	320	2.0	700	95	5	265	34	580	45	6.8
Pregnancy																			
14–18 years	135	50	530	66	12	1.2	1.2	14	1.6	520	2.2	785	160	23	335	40	1055	49	10.5
19–30 years	135	50	550	70	12	1.2	1.2	14	1.6	520	2.2	800	160	22	290	40	580	49	9.5
31–50 years	135	50	550	70	12	1.2	1.2	14	1.6	520	2.2	800	160	22	300	40	580	49	9.5
Lactation																			
14–18 years	160	60	885	96	16	1.2	1.3	13	1.7	450	2.4	985	209	7	300	35	1055	59	10.9
19–30 years	160	60	900	100	16	1.2	1.3	13	1.7	450	2.4	1000	209	6.5	255	36	580	59	10.4
31–50 years	160	60	900	100	16	1.2	1.3	13	1.7	450	2.4	1000	209	6.5	265	36	580	59	10.4

(Continued)

Appendix 27
(Continued)

Note: This table presents estimated average requirements (EARs), which serve two purposes: for assessing adequacy of population intakes and as the basis for calculating recommended dietary allowances (RDAs) for individuals for those nutrients. EARs have not been established for vitamin D, vitamin K, pantothenic acid, biotin, choline, calcium, chromium, fluoride, manganese, or other nutrients not yet evaluated via the DRI process.

[a] For individual at reference weight (Table 1-1).

[b] As retinol activity equivalents (RAEs): 1 RAE = 1 µg β-carotene, 24 µg α-carotene, or 24 µg β-cryptoxanthin. The RAE for dietary provitamin A carotenoids is twofold greater than retinol equivalents (REs), whereas the RAE for preformed vitamin A is the same as RE.

[c] As α-tocopherol: α-tocopherol includes RRR-α-tocopherol, the only form of α-tocopherol that occurs naturally in foods, and the $2R$-stereoisomeric forms of α-tocopherol (RRR-α-tocopherol, RSR-α-tocopherol, RRS-α-tocopherol, and RSS-α-tocopherol) that occur in fortified foods and supplements. It does not include the $2S$-stereoisomeric forms of of α-tocopherol (SRR-α-tocopherol, SSR-α-tocopherol, SRS-α-tocopherol, and SSS-α-tocopherol), also found in fortified foods and supplements.

[d] As niacin equivalents (NE): 1 mg of niacin = 60 mg of tryptophan.

[e] As dietary folate equivalents (DFEs): 1 DFE = 1 µg food folate = 0.6 µg of folic acid from fortified food or as a supplement consumed with food = 0.5 µg of a supplement taken on an empty stomach.

Sources: Dietary Reference Intakes for Calcium, Phosphorus, Magnesium, Vitamin D, and Flouride (1997); *Dietary Reference Intakes for Thiamin, Riboflavin, Niacin, Vitamin B_6, Folate, Vitamin B_{12}, Pantothenic Acid, Biotin, and Choline* (1998); *Dietary Reference Intakes for Vitamin C, Vitamin E, Selenium, and Carotenoids* (2000); *Dietary Reference Intakes for Vitamin A, Vitamin K, Arsenic, Boron, Chromium, Copper, Iodine, Iron, Manganese, Molybdenum, Nickel, Silicon, Vanadium, and Zinc* (2001), *and Dietary Reference Intakes for Energy, Carbohydrate, Fiber, Fat, Fatty Acids, Cholesterol, Protein, and Amino Acids* (2002). These reports may be accessed via http://www.nap.edu. Copyright 2002 by the National Academy of Sciences. All rights reserved.

APPENDIX 28 DIETARY REFERENCE INTAKES (DRIs): RECOMMENDED INTAKES FOR INDIVIDUALS, ELEMENTS (FROM FOOD AND NUTRITION BOARD, INSTITUTE OF MEDICINE, AND NATIONAL ACADEMIES)

Life-stage group	Calcium (mg/day)	Chromium (mg/day)	Copper (mg/day)	Flouride (mg/day)	Iodine (mg/day)	Iron (mg/day)	Magnesium (mg/day)	Manganese (mg/day)	Molybdenum (mg/day)	Phosphorus (mg/day)	Selenium (mg/day)	Zinc (mg/day)	Potassium	Sodium	Chloride
Females															
9–13 years	1300*	21*	700	2*	120	8	240	1.6*	34	1250	40	8	4.5*	1.5*	2.3*
14–18 years	1300*	24*	890	3*	150	15	360	1.6*	43	1250	55	9	4.7*	1.5*	2.3*
19–30 years	1000*	25*	900	3*	150	18	310	1.8*	45	700	55	8	4.7*	1.5*	2.3*
31–50 years	1000*	25*	900	3*	150	18	320	1.8*	45	700	55	8	4.7*	1.5*	2.3*
51–70 years	1200*	20*	900	3*	15	8	320	1.8*	45	700	55	8	4.7*	1.3*	2.0*
>70 years	1200*	20*	900	3*	150	8	320	1.8*	45	700	55	8	4.7*	1.2*	1.8*
Pregnancy															
14–18 years	1300*	29*	1000	3*	220	27	400	2.0*	50	1250	60	12	4.7*	1.5*	2.3*
19–30 years	1000*	30*	1000	3*	220	27	350	2.0*	50	700	60	11	4.7*	1.5*	2.3*
31–50 years	1000*	30*	1000	3*	220	27	360	2.0*	50	700	60	11	4.7*	1.5*	2.3*
Lactation															
14–18 years	1300*	44*	1300	3*	290	10	360	2.6*	50	1250	70	13	5.1*	1.5*	2.3*
19–30 years	1000*	45*	1300	3*	290	9	310	2.6*	50	700	70	12	5.1*	1.5*	2.3*
31–50 years	1000*	45*	1300	3*	290	9	320	2.6*	50	700	70	12	5.1*	1.5*	2.3*

Note: This table presents recommended dietary allowances (RDAs) in **bold** face type and adequate intakes (AIs) in ordinary type followed by an asterisk (*). RDAs and AIs asterisk (*). RDAs and AIs may both be used as goals for individual intake. RDAs are set to meet the needs of almost all (97–98%) individuals in a group. For healthy breastfed infants, the AI is the mean intake. The AI for other life stage and gender groups is believed to cover needs of all individuals in the group, but the lack of data or uncertainty in the data prevent one from being able to specify with confidence the percentage of individuals covered by this intake.

Source: Dietary Reference Intakes for Calcium, Phosphorous, Magnesium, Vitamin D, and Fluoride (1997); Dietary Intakes for Thiamin, Riboflavin, Niacin, Vitamin B₆, Folate, Vitamin B₁₂, Pantothenic Acid, Biotin, and Choline (1998); Dietary Reference for Vitamin C, Vitamin E, Selenium, and Carotenoids (2000); Dietary Reference Intakes for Vitamin A, Vitamin K, Arsenic, Boron, Chromium, Copper, Iodine, Iron, Manganese, Molybdenum, Nickel, Silicon, Vanadium, and Zinc (2001); and Dietary Reference Intakes for Water Potassium, Sodium, Chloride and Sulfate (2005). These reports may be accessed via http://www.nap.edu.

APPENDIX 29 DIETARY REFERENCE INTAKES (DRIs): RECOMMENDED INTAKES FOR INDIVIDUALS, VITAMINS (FROM THE FOOD AND NUTRITION BOARD, INSTITUTE OF MEDICINE, AND NATIONAL ACADEMIES)

Life-stage group	Vitamin A (μg/day)[a]	Vitamin C (mg/day)	Vitamin D (μg/day)[b,c]	Vitamin E (mg/day)[d]	Vitamin K (μg/day)	Thiamin (mg/day)	Riboflavin (mg/day)	Niacin (mg/day)[e]	Vitamin B$_6$ (mg/day)	Folate (μg/day)[f]	Vitamin B$_{12}$ (μg/day)	Pantothenic acid (mg/day)	Biotin (μg/day)	Choline[g] (mg/day)
Females														
9–13 years	**600**	**45**	5*	**11**	60*	**0.9**	**0.9**	**12**	**1.0**	**300**	**1.8**	4*	20*	375*
14–18 years	**700**	**65**	5*	**15**	75*	**1.0**	**1.0**	**14**	**1.2**	**400**[j]	**2.4**	5*	25*	400*
19–30 years	**700**	**75**	5*	**15**	90*	**1.1**	**1.1**	**14**	**1.3**	**400**[j]	**2.4**	5*	30*	425*
31–50 years	**700**	**75**	5*	**15**	90*	**1.1**	**1.1**	**14**	**1.3**	**400**[j]	**2.4**	5*	30*	425*
51–70 years	**700**	**75**	10*	**15**	90*	**1.1**	**1.1**	**14**	**1.5**	**400**	**2.4**[h]	5*	30*	425*
>70 years	**700**	**75**	15*	**15**	90*	**1.1**	**1.1**	**14**	**1.5**	**400**	**2.4**[h]	5*	30*	425*
Pregnancy														
14–18 years	**750**	**80**	5*	**15**	75*	**1.4**	**1.4**	**18**	**1.9**	**600**[j]	**2.6**	6*	30*	450*
19–30 years	**770**	**85**	5*	**15**	90*	**1.4**	**1.4**	**18**	**1.9**	**600**[j]	**2.6**	6*	30*	450*
31–50 years	**770**	**85**	5*	**15**	90*	**1.4**	**1.4**	**18**	**1.9**	**600**[j]	**2.6**	6*	30*	450*
Lactation														
14–18 years	**1200**	**115**	5*	**19**	75*	**1.6**	**1.6**	**17**	**2.0**	**500**	**2.8**	7*	35*	550*
19–30 years	**1300**	**120**	5*	**19**	90*	**1.6**	**1.6**	**17**	**2.0**	**500**	**2.8**	7*	35*	550*
31–50 years	**1300**	**120**	5*	**19**	90*	**1.6**	**1.6**	**17**	**2.0**	**500**	**2.8**	7*	35*	550*

Note: This table (taken from the DRI reports; *see* http://www.nap.edu) presents recommended dietary allowances (RDAs) in **bold** face type and adequate intakes (AIs) in ordinary type followed by an asterisk (*). RDAs and AIs may both be used as goals for individual intake. RDAs are set to meet the needs of almost all (97–98%) individuals in a group. For healthy breastfed infants, AI is the mean intake. The AI for other life stage and gender groups is believed to cover needs of all individuals in the group, but the lack of data or uncertainty in the data prevent one from being able to specify with confidence the percentage of individuals covered by this intake.

(Continued)

Appendix 29
(Continued)

[a] As retinol activity equivalents (RAEs): 1 RAE - 1 μg β-carotene, 24 μg α-carotene, or 24 μg β-cryptoxanthin. The RAE for dietary provitamin A carotenoids is twofold greater than retinol equivalents (REs), whereas the RAE for preformed vitamin A is the same as RE.

[b] As cholecalciferol: 1 μg cholecalciferol = 40 IU vitamin D.

[c] In the absence of adequate exposure to sunlight.

[d] As α-tocopherol: α-tocopherol includes RRR-α-tocopherol, the only form α-tocopherol that occurs naturally in foods, and the 2R-steroisomeric forms of α-tocopherol (RRR-α-tocopherol, RSR-α-tocopherol, RRS-α-tocopherol, and RSS-α-tocopherol) that occur in fortified foods and supplements. It does not include the 2S-steroisomeric forms of α-tocopherol (SRR-α-tocopherol, SSR-α-tocopherol, SRS-α-tocopherol, and SSS-α-tocopherol), also found in fortified foods and supplements.

[e] As niacin equivalents (NE): 1 mg of niacin = 60 mg of tryptophan; 0–6 months = preformed niacin (not NE).

[f] As dietary folate equivalents (DFEs): 1 DFE = 1 μg food folate = 0.6 μg of folic acid from fortified food or as a supplement consumed with food = 0.5 μg of a supplement taken on an empty stomach.

[g] Although AIs have been set for choline, there are few data to assess whether a dietary supply of choline is needed at all stages of the life cycle, and it may be that the choline requirement can be met by endogenous synthesis at some of these stages.

[h] Because 10–30% of older people may malabsorb food-bound B_{12}, it is advisable for those older than 50 years to meet their RDA mainly by consuming foods fortified with B_{12} or a supplement containing B_{12}.

[i] In view of evidence linking folate intake with neural tube defects in the fetus, it is recommended that all women capable of becoming pregnant consume 400 μg from supplements or fortified foods in addition to intake of food folate from a varied diet.

[j] It is assumed that women will continue consuming 400 μg from supplements or fortified food until their pregnancy is confirmed and they enter prenatal care, which ordinarily occurs after the end of the periconception period—the critical time for formation of the neural tube.

Copyright 2004 by the National Academy of Sciences. All rights reserved.

REVIEW QUESTIONS
THE ACTIVE FEMALE:
HEALTH ISSUES THROUGHOUT THE LIFESPAN

Part I- Focusing on Active Female's Health Issues: Unique Gender-Related Psychological and Physiological Characteristics of Females

Chapter 1: Body Image Concerns Throughout the Lifespan

1. According to Thompson et al. (2001), the term body image can *best* be defined as _____.

   a. an internal dissatisfaction with society's view of the ideal body
   b. the internal representation of one's own outer appearance
   c. a negative subjective evaluation of one's physical body
   d. an image that compares height and weight to a standardized norm

2. Body image concerns are best conceptualized as occurring along a continuum. Which of the following statements with respect to body image concerns represents a clinical condition?

   a. dissatisfaction with one's body
   b. negative subjective evaluation of weight
   c. negative body image
   d. body image disturbance

3. Evidence suggests that African-American females and Caucasian females internalize media representation of the female body in the same way.

   a. true
   b. false

4. The statement that body image construct has considerable lability means _____.

   a. body image is a reliable measure of psychological disturbance
   b. body image may change from setting to setting
   c. female athletes may experience positive effect in a sports environment yet experience negative body image within a larger social environment
   d. both b and c are acceptable answers

5. Which life stage is viewed as the stage of greatest risk in the development of body image concerns?

   a. adolescence
   b. young adulthood

c. pregnancy
d. old age

6. The perceptual component of body image refers to _____.

    a. an individual's attitudes about their body
    b. an individual's thoughts about their body
    c. an individual's ability to accurately estimate body size
    d. an individual's emotional reaction to their body

7. Which scale(s) has (have) an internal consistency and test–retest reliability score above 0.70?

    a. body Image Quality of Life
    b. body Image Concern Inventory
    c. physical Appearance State and Trait Anxiety Scale
    d. all of the above

8. Current/proximal factors to consider in body image assessment are _____.

    a. impact of body image concerns, external clues, and personality attributes
    b. sociodemographic factors, comorbidity, and physical characteristics of attribute
    c. patient's investment in appearance, cognitive and emotional processing, and impact of body image concerns
    d. all of the above

9. According to Wardle and Griffin (2001), socioeconomically advantaged women are more dissatisfied with or concerned about their bodies than socio-economically disadvantaged women.

    a. true
    b. false

Answers to Chapter 1 Questions
1. b, 2. d, 3. b, 4. d, 5. a, 6. c, 7. d, 8. d, 9. a

Chapter 2: Reproductive Changes in the Female Lifespan

10. What is the average age of menarche in the United States?

    a. 8 years
    b. 10.5 years
    c. 12.5 years
    d. 15 years

11. What was the legislative act that prevents sex discrimination in any educational program or activity in institutions from elementary school through college that receives federal funding?

    a. equal rights amendment
    b. title IX act
    c. civil rights act
    d. second amendment to the constitution

12. The hormone from the hypothalamus that stimulates LH release from the pituitary is _____.

    a. estrogen
    b. ACTH
    c. progesterone
    d. GnRH

13. The hormones primarily responsible for breast development at puberty in girls are _____.

    a. estrogen and progesterone
    b. LH and FSH
    c. cortisol and thyroxine
    d. GnRH and TRH

14. Menopause is caused by _____.

    a. decreased GnRH release from the brain
    b. depletion of ovarian follicles
    c. pituitary malfunction
    d. lack of responsiveness of the uterus

15. The most debilitating effect of low estrogen levels after menopause is _____.

    a. Alzheimer's disease
    b. cardiovascular disease
    c. osteoporosis
    d. stroke

Answers to Chapter 2 Questions
10. c, 11. b, 12. d, 13. a, 14. b, 15. c

Chapter 3: Considerations of Sex Differences in Musculoskeletal Anatomy

16. Compared with other species, sexual dimorphism in humans is _____.

    a. more prominent
    b. more subtle
    c. no different

17. Compared with men, women have _____.

    a. greater femoral anteversion
    b. greater femoral intercondylar notch width
    c. less tibial torsion
    d. more medial patellar alignment

18. Which two bones appear to be the best predictors of sex?

    a. calcaneus and talus
    b. femur and tibia
    c. humerus and pelvis
    d. pelvis and femur

19. In general, women have which one of the following cartilage characteristics compared with men?

    a. greater cartilage surface area
    b. greater cartilage thickness
    c. less cartilage degradation in older age
    d. slower cartilage accrual rate in youth

20. In general, women have which one of the following bone characteristics compared with men?

    a. faster bone turnover in adulthood
    b. greater compressive and bending strength
    c. less peak bone mass in adulthood
    d. less risk of bone fracture

21. Based on contemporary literature, which of the following is NOT sex-related with regard to afflictions at the shoulder?

    a. acromiohumeral interval space at rest
    b. glenoid fossa inclination
    c. inferior acromion changes
    d. anterior capsular laxity

22. All of the following are considered predilection sites for a woman's tendinosis of the lateral elbow extensor group, EXCEPT FOR _____.

    a. brachioradialis insertion at the supracondylar ridge
    b. extensor carpi radialis brevis insertion at the lateral epicondyle
    c. extensor carpi ulnaris insertion at the lateral epicondyle
    d. extensor digotorum communis insertion at the lateral epicondyle

23. All of the following factors appear to contribute to the woman's increased predisposition for carpal tunnel syndrome at the wrist/hand, EXCEPT FOR _____.

    a. body mass index
    b. carpal bone size
    c. hand length ratio
    d. repetitive hand movements

24. Which of the following characterizes a woman's distinctive clinical symptom profile associated with carpal tunnel syndrome?

    a. autonomic nervous system disturbances
    b. decreased intratunnel pressure from movement
    c. median nerve motor amplitude test outcomes
    d. median nerve sensory latency test outcomes

25. Compared with men, which one of the following appears to increase women's risk for macrotraumatic fracture at the hip?

    a. increased estrogen production
    b. decreased femoral neck strength
    c. deeper acetabular width
    d. narrower coxadiaphyseal angle

26. Compared with men, which one of the following appears to specifically increase women's risk for labral afflictions at the hip?

    a. decreased femoral neck thickening
    b. decreased labral sensory nerve supply
    c. decreased labral tissue tensile strength
    d. decreased coxadiaphyseal (CD) angulation

27. Which one of the following factors is LEAST likely going to predispose women to a symptom-producing condition at the cervical spine?

    a. cartilage coverage of the cervical facet articular surfaces
    b. segmental anterior–posterior translation during whiplash
    c. spinal canal narrowing that occur with the whiplash event
    d. vertebral canal-to-body diameter ratio changes

28. Which one of the following influences the development and progression of adolescent idiopathic scoliosis (AIS)?

    a. changes in the premenarchal kyphosis and lordosis
    b. curve magnitude measured after 15 years of age
    c. increased premenarchal skeletal growth rate
    d. leg-length discrepancy measured after 15 years of age

29. Contemporary literature has shown that all of the following are considered to be factors that contribute to the greater incidence of degenerative anterolisthesis at L4 in women compared with their male counterparts, EXCEPT FOR _____.

    a. pedicle facet angle
    b. pedicle structural thickness
    c. zygapophyseal facet shape
    d. zygapophyseal facet size

30. Compared with men, which one of the following sacroiliac joint characteristics is more commonly expected in women?

    a. equal thickness found in the sacral cartilage
    b. increased incidence of periarticular osteophytes
    c. increased pliability of the periarticular soft tissues
    d. reduced joint mobility by the 5th decade in life

Please use the scenario below to answer questions 31–35

Carrie is a 16-year-old female basketball player with a long history of knee problems. When she was 12, she complained of left anterior knee pain that began at the beginning of track season and lasted throughout the season. She was treated with quadriceps strengthening exercises, without improvement. The symptoms resolved after track season ended. The symptoms returned during track season the next year, and she reported the same complaints. She repeated the exercises without improvement.

When she was 14, she sustained a noncontact rupture of her left anterior cruciate ligament (ACL) when landing from a rebound in a basketball game. She was treated surgically with a patellar tendon autograft, with the graft material harvested from the injured knee. She completed rehabilitation without complications and returned to full participation the next year. At the time that she returned to play, her strength, balance, and control of the injured knee were symmetrical with those of the uninvolved side. She did not have any other complaints until 2 weeks ago, when she twisted her right knee when attempting to save the ball from going out of bounds during a conference championship game. The game was in the final minute of play, and her team was behind by 3 points. She had played the entire game and was attempting to keep the ball in play, when she planted on the right foot, turned to the left, and then felt a sharp pain accompanied by a loud pop. Her knee immediately swelled, and she could not walk on it but had to be carried off the court. She iced the knee that night, but the pain and swelling persisted. She reported to the team orthopedist the next day, where she was examined, and she received an MRI. The surgeon told her that she had completely torn her right ACL and would need surgery. The patient is presently very worried because she will be a senior next year and hopes to acquire a basketball scholarship at a prominent NCAA Division 1 university.

31. What is the percentage of all ACL injuries that are considered to be noncontact in nature?

    a. 20%
    b. 45%
    c. 70%
    d. 90%

32. Which of the following MOST likely contributed to your patient's increased risk for noncontact anterior cruciate ligament (ACL) injuries?

    a. decreased femoral notch width
    b. decreased ligament creep
    c. increased knee valgus
    d. increased thigh-foot (TF) angle

33. Which of the following muscle control strategy characteristics most likely contributed to her ACL noncontact injury risk during her landing or cutting sequences?

    a. decreased gastrocnemius activation
    b. decreased soleus activation
    c. increased hamstring activation
    d. increased quadriceps activation

34. Which of the following kinetic characteristics most likely contributed to Carrie's noncontact injury response?

    a. decreased peak posterior knee force
    b. increased leg stiffness
    c. increased peak knee flexor moments
    d. increased peak vertical knee force

35. All of the following physiological characteristics likely contributed to Carrie's ACL injury, EXCEPT FOR _____.

   a. increased fibroblastic activity within the ligament
   b. increased metallomatrix protease activity in the ACL
   c. increased population of estrogen receptors on the ACL
   d. menstrual fluctuations in estrogen concentration

Answers to Chapter 3 Questions

16. a, 17. a, 18. b, 19. d, 20. c, 21. b, 22. c, 23. b, 24. b, 25. b, 26. c, 27. d, 28. c, 29. b, 30. c, 31. c, 32. c, 33. d, 34. d, 35. a

Part II – Preoccupation with Body Image Issues and Disordered Eating Issues in the Active Female

Chapter 4: Body Image and Eating Disturbances in Children and Adolescents

36. Which of the following is not connected with body image disturbances?

   a. low self-esteem
   b. depression
   c. eating disorders
   d. self-efficacy

37. Numerous assessments that measure dietary restraint and body concerns have been developed with adult samples. What can be developed to use with these instruments to improve their reliability for use with adolescents?

   a. a new instrument
   b. a glossary
   c. a readability scale
   d. an instructional video

38. Which of the following is not considered a risk factor for body image and eating disturbances among children and adolescents?

   a. mood disorders
   b. self-directedness and assertiveness
   c. glamorization of the thin ideal
   d. perfectionism

39. Which of the following is not considered a protective factor for preventing body image and eating disturbances among children and adolescents?

   a. early puberty
   b. positive relationships with parents
   c. high self-esteem
   d. peers who do not overemphasize body weight and size

40. Primary prevention strategies include all of the following EXCEPT FOR _____.

   a. development of life skills, including media literacy and effective coping
   b. enhancing self-esteem

c. opportunities to build positive relationships with others
d. receiving a diagnosis and referral to a treatment facility

41. School-based prevention programs may be more effective if they possess all of the following characteristics EXCEPT THAT _____.

    a. they are based on experiences of fictional characters.
    b. they are part of a coordinated school health program
    c. they coincide with developmental issues
    d. they incorporate active, experiential, and peer-mediated formats

42. Which of the following describes how health care providers can play a role in preventing eating disorders among youth?

    a. communicate strategies for healthy eating and physical activity for health and quality of life.
    b. help girls and parents understand the normal developmental changes that occur during puberty.
    c. participate in advocacy efforts aimed at changing cultural norms aimed at children and adolescents.
    d. all of the above.

43. Which of the following statements about media literacy education is false?

    a. media literacy education can teach girls how to effectively counter messages that promote unrealistic body image and unhealthy eating.
    b. media literacy education can build self-esteem and foster resilience.
    c. most media literacy education efforts have targeted preadolescents.
    d. media literacy education should hold relevance in terms of the types of media the target group uses.

44. Which of the following is not one of the four "toxic" sociocultural myths that can increase unhealthy body image among children and adolescents?

    a. valuing image over substance
    b. denial of the harmful effects of restrictive eating for weight loss purposes
    c. acceptance of diverse body sizes and shapes
    d. discounting the value of healthy eating and regular physical activity to enhance health and wellness

45. Social Development Strategy focuses on the outcome of health-enhancing behaviors through exposing children and adolescents to two critical protective factors:

    a. pro-social bonding and healthy beliefs and standards for behavioral norms
    b. opportunities for meaningful participation and development of self-efficacy
    c. recognition for positive behavior and development of a positive identity
    d. development of social–emotional competencies and positive reinforcement

Answers to Chapter 4 Questions
36. d, 37. b, 38. b, 39. a, 40. d, 41. a, 42. d, 43. c, 44. c, 45. a

Review Questions 413

Chapter 5: The Female Athlete Triad: Disordered Eating, Amenorrhea, and Osteoporosis

46. The Female Athlete Triad is a term that describes _____.

    a. muscular endurance, cardiovascular endurance, and muscular strength
    b. amenorrhea, oligomenorrhea, and eumenorrhea
    c. speed, agility, and power
    d. disordered eating, amenorrhea, and osteoporosis

47. Athletes at greatest risk for developing signs and symptoms associated with the triad are _____.

    a. female basketball players under tremendous pressure to perform
    b. female hockey players at the peak of their season
    c. athletes competing in sports that emphasize leanness or a low body weight
    d. female softball players, particularly pitchers

48. Women deposit most of their bone mass between the ages of _____.

    a. 9 and 12
    b. 15 and 18
    c. 18 and 21
    d. 11 and 14

49. Which of the following statements is false with regard to athletic amenorrhea and bone loss?

    a. research has shown that in athletic amenorrhea, estrogen replacement alone normalized low bone density.
    b. mechanisms other than estrogen deficiency may account for low bone mineral density in women with amenorrhea.
    c. women who consistently diet may slow their metabolic rates, which could affect bone metabolism through decreased leptin levels.
    d. leptin receptors have been reported to be found in bone and may be important to osteoblastic function.

50. In this particular eating disorder, even though the drive for thinness persists, the individual's body weight is typically normal throughout the disorder.

    a. anorexia nervosa
    b. bulimia nervosa
    c. binge eating disorder
    d. compulsive overeating

51. Even though the profile of menstrual dysfunction varies considerably in athletes, there is a progression that develops ranging from the least severe form of menstrual dysfunction to the most severe form of menstrual dysfunction. [1. irregular cycles but still ovulating; 2. regular cycles with failure to develop and release an egg (ovulation); 3. absence of menses and anovulation; 4. irregular cycles and anovulation;

5. regular cycles with inadequate progesterone production; 6. regular cycles with a shortened luteal phase—progesterone production stops early]. The typical progression of menstrual dysfunction is

   a. 3, 6, 2, 1, 4, 5
   b. 6, 5, 2, 1, 4, 3
   c. 5, 6, 1, 2, 3, 4
   d. 1, 2, 4, 3, 6, 5

52. The World Health Organization defines osteoporosis as _____.

   a. 2.5 SD below the mean of young adults
   b. 2 SD below the mean of young adults
   c. 1–2.5 SD below the mean of young adults
   d. 2–2.5 SD below the mean of young adults

Answers to Chapter 5 Questions
46. d, 47. b, 48. d, 49. a, 50. b, 51. b, 52. a

Chapter 6: Disordered Eating in Active Middle-Aged Women

53. When questioning middle-aged women about their bodies, the most common concern is _____.

   a. a small chest size
   b. a protruding stomach
   c. sagging skin
   d. body weight

54. Important trigger events in the development of eating disorders in middle-aged women are _____.

   a. death of a loved one, divorce, a loss of youthfulness
   b. changing appearance and physical deterioration
   c. the empty nest syndrome and a changing lifestyle
   d. all of the above factors have been implicated as the precipitant of an eating disorder in middle-aged women

Answers to Chapter 6 Questions
53. d, 54. d

Chapter 7: Eating Disorder and Menstrual Dysfunction Screening Tools for the Allied Health Professional

55. Target groups for screening should include which group(s) of women?

   a. young women with low body mass index (BMI) compared with age norms
   b. patients with gastrointestinal symptoms
   c. women with menstrual disturbances or amenorrhea
   d. all of the above

56. An instrument that is routinely used in the primary care setting and serves as one of the most powerful and useful tools to aid in the diagnosis of an eating disorder is

    a. the EAT Inventory
    b. the BULIT-R Inventory
    c. the FAST Inventory
    d. a thorough medical history

57. If an eating disorder is suspected by the primary care physician, one of the most practical screening tools to use in immediate health care setting (physician's office) due to the brevity of the instrument is the _____.

    a. the SCOFF questionnaire
    b. the EAT Inventory
    c. the BULIT-R Inventory
    d. the FAST Inventory

58. National Screening Programs such as The National Eating Disorder Screening Program routinely use this instrument as a screening tool for eating disorders.

    a. the SCOFF questionnaire
    b. the EAT-26 Inventory
    c. the BULIT-R Inventory
    d. the FAST Inventory

59. This instrument has been shown to be a reliable and valid measure for identifying individuals who may suffer from bulimia nervosa both in a clinical and in a nonclinical population.

    a. the SCOFF questionnaire
    b. the EAT-26 Inventory
    c. the BULIT-R Inventory
    d. the FAST Inventory

60. This screening tool has been developed by McNulty et al. to identify eating pathology in female athletes.

    a. the SCOFF questionnaire
    b. the EAT-26 Inventory
    c. the BULIT-R Inventory
    d. the FAST Inventory

61. In 2006, National Collegiate Athletic Association Division 1 Schools adopted a standardized eating disorder and menstrual dysfunction screening tool to be used for all female athletes.

    a. true
    b. false

Answers to Chapter 7 Questions
55. d, 56. d, 57. a, 58. b, 59. c, 60. d, 61. b

Chapter 8: Education and Intervention Programs for Disordered Eating in the Active Female

62. The premise of this behavioral change model is that individuals vary in motivation and readiness to change their behavior.

    a. Transtheoretical Model
    b. Self-Efficacy Model of Change
    c. Self-Help Concept Model
    d. Freud Model of Change

63. The health behavior change model developed by Prochaska and DiClemente contains four related concepts considered central to health behavior change. These concepts are _____.

    a. stage of change, self-efficacy, decisional balance, and the processes of change
    b. personal dispositions, preferences, response tendencies, and social identities
    c. physical attractiveness, self-control, decisional balance, and self-esteem
    d. personal dispositions, self-control, decisional balance, and the processes of change

64. In this stage of change (TTM), self-control of the behavior has been established for 6 months.

    a. action
    b. maintenance
    c. commitment
    d. preparation

65. If an individual diagnosed with an eating disorder weighs less than 75% of average weight, most clinicians will recommend _____.

    a. treatment in an outpatient setting
    b. treatment in an inpatient setting
    c. hospitalization but no treatment until the patient reaches 85% of average weight
    d. a feminist approach to the treatment of eating disorders in an outpatient setting

Answers to Chapter 8 Questions
62. a, 63. a, 64. b, 65. b

Part III- Reproductive Health

Chapter 9: The Human Menstrual Cycle

66. The phase of the cycle when the lining of the uterus is growing is

    a. menstrual
    b. luteal
    c. follicular
    d. secretory

67. The hormone that causes ovulation is

    a. FSH
    b. LH

c. prolactin
d. GnRH

68. Hormonal birth control pills work on the basis of what relationship between the ovarian and pituitary hormones?

    a. negative feedback
    b. positive feedback

69. What hormone prepares the lining of the uterus for implantation of the early embryo?

    a. LH
    b. FSH
    c. GnRH
    d. progesterone

70. The most common cause of amenorrhea in women of child-bearing age is _____.

    a. nursing
    b. low energy availability
    c. pregnancy
    d. psychological stress

71. How many follicles normally mature and ovulate in the human?

    a. 1
    b. 2
    c. 8–10
    d. many

72. Which of the following are steroid hormones?

    a. FSH and LH
    b. inhibin and activin
    c. GnRH and prolactin
    d. estrogen and progesterone

73. The hormone progesterone is primarily produced by the _____.

    a. corpus luteum
    b. follicle
    c. pituitary
    d. hypothalamus

Answers to Chapter 9 Questions
66. c, 67. b, 68. a, 69. d, 70. c, 71. a, 72. d, 73. a

Chapter 10: Abnormal Menstrual Cycles

74. The term used to describe regular menstrual cycles of 25–38 days is _____.

    a. amenorrhea
    b. oligomenorrhea
    c. eumenorrhea
    d. dysmenorrhea

75. Primary amenorrhea is when _____.

   a. a young woman has her first menstrual period
   b. a young woman has not had her first menstrual period by the age of 16
   c. a young woman is infertile

76. Which of the following conditions can cause menstrual cycle disturbances?

   a. PCOS
   b. hyperprolactinemia
   c. Crohn's disease
   d. all of the above

77. The Female Athlete Triad is a recognized set of symptoms that was first reported in _____.

   a. 1944
   b. 1972
   c. 1993
   d. 2002

78. The physiological deficit in functional hypothalamic amenorrhea (FHA) is _____.

   a. decreased GnRH release from the hypothalamus
   b. ovarian failure
   c. hyperprolactinemia

Answers to Chapter 10 Questions
74. c, 75. b, 76. d, 77. c, 78. a

Chapter 11: Psychological Stress and Functional Amenorrhea

79. Which of the following are thought to inhibit reproductive function?

   a. psychological stress
   b. nutritional stress
   c. physical stress without accompanying energy needs
   d. all of the above

80. Suppressed LH secretion can sometimes be reversed with naloxone. What substance does this finding implicate in the inhibition of LH release?

   a. dopamine
   b. estrogen
   c. endogenous opiates
   d. GnRH

81. The individual who first recognized that stress might impair reproductive function was

   a. seyle
   b. refeinstein
   c. ferin
   d. loucks

82. The long-term consequences of FHA include _____.

   a. depression
   b. osteoporosis
   c. cardiovascular disease
   d. all of the above

83. Effective treatment of the underlying cause of FHA would be _____.

   a. hormonal therapy
   b. psychological therapy
   c. chemotherapy
   d. radiation therapy

Answers to Chapter 11 Questions
79. d, 80. c, 81. a, 82. d, 83. b

Chapter 12: Effects of the Menstrual Cycle on the Acquisition of Peak Bone Mass

Please use the scenario below to answer questions 84–88

You are a primary care sports medicine physician in a multispecialty group practice consisting of pediatrics, internal medicine, OB/GYN, and family physicians. For the past few years, you have been the team physician for local high school athletes, including track and field events and other court sports. Lately, the team athletic trainer has brought in a few of the female long-distance runners with similar complaints of vague, aching pain in their feet and legs, especially after a couple of weeks' practice, for example, running around the track. Upon further questioning, these female athletes also reveal that their menses have been very irregular and/or absent over the last year or so.

84. What should be your next action in terms of history taking, physical assessment, and diagnostic testing?

   a. no need for diagnostic studies yet.
   b. how regular are their menstrual periods occurring?
   c. what foods and how frequent are they eating?
   d. b and c

85. What are some of the causes of menstrual irregularity/absence in young female athletes?

   a. excessive gymnastics
   b. pregnancy
   c. long-distance running
   d. a and b

86. What are the consequences of dysmenorrheal or amenorrhea on bone?

   a. stress reactions
   b. osteopenia
   c. osteoporosis
   d. b and c

87. What are your recommendations for restoration of menses?

   a. resume normal eating patterns and decrease exercise
   b. consume foods with calcium and vitamin D
   c. begin hormonal replacement therapy
   d. all of the above

88. How would you go about treating osteopenia/osteoporosis?

   a. prescribe birth control pills
   b. prescribe an appropriate diet
   c. consult a gynecologist
   d. prescribe an appropriate workout regimen

Answers to Chapter 12 Questions
84. d, 85. b, 86. d, 87. a, 88. c

Part IV –Prevention and Management of Common Musculoskeletal Injuries in Active Females

Chapter 13: Prevention and Management of Common Musculoskeletal Injuries in Preadolescent and Adolescent Female Athletes

Please use the scenario below to answer questions 89–83

You are a pediatrician who has just recently completed a primary care sports medicine fellowship and decided to join a multispecialty group practice in a suburban town. One of your partners' physicians assistant (PA) has been involved with taking care of a local youth girls' soccer team. These young female athletes are under tremendous pressure to win because their team will disband if they do not make the playoffs this season; as a result, these female soccer players have been required to "train" extra hard every weekday before and after practice. Their training program consists of running for a couple of miles around the track, then lifting weights for strength and power, then run again for another mile or so. Evidently, their coach used to work with male football players, and he is incorporating the majority of football drills into the regular sports workout during the competitive season for these young female athletes. As far as he is concerned, the heavier weight moved the better. "Work through the pain!" echoes his motto. As much as these teenage girls want a good chance at the championship trophy, they are beginning to get discouraged, mainly because they cannot seem to get rid of their muscle aches and pains despite resting on the weekends and taking over-the-counter medications. In fact, their performances during the games have also started to deteriorate a bit, which makes their coach exercise them even harder. The PA is coming to you for advice concerning this group of adolescent females.

89. What do you think is happening to these young female athletes?

   a. they are out of shape
   b. they are overtrained and inappropriately trained

c. they are not motivated to work that hard
d. a and b

90. What questions should you ask of these athletes and how will you go about approaching the coach about his female soccer players' problems?

   a. are they involved in other sort of physical activity besides soccer?
   b. have they ever trained with weights or worked out before?
   c. confront the coach directly and ask him to "let up" a bit
   d. a and b

91. What elements of the history, physical examination, and diagnostic studies should you obtain to help you evaluate and treat these young female athletes?

   a. have they been involve with physical fitness before now?
   b. is the pain activity related or occurs even at rest?
   c. obtain radiographs, bone scan, and laboratory work
   d. all of the above

92. What initial measures should you institute?

   a. none, they need to just "gut it out" and keep going
   b. Stop any offending activity and cross train
   c. institute "rice" measures
   d. b and c

93. Who else should you consult for assistance in the management of their musculoskeletal concerns?

   a. an orthopedist or a surgical sports specialist
   b. no one else, you could handle it yourself
   c. an endocrinologist
   d. a and c

Answers to Chapter 13 Questions
89. b, 90. d, 91. d, 92. d, 93. a

Chapter 14: Prevention and Management of Common Musculoskeletal Injuries in the Adult Female Athlete

Please use the scenario below to answer questions 94–98

You are a fairly new athletic trainer for a collegiate-level female basketball team in a midsize town. This is your first season working with these women athletes. One half is composed of rookies (junior varsity) and the other half veterans (varsity) players. Recently you have noticed that not all of these females show up every day for practice and that several have had more "off" than "on" days in terms of "scrimmage" games. You have also been providing daily "treatments" as far as local modalities and physical therapy rehab exercises in the training room for over one-third of these female athletes for various musculoskeletal complaints. In fact, the majority of the problem lies in the lower extremity and most notably the tibia and knee regions. A couple of females are improving in terms of their symptoms, but the rest are still not making any appreciable gains. Alarmingly, some are actually getting worse. The regular sports season will begin in about 2 months, and you are quite concerned that some players may not be able to "start" in their best physical condition.

94. Initially, how should you go about discovering the root of these female athletes' problem?

   a. confront the whole team as group
   b. confront each player individually
   c. ask other players to "fill you in"
   d. befriend each player and allow her to confide in you privately

95. What specific questions should you pose to each woman about her complaints/symptoms?

   a. Where the pain is located and when does it occur?
   b. How severe is the pain and how long does it last?
   c. What other physical activity is she doing?
   d. All of the above

96. What physical sign(s) of overuse injury should you look for to point you toward a potential stress fracture?

   a. easy fatigability
   b. pain not necessarily associated with activity
   c. pain at rest
   d. pain after exercise
   e. b and c

97. What further actions, if any, should be taken to manage these athletes' problems?

   a. use the "no pain, no gain" rule so no quitting!
   b. make them go through extra drills since they are "out of shape."
   c. institute first-aid-type measures, that is, treat them symptomatically
   d. stop all physical activity

98. How can you ensure that these females will recover in time to start playing basketball once the regular sports season begins?

   a. enroll them in a focused strength and conditioning program
   b. let them train on their own, that is, give them a "home" program
   c. incorporate sports specific drills
   d. a and c

Answers to Chapter 14 Questions
94. d, 95. d, 96. c, 97. c, 98. d

Review Questions 423

Chapter 15: Prevention and Management of Common Musculoskeletal Injuries Incurred Through Exercise During Pregnancy

Please use the scenario below to answer questions 99–103

You are a family practice physician working in a small town where the closest major medical center is a couple of hours away. An emergency room is available nearby, but medical resources are scarce. You have been the "family doctor" for a few decades in the community. In fact, you have made "house calls" in the middle of the night to tend to extremely ill patients or to deliver an unexpected baby! Currently, you have five different women who are in various stages of pregnancy, three of whom are fairly healthy and all desire an exercise program. One woman, unfortunately, has gestational diabetes along with being extremely overweight. In fact, she has already gained over 60 pounds nearing the end of her second trimester. She has been giving in to her "food cravings" and wants to start exercising to lose as much weight as possible. The second woman just found out she is pregnant and has been working out for several years along with competing in triathlons. She wants to keep up with her training and competitions plus is very afraid of "putting on weight." The third woman is near the end of the first trimester, is gaining weight appropriately, and has begun a workout program. She needs to know whether she can continue to exercise and when she should stop working out. The fourth mother-to-be, although she has been involved with a training regimen for a few years, is so afraid that she will damage the fetus since it is her first baby, so she decides that she needs to quit exercising altogether. The fifth female patient is complaining of low back spasms and abdominal cramping when she attempts to work out, especially lower extremity exercises but she does not want to slow down.

99. For the obese, diabetic patient who has never exercised before, what other concerns do you have and how will you go about helping her to control further unwanted weight gain safely?

   a. is her diabetes under control?
   b. consult a nutritionist for the best type of diet
   c. do not let her workout at all
   d. a and b

100. As far as the competitive triathlete, what other issues must you address prior to advising her about training and nutrition?

   a. make her stop competing altogether
   b. ensure that she takes in extra calories to feed the fetus
   c. have her switch to cross training activities
   d. b and c

101. In terms of the third patient, what guidelines would you recommend in order for her to keep on exercising?

   a. she need not stop working out
   b. she can exercise in any environmental condition

c. she should stay away from risky-type sports
d. she can continue the same lifting program as before

102. As for the fourth patient, what advise do you have for her about potential fetal damage with exercising?

    a. reassure her that absolutely no fetal damage will occur with exercise
    b. modify her exercise routine to follow appropriate medical guidelines
    c. agree with her about stopping all workouts
    d. a and b

103. And finally with the last patient, what do you think is happening with her body and should she continue with the same workout program?

    a. abdominal cramps should be okay because they are probably because of the natural GI effects from the enlarged uterus
    b. disregard her symptoms—probably because of "aches and pains" of pregnancy
    c. see her obstetrician before resuming exercise
    d. give her some medications, tell her to keep working out as long as she feels okay

Answers to Chapter 15 Questions
99. d, 100. d, 101. c, 102. b, 103. c

Chapter 16: Prevention and Management of Common Musculoskeletal Injuries in the Aging Female Athlete

Please use the scenario below to answer questions 104–108

You are a nurse practitioner for an internist in a midsize town. Your adult patient population ranges from young adults to senior citizens. You see an array of patients with different clinical entities ranging from the simple common cold to complicated cardiopulmonary disease. One of your hobbies is playing recreational tennis in a "female over fifty" league at the YWCA. You meet with your teammates twice a week and you serve as the team captain for these female athletes. Another one of your pastimes is working out at a local health club three to five times a week. Your doubles tennis partner is 55 years of age and is an avid runner and climber, as well. She pretty much participates in some sort of sporting activity every single day of the week. Lately, she has been frequently absent, sending in a substitute player and complaining of vague, achy shoulder, and deep knee pain. Unfortunately, before you could help her sort things out, she suffered a fall, sustaining a wrist fracture on her dominant side (opposite of her involved shoulder) and finally made it to your clinic for evaluation and treatment. You have known this woman for several years and she refuses to let anything slow her down in terms of exercise. She is extremely frustrated with her injuries and, wants to heal really fast and get on with her active life again as soon as possible.

104. First of all, what pertinent questions should you ask about your patient's medical history and what modalities should you utilize for diagnosis?

    a. Is she postmenopausal and how did she fall?
    b. How often and how long does she work out and using what program?
    c. Obtain x-rays and DEXA scan.
    d. All of the above

105. What should you advise her about the shoulder/knee complaints and wrist injury?

    a. work through the pain
    b. ice and elevate all extremities
    c. exercise the wrist, shoulder, and knee
    d. lay off all three

106. What other medical specialist(s) should you consult and why?

    a. no need to consult, you can take care of everything!
    b. consult an orthopedist for the wrist
    c. consult an endocrinologist for presumed osteoporosis
    d. b and c

107. What other factors should you consider prior to sending her back to athletic activity after healing?

    a. her future osteoporotic fracture risk
    b. is she cleared from an orthopedist's or endocrinologist's standpoint?
    c. no need to worry about other factors
    d. a and b

108. How soon after she heals from her injury should you allow her to return to tennis? Will you let her go back to running; how about rock climbing?

    a. does not matter about the time; let her go back when she wants to
    b. 1 month to tennis, 6 weeks to running, and 2 months to climbing
    c. when she regains range of motion/strength of the involved extremities and cleared by her physical therapist
    d. do not let her return to any sport

Answers to Chapter 16 Questions

104. d, 105. b, 106. d, 107. d, 108. c

Chapter 17: Osteoporosis and Current Therapeutic Management

109. What does a *T*-score represent?

    a. the absolute measure of bone mineral density (BMD) by dual-energy x-ray absorptiometry
    b. the number of standard deviations above or below normal trabecular mass
    c. the number of standard deviations above or below average peak bone mass in young adults
    d. the number of standard deviations above or below the mean BMD of women the age of the patient

110. What is the World Health Organization (WHO) criterion to establish a diagnosis of osteoporosis?

    a. a T-score lower than −2.0
    b. a T score lower than −2.5
    c. a Z-score lower than −2.0
    d. a Z-score lower than −2.5

111. What is the pathophysiology behind postmenopausal osteoporosis?

    a. inadequate osteoclast expression producing large resorption pits
    b. inadequate osteoclast activity preventing bone formation
    c. inadequate bone formation as a result of osteoblast inhibition
    d. preferentially increasing cortical bone loss

112. To what does Type 1 primary osteoporosis refer?

    a. osteoporosis characterized principally by trabecular bone loss
    b. osteoporosis associated with the process of aging
    c. osteoporosis associated with vitamin D deficiency
    d. osteoporosis caused by failure to achieve adequate peak bone mass

113. What is the most common form of secondary osteoporosis?

    a. hypercalciuria and vitamin D deficiency
    b. glucocorticoid-induced osteoporosis
    c. hyperparathyroidism
    d. inflammatory bowel disease

114. Which of the following biochemical tests is most likely to be abnormal in a patient with glucocorticoid-induced osteoporosis?

    a. creatinine
    b. liver enzymes
    c. urinary calcium excretion
    d. parathyroid hormone (PTH) level

115. BMD measurements at which sites are considered the gold standard for establishing a diagnosis of osteoporosis?

    a. heel and hip
    b. femoral neck and spine
    c. femoral neck and total hip
    d. spine and hip

116. Which of the following agents has been shown to stimulate new bone formation in women with osteoporosis?

    a. alendronate
    b. risedronate
    c. raloxifene
    d. teriparatide

117. Alendronate and risedronate are examples of which class of medications for the treatment of osteoporosis?

    a. PTH
    b. selective estrogen receptor modulators

c. bisphosphonates
d. estrogens

118. What is the mechanism by which bisphosphonates reduce fracture risk in women with osteoporosis?

   a. promotion of osteoblast formation
   b. deposition of protein collagen matrix
   c. improvement in cortical bone mass
   d. inhibition of osteoclast activity

Answers to Chapter 17 Questions
109. c, 110. b, 111. b, 112. a, 113. b, 114. c, 115. d, 116. d, 117. c, 118. d

Part V-Safe Exercise Guidelines Throughout the Lifespan

Chapter 18: Physical Activity Recommendations and Exercise Guidelines Established by Leading Health Organizations

119. Moderate exercise can be defined as _____.

   a. any activity that burns 3.5–7 kcal/min.
   b. 40–59% of oxygen uptake reserve (VO_2R) or heart rate reserve (HHR)
   c. some increase in the breathing or heart rate or a "perceived exertion" of 11–14 on the Borg rate of perceived exertion (RPE) scale
   d. all of the above

120. Physical activity recommendations for health-related fitness from leading health organizations recommend that _____.

   a. people of all ages include a minimum of 30 min of physical activity of vigorous intensity (such as running) on most, if not all, days of the week
   b. people of all ages include a minimum of 30 min of physical activity of moderate intensity (such as brisk walking) on most, if not all, days of the week
   c. people of all ages include a minimum of 60 min of physical activity of vigorous intensity (such as running) on most, if not all, days of the week
   d. people of all ages include a minimum of 60 min of physical activity of moderate intensity (such as brisk walking) on most, if not all, days of the week

121. Health-related components of physical fitness include _____.

   a. balance, muscular strength, coordination, and cardiovascular endurance
   b. cardiovascular endurance, agility, coordination, and flexibility
   c. cardiovascular endurance, muscular strength and endurance, flexibility, and body composition
   d. balance, agility, coordination, speed, power, and reaction time

122. One MET (metabolic equivalent unit) is equal to approximately _____.

   a. 3.5 mL O_2 kg min
   b. 7 kcal min
   c. .5 mL O_2 kg min
   d. 3.5 L O_2 kg min

123. According to the American College of Sports Medicine, in order to achieve cardiovascular fitness, one should exercise in this prescribed intensity.

    a. 64%/70–94% of maximum heart rate (HR_{max})
    b. 40%/50–85% of heart rate reserve (HRR)
    c. 40%/50–85% of oxygen uptake reserve (VO_2R)
    d. all are acceptable intensities to achieve cardiovascular fitness

124. Which of the following statements is false with regard to the exercise sequence and progression guidelines for resistance training?

    a. a 12–15% increase in load should be applied when one to two repetitions can be performed over the current workload
    b. exercise large-muscle groups before small-muscle groups
    c. perform multiple-joint exercises before single-joint exercises
    d. perform higher intensity exercises before lower intensity exercises

125. According to the American College of Sports Medicine, flexibility guidelines state that stretches should be held for _____ to a position of mild discomfort and repeated for _____ repetitions for each stretch.

    a. 1 min, 2–4
    b. 10–20 s, 3–5
    c. 15–30 s, 2–4
    d. 4–60 s, 3–5

126. Although national standards have been developed and accepted for body mass index and waist circumference, there are no national standards for body composition. However, Lohman et al. (1997) proposed a set of standards for women using data from the National Health and Nutrition Examination Survey. Which of the following statements is false according to Lohman's recommendations?

    a. essential body fat needed for normal physiological functions for women is 8–12%
    b. recommended minimal body fat level is 10–12% for women
    c. body fat recommendation for women 34 years or less who are not athletic is 20–35%
    d. all of the above statements reflect Lohman's recommendations

Answers to Chapter 18 Questions
119. c, 120. b, 121. c, 122. a, 123. d, 124. a, 125. c, 126. d

Chapter 19: Specific Exercise Guidelines and Recommendations for Children and Adolescence

127. What is the highest load that should be used when having children engage in resistance training?

    a. 1 RM
    b. 6 RM
    c. 8 RM
    d. 0 RM

128. Which of the following is not a possible cause for increased metabolic costs in children?

   a. increased resting BMR
   b. relative VO_{2max} is higher than adults
   c. higher stride frequency
   d. biomechanically inefficient locomotion style

129. Which of the following characteristics can be significantly increased with resistance exercise in children?

   a. muscular hypertrophy
   b. VO_{2max}
   c. muscular strength
   d. fiber-type distribution in the muscles

130. Which of the following factors is not an important aspect of working with children when exercising?

   a. concentrate on sport-specific exercises
   b. maintain proper form
   c. select appropriate loads
   d. supervise all exercise sessions

131. Which of the following would fall under the category of "exercise" for children?

   a. playing tag
   b. swimming at a birthday party
   c. performing 20 pushups
   d. riding a bike to a neighbor's house

Answers to Chapter 19 Questions
127. c, 128. b, 129. c, 130. a, 131. c

Chapter 20: Exercise Precautions of the Female Athlete: Signs of Overtraining

132. The overtraining syndrome (OTS) is defined as _____.

   a. persistent underperforming, with or without other accompanying psychological and physical symptoms, despite weeks of lighter training or complete rest
   b. a brief period of heavy overload without adequate recovery
   c. the point where the athlete starts to experience physiological maladaptations and chronic performance decrements
   d. a period in which the body begins a regenerative process beyond the previous level, thus resulting in supercompensation

133. Symptoms of sympathetic staleness are _____.

   a. mood disorders, increased resting heart rate, decreased appetite, and weight loss
   b. depression, low pulse and fatigue very early in exercise, and excessive amounts of sleep.

c. anxiety, decreased maximal plasma lactate during exercise, and decreased catecholamine levels
   d. apathy, low resting heart rate, rapid heart rate recovery post exercise, and hypoglycemia during exercise

134. Signs and symptoms of overtraining have been categorized into four classes by Fry et al. (1991). Increased susceptibility to bacterial infections, reactivation of herpes viral infections, and decreased functional activity of neutrophils would be classified into which category?

   a. alteration in physiological functions and adaptations to performance
   b. psychological symptoms
   c. immunological dysfunction
   d. biochemical alterations

135. Hormonal responses to overtraining and staleness consistently shows an elevated cortisol response during recovery from exercise.

   a. true
   b. false

Answers to Chapter 20 Questions
132. a, 133. a, 134. c, 135. b

Chapter 21: Specific Exercise Guidelines and Recommendations During Pregnancy and Lactation

136. Which of the following recommendations regarding resistance training during pregnancy is false?

   a. women who have never participated in resistance training should not initiate a training program during pregnancy
   b. heavy resistance should be avoided because it may expose the joints, connective tissue, and skeletal structures of an expectant woman to excessive forces
   c. an exercise set consisting of at least 12–15 repetitions without undue fatigue is recommended
   d. as training occurs, overload initially by increasing resistance, and subsequently by increasing the number of repetitions

137. Resistance training on machines during pregnancy is preferred to free weights because machines can be more easily controlled and require less skill.

   a. true
   b. false

138. Sport activities that are encouraged during pregnancy include _____.

   a. walking, stationary cycling, low-impact aerobics, and swimming
   b. contact sports such as hockey, wrestling, football, and soccer
   c. gymnastics, horseback riding, skating, and skiing (snow and water)
   d. hang gliding, vigorous racquet sports, weightlifting, and scuba diving

139. According to the American College of Obstetricians and Gynecologists, an absolute contraindication to aerobic exercise during pregnancy is _____.

   a. severe anemia
   b. poorly controlled type 1 diabetes
   c. history of extremely sedentary lifestyle
   d. preeclampsia/pregnancy-induced hypertension

140. In the absence of either medical or obstetric complications, the American College of Obstetricians and Gynecologists suggests that pregnant women should _____.

   a. accumulate 30 min or more of moderate exercise on most days, if not all, days of the week
   b. accumulate 60 min or more of moderate exercise on most days, if not all, days of the week
   c. accumulate 60 min or more of vigorous exercise on most days, if not all, days of the week
   d. accumulate 30 min or more of vigorous exercise on most days, if not all, days of the week

141. The American College of Obstetricians and Gynecologists advice is to avoid exercise in this position as much as possible. This position should especially be avoided after the first trimester.

   a. prone
   b. upright
   c. semi-inclined
   d. supine

Answers to Chapter 21 Questions
136. d, 137. a, 138. a, 139. d, 140. a, 141. d

Chapter 22: Mindfulness, Quality of Life, and Cancer: A Mindfulness-Based Exercise Rehabilitation Program for Women with Breast Cancer

Please use the scenario below to answer questions 142–144

Your mother was diagnosed with breast cancer 8 months ago and underwent surgery as well as chemotherapy. Physically, she has poor muscle tone and strength and has been plagued by fatigue since she left the hospital. You visited home last weekend and tried talking with her about your observations of her "being stuck" in cancer and not moving on with life. She vehemently disagrees and insists her that breast was removed simply as a precaution and that she did not have cancer so she is not "stuck" at all. Yet, she refuses to look at or feel the front of her upper torso and refuses to look at herself in the mirror. Your observations lead you to think that she is seriously in denial and afraid to face the reality of her recent health condition and experience.

142. From the MBSR program, which core technique do you think would help to transform the mother's cognitive or mental denial into acceptance of reality?

   a. yoga
   b. sitting meditation

c. walking meditation
d. body scan
e. moderate exercise program (daily brisk walking, etc.)

143. From the MBSR program, which core technique do you think would assist in the mother's acceptance of the physical/surgical changes to her body?

   a. yoga
   b. sitting meditation
   c. walking meditation
   d. body scan
   e. moderate exercise program (daily brisk walking, etc.)

144. What would you recommend to improve the mother's continued fatigue and poor muscle tone and strength?

   a. yoga
   b. sitting meditation
   c. walking meditation
   d. moderate exercise program (daily brisk walking, etc.)
   e. both a and d

Answers to Chapter 22 Questions
142. b, 143. d, 144. e

Chapter 23: Resistance Training Guidelines for the Postmenopausal Woman

145. At what rate does the $VO_{2\,max}$ decline after the age of 25 in sedentary individuals?

   a. 5% each year
   b. 5% each decade
   c. 10% each year
   d. 10% each decade

146. Which of the following statements is true ?

   a. type I fibers, which are responsible for explosive exercise, increase with age
   b. type II fibers, which are responsible for endurance exercise, increase with age
   c. type I fibers, which are responsible for endurance exercise, decreases with age
   d. type II fibers, which are responsible for explosive of exercise, decrease with age

147. All of the following decline with age except for _____.

   a. muscle mass
   b. vo_2 max
   c. elastin in cardiac muscle tissue
   d. maximum heart rate

148. Which of the following is not a common location of fractures in women with low bone mass density?

   a. vertebrae
   b. ankle

c. wrist
d. hip

Answers to Chapter 23 Questions
145. d, 146. b, 147. c, 148. b

Part VI- Nutrition, Energy Balance, and Weight Control
Chapter 24: Estimating Energy Requirements

149. In the majority of individuals, _____ of the total energy expenditure (TEE) constitutes the largest portion (60–75%) of TEE.

   a. resting energy expenditure (REE)
   b. the thermic effect of food (TEF)
   c. energy expended in physical activity also referred to as EEPA
   d. metabolic energy expenditure (MEE)

150. The main determinant of resting energy expenditure (REE) is _____.

   a. body composition (lean body mass)
   b. sex
   c. hormonal status
   d. gender

151. There are many available methods to measure energy expenditure. Which method would yield the highest accuracy?

   a. direct calorimetry
   b. indirect calorimetry
   c. doubly labeled water (DLW) technique
   d. seven-Day Recall Physical Activity Questionnaire

Answers to Chapter 24 Questions
149. a, 150. a, 151. c

Chapter 25: Nutritional Guidelines and Energy Needs for Active Children

152. The MyPyramid Food Guidance bases their calorie intake on what three criteria?

   a. height, weight, and sex
   b. physical activity, sex, and age
   c. age, sex, and height
   d. weight, age, and physical activity

153. _____ % of total calories in a child's diet should come from carbohydrates.

   a. 40–55
   b. 35–50
   c. 50–70
   d. 45–65

434   Review Questions

154. All of the following are symptoms that are caused because of a lack of protein in the diet EXCEPT _____?

   a. stunted growth
   b. increased thyroid stimulating hormone
   c. immune system deficiencies
   d. edema

155. Amino acids are important in the maintenance of the body because they _____

   a. help release body fluids
   b. are the primary energy provider to the body
   c. provide fuels to the nervous system and the brain
   d. are used as building blocks for the cells of the body

156. Children should consume _____ % of their calories from fat in order to prevent detrimental health diseases and weight gain.

   a. 25–35
   b. 30–40
   c. 20–30
   d. 10–20

157. _____ fatty acids lower LDL cholesterol levels and help maintain good cardiovascular health.

   a. Hydrogenated oils
   b. Polysaturated fatty acids
   c. Transfatty acids
   d. Polyunsaturated fatty acids

158. What is the overall purpose of the 2005 Dietary Guidelines for Americans?

   a. propose a new health system that incorporates dietary requirements
   b. educate individuals about the dietary and physical needs of the body and how to meet those needs
   c. examine the current dietary habits of American citizens
   d. evaluate the current school lunch plan for American children.

Answers to Chapter 25 Questions
152. b, 153. d, 154. b, 155. d, 156. a, 157. b, 158. b

Chapter 26: Nutritional Guidelines and Energy Needs for the Female Athlete- Determining Energy and Nutritional Needs to Ameliorate Functional Amenorrhea Caused by Energy Imbalance

159. The adult female human brain oxidizes approximately _____ of glucose each day at a continuous rate, and this must be provided daily by dietary _____, because the brain's rate of energy expenditure can deplete liver glycogen stores in less than 1 day.

   a. 50 g, carbohydrate
   b. 30 g, protein

c. 80 g, carbohydrate
d. 40 g, fat

160. According to Loucks et al. (1998, 2003), functional amenorrhea in young exercising women occurs primarily because of _____.

   a. excessive exercise stress
   b. excessive psychological stress related to competition and performance
   c. dietary energy restriction or "energy drain"
   d. protein availability

161. Reproductive function in women critically depends on the pulsatile release of _____ from neurons in the arcuate nucleus of the hypothalamus and on the consequent pulsatile release of _____ from the pituitary.

   a. follicle-stimulating hormone (FSH), progesterone
   b. gonadotropin-releasing hormone (GnRH), luteinizing hormone (LH)
   c. luteinizing hormone (LH), gonadotropin-releasing hormone (GnRH)
   d. follicle-stimulating hormone (FSH), estrogen

162. A joint position statement by the American College of Sports Medicine, the American Dietetic Association, and the Dietitians of Canada stated that a diet substantially different from that recommended in the Dietary Guidelines for Americans or the Nutrition Recommendations for Canadians is needed for athletes.

   a. true
   b. false

Answers to Chapter 26 Questions
159. c, 160. c, 161. b, 162. a

Chapter 27: Sports Supplements and Ergogenic Aids

163. Which one of these supplements has a structure similar to that of adenosine?

   a. creatine
   b. caffeine
   c. ginseng
   d. soy protein

164. Which one of these is not a reason female athletes take protein/amino acid supplements?

   a. tasty
   b. enhanced recovery
   c. hypertrophy
   d. to meet nutritional needs

165. Ephedra has been replaced by which substance in many fat-burning/thermogenesis products?

   a. Ma Huang
   b. Siberian Ginseng

c. Creatine
d. Bitter Orange

166. Choose the supplement that does not require a loading phase to yield an ergogenic effect.

   a. creatine
   b. ginseng
   c. caffeine

167. Adverse effects of anabolic-androgenic steroids (AASs) include all EXCEPT FOR _____.

   a. voice deepening
   b. menstrual irregularities
   c. increased protein synthesis
   d. male pattern baldness

Answers to Chapter 27 Questions
163. b, 164. c, 165. d, 166. c, 167. c

Chapter 28: Nutritional Guidelines and Energy Needs During Pregnancy and Lactation

168. A woman of normal weight should consume an additional _____ kcal/day, and those calories should be concentrated in foods of high-nutrient density, a value based on the percent protein, vitamins, and minerals per 100 kcal.

   a. 500
   b. 150
   c. 300
   d. 600

169. For normal-weight women, the recommended total weight gain ranges from _____. _____ This translates to _____ a week for normal weight women.

   a. 11 to 15 kg (25–35 lb), 0.4 kg
   b. 12.7 to 18 kg (28–40 lb), 0.5 kg
   c. 7 to 11 kg (15–25 lb), 0.3 kg
   d. 18 to 23 kg (40–50 lb), 0.7 kg

170. The impact of maternal weight gain in the _____ is most important for fetal development and is protective of fetal growth even if overall weight gain is poor (7).

   a. first trimester
   b. second trimester
   c. third trimester
   d. 9th month

171. Protein should comprise approximately _____ of a normal pregnancy diet. Fat should consume approximately _____ of diet in pregnancy, and carbohydrates the remaining _____.

   a. 35%, 15%, 50%
   b. 20%, 10%, 70%

c. 30%, 30%, 40%
d. 20%, 30%, 50%

172. The Institute of Medicine recommends that all pregnant females who are consuming a well-balanced diet should take _____ mg in divided doses of ferrous iron supplementation daily during the second and third trimesters to prevent anemia.

a. 50
b. 30
c. 60
d. 70

Answers to Chapter 28 Questions
168. c, 169. a, 170. b, 171. d, 172. b

Chapter 29: Nutritional Guidelines, Energy Balance, and Weight Control: Issues for the Mature Physically Active Woman

173. Which of the following predictive equations most accurately estimates resting metabolic rate (RMR) in normal-weight, overweight, obese, and severely obese individuals ages 19–78.

a. Mifflin–St. Jeor
b. Harris–Benedict
c. Owen
d. World Health Organization/Food and Agriculture Organization/United Nations University

174. Caloric expenditure _____ because of the aging process, this is due in part to _____.

a. increases, the decrease in FFM
b. decreases, the increase in FFM
c. decreases, the decrease in FFM
d. increases, the increase in FFM

175. It is recommended that older adults, people with dark skin, and people exposed to insufficient ultraviolet band radiation (i.e., sunlight) consume extra _____.

a. vitamin D from vitamin D-fortified foods and/or supplements
b. vitamin C from vitamin C-fortified foods and/or supplements
c. vitamin A from vitamin A-fortified foods and/or supplements
d. vitamin K from vitamin K-fortified foods and/or supplements

176. The need for this macronutrient increases because of the physiological changes associated with the aging process.

a. fat
b. carbohydrate
c. lipids
d. protein

177. Presently, the DRI for calcium for adult women is _____ of calcium, but there have been suggestions that daily intakes of _____ may by appropriate for postmenopausal women or women over age 65.

   a. 600 mg/day, 800 mg/day
   b. 1200 mg/day, 1500 mg/day
   c. 1500 mg/day, 1200 mg/day
   d. 500 mg, 900 mg

Answers to Chapter 29 Questions
173. a, 174. c, 175. a, 176. d, 177. b

INDEX

A

abnormal menstrual cycles, 131. *See also* amenorrhea; oligomenorrhea
   acyclic, 132
   irregular, 132
Achilles tendonopathy, 37, 38, 43. *See also under* ankle foot
ACL. *See* anterior cruciate ligament (ACL) injury
acromion process. *See also* sex differences
   Type I, 31
   Type II, 31
   Type III, 31
ACSM
   exercise recommendations
      CR fitness, 230
      for flexibility, 236
      for healthy body composition, 235
      muscular fitness, 230
   Female Athlete Triad, position stand on, 82–83
ACTH. *See* adrenocorticotrophic hormone
activin, 124
acyclic menstrual cycles, 132. *See also* amenorrhea; irregular menstrual cycles
adipose tissue
   deposition, 157
   musculoskeletal injuries in adult female athletes, 171
adolescent growth spurt, 145, 170. *See also* musculoskeletal injuries in adult female athletes; peak bone mass
adolescent idiopathic scoliosis (AIS), 39
adolescents
   body image concerns, 6
   exercise guidelines for, 241, 374
   eating disturbances in, 57–60
adrenal gland, 19
adrenocorticotrophic hormone (ACTH)
   functional hypothalamic amenorrhea (FHA), 138
adult body image concerns
   body dissatisfaction and, 7
   older adult, 6
   young adult, 6
aerobic
   capacity for children and adolescence, 242
   exercise, pregnancy and, 256
aging female athlete. *See* musculoskeletal injuries in ageing female athletes
alendronate, osteoporosis management and, 218

alkaloids, sympathomimetic, 315
allied health professional
   eating disorder and, 99
   Female Athlete Triad and, 88, 89
   menstrual dysfunction screening tool, 99
alternative therapy, 221. *See also* osteoporosis
Alzheimer's disease
   estrogen and, 22
   hormone replacement therapy, 22
   postmenopause and, 22
amenorrhea, 81, 132
   anorexia nervosa and, 134
   defined, 134
   estrogen, 135
   estrogen replacement and, 87
   Female Athlete Triad, 81, 82, 87, 88
   gonadotropin levels, 135
   hypothalamus and, 82, 89
   menstrual disturbances in athletes and, 86
   peak bone mass and, 146
   prolactin or hyperprolactinemia level, 134
   reproductive changes scenario, 23
   reversible, 135
   types
      functional, 135, 137, 299
      primary, 19, 23, 81, 132
      secondary, 81, 132
amino acids, 19
   creatine, 314
   ergogenic aids for athletes, 312
   essential, 313
   glutamine, 312
   nonessential, 313
anabolic-androgenic steroids (AAS), 317, 318
   adverse effects, 318
   dosage, 318
anaerobic capacity for children and adolescence, 243
androgens, 19
   osteoporosis management, 220
   puberty and, 19
anemia, pregnancy and, 328
Angelicia sinesis, 315
ankle foot. *See also* lower extremity based sex differences
   Achilles tendonopathy, 37, 38
   ankle inversion trauma
      grade I, 37
      in athletes, 37

based sex differences, 37
structural differences in, 37
anorexia, 4, 5. *See also* bulimia
anorexia nervosa, 3, 5, 6, 84, 134. *See also* body dissatisfaction
   diagnostic criteria for, 84
   eating disorders (ED) in children and, 59, 61
   obesity and, 3
   treatment principles, 111, 113, 114
anterior cruciate ligament (ACL) injury, 35–37, 43. *See also* knee complex
   in adult female athletes, 172
   in athletes, 36
   jump training and, 36
   prevention measures, 163
   tearing, 161
anxiety scale, 351
apophysis, 158
appearance-related beliefs
   body image and, 7
   body image disturbances management and
      cognitive approach, 11
      feminist approach, 11
   middle-aged women, 94
   physical appearance state, 351
arginine, 314
atherosclerosis, 272
athletes. *See also* Female Athlete Triad
   ACL injury, 36
   aging female injuries, 199–208
      activities of daily living (ADL), 206
      differences and similarities between older males and females, 202–205
      fragility fracture, 206
      hormonal, physiological, and anatomical changes from young-to-adult-into-middle-and-beyond transition, 200–201
      orthopedic conditions and, 205
      prevention measures, 206, 207
      treatment, 208
   ankle inversion trauma, 37
   body dissatisfaction, 5, 6
   energy intake for, 307
   ergogenic aids, 312–319
   exercise precautions
      overtraining, 247–250
      staleness, 248–250
   hormonal changes during childhood, 18
   hydration before, during, and after exercise, 307
   knee injury, 36
   menstrual disturbances, 86
   musculoskeletal injuries
      in adult female athletes, 169–181
      in preadolescent and adolescent female athletes, 155
      treatment in aging female athletes, 208–209
   nutritional guidelines, 303, 304, 305, 306
      energy needs and, 299
      nutritional health questionnaire, 364

   screening tools for eating disorder detection, 103, 366–368
Athletes@Risk® Program, 111–113, 118
ATP-phosphocreatine (ATP-PC) energy system, 314
attractiveness factor, 7. *See also* body image assessment
avulsion fractures, 160

B

basal energy expenditure (BEE), 280
basal metabolic rate (BMR), 280
BED. *See* binge eating disorder
behavioral strategies/self-regulatory behaviors, 7. *See also* body image assessment
BID. *See* body image disturbances
Binge eating disorder (BED), 85. *See also* eating disorder not otherwise specified (EDNOS)
biological risk factors (body image and eating disturbances), 62
birth control
   FSH release, 127
   LH release, 127
bisphosphonates, 22, 218. *See also* osteoporosis
BMD. *See* bone mineral density
body composition, exercise recommendations for, 235
body dissatisfaction. *See also* body image; body image concerns; body image disturbances (BID)
   adult women and, 7
   anorexia nervosa to obesity and, 3
   athletes and, 5, 6
   body image distortion, difference with, 5
   body mass index (BMI) and, 7
   defined, 4
   eating disorder and, 59
   social factors, 6
   weight concerns and, 59
body dysmorphic disorders, 5
body image. *See also* body image assessment (BIA); body image concerns; body image disturbances (BID)
   clinical assessment tools for, 7
   eating disturbances and. *See* body image and eating disturbances
   education, 8–11
   enhancement activities, 11
   exposure and desensitization, 11
   negative, 5
   quality of life inventory (BIQLI), 347
body image and eating disturbances
   future scope, 72
   positive youth development, 70, 71, 72
   primary prevention, 62
      family, 64
      health care providers, 67
      health education, 63
      individual aspects, 63
      media, 67
      peer, 66

Index

schools, 65
self-efficacy aspects, 64
self-esteem aspects, 63
society, 68, 69, 70
sociocultural groups, 65
protective factors
positive family relationships, 62
sociocultural, 62
risk factors, 61
biological, 62
familial, 62
sociocultural, 62
body image assessment (BIA)
BIA-C (children), 60
BIA-P (preadolescents), 60
cognitive behavioral approach, 7
procedure, 60
scales and questionnaires, 8
without psychometric properties, 8
body image concerns, 3
adolescent, 6
adult
body dissatisfaction and, 7
older adult, 6
young adult, 6
body dissatisfaction and, 4
body dysphoria, 4
body esteem, 4
body image
defined, 3
distortion, 4
disturbance, 4
bulimia, 5
cognitive behavioral approach, 7
culture-related aspects and, 5
family attitudes and beliefs, 6
inventory, 349
menarche and, 12
metacommunion system and, 5
midlife, 6
negative body image, 4
older adult, 6
prepubertal, 6
pro-anorexia, 5
psychiatric disorders, 4
social factors and, 6
weight and shape, 5
young adult, 6
body image distortion, body dissatisfaction and, 5
body image disturbances (BID). *See also* body dissatisfaction; eating disturbances
cognitive behavioral management of, 11
defined, 4
dysmorphic disorders, 5
eating disorders, 5
body image assessment (BIA) procedure, 60
in children and adolescents, 59, 60
eating disturbances in children and adolescents, 57–59

perceptual body-size distortion, 59
weight concerns and, 59
body image disturbances management
cognitive behavioral strategy
appearance-related beliefs, 11
body image education, 11
body image enhancement activities, 11
body image exposure and desensitization, 11
feminist approach, 11
body image quality of life inventory (BIQLI), 347
body mass index (BMI)
body changes during childhood, 18
body dissatisfaction and, 7
body surface area (BSA), 280, 281
bone
female sex hormones effect on, 143
loss, 146, 147, 149
bone mass
bone composition, 144
female sex hormones effect on, 143
lumbar and, 39–40
osteoclasts, 144
ovarian hormones and, 144
peak bone mass, 143, 145
abnormal menstrual cycle and, 146
adolescent growth spurt, 145
bone loss, 146, 147
bone mineral density, 147
hormonal replacement therapy, 147
mechanical loading aspects, 146
osteopenia/osteoporosis management, 147
secondary amenorrhea, 146
physiology and function, 144
progesterone and, 144
bone mass density, postmenopausal woman, 272
bone mineral density, 82, 147
measurements, 216
osteoporosis, 214, 216
peak bone mass and, 147
bone mineralization, 87
bone thinning, 206
bone tissue. *See also* osteoporosis, 214
in adults, 30
in children, 30
sex differences in, 30
sex hormones and, 30
BORG rating of perceived exertion (RPE) scale. *See also* heart rate reserve (HRR) method
moderate exercise intensity and, 372
vigorous exercise intensity and, 372
brain, reproductive health and, 124
branched-chain amino acids (BCAAs), 312
breast and endometrial cancer
estrogen and, 22
hormone replacement therapy, 22
postmenopause and, 22
progesterone and, 22

breast cancer, exercise program and, 261
breast development, 19
bulimia, 5. *See also* anorexia
bulimia nervosa, 4
    cognitive behavioral therapy (CBT), 113, 114
    diagonistic criteria for, 84, 85
    eating disorders (ED) in children, 59, 61
    treatment principles, 111, 113–114
BULIT-R (bulimia test—revised), 102, 358–363. *See also* eating disorder detection

C

caffeine
    adverse effects, 317
    dosage, 317
    endurance and, 317
    high-intensity exercise and, 316
    strength, effect on, 317
calcitonin
    calcitonin generelated polypeptide (CGRP), 32
    osteoporosis management, 220
calcium. *See also* bone mass
    deposition, 170
    ergogenic aids for athletes, 319
    for mature women, 342
    homeostasis, 22
    metabolism, 144
    musculoskeletal injuries in adult female athletes and, 170
    osteoporosis management and, 217
calories
    discretionary, 396
    discretionary calorie allowance, 397
    essential, 289
    levels, 396
    requirements, 389
cancer
    ACS's physical activity guidelines, 263
    breast, 22
    endometrial, 22
    exercise therapy, 263
        ACS's physical activity guidelines, 263
        quality of life aspects, 264
        therapy-related fatigue, 264
    MBSR and, 263
    quality of life (QOL) factors, 262, 264
cancer-related fatigue, 264
    defined, 264
    exercise and, 264
carbohydrates, 291. *See also* nutrients for children
cardiac output, exercise guidelines during pregnancy and, 254
cardiopulmonary physiology, exercise guidelines for postmenopausal woman and, 271
cardiorespiratory (CR) fitness
    ACSM's exercise recommendations, 230
    exercise guidelines for children and adolescence, 243
    FIT principles, 230
    moderate activity for, 230
    physical activity guidelines, 232
    vigorous activity for, 230
carnitine, 313. *See also* ergogenic aids for athletes
carpal tunnel syndrome (CTS), 33, 34
collagen tissue, 28. *See also* bone tissue
cartilage morphology, 29
cartilage tissue
    sex differences in, 29
    sex hormones and, 29
casein protein, 312
catecholamine, 315
CBT. *See* cognitive behavioral therapy
cervical based sex differences, 38
    cervical disc failure aspects, 39
    whiplash-related disorders, 38
childhood, hormonal changes during, 18
children
    eating disturbances in, 57–60
    exercise guidelines for, 241, 374
    resistance training with, 375
    nutritional guidelines for, 287
        Dietary Guidelines for Americans 2005, 288
        discretionary calories, 289
        family meals, 290
        MyPyramid, 289
        nutrients that provide energy, 291
Chinese ginseng, 315
cholesterol, 292
    bad, 293
    dietary guidelines for mature woman, 338
    good, 293
cognitive approach for body image disturbances management. *See also* feminist therapy
    appearance-related beliefs, 11
    body image
        education, 11
        enhancement activities, 11
        exposure and desensitization, 11
cognitive behavioral approach, body image assessment and, 7
    attractiveness factor, 7
    behavioral strategies/self-regulatory behaviors, 7
    comorbidity, 8
    cultural/socialization factors, 7
    historical/developmental factors, 7
    personality attributes, 8
    self-monitoring, 8
    sociodemographic factors, 7
    treatment attempts history, 8
cognitive behavioral therapy (CBT), 113–114
collagen
    ligament, 28, 29
    skin, 28, 29
    stress patterns, 29
    tendon, 28, 29
collagen tissue. *See also* bone tissue; cartilage tissue
    diseases, 29
    disorders of, 29

Index

sex differences in, 29
sex hormones and, 29
cartilage tissue, 28. *See also* bone tissue
combination therapy, osteoporosis management and, 221
comorbidity, 8. *See also* body image assessment
complementary and alternative therapy, 221. *See also* osteoporosis
compulsive overeating disorder, 85. *See also* eating disorders
concentric muscle actions, 235. *See also* muscular fitness
conditioning programs, musculoskeletal injuries in adult female athletes and, 178–179
coordinated school health program (CSHP), 66
coronary artery disease and stroke
  estrogen and, 22
  hormone replacement therapy, 22
  postmenopause and, 22
corticotropin-releasing hormone (CRH), 89, 138. *See also* functional hypothalamic amenorrhea
coupling effect, 143. *See also* bone mass
coxadiaphyseal (CD) angle, 34
creatine
  arginine, 314
  ergogenic aids for athletes, 313, 314
  glycine, 314
  methionine, 314
  phosphocreatine (PC), 314
  supplementation benefits, 314
cultural/socialization factors, 7. *See also* body image assessment (BIA)

D

degenerative spondylolisthesis, 189
delayed menarche, 82. *See also* amenorrhea; Female Athlete Triad
dietary energy intake values for active individuals, 283
Dietary Guidelines for Americans 2005, 288, 293, 395
dietary guidelines for children. *See also* enery needs for children
  Dietary Guidelines for Americans 2005, 288
  discretionary calories, 289
  family meals, 290
  MyPyramid, 289
  nutrients that provide energy, 291
dietary guidelines for mature woman, 337, 338
  calcium, 342
  folic acid, 339
  iron, 342
  thiamin, 339
  vitamin B_{12}, 339
  vitamin C, 341
  vitamin E, 341
dietary intake
  for athlete, 303
  peak bone mass, 148
Dietary Reference Intakes (DRIs), 398
  elements intake, 402

estimated average requirements for groups, 400
  for athletes, 304
  for children, 292
  for mature women, 338
  for pregnancy, 324
  macronutrient
    distribution ranges, acceptable, 399
    recommended intakes, 398
  vitamin intake, 403
dimorphism. *See* sexual dimorphism
direct calorimetry, 282. *See also* doubly labeled water (DLW) technique
disaccharides, 291
discretionary calorie allowance, 397
discretionary calories, 289, 396. *See also* dietary guidelines for children
disordered eating, 81. *See also* eating disorders (ED)
  cognitive behavioral therapy (CBT), 113, 114
  defined, 84
  education and intervention programs
    anorexia nervosa treatment principles, 111, 113, 114
    Athletes@Risk® Program, 111–113
    bulimia nervosa treatment principles, 111, 113, 114
    hospitalization criteria, 115
    Transtheoretical Model (TTM), 110
  education and intervention programs, 109–111
  Female Athlete Triad
    anorexia nervosa diagnostic criteria, 84
    BED criteria, 85
    bulimia nervosa diagnostic criteria, 85
    disordered eating prevalence, 88
    disordered eating, amenorrhea, and osteoporosis, interrelation between, 87
    EDNOS criteria, 85
  in active middle-aged women, 93, 96
    risk factors for, 94
    stress factors for, 95
    TSAB treatment program, 95
  organization and resources, 115, 116, 117
  self-esteem aspects, 109, 110
doubly labeled water (DLW) technique, 279. *See also* total energy expenditure (TEE)
  EEPA measurement, 282, 283
  REE measurement, 282
  TEE measurement, 282
  TEF measurement, 282
  tri- and uni-axial monitor usage, 283
dumbbell arm curl, 383

E

eating attitudes test, 353
Eating Attitudes Test (EAT), 102, 356
eating disorder not otherwise specified (EDNOS), 61, 85. *See also* binge eating disorder (BED)
eating disorders (ED), 5. *See also* disordered eating; eating disorders detection
  allied health professional and, 99
  and negative body image scenario, 73

anorexia nervosa (AN), 59, 61, 84
binge eating disorder (BED), 85
body dissatisfaction and, 59
bulimia nervosa (BN), 59, 61, 84
compulsive overeating, 85
eating disorder not otherwise specified (EDNOS), 85
Female Athlete Triad, 88
in active middle-aged women, 93, 94
in children and adolescents
   body image assessment (BIA) procedure, 60
   EDNOS criteria, 61
organization and resources, 115, 116, 117, 369
primary care provider's role in detecting, 100
psychological disorders and, 61
eating disorders detection
   behavioral/emotional aspects, 100
   BULIT-R, 102
   Eating Attitudes Test (EAT), 102
   Female Athlete Screening Tool (FAST), 103
   NCCA Division I program, 103
   physical aspects, 100
   preparticipation physical examination (PPE), 103
   primary care provider's role, 100
   SCOFF questionnaire, 102
   screening tools
      for female athlete, 103
      scenario, 105, 106
   self-assessment and educational Web tools, 103, 104
eating disturbances, 5. *See also* body image disturbances (BID)
   future scope, 72
   positive youth development, 70, 71, 72
   primary prevention
      family, 64
      health care providers, 67
      health education, 63
      individual aspects, 63
      media, 67
      peers, 66
      schools, 65
      self-efficacy aspects, 64
      self-esteem aspects, 63
      society, 68, 69, 70
      sociocultural groups, 65
   primary prevention, 62
   protective factors
      positive family relationships, 62
      sociocultural, 62
   risk factors, 61
      biological, 62
      familial, 62
      sociocultural, 62
eating disturbances in children and adolescents, 57–60
   body image assessment (BIA) procedure, 60
   perceptual body-size distortion, 59
   weight concerns and, 59
Eating Smart, Eating for Me prevention program, 65
eccentric muscle actions, 235. *See also* muscular fitness
EDNOS. *See* eating disorder not otherwise specified

Education Amendments Act, 173. *See also* musculoskeletal injuries in adult female athletes
education and intervention programs, 109. *See also* disordered eating
   anorexia nervosa treatment principles, 111, 113, 114
   Athletes@Risk® Program, 111–113, 118
   bulimia nervosa treatment principles, 111, 113, 114
   cognitive behavioral therapy (CBT), 113
   hospitalization criteria, 114, 115
   organization and resources, 115–117
   self-esteem aspects, 109, 110
   self-help (SH) aspects, 114, 115
   Transtheoretical Model (TTM), 110, 111, 118
EEPA. *See* energy expended in physical activity
EER. *See* estimated energy requirements
egalitarian relationship, 11. *See also* feminist therapy
eggs, 142. *See also* menstruation; ovulation
elbow
   tendinosis, lateral, 32
   tennis elbow, 32
elbow based sex differences. *See also* upper extremity based sex differences
   extensor carpi radialis longus (ECRL), 32
   extensor digitorum communis (EDC), 32
   metacarpophalangeal (MCP) joints, 32
Eleutherococcus ginseng, 315
endometrial cancer, postmenopause and, 22
endometrium, 125. *See also* menstruation
endurance, caffeine and, 317
energy balance. *See also* energy imbalance
   energy needs for athlete, 302
   in children, 288
   resting metabolic rate (RMR) and, 336
   weight control, 336
energy demand, pregnancy and, 255
energy expended in physical activity (EEPA), 279, 282. *See also* doubly labeled water (DLW) technique; total energy expenditure (TEE)
energy expenditure
   measurement, 282
      direct calorimetry, 282
      DLW technique, 282, 283
   prediction equations, 283
      EER, 283
      Harris–Benedict formula, 283
      Mifflin–St. Jeor, 283
      physical activity level (PAL), 283
      RMR, 283
energy expenditure components
   energy expended in physical activity (EEPA), 282
   resting energy expenditure, 280
      basal energy expenditure (BEE), 280
      basal metabolic rate (BMR), 280
      body surface area (BSA), 280, 281
      fat-free mass (FFM), 281
      hormonal aspects, 282

Index

lean body mass (LBM), 281
    resting metabolic rate (RMR), 280
    sex difference in metabolic rates, 281
  thermic effect of food, 282
energy imbalance. *See also* energy balance
  functional amenorrhea and, 299
  in children, 288
energy intake for athletes, 307
energy needs during lactation, 324
energy needs during pregnancy, 323, 324
  folic acid supplementation, 329
  iron reserves prior to conception, importance of, 328
  iron–folic acid supplementation, 328
  minerals, 328
  nutrients, 328
  vitamin D deficiency, 329
  vitamins, 328
energy needs for athlete
  energy balance, 302
  estimating energy and nutritional intake, 301, 303
  functional amenorrhea and, 299, 300
energy needs for children, 287. *See also* dietery guidelines for children
  energy balance, 288
  energy imbalance, 288
energy requirements estimation, 279, 280. *See also* estimated energy requirements (EER)
energy supplements
  anabolic-androgenic steroids, 317
  caffeine, 316, 317
  calcium, 319
  ephedra, 315, 316
  ergogenic aids for athletes, 315
  ginseng, 315
  iron with normal ferritin levels, 319
  micronutrient, 335
  multivitamins, 319
ephedra, 315, 316
ephedra, side effects, 316
ephedrine, 315
epinephrine, 282
ergogenic aids for athletes
  anabolic-androgenic steroids, 317, 318
  caffeine, 316, 317
  calcium, 319
  carnitine, 313
  creatine, 313, 314
  energy supplements, 315
  ephedra, 315, 316
  fat burners, 315
  ginseng herbs, 315
  glutamine, 312
  iron, 319
  multivitamins, 319
  protein and amino acids, 312
  whey versus soy protein, 312
essential amino acids, 313

estimated calorie requirements, 389
estimated energy expenditure, 386
estimated energy expenditure prediction equations
  EER, 283
  Harris–Benedict formula, 283
  Mifflin–St. Jeor, 283
  physical activity level (PAL), 283
  REE, 283
  RMR, 283
estimated energy requirements (EER), 280. *See also* total energy expenditure (TEE)
  average dietary energy intake values for active individuals, 283
  BMI and, 386
  energy definition, 279
  estimated energy expenditure prediction equations, 283
  for lactating women, 387
  pregnant women, 387
  TEE and, 386
estradiol, 20
estrogen. *See also* hormone release
  Alzheimer's disease and, 22
  amenorrhea, 135
  athletic amenorrhea and, 87
  bone mass and, 144
  breast and endometrial cancer, 22
  coronary artery disease and stroke, 22
  functional hypothalamic amenorrhea (FHA), 137
  menopause, 21
  menstruation and, 123, 125, 126, 143
  musculoskeletal injuries in adult female athletes, 170, 171
  osteoporosis and, 22, 214
  osteoporosis management, 219, 220
  postmenopause, 21
  pregnancy and, 126
  puberty and, 19
etiological factors, 11. *See also* feminist therapy
exercise, 261. *See also* mindful exercise
  and physical fitness, 229
  defined, 229
  for athletes, 307
  hydration before, during, and after exercise, 307
  physical activities and, difference between, 229
  musculoskeletal injuries
    during pregnancy, 183–196
    in adult female athletes, prevention measures for, 177
  resistance program for postmenopausal women, 379
exercise guidelines. *See also* physical activity recommendations
  for children and adolescents, 374
  for resistance training, 376
  resistance-training guidelines, 233
exercise guidelines and recommendations during pregnancy, 253
  activities to encourage and discourage, 255
  aerobic exercise, 256

benefits, 253
cardiac output, 254
effect on baby, 255
energy demand, 255
gestational diabetes mellitus (GDM), 254
guidelines for exercising, 255
hyperthermia, 254
moderate-intensity exercise, 254, 255
oxygen demand, 254
physiological changes, 254
postpartum exercise, 258
precautionary guidelines, 257
resistance training, 259
RPE, 257
exercise guidelines for children and adolescents, 241
aerobic capacity, 242
anaerobic capacity, 243
approach aspects, 242
cardiorespiratory exercise, 243
exercise aspects, 242
exercise capacity characteristics, unique, 242
metabolic costs, 243
resistance training, 243, 244
$VO_{2\ max}$, 242
exercise guidelines for postmenopausal women, 271
bone mass density changes, 272
cardiopulmonary physiology for older women, 271
conformational changes
muscle, 273
of strength, 273
guidelines for appropriate exercise, 273
resistance training, 273
training technique, 274
exercise intensity
moderate, 371, 372
vigorous, 371, 372
exercise precautions for female athlete, 247
overtaining, 247
hormonal responses, 250
mood variations in response to, 249
signs, 250
staleness, 248
hormonal responses, 250
mood variations in response to, 249
parasympathetic, 249
sympathetic, 249
exercise programs, lean body mass, and 236
exercise recommendations, 228
cardiorespiratory (CR) fitness, 230
for flexibility, 236
for healthy body composition, 235
muscular fitness, 230
exercise resistance program for postmenopausal women, 377
dumbbell arm curl, 383

machine arm curl, 384
machine chest press, 382
wall squats, 379
wall squats with ball, 380
exercise therapy for cancer, 263
ACS's physical activity guidelines, 263
quality of life aspects, 264
related fatigue, 264
extensor carpi radialis brevis (ECRB), 32
extensor carpi radialis longus (ECRL), 32
extensor digitorum communis (EDC), 32

F

fallopian tubes, 125. *See also* menstruation
familial factors of body image and eating disturbances, 62, 64
familial risk factors (body image and eating disturbances), 62
family meals, 290
fat-free mass (FFM), 281
fatigue
cancer-related, 264
fractures, 160
fats
burners
ergogenic aids for athletes, 315
ginseng, 315
cholesterol, 292, 293
deposition aspects, 170
for children, 292
HDL, 293
LDL, 293
lipids, 292
monounsaturated, 293
musculoskeletal injuries in adult female athletes, 170
polyunsaturated, 293
saturated, 293
trans, 293
unsaturated fatty acids, 293
fatty acids
saturated, 338
trans, 338
Female Athlete Screening Tool (FAST), 103
Female Athlete Triad, 134
ACSM Position Stand on, 82, 83
amenorrhea, 81, 87, 88
bone mineralization aspects, 87
disordered eating, 81, 87
amenorrhea, and osteoporosis, interrelation between, 87
prevalence, 88
disordered eating and eating disorders aspects
anorexia nervosa diagnostic criteria, 84
BED criteria, 85
bulimia nervosa diagnostic criteria, 85
EDNOS criteria, 85
eating disorder

Index

detection, 103
prevalence, 88
health concerns, 88
hormonal release and, 87
hypothalamic amenorrhea, 89
menstrual disturbances. *See* menstrual disturbances in athletes, 86
menstrual history scenario, 90
musculoskeletal injuries in adult female athletes, 172, 176
osteoporosis, 81, 82, 87, 88
osteoporosis from osteopenia, differentiating, 86
peak bone mass and, 146
feminist therapy. *See also* cognitive approach for body image disturbances management
appearance-related beliefs, 11
egalitarian relationship characteristics, 11
etiological factors, 11
femoral anteversion, 42
ferritin, 319
fertilization, 125, 128. *See also* implantation; menstruation
FHA. *See* functional hypothalamic amenorrhea
fibers for children, 294
first menstruation. *See* menarche
FIT. *See* Fracture Intervention Trial (FIT)
flexibility. *See also* ACSM, exercise recommendations
alternative stretches, 237
high-risk stretch, 237
folate for mature women, 339
folic acid supplementation, 329
follicle stimulating hormone (FSH), 19, 124
birth control and, 127
functional hypothalamic amenorrhea (FHA), 137, 138
menstruation, 126, 142
reproductive health and, 124
follicular development, 21
food frequency questionnaire, 393
foot. *See also* musculoskeletal injuries in adult female athletes
ankle foot, 37
problems, 175
Fracture Intervention Trial (FIT) principles, 192, 218
cardiorespiratory (CR) fitness, 230
musculoskeletal injuries
during pregnancy. *See* musculoskeletal injuries during pregnancy, exercise-related
treatment in ageing female athletes, 208
fractures, 30, 31. *See also* bone tissue; osteoporosis
avulsion fractures, 160
fatigue, 160
fragility, 206, 213
hip joint, 34
macrotraumatic frank, 34
stress, 160
fruit group, 396
FSH. *See* follicle stimulating hormone
functional amenorrhea, 135, 299. *See also* amenorrhea
energy imbalance and, 299

energy needs for athlete and, 300
psychological stress and, 137
functional hypothalamic amenorrhea (FHA), 137
ACTH hormone, 138
corticotropin-releasing hormone (CRH), 138
cortisol hormone, 138
defined, 137
estrogen, 137
FSH, 137, 138
GnRH hormone release, 137, 138
HPO axis, improper functioning of, 137
LH, 137, 138, 139
psychological stress, 138

G

genital development, 19
gestational diabetes mellitus (GDM), 254
ginseng, 315
Chinese, 315
Siberian, 315
glutamine, 312. *See also* ergogenic aids for athletes
glycine, 314
glycogen sparing, 317
GnRH. *See* gonadotropin-releasing hormone
gonadotroph, 124
gonadotropic releasing hormone (GnRH), 19, 87. *See also* puberty
amenorrhea and, 135
functional hypothalamic amenorrhea (FHA) and, 137, 138
menopause and, 21
menstruation and, 126, 142
reproductive health and, 124
grains group, 397
granulomatous tissue, 29
gravid uterus, musculoskeletal structures and, 184
growth plate injury, 158

H

hand. *See* wrist and hand based sex differences
Harris–Benedict formula, 283. *See also* estimated energy expenditure prediction equations
HCG, pregnant woman's ability to exercise and, 185–186
health care providers
body image and eating disturbances
health education, 63
prevention and, 63, 67
health professionals, allied, 88, 89
health-related fitness, 229, 230
cardiorespiratory (CR) fitness, 230
physical activity recommendations, 231
health-related quality of life (HRQoL), AIS impact on, 39
Heart and Estrogen/Progestin Replacement Study (HERS), 22
heart rate reserve (HRR) method, 371. *See also* BORG rating of perceived exertion (RPE) scale

moderate exercise intensity, 229, 371
vigorous exercise intensity and, 229, 371
hematopoiesis, 328
hemodynamics changes, exercise guidelines and recommendations during pregnancy, 254
HERS. *See* Heart and Estrogen/Progestin Replacement Study
high-density lipoprotein (HDL), 293, 318
high-intensity exercise, 316. *See also* caffeine
high-risk stretch, flexibility and, 237
hip joint based sex differences. *See also* lower extremity based sex differences
  coxadiaphyseal (CD) angle, 34
  labral trauma, 34
  stress reactions, 34
historical/developmental factors. *See also* body image assessment (BIA)
homeostasis, 144
hormonal birth control, 127. *See also* reproductive health
hormonal changes during childhood, 18. *See also* reproductive changes
hormonal responses
  exercise precautions for female athlete overtraining, 250
  pregnant woman's ability to exercise and, 187
  REE and, 282
  staleness, 250
hormone replacement therapy (HRT)
  Alzheimer's disease, 22
  breast and endometrial cancer, 22
  coronary artery disease and stroke, 22
  osteoporosis and, 22
  peak bone mass and, 147
hormone therapy (HT), osteoporosis management and, 218
  androgen, 220
  calcitonin, 220
  estrogen, 219, 220
  progesterone, 220
  teriparatide, 221
hospitalization criteria, 114–115. *See also* education and intervention programs
hot flashes, 21
HPO. *See* hypothalamic-pituitary-ovarian (HPO) axis
HRT. *See* hormone replacement therapy
hydration, exercise and, 307
hypertension, 338
hypertrophy, 234, 235. *See also* muscular fitness
hypothalamic amenorrhea, 82, 89, 137. *See also* amenorrhea
hypothalamic-pituitary-adrenal (HPA) axis, 82
hypothalamic-pituitary-gonadal (HPG) axis, 82
hypothalamic-pituitary-ovarian (HPO) axis, 82
  functional hypothalamic amenorrhea (FHA) and, 137
  menstruation and, 123, 142
hypothalamus hormone secretion, 126. *See also* reproductive health
hypothyroidism, 134

I

ibandronate, 219. *See also* osteoporosis
iliolumbar ligament, 42
impingement syndrome, 176. *See also* musculoskeletal injuries in adult female athletes
implantation, 125. *See also* fertilization, menstruation
inhibin, 124
insulin-like growth factor 1, 87
integrated care, 262
iron
  energy needs during pregnancy, 328
  ergogenic aids for athletes, 319
  for mature women, 342
  normal ferritin levels, 319
iron–folic acid supplementation, 328
irregular menstrual cycles, 132. *See also* oligomenorrhea

J

jump training and anterior cruciate ligament (ACL) injury, 36

K

knee complex. *See also* lower extremity based sex differences
  anterior cruciate ligament (ACL) injury, 35, 36, 37
  based sex differences, 35
  patellofemoral complex, 35
knee injury
  musculoskeletal injuries treatment in adult female athletes, 179
knee/ACL injury prevention measures, 163
Kraemer's age-specific exercise guidelines for resistance training, 376
kyphosis, thoracic, 39

L

labral trauma, 34. *See also* hip joint based sex differences
lactation, 126. *See also* reproductive health
  EER for lactating women, 387
  energy needs during, 323, 324
  nutritional guidelines during, 327
LBP. *See* low back pain
lean body mass (LBM), 236, 281
leptin, 87
LH
  birth control, 127
  energy needs for athletes and, 300, 301
  functional hypothalamic amenorrhea (FHA), 137–139
  menstruation, 126, 142
LH release
  menopause and, 21
LH. *See* luteinizing hormone
LHRH. *See* luteinizing hormone releasing hormone
ligament system, 41. *See also* sacroiliac joint (SIJ)

Index

lipids for children, 292
long dorsal sacroiliac ligament, 41
lordosis, thoracic, 39
low back pain (LBP), 40, 188
   musculoskeletal injuries during pregnancy, exercise-related, 188–189
   musculoskeletal injuries treatment during pregnancy, exercise-related, 193
low-density lipoprotein (LDL), 293
lower extremity based sex differences. *See also* upper extremity based sex differences
   ankle foot, 37
   hip joint, 34
   knee complex, 35
lower extremity musculoskeletal injuries, 175
lumbar
   bone mass differences and, 39, 40
   load tolerances and, 40
   low back pain (LBP), 40
   muscle cross-sectional areas and muscle geometry, 40
   muscle reflexes, 40
lumbar based sex differences, 39–40. *See also* spine based sex differences
luteal phase deficiency, 86
luteinizing hormone (LH), 87
   birth control, 127
   energy needs for athletes and, 300, 301
   functional hypothalamic amenorrhea (FHA), 137–139
   menopause and, 21
   menstruation, 126, 142
   puberty and, 19
luteinizing hormone releasing hormone (LHRH), 124. *See also* reproductive health
lysine, 313

M

machine arm curl, 384
machine chest press, 382
macronutrients
   for mature women, 338
   recommendations for athletes, 307
macrotraumatic frank fracture, 34
maternal anemia, 328
MBSR. *See* mindfulness-based stress reduction
meat and beans group, 397
mechanical loading, 146. *See also* peak bone mass
media, body image and eating disturbances prevention and, 67
menarche, 17, 19, 20. *See also* menstruation; puberty
   body image concerns scenario and, 12
   reproductive changes scenario, 23
menopause, 18, 20. *See also* menstruation
   estrogen relsease and, 21
   GnRH release, 21
   hot flashes, 21
   LH release, 21
   norepinephrine activity, 21

menstrual cycles, 82. *See also* amenorrhea; eumenorrhea; menstruation; oligomenorrhea
   abnormal, 131, 135
      acyclic, 131, 132
      irregular, 131
   bone loss, 149
   estrogen, 143
   FSH, 124, 126, 142
   GnRH, 126, 142
   HPO axis, 123, 142
   LH, 124, 126, 142
   ovaries and, 142
   peak bone mass and, 141, 146
   progesterone, 143
   regular, 131
   reproductive health scenario, 128
menstrual disturbances
   allied health professional's role and, 99
   amenorrhea and, 82, 86, 131, 132, 137
   in athletes, 86
menstruation, 17, 20. *See also* menopause; menstrual cycles; puberty; reproductive changes
   amenorrhea, 134, 137
   brain involvement, 124
   defined, 123
   estrogen hormone secreation and, 123, 125, 126
   fallopian tubes, 125
   first menstruation (menarche), 17
   FSH, 124, 126, 142
   functional amenorrhea, 137
   GnRH, 126, 142
   hypothalamic-pituitary-ovarian axis and, 123, 142
   hypothalamus, 125, 126
   hypothyroidism, 134
   implantation, 125
   LH, 124, 126, 142
   menopause and, 17, 18
   oligomenorrhea, 132, 133
      low energy availability, 133
      PCOS, 133
   ovary, 125, 126
   pituitary hormones secretion, 125, 126
      follicle stimulating hormone, 124
      luteinizing hormone, 124
   primary amenorrhea and, 132
   progesterone secretion and, 123, 125, 126
   puberty, 17
   secondary amenorrhea and, 132
   steroids and ova release, 125
   uterus, 125
metabolic costs for children and adolescence, 242
metabolic rate. *See also* total energy expenditure (TEE)
   basal, 280
   resting, 280
   sex difference in, 281
metacarpal septa, 32
metacarpophalangeal (MCP) joints, 32
metacommunion system, 5
methionine, 313, 314

micronutrients, nutritional guidelines and, 335
middle-aged women. *See* disordered eating in active middle-aged women, 95
midlife body image concerns, 6
Miffin–St. Jeor equation
  estimated energy expenditure prediction equations, 283
  RMR estimation and, 336
milk group, 397
mind-body medicine, 262
mindful exercise, 261. *See also* breast cancer
mindfulness
  concept of, 263
  MBSR program, 263
mindfulness meditation (MM), 262, 263
Mindfulness-Based Exercise Rehabilitation (MBER) program, 265
  cancer scenario, 267
  format and structure, 265
  session contents, 265, 266, 267
mindfulness-based stress reduction (MSBR), 261, 262
  cancer and, 263
  MBSR Program and mindfulness, 263
mineralization, bone, 87
minerals
  energy needs during pregnancy, 328
  for children, 294
  for mature women, 338
  osteoporosis management, 217
miserable malalignment syndrome, 161
MM. *See* mindfulness meditation
Model for Healthy Body Image (MHBI), 70
moderate exercise intensity
  BORG rating of perceived exertion (RPE) scale, 372
  heart rate reserve (HRR) method, 371
  pregnancy and, 254, 255
moderate-intensity physical activity, 229
  defined, 229
  heart rate reserve (HRR), 229
  pregnancy, 255
  rate of perceived exertion (RPE), 229
monosaccharides, 291
monounsaturated fats, 293
mood variations
  overtraining and, 249
  staleness and, 249
multiple-joint exercise, 235. *See also* muscular fitness
multivitamins, 319
muscle mass, postmenopausal woman, 273
muscle reflexes, lumbar and, 40
muscle strength, 172. *See also* musculoskeletal injuries in adult female athletes
muscoloskeletal alterations, pregnancy and, 254
muscular endurance training, 234
muscular fitness, 230
  ACSM's exercise recommendations, 230
  concentric muscle actions, 235
  eccentric muscle actions, 235
  exercise sequence, 234
  hypertrophy, 235
  hypertrophy training, 234
  intermediate to advanced training, 234
  multiple-joint exercise, 235
  muscular endurance training, 234
  novice training, 234
  overload principle, 233
  periodization, 235
  power, 235
  power training, 234
  principle of progression, 233
  progressive overload, 233
  repetition maximum (RM), 234, 235
  repetitions, 235
  resistance exercise, 231
  resistance training, 234
  resistance-training guidelines, 233
  RPE, 233
  set, 235
  training frequency, 234
musculoskeletal anatomy. *See also under* sex differences
  musculoskeletal tissues-based sex differences, 28, 29
    bone, 28, 30
    cartilage, 28, 29
    collagen, 28
  sexual dimorphism and, 25, 26
  skeletal geometry-based sex differences, 27, 28
musculoskeletal injuries during pregnancy, exercise-related, 183
  degenerative spondylolisthesis, 189
  exercising issues, 187
    anatomical, 187
    biomechanical aspects, 188
    structural, 187
  exercise affecting factors
    physiological, 185
    systemic factors, 185
  gravid uterus and organ growth, 184
  HCG secretion, 185, 186
  hormonal secretion, 187
  low back pain (LBP), 188, 189
  osteoporotic hip, 189
  patellofemoral pain, 190
  pelvic girdle pain (PGP), 189
  pelvis morphology, 185
    ilium/pubis, 184
    sacrum/coccyx, 184
  prevention measures, 191, 192
    exercise regimen, 192
    FITT principle, 192
  treatment modes
    correct posture, 193
    for LBP, 193
    for PGP, 193
    NSAIDs, 193
    physical therapy program, 193

Index

musculoskeletal injuries in adult female
  athletes, 169, 180
  ACL injuries, 173
  adipose tissue, 171
  adolescent growth spurt, 170
  anterior cruciate ligament, 172
  calcium deposition, 170
  differences in anatomy, physiology, and body
    composition between adolescent and adult
    females, 170
  Education Amendments Act, 173
  estrogen, 170, 171
  fat deposition aspects, 170
  Female Athlete Triad, 172, 176
  foot problems, 175
  impingement syndrome, 176
  lower extremity, 175
  muscle strength and, 172
  Olympic sporting events, 172
  orthopedic injuries, 172
  patellofemoral joint, 175
  patellofemoral maltracking and pain (PFPS), 174
  PFPS, 175
  prevention measures
    conditioning programs, 178, 179
    exercises, 177
    extrinsic factors, 177, 179
    for stress injury, 178
    intrinsic factors, 177
    muscular contraction aspects, 178
  risky landing positions, 174
  similarities and pertinent differences between adult
    males and females concerning anatomy, body
    composition, and biomechanics, 171
  skeletal density, 170
  testosterone, 170, 171
  treatment
    knee injury, 179
    NSAID, 179
    PRICE, 179
    ROM, 179
  upper extremity, 175, 176
musculoskeletal injuries in aging female athletes, 199
  activities of daily living (ADL), 206
  differences and similarities between older males and
    females in terms of body composition,
    musculoskeletal components, and athletic
    performance, 202, 203, 204, 205
  fragility fracture, 206
  hormonal, physiological, and anatomical changes from
    young-to-adult-into-middle-and-beyond
    transition, 200, 201
  orthopedic conditions and, 205
  prevention measures, 206, 207
  treatment, 208
    FIT (T) activity, 208
    NSAIDs, 208
    PRICE, 208
    stretching, 209

musculoskeletal injuries in preadolescent and adolescent
  female athletes, 155
  adipose tissue deposition and, 157
  adolescent growth spurt and, 158
  anterior cruciate ligament (ACL) tearing, 161
  fatigue fractures, 160
  growth plate injuries, 160
  growth plate injury and, 158
  knee injury, 161
  knee joint alingment, 162
  landing positions, risky and safe, 162
  limb landing, 159
  miserable malalignment syndrome, 161
  pelvic apophyses and, 159
  physiological aspects, 156
  prepubescence to postpubescence changes, 156
  prevention measures, 161
    conditioning program, 163, 164
    knee/ACL injuries, 163
    orthopedic treatment, 163
    PRICE, 163
    range of motion (ROM) aspects, 163
    six S's institutionalization/modification, 163
  stress fractures, 158, 160
  tendinous tightness, 161
  training techniques scenario, 165, 166
  young females and males anatomy, physiology, and
    biomechanics, differences and similarities
    between, 157
musculoskeletal injuries treatment in aging female
  athletes
  FIT (T) activity, 208
  NSAIDs, 208
  PRICE, 208
  stretching, 209
musculoskeletal tissues-based sex differences, 28, 29
  bone, 28
  bone tissues, 30
  cartilage, 28, 29
  collagenous tissues, 28, 29
myometrium, 125. *See also* menstruation
MyPyramid, 289, 396. *See also* dietary guidelines
  for children
MyPyramid food intake patterns
  discretionary calorie allowance, 397
  fruit group, 396
  grains group, 397
  meat and beans group, 397
  milk group, 397
  oils, 397
  vegetable group, 397

N

NCCA Division I programs, 103
negative body image, 4, 5, 64. *See also* body image
  concerns
  and eating disorder scenario, 73
  prevention, 64

neural tube defects (NTD), 323
nonessential amino acids, 313
norepinephrine, 21, 282. *See also* menopause
NSAIDs for musculoskeletal injuries treatment
   in adult female athletes, 179
   in aging female athletes, 208
   during pregnancy, 193
nutrients
   energy needs during pregnancy, 328
   for athletes, 307
   for mature women, 338
nutrients for children. *See also* dietary guidelines
     for children
   carbohydrates, 291
   fats/lipids, 292
   fibers, 294
   minerals, 294
   nutrients that provide energy, 291
   proteins, 292
   vitamins, 294
   water, 295
nutrition questionnaire, 390
nutritional guidelines
   during lactation, 327
   during pregnancy, 323, 327
   for athletes, 299
   for children, 287. *See also* dietery guidelines
     for children
   for female athlete, 303–306
   for functional amenorrhea, 299
nutritional health questionnaire, student-athlete, 364
nutritional intake for athlete, 301, 303

O

oils, 397
older adult body image concerns, 6
oligomenorrhea
   defined, 132
   hypothyroidism and, 134
   low energy availability and, 133
   PCOS, 132, 133
Olympic sporting event, 172. *See also* musculoskeletal
     injuries in adult female athletes
orthopedic conditions
   adult female athletes
     musculoskeletal injuries in, 172
     musculoskeletal injuries treatment, 179
   aging female athletes
     musculoskeletal injuries in, 205
     musculoskeletal injuries treatment, 208
   preadolescent and adolescent female athletes, 163
osteoblasts, 144, 214
osteoclasts, 144
osteopenia, 86, 134, 147. *See also* Female Athlete Triad;
     peak bone mass
osteoporosis, 213
   bone mineral density (BMD) and, 82,
     214, 216

classifications, 214
defined, 214
diagnosis, 215
estrogen deficiency and, 214
estrogen hormone and, 22
Female Athlete Triad and, 81, 82, 87, 88
fragility fractures, 213
hormone replacement therapy, 22
hormone therapy
   androgen, 220
   calcitonin, 220
   estrogen, 219, 220
   progesterone, 220
   teriparatide, 221
management
   alendronate, 218
   bisphosphonates, 218
   calcium, 217
   combination therapy, 221
   complementary and alternative
     therapy, 221
   Fracture Intervention Trial (FIT), 218
   herbal remedies, 222
   hormone therapy (HT), 218
   ibandronate, 219
   pharmacologic therapy, 217
   phytoestrogens, 221
   raloxifene, 219
   risedronate, 218
   selective estrogen receptor modulators,
     218, 219
   vitamin D, 217
   vitamins and minerals, 217
osteopenia and, 86
parathyroid hormone (PTH), 214
pathophysiology, 214
peak bone mass, 147
postmenopause and, 21
prevention, 216
primary, 214
   Type 1, 214
   Type 2, 214
risk factors
   modifiable, 215
   nonmodifiable, 215
secondary, 215
T-score, 216
Z-score measurement, 216
osteoporotic hip, 189. *See also* musculoskeletal injuries
     during pregnancy, exercise-related
OTS. *See* staleness, 248
ova, 125. *See also* menstruation
ovarian hormones, 19. *See also* puberty
   bone mass and, 144
   reproductive health, 126
ovaries, 19, 142
overload principle, 233. *See also* muscular fitness
overreaching, 248

Index

overtraining, 247, 248. *See also* exercise precautions for female athlete
  hormonal responses, 250
  mood variations in response to, 249
  parasympathetic, 251
  signs and symptoms, 250
    biochemical, 251
    immunological, 251
    physiological, 250
    psychological, 251
  sympathetic, 251
overweight, 388. *See also* physical activity level (PAL)
ovulation, 20, 125
ovulation, PCOS syndrome and, 133
oxygen demand, exercise guidelines and recommendations during pregnancy, 254

P

Panax ginseng, 315
Panax notoginseng, 315
Panax pseudoginseng, 315
Panax quinquefolium, 315
parasympathetic
  overtraining, 251
  staleness, 249, 251
parathyroid hormone (PTH), 214. *See also* osteoporosis
patellofemoral complex of knee, 35
patellofemoral joint (PFJ), 174, 175. *See also* musculoskeletal injuries in adult female athletes
patellofemoral maltracking and pain (PFPS), 174, 175
patellofemoral pain
  musculoskeletal injuries during pregnancy, exercise-related, 190
  PFPS, 174
peak bone mass, 141, 145
  abnormal menstrual cycle and, 146
  adolescent growth spurt, 145
  bone loss, 146, 147
  bone mineral density, 147
  hormonal replacement therapy, 147
  mechanical loading aspects, 146
  osteopenia/osteoporosis management, 147
  secondary amenorrhea, 146
peers, body image and eating disturbances prevention and, 66
pelvic girdle pain (PGP), 189, 193. *See also* musculoskeletal injuries during pregnancy, exercise-related
perceptual body-size distortion, 59. *See also* body dissatisfaction
periodization, 235. *See also* muscular fitness
personality attributes, 8. *See also* body image assessment
PFPS. *See* patellofemoral maltracking and pain
PGP. *See* pelvic girdle pain
pharmacologic therapy for osteoporosis management, 217
phosphocreatine (PC), 314. *See also* creatine
physical activity, 287
  and exercise, difference between, 229
  and fitness objectives, 228
  and physical fitness, 229
  benefits of, 228
  cancer guidelines, 263
  cardiorespiratory (CR) fitness, 230
  defined, 229
  estimated calorie requirements and, 389
  exercise guidelines for children and adolescents and, 242
  health-related fitness recommendations and, 230
  moderate-intensity, 229
  physical activity readiness questionnaire (PAR-Q), 373
  vigorous-intensity, 229
physical activity levels (PAL)
  energy needs during pregnancy, 324
  estimated energy expenditure and, 386
  estimated energy expenditure prediction equations, 283
  walking equivalence and, 385
  weight maintenance and, 387
physical activity readiness questionnaire (PAR-Q), 373
physical activity recommendations, 227, 231. *See also* exercise guidelines
physical appearance state, 351
physical fitness, 227
  exercise and, 229
  health-related, 230
  physical activities and, 229
  skill-related components of, 229
physiological changes, exercise guidelines and recommendations during pregnancy, 254
physis, 158
phytoestrogens, 221
pituitary hormones
  follicle stimulating hormone, 124
  luteinizing hormone, 124
  reproductive health and, 124
polycystic ovary syndrome (PCOS), 132, 133, 137
polyunsaturated fats, 293
porous bones. *See* osteoporosis
positive youth development, 70–72. *See also* body image and eating disturbances
posterior superior iliac spine (PSIS), 41
postmenopausal state, 18. *See also* menopause
postmenopausal woman
  exercise guidelines for, 271
  resistance program for, 377, 379
postmenopause, 21
  Alzheimer's disease issues, 22
  breast and endometrial cancer issues, 22
  coronary artery disease and stroke issues, 22
  estrogen and, 21
  osteoporosis issues, 21
postpartum exercise, 258. *See also* exercise guidelines and recommendations during pregnancy
power, 235. *See also* muscular fitness
power training, 234

pregnancy, 126. *See also* reproductive health
  aerobic exercise during, 256
  body temperature, 254
  cardiac output, 254
  energy demand, 255
  energy needs
    folic acid supplementation, 329
    iron reserves prior to conception, importance of, 328
    iron–folic acid supplementation, 328
    minerals, 328
    nutrients, 328
    vitamin D deficiency, 329
    vitamins, 328
  exercise guidelines
    activities to encourage and discourage, 255
    aerobic exercise, 256
    benefits, 253
    cardiac output, 254
    effect on baby, 255
    energy demand, 255
    gestational diabetes mellitus (GDM), 254
    guidelines for exercising, 255
    hyperthermia, 254
    moderate-intensity exercise, 254, 255
    oxygen demand, 254
    physiological changes, 254
    postpartum exercise, 258
    precautionary guidelines, 257
    resistance training, 259
    RPE, 257
  folic acid supplementation, 329
  hyperthermia, 254
  iron reserves, 328
  iron–folic acid supplementation, 328
  maternal anemia, 328
  musculoskeletal alterations, 254
  musculoskeletal injuries incurred during. *See* musculoskeletal injuries during pregnancy, exercise-related
  nutritional guidelines, 323
  oxygen demand, 254
  resistance training, 259
  resistance training and, 257
  RPE, 257
  vitamin D deficiency, 323, 329
  weight gain ranges, 325
pregnant women, EER for, 387
Premarin, 22
Prempro, 22
preparticipation physical examination (PPE), 103
prepubertal body image concerns, 6
PRICE, 181
  measures, 163
  musculoskeletal injuries treatment in adult female athletes, 179
  aging female athletes, 208
primary amenorrhea, 19, 23, 81, 82, 132. *See also* menstruation

primary care provider, 100, 101. *See also* allied health professional; eating disorder detection
  behavioral/emotional aspects, 100
  BULIT-R, 102
  Eating Attitudes Test (EAT), 102
  in adolescent and pre-adolescent patients, 100
  physical aspects, 100
  SCOFF questionnaire, 102
  screening tools for female athlete, 103
primary osteoporosis, 214
principle of progression, muscular fitness and, 233
Profile of Mood States (POMS)
  overtraining and, 249
  staleness and, 249
progesterone
  bone mass and, 144
  breast and endometrial cancer, 22
  menstruation and, 123, 125, 126, 143
  osteoporosis management, 220
  pregnancy and, 126
  puberty and, 19
progressive overload, 233. *See also* muscular fitness
prophylactic protective effect, 164
protein
  ergogenic aids for athletes and, 312
  for children, 292
  hormones, 19
psychological stress, functional hypothalamic amenorrhea (FHA) and, 138
psychometric properties, body image assessment and, 8
puberty, 17, 19, 20, 42. *See also* menopause; menstrual cycles; menstruation; reproductive changes; sexual maturity
  androgen hormones and, 19
  estrogen hormone secretion, 19
  follicle stimulating hormone and, 19
  gonadotropin-releasing hormone secretion, 19
  luteinizing hormone, 19
  menarche, 19
  ovarian hormones secretion, 19
  progesterone hormone, 19
  secondary sexual characteristics, 19
  sexual maturation and, 19
  steroid hormone secretion, 19

Q

Quality of life (QOL)
  cancer and, 262
  exercise therapy for cancer, 264

R

raloxifene, 219. *See also* osteoporosis management
range of motion (ROM), 163
  musculoskeletal injuries treatment in adult female athletes, 178, 179
  musculoskeletal injuries treatment in aging female athletes, 208

Index

rate of perceived exertion (RPE), 316
  physical activity
    moderate-intensity, 229
    vigorous-intensity, 229
  muscular fitness and, 233
  pregnancy and, 257
recommended dietary allowance (RDA), 292
REE. *See* resting energy expenditure
regular menstrual cycles, 132. *See also* eumenorrhea
repetition maximum, 234, 235. *See also* muscular fitness
repetitions, 235
reproductive changes
  hormonal changes during childhood, 18
  menopause, 18, 20
  menstrual cycles, 17
  post-menopause
    Alzheimer's disease and, 22
    coronary artery disease and stroke and, 22
    osteoporosis, 21
  puberty, 19
    hormonal changes, 19
    menarche, 19
  sexual maturation, 18, 20
reproductive health
  brain involvement, 124
  follicle stimulating hormone and, 124
  GnRH and, 124
  hormonal birth control, 127
  hormonal secretion and, 124
  hypothalamus hormone secretion, 125, 126
  luteinizing hormone releasing hormone and, 124
  menstrual cycle, 123
  menstrual period scenario, 128
  menstruation
    FSH secretion, 124
    LH secretion, 124
    steroids and ova release, 125
  ovarian hormone secretion, 126
  ovary, 125, 126
  pituitary hormones secretion, 124–126
  pregnancy and lactation, 126
  steroid hormone secretion, 126
resistance exercise, 231. *See also* muscular fitness
resistance program for postmenopausal women, 377, 379
resistance training
  exercise guidelines for, 376
    muscular fitness, 233
    postmenopausal woman, 273
  for children and adolescents, 243, 244, 375
  Kraemer's age-specific exercise guidelines, 376
  muscular fitness, 234
  pregnancy and, 257, 259
resting energy expenditure (REE), 279, 280, 336. *See also* total energy expenditure (TEE)
  basal energy expenditure (BEE), 280

basal metabolic rate (BMR), 280
body surface area (BSA), 280, 281
  defined, 280
  DLW techniques, 282
  estimated energy expenditure prediction equations, 283
  fat-free mass (FFM), 281
  hormonal aspects, 282
  lean body mass (LBM), 281
  measurement, 282
  resting metabolic rate (RMR), 280
  sex difference in metabolic rates, 281
resting metabolic rate (RMR), 280
  energy balance and, 336
  estimated energy expenditure prediction equations, 283
  Miffin–St. Jeor equation, 336
reversible amenorrhea, 135
risedronate, osteoporosis management and, 218
risky landing positions, 174. *See also* musculoskeletal injuries in adult female athletes
RMR. *See* resting metabolic rate
ROM. *See* range of motion
RPE. *See* rate of perceived exertion

S

sacroiliac (SI) joints, 184
sacroiliac joint (SIJ), 41, 42. *See also* spine based sex differences
  C-shape in men, 41
  ligament system, 41
    iliolumbar, 42
    long dorsal sacroiliac ligament, 41
    sacrospinous and sacrotuberous ligaments, 41
    thoracolumbar fascia, 41
  L-shape in men, 41
sacrospinous ligaments, 41
sacrotuberous ligaments, 41
saturated fats, 293
saturated fatty acids, 338. *See also* dietary guidelines for mature women
scales and questionnaires approach, body image assessment and, 8
schools, body image and eating disturbances prevention and, 65
  coordinated school health program (CSHP), 66
  Eating Smart, Eating for Me prevention program, 65
SCOFF questionnaire, 102, 352. *See also* eating disorders (ED)
screening tool
  athlete, 366, 367, 368
  for eating disorder, 103
secondary amenorrhea, 81, 82. *See also* functional amenorrhea; secondary amenorrhea
  menstruation and, 132
  peak bone mass, 146

secondary osteoporosis, 215
selective estrogen receptor modulators (SERMS), 218, 219. *See also* osteoporosis
self-efficacy aspects of body image and eating disturbances prevention, 63, 64
self-esteem, 63, 109. *See also* disordered eating
self-help (SH) aspects, 114–115. *See also* education and intervention programs
self-monitoring, 8. *See also* body image assessment
self-respect, 63
serum markers of bone, 30
set, 235. *See also* muscular fitness
sex differences
  lower extremity
    ankle foot, 37
    hip joint, 34
    knee complex, 35
  musculoskeletal anatomy and, 25, 27
    musculoskeletal tissues, 28–29
    skeletal geometry, 27–28
  musculoskeletal tissues
    bone tissues, 28, 30
    cartilage tissue, 28–29
    collagen tissue, 28–29
  spine
    cervical, 38
    lumbar, 39–40
    sacroiliac and pelvis, 41–42
    thoracic, 39
  upper extremity
    elbow, 32
    shoulder, 31–32
    wrist and hand, 33
sex hormones
  bone tissue and, 30
  cartilage tissue and, 29
  collagen tissue and, 29
sexual activity, reproductive change scenario, 23
sexual dimorphism, 25, 26, 43
sexual maturation, puberty and, 19, 20
shoulder, 31. *See also* upper extremity based sex differences
shoulder based sex differences, 31–32. *See also* acromion process
Siberian ginseng, 315
SIJ. *See* sacroiliac joint
skeletal density, 170. *See also* musculoskeletal injuries in adult female athletes
skeletal geometry-based sex differences, 27, 28
skill-related physical fitness, 229
Social Development Model, The, 70
society, body image and eating disturbances prevention and, 68, 69, 70
sociocultural aspects of body image and eating disturbances prevention, 65
  protective factors, 62
  risk factors, 62
sociodemographic factors, 7. *See also* body image assessment

soy protein, 312. *See also* ergogenic aids for athletes
spine based sex differences. *See also* lower extremity based sex differences; upper extremity based sex differences
  cervical, 38
  lumbar, 39, 40
  sacroiliac and pelvis, 41, 42
  thoracic, 39
spondylolysis, 160
sports medicine guidelines, 375
squats, 379, 380
staleness, 248. *See also* exercise precautions for female athlete
  hormonal responses, 250
  parasympathetic, 249, 251
  sympathetic, 251
steroid hormone secretion
  anabolic-androgenic, 317–318
  puberty and, 19
  reproductive health, 126
steroids and ova release, 125. *See also* menstruation
strength capabilities, postmenopausal women, 273
stress fractures, 158, 160
stretching, musculoskeletal injuries treatment in aging female athletes and, 209
sympathetic
  overtraining, 251
  staleness, 249, 251
sympathomimetic alkaloids, 315
systemic ligament laxity, 161

T

TEE. *See* total energy expenditure
tendinous tightness, 161
tendon stres, elbow and, 32
tennis elbow, 32
teriparatide, osteoporosis management and, 221
testosterone
  anabolic-androgenic steroids and, 317
  musculoskeletal injuries in adult female athletes, 170–171
thermic effect of food (TEF), 279, 282. *See also* total energy expenditure (TEE)
thiamin for mature women, 339
Thirty Something and Beyond (TSAB) treatment programs, 95
thoracic based sex differences, 39. *See also* spine based sex differences
thoracic kyphosis, 39
thoracic lordosis, 39
thoracic spine based sex differences, 39
thoracolumbar fascia, 41
total energy expenditure (TEE)
  doubly labeled water (DLW) technique, 279, 282
  energy expended in physical activity (EEPA), 279, 282
  estimated energy requirements (EER), 279, 386
  REF, 279
  resting energy expenditure, 279, 280

basal energy expenditure (BEE), 280
basal metabolic rate (BMR), 280
body surface area (BSA), 280, 281
fat-free mass (FFM), 281
hormonal aspects, 282
lean body mass (LBM), 281
resting metabolic rate (RMR), 280
sex difference in metabolic rates, 281
TEF, 279, 282
training, 243. *See also* resistance training
exercise guidelines for postmenopausal woman, 274
musculoskeletal injuries and
in preadolescent and adolescent female athletes, 166
prevention in adult female athletes, 178
trait anxiety scale, 351
trait scale, 351
trans fats, 293
trans fatty acids, dietary guidelines for mature woman and, 338
Transtheoretical Model (TTM), 110
action (A) stage, 110
contemplation (C) stage, 110
decisional balance aspects, 111
eating disorder scenario, 118
maintenance (M) stage, 110
precontemplation (PC) stage, 110
preparation (P) stage, 110

U

upper extremity based sex differences. *See also* lower extremity based sex differences
elbow, 32
shoulder, 31, 32
wrist and hand, 33
upper extremity musculoskeletal injuries in adult female athletes, 175–176
uterus, 125. *See also* menstruation

V

vegetable group, 397
vigorous exercise intensity
BORG rating of perceived exertion (RPE) scale, 372
heart rate reserve (HRR) method, 371
vigorous-intensity physical activity, 229
defined, 229
heart rate reserve (HRR), 229
rate of perceived exertion (RPE), 229
vitamin B12 for mature women, 339
vitamin C for mature women, 341
vitamin D. *See also* bone mass
deficiency, 323
pregnancy and, 323, 329
with dark complexions and limited sun exposure, 329
metabolism, 144
osteoporosis management, 217
vitamin E for mature women, 341
vitamins
Dietary Reference Intakes (DRIs), 403
energy needs during pregnancy, 328
for children, 294
for mature women, 338
osteoporosis management, 217

W

walking equivalence, physical activity level (PAL) and, 385
wall squats, 379
wall squats with ball, 380
water for children, 295
weight and shape concerns, 5. *See also* body image concerns
weight control, energy balance and, 336
weight dissatisfaction, 7. *See also* body dissatisfaction
weight gain ranges during pregnancy stages, 325
weight maintenance
PAL and, 387
TEE and, 387
whey protein, 312. *See also* ergogenic aids for athletes
whiplash-related disorders, 38
Women's Health Initiative (WHI), 22
wrist and hand based sex differences, 33, 34. *See also* upper extremity based sex differences

Y

young adult body image concerns, 6

Printed in the United States of America